A PLAY OF
INFINITE FORMS

Other Books by Tom Weil

Fiction

A Clearing in the Jungle (1979)

A Balance of Power (1981)

Travel (*published by Hippocrene Books*)

Last at the Fair: A Book of Travel (1986)

America's Heartland: A Travel Guide to the Back Roads of Illinois, Indiana, Iowa and Missouri (1989)

America's South: A Travel Guide to the Eleven Southern States (1990)

The Cemetery Book: Graveyards, Catacombs and Other Travel Haunts Around the World (1992)

The Mississippi River: Nature, Culture and Travel Sites Along the "Mighty Mississippi" (1992)

America's Heartland: A Travel Guide to the Back Roads of Illinois, Indiana, Iowa, Missouri and Kansas (2d edition, 1992)

America's South: The Atlantic States (2d edition, 1993)

America's South: The Gulf and Mississippi States (2d edition, 1994)

Hippocrene U.S.A. Guide to Civil War Sites (1994)

A PLAY OF INFINITE FORMS

The Ventures and Adventures
of A Curious Man and
What He Found Through the Years
and Around the World

TOM WEIL

HIPPOCRENE BOOKS, INC.
New York

*In memory of Paul Kalter Weil, who began it all for me,
and—now toward the end—
For Edith, as the wingèd chariot draws near*

Copyright © 2021 Tom Weil

All rights reserved

For information, address:
HIPPOCRENE BOOKS, INC.
171 Madison Avenue
New York, NY 10016
www.hippocrenebooks.com

ISBN 13: 978-0-7818-1418-8
ISBN-10: 0-7818-1418-9

Printed in the United States of America.

Imagination, not invention, is the supreme master of art as of life. An imaginative and exact rendering of authentic memories may serve worthily that spirit of piety towards all things human which sanctions the conceptions of a writer of tales, and the emotions of the man reviewing his own experience.

—Joseph Conrad

ACKNOWLEDGEMENTS

My thanks to Priti Chitnis Gress, Publisher and Editorial Director at Hippocrene Books, for her professional and good-natured assistance in bringing into existence yet another form—in the form of this book—in the "play of infinite forms."

My thanks also to Natalie Blagowidow, Chairman of Hippocrene, for her support—much appreciated—of the project.

My thanks to Hippocrene's founder, the late George Blagowidow, Natalie's father, for his many courtesies and his friendship over my nearly 40-year relationship with Hippocrene and the family.

My appreciation to Barbara Keane-Pigeon, who processed, formatted, and typeset the text for Hippocrene with skill that would make Gutenberg himself proud.

To my long-time close friend Bud Carlson, my digital adviser and facilitator, my thanks for advancing me from the pen and paper age to word-processing, truly an heroic accomplishment on his part. Thanks to Bud, I discovered that one great advantage of digital text is that it eliminates ink stains on a writer's fingers, and, what's more, I no longer have to keep sharpening the quill.

Finally, special thanks to my great friends of more than half a century, Kathrine and the late Henry Jackson in London and Giuseppina d'Amely Melodia in Rome. Although the three of them were not directly connected with this book, those friends frequently hosted me in their cities—giving me there a home away from home in a familial ambiance—when I went to visit them or stopped off in London or in Rome on my way to or from adventures around the world. My thanks also to many other friends around the world who offered me less frequent but no less appreciated hospitality over the years.

CONTENTS

Backward		viii
SECTION I	Plots and Counter-Plots	1
SECTION II	The Fatal Circle	41
SECTION III	Charmed Lives	109
SECTION IV	Twilights Far From Home	153
SECTION V	The Misty Chords of Memory	195
SECTION VI	Subjective Objectives	285
SECTION VII	Shadows and Smoke	331

BACKWARD

A forward view never offers as much information or utility as does a retrospective view. Looking back after the story is almost complete allows you to gain a perspective on the entirety. A review clarifies much of what seemed indistinct, random and bewildering at the time you experienced it. Time sharpens your focus and eventually brings a more integral comprehension of your previous seemingly theme-less experiences so widely scattered in time and place. From the ghostly shadows of the past begin to form toward the end patterns as the entire picture of a long life takes shape. A backward view helps to align disconnected components experienced in dispersed helter-skelter catch-as-catch-can confusion.

Only afterwards—when a life, a book—is complete do you write a forward, so it's not really prospective: "forward" is a misnomer. A forward perspective in life ever remains only theoretical, a vague guess subject to many contingencies. Until actually realized, all forward-looking views are hypothetical—mere potential until the tale is told. As the story draws to an end, I am now able to tell my tale. A fully rounded account becomes possible only when the fullness of time contains very nearly all the elements of an entire life. Toward the final phase you can assess how content you are with the content which filled your allotted time.

Over the years I've learned that looking at things from a different and perhaps even eccentric viewpoint can often greatly enhance understanding. I have thus inverted a forward into a backward to conform to my nearly concluded very late in life retrospective perspective. The views and reviews in my reminiscence serve to connect into something of a coherent story many of the varied and disparate happenings which largely by chance comprised my earthly existence.

In these pages the random infinite play of forms which became known to me I've ordered into a story, the plot and the counter-plot—my responses to what happened—imposed by my backward perspective recorded here shortly before the story ends. By the time you read it my story may be over, after I transition from existence to the ghostly other side of forever.

Section I
PLOTS and COUNTER-PLOTS

~ 1 ~

I am my own ghost writer. It may be odd to exclude from a personal account the person who writes it, but my intention is to remain largely removed from the pages, a phantom-like figure for the most part absent from the story. I remain ghostly, shadowy, indistinct, like an X-ray image. The book includes only the bare bones, the underlying experiences which structure the story. Few personal details appear because the book—in no way a memoir or autobiography—is not about me. Rather, it's in the nature of a reminiscence—an account of a keenly curious character, one who happened to exist at a certain time and place, and his encounter over the years with the outside world all around the world. As for every earthly mortal, my life contains no plot—other than the one which will eventually contain my remains. A person's life story is always a short story, with any theme only artificially and arbitrarily imposed. My role in the contrived plot is simply as an intermediary, a kind of go-between and facilitator connecting the reader with some selected experiences and adventures. It so happened that those happened to me, but the me—the particular person involved—is largely irrelevant. Some, most or even all of the events might never have taken place and any or all of them (or similar) could have occurred to someone else. I was only a receptor, and otherwise of no importance to the story. The specifics which pertain to me—personal details—have no pertinent part in the tale and so are omitted. My most appropriate role is to remain behind the scenes as the man who isn't there—an off-stage spectral figure hidden in the shadows.

As time goes by you become ever more gradually the man who isn't there. An ever-depleting resource, your earthly existence shrinks away as various random infinite forms fill out your life. Trading time for content—a forced trade, as you have no choice—your allotted span shrivels as it swells with experiences. As the views—the experiences—proliferated over the years, so decreased the viewer's life. The remembered residues of those disorganized happenings form the skeletal structure of this reminiscence. My enterprise was to gain, over time, original views and then near the end, which is now, to review them. My job now done, I'm something of a phantom figure, a fading and ghostly nearly disappeared presence mutating into an absence, the absence of a man who isn't there.

My absence from the account represents not a defect but a benefit. Removing the narrator eliminates inessential and uninteresting parts of the narrative, so enhancing it. The story is enlivened by eliminating my life. I've purged the text of its author so that the tale, rather than its teller, represents the essential factor. By largely eliminating myself from the account, it offers more substance. Removing myself from a featured role not only disembodies an unnecessary actor and factor but also brings the additional personal advantage of preparing me to adjust to my eventual complete severance from my existence. Writing a tale without the teller serves to remind me that the disassociated phantom elderly author will before long become removed from his earthly being. A slight something of my history will remain, but nothing of he who experienced, created and remembered it. Not being present in the story gives me some familiarity with not being at all. Just as a dying campfire's waning smoke evidences light and warmth but is not them—only an insubstantial vapor slowly rising and writhing to disappear into the air—so does this narrative offer embers of remembers but nothing of the once solid soon ashed logs. From spectator to specter, from ghost writer to ghost: that's the real story—for me and for everyone else.

~ 2 ~

The book's main character, then, is not me but the outside world—quite a character it is and, in truth, much more interesting than am I. The author in my account serves simply as an ancillary participant in the narrative, a behind-the-scenes auxiliary helper acting as a facilitator rather than as a leading man. In March 1768 at age fifteen Fanny Burney, later part of James Boswell's and Samuel Johnson's literary circle, began a journal "Addressed to a Certain Miss Nobody," as "to Nobody can I reveal every thought, every wish of my heart." My account is meant to be from "a certain Mister Nobody," as I wish to remain a disembodied ghost writer.

The play of infinite forms rather than the player who experienced them predominates in my reminiscence. Although I've tried to remain in the background, publishing a book obviously entails making a public statement. Anyone pretentious enough to emerge from obscurity by giving up his privacy to enter the public eye seeks some sort of notice. Even the most diffident author doesn't want to remain undiscovered. But my account is only descriptive and not proscriptive. There's no shortage of characters trying to tell other people how to think and behave, but there's a great shortage of observers who offer well-meaning intelligent and credible opinions. I fall into neither category, for my agenda is not to persuade or advise but only to describe how things were for me. Such outspoken self-seeking types as politicians, media personalities, entertainers, soft subject experts stating seemingly certain views which in fact are mere opinions, social pretenders, academic pontificators, exhibitionists, social media personalities, know-it-all "money talks" chattering billionaires, and other know-little or know-nothing self-promoters thrive on attention and often profit from it. Those publicity seekers are essentially like an organ grinder's monkey—just as are authors. By publishing a book I'm admittedly just another monkey rattling coins—my reminiscence—in a tin cup, making a racket to attract a little attention in the crowd. But that sort of self-seeking racket is not my purpose. I prefer to leave the stage to other performers, of which there is no shortage. The play of infinite forms includes a vast cast of characters. For my presentation the featured attraction is not the chimp but the organ, that insensate mechanical force—driven by luck, chance, fate, circumstances and by the operation of natural forces—which creates the mood music to which every humanimal dances. The organ is like the organs revealed by an X-ray—the systems which enable organic existence. To the music which chanced to reach me I performed my dance. In accordance with my arrangement of this reminiscence, the monkey-author remains only a minor factor who observes the cosmic spectacle in all its glory and oddity, and who then records his observations based on the mechanical chords and discords of the organ as day by day it grinds away.

By no means a memoir, this book is in essence simply an account of curiosity meeting spectacle. Although the great play of earthly forms was obviously not created to entertain a curious mind—in fact, not created for any transcendental purpose at all—nonetheless the interface of those two factors, curiosity and the infinite forms, can serve to animate many delightful adventures, experiences and memories. The great spectacle inspires curiosity, curiosity illuminates the spectacle. Because the outside world is far more varied and interesting than my own life, I've attempted to objectify myself as a character outside the narrative rather than one central to it. By removing my persona and presence I sought to deflect attention from the author in order to focus it on the world beyond. I am merely a kind of utility, both in life and in print—a camera through which are seen images and impressions viewed and now reviewed, as captured on film independent of the operative mechanism. I was an imaging machine—a kind of X-ray vision viewer looking behind the scenes—which now, in these pages, imagines a structure for what I saw.

In *Follow the Ecstasy: The Hermitage Years of Thomas Merton* John Howard Griffin observes that he sought to distance himself from the book by serving as a conduit rather than as a participant in the story. Griffin wanted to make "it purely the life of Tom and not the life of Tom as seen by anyone." Similarly, my aim as a ghost writer is to present some chosen fragments from the life of Tom but without intruding as the author. For my account I am not the principal but only an agent mediating between Tom's curiosity and how it encountered and viewed the outside world. Jane Welsh Carlyle, wife of the renowned historian Thomas Carlyle, felt somewhat upstaged by her famous husband Tom and sought to search for what she called her "I-ity"—her "self-seeking Me." My approach is the exact opposite: to include as little "I-ity" and "Me" as possible. Tom the phantom author should not overshadow the main character which lies outside him. Tom the traveler trumps the Tom writing about him. In the forward to *Titan*, the biography of John D. Rockefeller, Ron Chernow writes: "Rockefeller often seems to be missing from his own biographies, flitting through them like a ghostly disembodied figure." This is precisely my intention: to remain absent, participating only as an insubstantial ghostly presence.

For the twenty or more years that I contemplated writing my account I faced the challenge of how to produce a personal narrative with the person largely removed. I was finally able to imagine a way to tell the story as a kind of memoir with the "me" removed—something of a new category or (to use the fancy literary word) genre. In this format the author (a fancy word for writer) is in effect reduced from first person to third, removed from the limelight as was Harry Lime, the elusive and evasive *Third Man*. Like Graham Greene's shadowy Third Man, the distancing third person format which de-emphasizes Number One features primacy of the tale rather than the teller: less is more. I have not literally but only literarily eschewed the first person. It would be an affectation to refer to myself as "he." But the sense of my approach is to view my participation in the story somewhat as a distraction and, as such, not as the featured attraction but only as a surrogate for the reader.

I've always believed that writing a book is simply a means to an end, a vehicle like a taxi or a taxi dancer, a way to convey a reader to a certain destination or result. It's irrelevant who the author is. The essence is the account, not the accountant who balances the books (or the book) to reach the bottom line. A book of any kind is simply the residue of a person's selected experiences and his contemplation about them just as, when the writer is gone, his name survives as a phantom remnant of what he was once called but without defining his essential being. In his 1965 essay "The Writer as Detective Hero," detective story novelist Ross Macdonald wrote: "My narrator Archer is not the main object of my interest, nor the character with whose fate I am most concerned." He is, Macdonald continued, only "a consciousness" to delineate the meaning of other peoples' lives. Similarly, my sensory intake served to collect various experiences and impressions followed by contemplation of them—the views and reviews which I record in this book. I was merely a consciousness somehow fated to absorb and then to discern as best I could the meaning of what by chance happened to me. Apart from that fatalistic happenstance—the randomness not only of my very existence but of my experiences and my reactions to them—my own fate is irrelevant to the story. In 1878 when James McNeill Whistler's *Mother* was on display (the painting, not the lady) at the Royal Academy in London the artist observed that *Arrangement in Gray and Black No.1* (as the work was entitled) represents "what it is. To me it is interesting as a picture of my mother, but what can or ought the public to care about the identity of the portrait." In the same way, my black and white arrangement of type—a work on paper portraying the outside world—need not concern readers about the identity or personal aspects of the author.

My reticence arises not only to eliminate irrelevant material from the narrative but also from my disinterest in talking about myself. As a very private man—perhaps to a fault, but such am I—never have I felt a desire to reveal myself or comfortable doing so, other than to very close family and friends, and even then only partially and somewhat guardedly. Apart from the materials I've chosen to include in this book (and the many omitted adventures) my own life is truly of no interest. This is not false modesty but a true view, and in any case I have no interest in discussing my life, other than some of my experiences out in the world. I must admit that it's something of an anomaly for such a private person to produce a public writing, especially a personal reminiscence. The formats of my other books and of my articles did not require personal details and none were offered.

Although normally an account like this reminiscence would contain some quantum of details about the author's life, the few which appear here I included as relevant to the story rather than to talk about myself. I focus on general topics, like travel, culture, history, curiosities (perhaps myself included) which my curiosity led me to, personalities I encountered—some colorful, a few famous, all noteworthy—and other matters external to my personal existence. The contents are more in the nature of William (or Tom) Tell tales—derived from time and place beyond my inner being—rather than kiss-and-tell private life stories. Not my inner life but my outer one—interactions with the outside world—comprises the essence of the book. My goal is not to present a memoir or an autobiography but simply an account of some life experiences which may hold some interest and perhaps even offer some lessons for readers to enjoy or otherwise benefit from. To such readers, if any, I say: congratulations on your good taste in literature. I compliment you on your discernment in choosing this book to read. I hope you bought it rather than only borrowing it from the library.

The memoir category (or "genre" for my most cultivated and discerning readers) has in my view in recent years often mutated into an unappealing format, one uninteresting to me as a reader or as a writer. For this reason I've refrained from writing a memoir or describing the book as such. As a ghost written author-subordinated work, this narrative could instead perhaps be called a "moir"—a "me"-less memoir. Pursued by publishers to write his autobiography, Ezra Pound refused to produce his "meeemoirs," as he called them. Unlike for Pound, no publisher has asked for my reminiscence, but like him it's not a "meee" account of the kind in fashion these days. Such intensely personal inward looking me-me memoirs are littered with "I" and "me." The self-reflecting writer, preoccupied with his or her own sensitivities, stuffs the text with recollections of childhood memories, lifelong traumas and dramas, psychological disorders, repression and suppression and depression, dysfunctional families, too functional neuroses, melancholies, anxieties and "me-isms" which reveal any number of other personal downers and demons. Such problems of course often do shadow lives. Some readers, with a voyeristic tendency who identify with the problems or who are somehow comforted by knowing that other people suffer similar situations or for various other reasons, find such memoirs of interest. For my purposes, such a confessional and self-referential account would be too exhibitionist and narcissistic, too self-oriented and based on the viewpoint that the writer represents the all-important center around which the world revolves. Such a writer believes that what befalls him or her in an emotional or psychological way is somehow worth recording. Those benighted souls certainly deserve sympathy, but I'm not sure they deserve readers. The public display of highly personal experiences and feelings in the form of written "selfies" seems to me less worthy of note than more selfless works based on observation of and response to the outside world, with its vast play of infinite forms. The eye which focuses on the world at large beyond the self offers larger and more diverse sources of experiences and brings a wider perspective than an inward looking mind's eye gazing at the "I." In *Man's Search for Meaning*, Nazi prisoner Viktor E. Frankl wrote, "True meaning in life is to be discovered in the world rather than within man or his own psyche"—being in a concentration camp tends to concentrate your thoughts on the essentials. By a certain paradox, reticence and restraint regarding personal matters in a worldly work can speak more loudly—by focusing attention on external experiences—than an elaborate introverted memoir based on a person's personal and psychological life.

A me-moir author recounting his or her most personal and private matters seemingly supposes that such details hold enough interest to appeal to readers. For my part, I recognize that my life from first breath to now approaching death truly lacks any noteworthy "me" personal story. My regular colorless existence—childhood, growing up, private and family life, work-a-day endeavors, social activities, conventional suburban dailiness—contains barely a scintilla of interest and is quite drab compared to my Technicolor adventures out in the wide world. Apart from the horror that my mother at times attempted to murder me—behavior so disturbing that never could I read an Agatha Christie murder mystery novel—nothing much happened to me in my youthful home life. Alas, it was my misfortune that those dramatic infanticide attacks never occurred and so, regretfully, I can't even include such deathly episodes to enliven an otherwise nearly "me"-free account. I sincerely regret that I never suffered such life-threatening happenings like being murdered by mom. This would have given me great memoir (or "mom-oir") material, to be produced by a posthumous real rather than simply a virtual ghost writer.

My settled, normal, routine existence—apart from travel and a few other types of adventures—in fact couldn't be more boring. Well, maybe it could but my regular life was for sure quite boring enough to lack interest for anyone but me and, in truth, even for me my mundane home-based routines weren't all that interesting. For want of pertinent material and because of my innate sense of privacy this book, then, omits the kind of self-focused, inward looking accounts so common in contemporary me-moirs. I instead concentrate on the outward looking experiences and adventures I encountered in the world at large rather than the personal interior world "at small." The book deals with motion, not emotion, and with my travels rather than my travails. The text includes a few selected reminiscences, out of many I might have chosen, which seem to me to represent experiences that typify others not mentioned. Included are not only those which are representative but also selections pertaining to people, places, adventures and happenings I found most memorable and noteworthy. My curiosity, facilitated by my bank account, took me far and wide over many years, sparking many encounters, relationships, connections, adventures and other experiences which became treasured memories. Now, with a nearly complete lifetime perspective of the whole, I'm able to review the almost concluded content of my experiment in living and so give the entirety a sort of shadowy shape which enables me to draw some conclusions from all the disparate infinite forms.

As for every mortal soul, both my individual existence as a humanimal and my encounter with the world, my life experience, are experiments permeated with randomness. A capricious and arbitrary natural regimen, teeming with random unpredictable forces operating in a world fraught with uncertainty, largely governs how each individual experiment turns out. Only toward the end, a stage I've now reached, can a semblance of a sketchy pattern or the vague outlines of a design begin to form. A late in life perspective allows you to begin to perceive how some of the pieces of the mosaic start to fit together. Moreover, in your mind and then on the page (or screen) you can rearrange the tesserae which in reality are ineluctably forever set in time and place. The bookish exercise allows the writer, by his reviews, to create with his own hand artificial arrangements, associations and formats unavailable with the original views, all individually unalterable after the fact but later subject to authorial reordering. The experiences randomly came to you, malleable up to the moment they occurred, and thereafter eternally ineluctable. The passing moments disappear and then last forever—gone and eternal. Words can arrange the infinite forms into an imposed order, one not necessarily chronological, and such has been my intention for this reminiscence.

Only near the end, then, can there be the beginning of some slight understanding of one's past. At that late date, which for me is now, a certain shadowy comprehension of the whole starts to take shape. The long span of years brings you ever more deposits into your memory bank, so enriching your resources and allowing you to refine and define the treasure trove inherited from elapsed time. The congeries of memories deposited by accretion serve as reference points to facilitate connections which at the time of happening seemed to be unrelated events. The longer you live the more the interstices fill in, eventually becoming dense enough to create patterns. Of course, as your allotted time fills with content your life empties of potential. As the tenure of your earthly existence drains away, near the end the capacity to absorb further impressions becomes limited. At that late date the interval between the now and the never offers little space for additional accretions to your memory supply. But what you've lost by the end in terms of time and opportunity you've gained by the ability to perceive a more complete picture of the whole. Whether or not this is a viable bargain, a fair exchange of time for lived experiences, depends of course on what luck, chance and fate have brought to form the pieces of your mosaic. In my own case, largely by luck and only slightly by will and effort, I enjoyed good fortune of many kinds. I would otherwise not have written this account because, as I've explained, I disfavor confessionals based on negative experiences. I leave it to others to describe, in the way of many latter day me-moirs, personal miseries and woes. My reminiscence deals with the impersonal outside world, indifferent to humanimals and their concerns, as I happened to encounter that world.

To produce my "moir" I waited until the twilight darkening which now shadows my earthly tenure. Because of my advanced age, even if one so far unencumbered by old age ailments, I felt in the last few years as if I were sitting on a time bomb loaded with the shrapnel of mortality. My life could destruct at any moment, thus exploding my memories and destroying the possibility of memorializing some of them. I felt that before it became too late it might be worth recording something of how one keenly curious character experienced and perceived the world in his time. Toward the end of my time on the lonely planet, then, I began my account, which hopefully I can complete before I transition to the type of ghost writer who can no longer write. In any case, this book will before long become transmuted from a living document—produced by a humanimal still able to remember, reflect and recall some highlights which illuminated his life—to a posthumous work left behind to anyone with a passing interest in my theme: a curious man's encounter with the world's play of infinite forms during a particular era. As a practical and realistic fellow—traits perhaps not always beneficial, as they sometimes diminish the poetry of life (I've always been more of a prose type)—I of course well realize that few people or even none will ever be interested in my tales and details, and such indifference I completely understand and accept. Those who never become aware of my existence or of my reminiscence—the millions alive now and those in the future forever oblivious to this book and my vanished being—will by far outnumber the very few who happen upon my account. To those few I express my appreciation of your passing interest and my hope that you enjoy and in some way manage to benefit from your encounter, most likely by chance, with my "moir."

Younger readers might not realize this yet—if ever they do, even when they're older—but a late in life retrospective reminiscence can rouse pleasing phantasmagoric memories which in a way console you for the fugitive long gone years during which were created such a store of personally pleasurable experiences. One by-product from producing this account is how the process activated my mind to revive so many delightful recollections from forever gone bygone days. Often somewhat at random did those long buried old views come to mind, momentarily illuminating my remembered life before once again reverting to the obscurity of the distant forgotten past. Sometimes every now and then by a sort of spontaneous combustion, as with stirring a dying log fire's last embers, a particular collection of recollections light up to spark other yesteryear images. They flare up, momentarily glow and then the brightness disappears into the darkness of elapsed time. There's no telling what will suddenly illuminate a random memory. A stray glimpse, a momentary sound, a brief whiff of a certain aroma, a dream, a fragment of something seen, a phantom image, a fleeting feeling, a momentary taste or touch—any of these passing impressions might suddenly bring to mind a reminder of something experienced long ago, far away. Certain faces or places; a river flowing by and flowing on as it winds its way to the distant sea; a photo; an unexpected image—mind or matter; oddly shaped shadows; a long forgotten memento; random patterns of windblown fallen autumn leaves; birds winging into the dusk of a waning summer's eve; a name—person or place; the fragment of a song; an echo; a reflection; words written, words spoken; haunting thoughts: all those, and much else, can inspire a kind of epiphany based on the long lost past.

At times I don't exactly remember the where or the when or the whom of the original happenings, but I know that once upon a time there was one. You sense a certain deeply embedded even if sometimes indistinct realization that something from your past has somehow suddenly revived in your memory in the here and now. Those unsummoned random and at times melancholy rememberings—an admixture of pleasurable past events tinctured with regret for their now forever vanished nature—recall the vague memories, the poetry of the past, described in Archibald MacLeish's "Memory Green," with the autumn smelling of rain and summer stripping "branches of their dying leaves" as "you in Paris along the windy quai" shuffle the "fallen leaves before you," then sweetness suddenly fills your heart and "tears come to your eyes," but why? When was it like this before, and with whom? This you will not remember as you feel the wind on your throat and "smell the dead leaves." You try to remember: "With whom, you will say, Ah where?"

~ 3 ~

Every written work should of course speak for itself. Sometimes, however, a few additional words—especially by an elusive ghost writer from his shadowy haunts behind the scene—can serve to illuminate the narrative. Anthony Trollope begins his *Autobiography* by noting that "it will not be so much my intention to speak of the little details of my private life," even if "the garrulity of old age" tempts him to include such information. My reminiscence may be somewhat garrulous, but (with very few exceptions) nowhere have I been tempted to include any private details. I refer to my presentation as a reminiscence, rather than an autobiography or a memoir, because the designation I've chosen eliminates the need to include personal references. An old-timer can reminisce about the old days as he chooses without revealing anything personal at all. My account is thus not a memoir but a "moir," a truncated version with the "me" removed. This invented word to designate my reminiscence happens to evoke "moiré," which refers to a substance with a "watered, clouded or frosted appearance." This definition by chance well describes the clouded, shadowy nature of an account like mine from which the author remains hidden. In contrast to many contemporary confessional memoirs as transparent as an open book—filled with revealing personal details—my "moir" lacks the "me" element. My appearance in these pages is more of a disappearance, leaving them in that way cloudy, frosty, ghostly

Similarly, looking back now near the end on my entire life experience it somehow seems rather ghostly and insubstantial, like the vanished smoke from a smoldering nearly burnt out campfire. Indistinct random events crowded themselves fleetingly into my years, an amorphous mass of raw material from which I've now chosen a few fragments to arrange into an arbitrary sort of limited structure. Two limitations confine the structure to a relatively narrow range. First, of all the possible infinite forms which might have given me sensory impressions only very few happened, mostly by chance, to reach me. Second, from those which came to my attention to comprise my life I've selected only some samples to represent the entirety. My reminiscence thus records only one person's limited view based on his individual life experiences and is not meant to present any transcendental principles. My subjective perceptions and my responses to them pertain only to what happened to me. Observers more confident, opinionated and certain than am I would no doubt attempt to generalize what they perceived and believe into sweeping authoritative and definitive pronouncements. My views and reviews, however, are not meant to be didactic. I am not a pundit seeking center stage but only a ghost writer taking a fond last look back on what befell me during my brief even if long-lived and by now very short remaining earthly existence.

No theme or thread shaped the various random sensory impressions I experienced. They were accumulated, not collected, as no presiding consciousness operated to arrange what happened into a coherent narrative. What befell me survives only as a store of confused memories. Everything came to me in a disorderly way, a disorganized and even chaotic jumble of impressions, feelings, ideas, thoughts, realizations, nuances, memories, all filtered through my mind, now to emerge as writings which transmute the confusion of raw material into a certain order. No pre-existing organizing structure offers a way to explain the discordant random happenings. They become slightly meaningful only in an artificial way, one of many arrangements which might have been chosen, by the arbitrary construct I've imposed on them.

In order to suggest the disconnected nature of the random happenings my account is time-scrambled. This discontinuity reflects the lack of coherence in the way experiences reached me. But the narrative is not simply episodic, as some underlying ligatures hold the text together. Although not chronological, the narrative is logical as I've arranged it into an order to produce a comprehensible perspective on the entirety. When things happen is not necessarily meaningful. Patterning stray events from their original temporal order into a more symmetrical design helps to endow them with enhanced significance. Because the contents of my reminiscence derive from such random confusions I avoided not only a chronological order but also the usual designation of each part as a "chapter." So much of what happened seemed sectioned off into scattered individual events which didn't cohere into cohesive chapter-like categories. Somewhat similar happenings arose in a discursive way separated in time and place until finally connected only in my imposed narrative. I thought—but I may be wrong—that "Section" evokes better than the more bookish designation "Chapter" the random, disparate and eclectic nature of the real-world happenings. I thus chose to organize the book into sections rather than chapters.

In contemplating the past, it was clear to me that no one lives a life in chapters, with each episode or phase neatly arranged into a beginning, middle and end. A chapter is an arbitrary unit, a fictional literary convention shaped, or misshaped, to fit a theme, a contortion quite contrary to the episodic and for the most part disconnected random happenings typical of how the experiment of being alive really occurs. "Section" better describes the different units of the book more accurately than does "chapter," which seems to me to endow the material with a logical structure which is in fact only arbitrary. This approach conforms to the spirit of Robert Burton's 1628 classic, *The Anatomy of Melancholy*, which he divided into Partitions, Sections, Members and Subsections. Although I include many remembers, I do not label the units "Members." Moreover, unlike Burton's book mine deals not with the anatomy of melancholic states but with the melancholy of anatomy—the eventual inevitable disintegration of every humanimal when it gives up the ghost. A ghost writer is already part way to that final departure.

If you look beyond the surface of things, as if gazing at an X-ray of the skeleton beneath the skin, you see a ghostly shadowy image which reflects the underlying reality of the humanimal. Even apart from X-rays, early-day "photography was almost uncannily predisposed to the creation of ghostly images," notes Susan Owens in *The Ghost: A Cultural History*. As such, those primitive spooky pictures depict more vividly than contemporary ones the truth about human existence. It's not a rehearsal in a "playhouse of infinite forms" (Rabindranath Tagore's phrase) but a hearse-al. Because you don't remain on stage for the reviews after the curtain drops to shroud the scene, a reminiscence such as mine is an auto-review written to meet a deadline as the production nears the last scene, after which you exit in a hearse. Every mortal is a phantom of the opera.

If my melancholy observations tend to demoralize the reader, feel free to ignore them. In fact, any and all of my comments in this book can be ignored without diminishing the delightful (this may be the only fictional word in the entire text) experience of reading my reminiscence, for whatever I say is only one person's viewpoint based on his particular life experience. In truth, both the format and the content of a book are not all that important as long as a few basic benefits for the reader exist. Those who chance to read my account may be somewhat surprised at my view that it really doesn't make all that much difference what an author says as long as the work entertains, pleases, delights, informs, amuses or otherwise offers some modicum of pleasure or usefulness to whomever happens to read the work. Apart from scientific works with provable facts, whatever a writer propounds is not definitive or ultimate and can be contradicted by any number of other writings The experience and experiment of being alive creates all manner of beliefs, dogmas, doctrines, perceptions, responses, reactions, value-systems, solutions and other conclusions, none of them conclusive but which are personalized and unique only to the individual and not definitive in a general way. The important thing for someone who purports to be a writer—that is, someone who views himself as having a reasonably valid claim to saying something of value in a clear way to a transient reader—is to produce some sort of functional and useful word work without overestimating the transcendental nature of the content, which in the end represents only how the author saw it. As Henry James wrote: "Live all you can. It doesn't so much matter what you do in particular so long as you have your life." Similarly, it doesn't much matter what an author produces as long as he says something worthwhile for the reader. Whatever a writer says is ever subject to revision by readers: the author's views can always be re-reviewed. As much as a writer aspires to entice and please a reader, some works—this one, among others—may fail to please. Many books are in some ways uninteresting, uninspiring, uninformative, boring, fail to entertain or amuse, or are otherwise unworthy of the valuable and ever depleting time of a reader, who can at any moment abandon the work in favor of more compatible books or activities which better conform to his or her interests, needs, tastes or sentiments. Please do not abandon me—at least not just yet. Hear me out—read me out.

My hope, of course, is that the essentially random happenings which became the content of my life along with my selection and responses to those events as contained in my reminiscence are in some way sufficiently worthy to retain a reader's attention. The book mediates between those two different components: the long ago "views"—what I actually experienced—and the more recent "reviews" selected for this reminiscence. Before the chance happenings which ripened into reality occurred they remained pure possibility and then, from moment to moment, became transmuted into the unalterable facts which comprise my "views." My latter-day "reviews," much more malleable, represent only an arbitrary selection of material I've integrated into an artificially imposed order. The long settled fixed facts furnish the basis for a discretionary composition: immutable realities arranged into a flexible forced but not conclusive finality, definitive only in a provisional way.

From the chaos of disparate experiences I could well have chosen any number of other examples to serve as surrogates for the whole. But somehow the ones recorded in my reminiscence came to mind and serve to form a picture—not a definitive one, but only kind of a ghostly image—of the piecemeal jigsaw puzzle of the past. Such highly contingent content, imposed on unalterable facts, could all have been different, just as my experiences before they occurred might have happened in ways other than how they actually occurred. Empedocles of Agrigentum (493-433?) explained the origin of living beings by the separate creation of individual human components which by chance joined together in all possible combinations. Only the individuals which happened to survive the many experiments were suitable to exist. Similarly, chance operated to bring me out of all possible combinations those somehow meant to become my reality, and from those randomly created views I selected out of many possibilities for review the relatively few which comprise my account.

I wrote the reminiscence entirely from memory without consulting my extensive and detailed Journals in which I recorded my travels as they occurred day by day over more than half a century. As for everyone, my memory is selective—but, I believe, not defective—so only certain episodes from the past came to mind while many others remained interred in dead time, never to be remembered. It's of course impossible to know just why some vintage views happened to have entered my consciousness, either somehow wafting to the surface on their own by some mysterious process or forced to the fore by more willful thought. But some events and impressions remain so vivid and telling they naturally form a permanent part of my memories, while more obscure ones buried in the depths arrived unsummoned. In any case, reliving the past in my mind's eye to view again what my original sight-seeing brought me represented something of a capricious exercise in recollection. Writing a "moir" such as this one in a way involves an odd and inexplicable process, and as such quite resembles each life and human existence itself: a phenomenon to be experienced, not one to be understood.

~ 4 ~

From my early days I realized that creating experiences and relationships rather than possessions represented for me the highest and best use of my allotted time. Those priorities aren't particularly unusual or distinctive, and I mention them only to show the frame of reference which governed my experiment in living. Never did I covet material objects, even ones which would greatly improve my standard of living. Luxury and frivolous frills were for me redundant and irrelevant; basic functionality sufficed. Only things of sentimental value or those—none of any monetary value—acquired the world over during my far-flung travels interested me. Such treasured but financially worthless modest items serve to furnish both my home and my memory. The objects evoke trips of long ago and far away while also adding color and meaning to my surroundings. The keystones of my existence were not gemstones and similar baubles but the intangibles of experiences and relationships—the opposite of possessions. As objects in the world, things are at hand to see, touch, feel, sell, move. Happenings, feelings, experiences, personal connections can be moving and touching but, no matter how worldly, remain ever evanescent. If such invisible life components are to assume even a shadow of a trace of ethereal being in the form of lasting descriptions, such fleeting, will-o'-the-wisp, gossamer nothings must be recorded. Those misty memories are special-needs phenomena, orphaned until adopted by being remembered and then transmuted into text so they can find a place before the rememberer loses his. Possessions need no such help as they exist on their own in one place or another. Objects define themselves; intangibles need someone to speak for them. The descriptions are of course not the events or the relationships as such but only shadows of the original reality—the best anyone can do to capture something of the past. In things inhere their essence; by way of contrast, intangibles gain a sort of substance, a weak version of what once was, only by how they happen to be remembered and then described. They otherwise disappear when the life which experienced them ends.

Writing about experiences after the fact is one thing, but creating them is quite another process. I didn't infuse my life with experiences and relationships in order to write about them but only in order to live in accordance with my interests and values. French poet Stéphane Mallarmé claimed that "Everything in the world exists in order to end in a book," the antithesis of my view. Nothing at all exists to furnish material for word-processing. The world exists for no reason, least of all to offer thought workers stuff they can put into books. Never did I contour or contort my life to gain experiences to write about. I lived so as to live, not to put what happened into a book. The lonely planet does not serve as a festival of images, a play of infinite forms created to furnish book, chapter and verse for creative types.

To enhance my ability to shape my existence in the desired ways—not for writing but for living—I was conscious of the need to address the intangible elements I favored, experiences and relationships, with some sort of connective tissue by way of attitude and preparation. I attempted to remain alert and aware both of the opportunities and the happenings themselves as well as bringing some thoughtful perspectives on what chanced to befall me as my life unfolded. I aspired to let nothing go to waste. This mind-set included two dimensions. My observations, reactions, sentiments and feelings, contemplations and speculations were based both on an intellectual framework for how I viewed the world—forethought to establish a perspective early on—and an on-going meditative state of mind which served to curate the experiences. I tried to fit the passing events into the prefabricated mind-set, which from time to time I changed in response to what happened. I adjusted and revised my general outlook based on what I learned as my experiences accumulated.. The two elements—my pre-existing framework and then the responses—represented an attempt to benefit from the flow of events. I didn't want them to occur and then pass me by without making use of what happened to grasp from the random flow at least a few durable realizations. A synthesis of mind and matter—not stuff, but what to me mattered—served to combine that which happened with contemplation about what by chance happened to have occurred. This sort of adhesive conjunction helped to bind events and my responses to them, so rescuing such transient happenings from insignificance and probable oblivion. Binding the one with the other—experiences and my reaction to them—enabled me to grasp a fragment of meaning from the passing events rather than just letting them wash over me and drain away. I could at least try to capture a few drops of the flow and freeze them into remnants more solid and lasting than the liquid moments. If not properly stored in the deep freeze of a memory vault, some of the more fragile and melty remembrances would dribble away. I tried to preserve as many as possible, and over the years they accumulated into a glacier of memories.

A precondition of recollection and reflection requires access to the underlying reality of actual experience to support the later derivative memory image which mirrors the original. To imagine into being a valid tale, be it fact or fiction, requires some experience. Mere invention of a story represents a lesser enterprise, one more removed from reality than imaginings based on the real world. Theory, speculation and invention can serve as aides to comprehension but not as its essence. As Jacob Bronowski observed, "The world can only be grasped by action, not by contemplation." This maxim indeed contains a great truth but not a complete one. In fact, you need a synthesis of both elements to be able to discern some small amount of meaning from what chance and randomness conspire to activate that which befalls you. This is ultimately a mysterious and baffling phenomenon, and without the vital components of both action and reaction you remain in a hermetic (or himetic) and theoretical limbo. Lack of experience leaves you empty of content, while a failure to contemplate what happens strands you with just a jumble of essentially random events deprived of connective tissue.

Andrew Grove, one of Intel's founders, made a practice of questioning the company's middle managers as he thought they could better recognize changes in the marketplace than could senior management more remote from the front lines. The lower echelons were more attuned to developments because those employees, Grove noted, "spent so much time 'out doors' where the winds of the real world blow in their faces." Many academics prefer to remain indoors where the scholars can enjoy climate-controlled conditions. But every so often humanities professors happen to offer some useful outside learning. In *Twisted Tales From Shakespeare* Richard Armour notes that on-stage doors, "according to scholars who have given their lives to a study of the Elizabethan theater, were used to go in and out." Thanks to the labors of some learned tenured academics, we now know how to follow Andrew Grove's recommendation to go "out doors" to gain real life experience--just go out the door (open it first). Sheltering from the winds in an ivory tower or settled in an armchair by a cozy fire is certainly more comfortable than venturing "out doors" into the storm, but outside actions and activities abroad, such as travel, are necessary for a grounded and well-rounded granular life experience and a valid perspective on it. After using naturalist and geologist Louis Agassiz's 1850 *Lake Superior* as a guide to the rich copper and iron deposits in the area William Butler Ogden, the locomotive force in establishing and building the North Western Railroad system, called on the famous scientist at the Lowell Institute in Boston. Agassiz told Ogden, "You can study nature in my book, Mr. Ogden. But unless you go out of doors yourself you cannot find her. This new country needs students far more than it needs text-books. The book of nature is always open." So it was that I subordinated books to looks. I aspired to become a student-at-large to read the encyclopedia of the world as it really was rather than as filtered through merely an observer's or an author's mind and transmuted to the printed page.

To combine both dimensions—the doing and the thinking—presents a great challenge for most people. They're usually restricted by circumstances, temperament, habit, personality as well as forcibly being attracted to and distracted by all sorts of practical demands and considerations, all of which limits more wide-ranging adventurous kinds of actions and contemplations about them. Those most mired in the world by force of circumstances seldom enjoy the opportunity to transcend their immediate context. Life presses in rather than opening out. Similarly, more philosophical and intellectual types often remain rather too cerebral and lack the practical experience necessary to infuse thought with reality. Bridging this gap by fusing the two elements yields a richer content than simply relying mostly on the one (doing) or the other (thinking). To paraphrase Ortega y Gasset, not a Spanish vaudeville act but just a single philosopher, those most able to think about things are the least experienced in the outside world, while people with the most experience are usually less inclined to contemplate their actions and reactions.

In my own case, from early days I aspired to grasp from the fleeting flow of passing events a few residues of impressions which might serve to enrich the experiences so as not to waste them. For such enrichment I needed to decide what I valued and how to arrange my life to focus on what was meaningful for me. Once I realized that experiences and relationships represented for me the prime factors, that straight-forward and focused perception served to outline my path. In addition, an even more fundamental and influential maxim, apart from Bronowski's, shaped my aspirations, choices and activities: Christopher Morley's observation that "There is only one success—to be able to spend your life in your own way." This wisdom served to set my agenda and to govern my every move and decision. Other than the momentous matter of what kind of dog (if any) to get and what to name it, Morley's desideratum represents one of the biggest challenges for someone who hopes to live on his own terms and—an essential "and"—who enjoys the possibilities to do so. To accomplish that desirable goal it's necessary to contemplate what "your own way" is, and then how to reach that route given your specific circumstances. To connect "want" with "can" entails realistic and workable assumptions and assessments of your situation and your ability to produce the desired results. This boils down to deciding what you want out of life and, given your circumstances, what you might be able to get from it. You begin with a given fatal circle, and then chance and luck and some volition add further opportunities and constraints. What you want depends on you; what you eventually get largely depends on random forces beyond your control.

The process requires not only some specific thought but also an awareness of the mix of considerations which come into play when addressing the matter of a life course. Many moving parts must mesh to produce a smoothly functioning operation. For most people, the gears don't engage sufficiently well to bring about Morley's definition of success. In my own case, by chance it happened that many of the factors facilitating such a life existed, and to those I added the necessary element of contemplating how best to live to attain that success. My engagement with life thus evolved over the years based on a number of favorable variables along with a conscious effort to exploit them.

Although I enjoyed access to sufficient advantages appropriate to facilitate my goal, I resolved not to contort my activities and procedures to live only on my own terms. My restructuring of my life to follow Morley's maxim was thus somewhat range-bound, as I willingly retained some elements which restricted my ability to pursue my "own way." Of course, from a practical and realistic standpoint no one can exist completely on their own terms; other considerations always limit options. My protocol was to flex but not to snap my existing situation, reshaping it enough to diversify without eliminating my regular way of life. I sought a balance, supplementing what was shaping up as a rather mundane, conventional and unadventurous but comfortable existence with some more exciting and varied experiences out in the wider world. I wanted to get "out doors," as Andy Grove put it.

With that basic principle in mind—to retain and extend—I could begin to focus on some of the factors I deemed necessary to facilitate living, at least in part, on my own terms. One essential is an appropriate temperament. This includes the desire, willingness and ability to think and act in an independent way. You should feel comfortable—and, even better, invigorated—to go against the grain of conventional wisdom and social convention. Such contrary behavior might produce a few splinters, but better that than rather wooden unthinking conformity. Living in accordance with your own terms by definition meant in my case that I'd eschew some, many or most of the structures and strictures which prevailed in my normal suburban context. I didn't pursue non-conformance purely for its own sake; it was a by-product of living in my own way. To determine pursuits based on your own terms demands a certain free-spirited attitude, one unfettered by custom, ritual, routine, ordinary practices, cohort beliefs and values, prevailing opinions, judgments by your familiars, and other such limiters which operate to constrain independence. To peer pressure I reacted with a disappear response, removing myself both by attitude and by distance, with extensive travel, from my ambient environment. But I always returned to my provincial native habitat, and when there happily functioned in those regular and normal circumstances which defined my home life. The balance I established—a synthesis of far-flung adventures and a settled existence—proved quite satisfying and brought me an equilibrium which enabled me to succeed in living as Morley deemed ideal.

Independence presupposes not only an appropriate attitude and mindset—better stated as a mind un-set—but also and obviously the benefit of monetary resources. Without the lubrication of money the gears, even if they otherwise mesh, won't turn. Financial freedom is a necessary but not a sufficient element for living according to your own terms. While essential, financial means are by no means the ultimate or governing factor. Although some people enjoy financial security, they never convert that fortunate endowment into a rich life beyond their money moat and their immediate context. It's in fact something of a challenge to overcome your advantages. For me, foreign countries rather than local country clubs represented more valuable terrains. Wealthy people indeed often use money to live on their own terms, but those sometimes entail a self-indulgent and materialistic way of life which gives wealth a bad name. Owning wealth rather than letting it own you requires using it well to seek experiences money can't buy but ones which require financial resources.

Gaining and maintaining the financial freedom to live a life of your own choice is facilitated by keeping your needs and wants limited and simple, an attitude somehow innate to me. Consumerism and possessions distract and confine a person; stuff is sticky and burdensome. A basic conservatism and frugality characterized my personality. Things, I realized, limit life; experiences and relationships expand it. Acquisitions I restricted to practical and necessary items with no frills or showy features extraneous to functionality. I aimed to reduce clutter and complications to a minimum. Apart from my home-based existence and its essentials, my main discretionary expenses consisted of travel and the relatively few and inexpensive mementos I acquired, some by purchase and others by gathering otherwise ignored bits of archaeological debris. Sentiment rather than financial worth underlies my attachment to those objects collected over the years in the far corners I visited. The objects now occupy near corners close at hand at home.

Those objects pleasurably allow me to recall where and when I added them to my life, so giving me from time time a review of pleasant experiences gone but not forgotten. But even those treasured items represent something of a burden, a clutter, an obligation, a claim on space and time. Even the most modest stuff has to be dusted. Although I'd hate to lose any of my travel mementos, I can well understand Virginia Woolf's reaction to the destruction of her house and its contents during the World War II London bombing raids. In *A Writer's Diary*, Woolf wrote of her relief when her house at 37 Mecklenburgh Square was destroyed:

"Exhilaration at losing possessions—save at times I want my books and chairs and carpets and beds. How I worked to buy them—one by one—and the pictures." She goes on: "it's odd—the relief at losing possessions. I should like to start life, in peace, almost bare—free to go anywhere."

Loss of my simple and meager possessions would not be a relief, nor did such loss occur to prompt me to start a new life bare of things. My old life was already rather bare of things. Even without such loss, however, I proceeded to arrange a satisfying equilibrium between two completely different ways of life—a settled routine existence at home, frequently punctuated by the counterpoint of transient, rootless travels "out doors" in the big wide world. To accomplish my goal to be "free to go anywhere," as Woolf desired, I cleared away some of the activities, restraints and obligations of daily life at home to free myself to go anywhere and everywhere, and in the course of time I very nearly did. It was a great adventure, one which lasted more than half a century.

So much of interest lay beyond the limited bounds of my narrow regular life I couldn't bear the thought of never seeing such alluring attractions. I wanted to experience as much of the world as possible. Of course, the world is far too vast and interesting for any one consciousness to absorb even a tiny fragment of the offerings. The infinite play of forms is indeed so multifarious a single life can experience only a few tiny fragments of the entirety. What time has jolted into being by natural forces and by the hand of man has created a world far beyond the capacity of any humanimal—or all of them as a totality who did, do and will exist—to comprehend, even in small part. Moreover, so much is in flux—constant proliferation and mutations, appearances and disappearances, as if a veritable jungle of ceaseless growth, change and decay which over time creates the past—there exists an overwhelming surfeit of sensory phenomena, a dazzling play of forms which spark, even if not much to enlighten, the life experience. But I wanted to experience my share. The wide world beckoned, and I responded.

~ 5 ~

My desire to be free to expand my horizons—go anywhere, experience as much as possible—arose from my most dominant and compelling trait, one which shaped and maybe even in a contorted way misshaped my entire existence: curiosity. Where this fundamental and driving characteristic came from I don't know, nor do I care. That's one thing I'm not curious about. Its origins are of no importance to me; only its existence is. The crucial element is that an overwhelming sense of curiosity somehow possessed me from my earliest days. A curious person both enjoys the benefits of an endless variety of possible phenomena to explore while at the same time suffering the frustrating disadvantage of being able to satisfy only a very small quantum of that curiosity. Only the merest mite of a minuscule microcosm of the whole can you examine or experience. Just as time must have a stop, so endless curiosity ultimately reaches its maximum point of limited satisfaction with all too much left unexplored. Within those limits, however, exist sufficient pleasurable opportunities to endow a life with enough adventures to satisfy even the most curious character. Well did I enjoy access over the years to ever more fragments of the world's overwhelming complexity—humanimals' doings and don'tings, the earthly spectacle in all its panoramic richness, the diversity of happenings and mishaps which happened to befall me, all an ever fascinating multifaceted experiment in the curious phenomenon of being alive as a keenly curious fellow.

Curiosity both fuels and facilitates an independent life you spend in your own way. Without at least a modicum of curiosity it may not even be possible to live in the Morley mode. Lacking curiosity, it's difficult to contemplate and then to activate alternative ways of life-enhancing experiences. Moreover, even with compulsive curiosity you can't be interested in everything. Because many vast categories, even ones which attract you, must be ignored, pursuing only the chosen few which most interest you will occupy just about all your time and attention. My own rather restless curiosity both bedeviled and delighted me, as seldom could I focus only on the topics or activities I favored. I was often diverted into exploring a wider range of subjects and experiences, a practice which both diluted and enhanced my efforts to get to know the world. One thing led to another as the play of infinite forms played on and I played along.

My inquisitive behavior permeated my existence, a pleasurable curse and an undoubted blessing. For someone who aspires to spend his life in his own way, curiosity is a highly desirable trait but also an unsettling one. This suited my purpose, however, because one of the points in revising my life by adding various ventures and adventures was precisely to become unsettled. Behavior activated by keen curiosity brought to me over more than half a century many enjoyable viewings and happenings which endowed my experiment in living with memorable experiences. To curiosity I owe a rich trove of cherished memories which now, in old age, comfort and divert me. So crucial in my view is the importance of curiosity in living a good life, I assess people's chances for such a life by observing their CQ as much as their IQ or EQ. I've noticed that a high Curiosity Quotient, in addition to Intelligence and Emotional Quotients, serves as one of the best indicators or predictive factors to assess if a person does or will live a successful and contented life. Of course, the vicissitudes of chance, the capricious nature of luck, the influence of randomness, the uncertainties of how potential will develop into reality, the impact of ill-fated and well-fated happenings, along with other such uncontrollable factors, can defeat the benefits of even the most ardent curiosity. But without that fundamental and life-enhancing trait, its origin in any particular person unknown, life chances are diminished. To be interested in that which transcended my immediate needs, context and surroundings seemed to me an appealing formula for living an enriched and more exciting, even if more exacting, existence. "Is there not at the root of every passion something of curiosity?" mused Miguel de Unamuno in *The Tragic Sense of Life*. My passion for experiencing the world grew from deeply rooted curiosity.

As an aside I'd note that my personal references in discussing my curiosity and its role in motivating my travels do not contradict my intention to refrain from talking about myself or my desire to remain off stage as a ghost writer. My comments are meant mainly to give some background information to explain in part what inspired me to restructure my life so as to add to my regular settled suburban existence an entirely different dimension motivated by my curiosity and aimed at satisfying it. Discussion of its role may serve to elucidate some of the dynamics of my life and of this reminiscence, which pertains only to the peripatetic phase of my existence. My un-dynamic sedentary provincial home life merits no discussion at all, nor am I interested in writing about my private, personal, family or affective constituents.

Being curious, I was interested in what others had to say about the trait. One of the most famous and exemplary cases pertained to a curious character whose entire existence was based on curiosity, from which he made both a life and a living. As Neal Thompson relates in *A Curious Man: The Strange and Brilliant Life of Robert "Believe It or Not!" Ripley*, O. B. Keeler, one of Ripley's friends, wrote in a column about him: "The wide-eyed, innocent curiosity that started Bob Ripley along the trail of fame more than 15 years ago [about 1908] burns as brightly today as ever ... [T]he freshness of life is an eternal dew in his heart." A profile writer early in Ripley's career referred to his "bottomless, off-kilter curiosity"—nothing was safe from his mind, as Keeler observed in his column. A tremendous gusto and interest in the world drove Ripley to explore it. He was much more than a globe-trotter: he cantered and galloped around the course, a course which of course no school could offer him.

Apart from all the oddities, curiosities and weird "believe its" in Ripley's cartoons, his "Believe It or Not!" rubric serves to suggest the strangeness of the entire play of infinite forms. Viewed from a cosmic perspective, the way things are arranged here on the lonely planet both by earthlings and by natural forces is very strange. It's truly a curious sort of place, one impossible to imagine if it did not exist. To Ripley's challenge, outside observers of the world from far-away planets would reply, "I don't believe it!" It's all too unbelievable for words, although every now and then a writer tries to capture in words something of the strangeness of the infinite play of forms.

Other curiosity commentators and characters seem to lack Ripley's delightful and somewhat innocent natural sense of wonderment about the world. As a thinker rather than a traveler, Alberto Manguel views curiosity more as an intellectual trait than as one which motivated worldly explorations such as Ripley pursued. In *Curiosity* Manguel discusses the characteristic with an erudite treatment, citing all sorts of sages but sighting and describing few exotic places in the world at large. Manguel's bookish approach certainly doesn't conform to Bronowski's view that only action and not mere contemplation can enable you to understand the world. Astrophysicist Mario Livio titles his book *Why?: What Makes Us Curious*. That interrogative seems to me exactly the wrong question. True curiosity is based not on "why" but on "what." To pursue why things are the way they are represents seeking a chimera, creating a diversion from the essential nature of pure curiosity—an attitude concerned not with how things work but what the play of infinite forms presents. The "why" such a play takes place is irrelevant unless, as Livio notes, you seek "results." While scientists may (must) ask "why?" adventurers possessed by curiosity focus on what exists without attempting to explain the reason for such forms or sensory phenomena.

Similarly, in *A Curious Mind: The Secret to a Better Life*, movie producer Brian Grazer (with Charles Fishman) gives us fully 300 pages about "the value and power of curiosity." There's no doubt that curiosity can bring life-enhancing advantages, but to a large degree the author views an inquisitive nature as a pragmatic, functional tool to help him network his way through the Hollywood labyrinth. This strikes me as something of a Hollywooden take—a rough cut on the matter. Being inquisitive can of course facilitate professional advancement and a lot of other good things, but to my view that's really not the purpose or the essence of curiosity. For me, being curious entails a kind of Ripley-like playful encounter with the play of infinite forms. I wanted to engage with what's out there—how the world works, how humanimals function, the creature's culture and civilizations in all their present and past vast variety and fascinating forms, and nature's as well. Why everything endlessly spins through space and time, a meaningless whirligig, never interested me. I saved a lot of time and thought avoiding that unanswerable question. Of course, Ripley used his curiosity-driven experiences and findings in a practical way to produce "Believe It or Not!" But his acute and searching curiosity seems less contrived and purposeful than for those who view it as a tool for one's livelihood. My curiosity wasn't motivated by the hope of gaining any professional advantages. For me, in fact, it was precisely the lack of utility of my enterprise which attracted me. Although Abraham Flexner in his 1939 article endorsed "The Usefulness of Useless Knowledge," I valued the uselessness of useless knowledge. For my purposes, the absence of functional practicality in my pursuit of curiosity-inspired adventures represented their very attraction. My travels served no useful purpose other than to expose me to a wider range of infinite forms, which in turn was one of the ways I tried to live in my own way. I emerged into the outside world with no ulterior motive or agenda other than to scratch my itching curiosity. I assumed the role of Mark Twain's "unbiased traveler seeking information" rather than someone intending to profit professionally. I wanted to enrich my life, not my bank account. No movie mogul hoping to make connections, I was simply a human "movie" moving about to far destinations to enjoy the world's disconnected variety of forms, an activity which helped me to spend my life in my own way.

Ian Leslie's *Curious: The Desire to Know and Why Your Future Depends On It* also discusses the topic with rather too much practicality. Contrary to his title, my future did not depend on curiosity but was shaped by it. For me the delights of curiosity didn't relate to its power to bring specific advantages in the future. Being inquisitive entailed only the pursuit of the unnecessary, with the main benefits memorable adventures and intellectual satisfaction rather than practical utility. Although it's not especially useful to approach curiosity in an overly analytical way, Leslie's distinction between "diversive" curiosity, which is unfocused and unchanneled, and the "epistemic" kind, which is disciplined and directed to some purpose or end, offers an enlightening contrast between the two quite different types of inquisitiveness. This sort of distinction Salvador de Madariaga discussed in a more concrete way in *Englishmen, Frenchmen, Spaniards* where he notes that the Russian national character exhibits "an uncontrollable appetite for all kinds of knowledge," while the English are "gifted with a vigorous defensive instinct against all unhealthy intellectual curiosity." By English standards, my wide-ranging rather discursive eclectic curiosity would be deemed unhealthy, but in fact quite the opposite was the case. I was more Russian than English. An unbridled inquisitive nature can lead you to discover many unexpected delights you're likely to miss with cramped defensive or with primarily purposeful curiosity. In some cases, practical collateral effects by chance brought me benefits unsought but welcome.

Favoring the "epistemic" form represents the wrong motivation, for true curiosity implies the pursuit of experiences and knowledge for their own sake. Both Grazer's and Leslie's books emphasize what might be called the utilitarian school of curiosity, its function primarily to help you in your career, improve your chances to age well, bring you practical benefits, make you thinner or richer or whatever. But once it becomes purposeful is it truly curiosity? Never did utility cross my mind. I didn't will, cultivate or activate curiosity as a facilitator, other than try to fashion an interesting life in my own way. My curiosity-inspired activities were only gratuitous, in no way forced or contrived in an effort to gain professional or personal benefits. The American Mars rover *Curiosity*, which has spent more than six years crawling around a crater called Gale, bears an inappropriate name. The robot isn't at all curious but only a mechanical automaton operating with a very narrow mono-focus and interested in only a specific practical activity.

Deeply embedded in my personality, curiosity was for me a force of nature, a characteristic unforced by me and one which ineluctably endowed me with a free-spirited attitude motivating me to explore what lay beyond my given conventional provincial existence. A sense of adventure, not advantage, activated me. The Chinese refer to a failure to focus as "a monkey mind." I do not refrain from describing myself that way—and thus, valued reader, you have finally encountered an honest author who compares himself to a monkey. Of course, I would prefer not to refer to myself as a monkey but, as noted earlier, I'm a realistic fellow. Moreover, it's better to be a discursive, curiosity-driven monkey-like creature than an organ grinder who day after day after day grinds out the same repetitive music. The music of the world, discordant as it often is, encompasses a vast repertoire far beyond the grinder's limited play list. Reader, treasure this paragraph, for where else will you ever find the literary phenomenon of an author who admits that in a way he's a monkey? (I hope this doesn't insult monkeys.) In truth, it's not such a far-fetched comparison, for me or for any other humanimal. As I swung my way around the world, high-tailing it hither and yon, people and simians somehow seemed similar. An organ grinder's monkey could well be my third cousin once removed, or—to ape the same thought as it might apply to you—perhaps your first cousin not removed.

It's perhaps somewhat curious that I have no curiosity about the origins of a personal trait so dominant as my curiosity. Apart from the reality that it's unlikely one can determine the source of such a controlling characteristic, my paradoxical attitude—acutely curious, but not about my own curiosity—conforms to my general disinterest about the reasons people are the way they are. Never did my curiosity extend to analytical observations based on psychology, psychiatry, personality traits, behavior "disorders" or orders, and other such mental and affective techniques or conditions. The "why" of human behavior, or of anything else, held much less interest for me than the "what." Why a person acts or reacts in a particular way, why the world is as it is, why things happen as they do, the reasons why luck and chance and circumstances manifest themselves one way rather than another all seem to me matters so opaque and inexplicable that attempting explanations only distracts from more accessible and enjoyable ways to satisfy curiosity. For me, pursuit of the "is" rather than the "why" offered more satisfactory and sportive possibilities. Addressing "whys" doesn't make you wise but only draws your attention away from enjoying the vast richness of the "what"—that which actually happened or exists and is known or knowable. Speculating on "why" adds little to the delights of what a curious adventurer encounters. Why both the totality, the cosmos, or each individual component—the world, people, all the rest—exist in the forms familiar to us is unknowable and irrelevant. Any "why" explanation is in any case not definitive as no transcendental explanation offers conclusive proof. Explaining why something is or occurs is usually just speculation unless subject to scientific proof, and even then no ultimate explanations are possible.

Omitting "whys" in no way diminished the range of experiences activated by my avid curiosity. The ways rather than the "whys" of the world represented what I sought. Because even a monkey mind comes with limits, I in fact eliminated not only "whys" but many other matters, for a highly curious mind can't be mindful of everything. Wide curiosity can't begin to encompass even a small fraction of all the infinite forms which enter and swirl around in a person's consciousness. We for sure have world enough but, alas, never enough time to explore more than a minuscule fragment of the totality. Each curious person must eschew almost everything in favor of concentrating on only a few selected matters. In my own case, my curiosity was never piqued by such things or activities or topics, to mention only a few, as digital technology and gadgets; professional sports (which includes many university varsity teams); toadstools, toads, frogs (except one in Calaveras County); bats (baseball and flying) and balls (in games and for debutantes); Olympic games, card games, computer games, games of chance (chance enough existed in life without adding more in the way of gambling); ballet dancing (belly kind okay); gourmet cuisine; politics and politicians; insects and reptiles; addictive substances, other than ice cream and M&M's; and much else. The residue which remains is of course so vast that a millennium would scarce serve to satisfy even the tiniest quantum of my curiosity. Human finitude in the face of cosmic infinity is rather bothersome for a keenly curious person. You wish that somehow the allocation of your time on earth would permit you to explore many topics, but most remain untouched for lack of time. I'd suggest that if anyone or anything in cosmic authority happens to read these words, the he or the she or the it charged with arranging the next planetary version of organic life and consciousness include a "BOGO"—Buy One Get One, an effective two-for-one marketing device. If such a promotion works for hamburgers or dog food it should work for more universal matters. If you happen to enter existence in that new, improved world you get a second life. In this way you enjoy not just one but two chances to engage with that world rather than the existing all-or-nothing arrangement. Either give the humanimal two lives or don't bother creating just a single one, a far too limited offering. Although I greatly enjoyed and treasure my adventures, at the same time I regret leaving so much undiscovered, but that's the fate of every earthling. Samuel Beckett observed that the whiskey resents the decanter. I don't resent the container's permanence but do lament how little of the booze I managed to imbibe. I would have enjoyed becoming more drunk on the play of infinite forms.

In addition to a humanimal's very limited time-constrained ability to absorb the world, yet another major limiting factor exists. What you are able to accomplish and experience and what happens to you are subject to so many uncertainties and contingencies you really have very little control over what occurs. Life comes at you, not from you. No matter how keen the curiosity, it might well remain largely unfulfilled. Not withstanding intensive and extensive forethought and planning and the exercise of the most willful intentions, luck and chance may operate willy-nilly to defeat your aspirations. Those factors influence outcomes even more than the circumstances which define the context of your fatal circle—your place in time, place, the scheme of things—which itself is happenstance. As for specific events which comprise your life, random intangibles like chance, luck, fate, destiny, coincidence, serendipity and other such uncontrollable influences govern the conditions which both facilitate and limit your ability to act. My reminiscence serves to illustrate the influence of luck in life—how mankind is "the slave of accident," the vivid phrase from Robert Graves' translation of Lucan's account of the civil war between Pompey and Caesar. In *The Drunkard's Walk: How Randomness Rules Our Lives*, Leonard Mlodinow discusses how a "conspiracy of random and minor factors—that is luck" determines outcomes. He notes how "seemingly inconsequential random events" lead to big changes and how "random forces and our reactions to them account for much of what constitutes our particular path in life." To cope with the capricious random influences it's better to "focus on the ability to react to events rather than relying on the ability to predict them, on qualities like flexibility, confidence, courage, and perseverance." As I read Mlodinow's book I imagined that he must have read my mind.

To pursue a Morley-type self-designed and executed life requires a number of unlikely and fateful factors to combine to enhance your chances, which in any case always remain hostage to uncertainty. For some reason—the "why" unimportant and anyway unknowable—chance happened to endow me with a set of circumstances favorable to my agenda. The situational elements coincided with my aspirations. This of course did not guarantee a favorable outcome but only made it somewhat more achievable. Somehow all the necessary factors coalesced to bring about my desired results. Even with a head-start you can end up with many setbacks which defeat your purposes. Often an advantaged person who enjoys favorable circumstances becomes a self-unmade man. Such an unfortunate type could be called Regla Oitarch–Horatio Alger in reverse. It's far more common for people to make nothing out of something than to produce something out of nothing. In fact, any life accomplishments at all essentially involve creating something out of nothing. The fateful downs and ups, outs and ins, the skelter-helter and catch-as-catch-can-or-can't nature of life imposes far too many uncertainties for even the most favored earthlings to attain their desired ends. Many such advantaged but benighted souls suffer only downs and downs, outs and outs. No need to ask why fortunate situations sometimes fail to produce fortunate results. That's just how things happen to work out for some people.

Because everyone no matter how fortunate by fatal circle, circumstances and later developments ever remains a hostage to fortune, misfortune always lurks in the shadows. One of the main reasons the lives of some people become prone to degradation arises because they lack realism. Unrealistic beliefs or expectations, exaggerated aspirations, misguided behavior based on incorrect premises, ill-judged decisions, misbegotten endeavors, dreamy assumptions, invention from theory or academic learning rather than fact-based imagination, and other infelicitous notions often originate from a failure to perceive the world as it really is. This defect is especially common with trust fund babies, inheritors who gain wealth without earning it. This bonanza, which comes with a cost, is laden with many risks. Such fortunate types often seem to lack comprehension of the real world, an understanding of peoples' motivations, the keen instincts sharpened by experience, a practicality and savvy of the kind street-smart operators learn early. In many ways, streets educate more than books. Sheltered trust fund heirs often trust too easily. Gratuitous inherited wealth in many cases results in the spoilation of one's agenda for life. The wealth interferes with living a rich life.

Although money doesn't represent the sole factor in the ability to arrange life in accordance with Morley's principle, it would be somewhat obtuse to omit discussion of the role of wealth as facilitating an independent existence. Contrary to what is no doubt general belief, wealth—especially the inherited kind—doesn't bring an unmixed blessing. It's in fact something of an odd hermaphrodite-like thing which encompasses at the same time two quite opposite potentialities. On the one hand, financial resources can be greatly corrosive, spoiling its possessor, stunting ambition, warping perceptions, creating an entitlement mentality, perverting behavior, contorting values, inducing arrogance and, above all, distorting reality. Handling and spending money in ill-advised ways obviously interferes with constructively spending your life in your own way. But on the other side of the balance sheet, wealth properly understood and managed provides an essential ingredient to enable you to pursue your own interests. If you by chance happen to enjoy some wealth by diligence or inheritance—each case laced with luck—along with other advantages, failure to respect and nurture your good fortune means that you will eventually no longer face the challenge of dealing properly with such benefits, as they will cease to exist. This will eliminate the problems of oversight while creating other worries.

You never know ahead of time how someone will handle wealth, whether unearned or self-generated. In some cases heirs behave as if they possess no inherited money and they pursue productive and constructive lives. In other cases such advantaged characters abuse their good fortune. Self-made successful types are also subject to mishandling their hard-earned wealth, dissipating it or abusing it by hubris, over-confidence, erratic behavior, euphoria, show-off conspicuous consumption, investing in unfamiliar areas (places or industries) or by various other defective behavior. A mercurial substance ever subject to volatility and dispersion, wealth—like luck—comes and goes. Money which burns a hole in your pocket can at any time evaporate into smoke, just as do a campfire's logs. As a practical and, I believe, realistic fellow, I recognized early that in order to live in my own way I'd need a certain level of financial support, and this I managed to establish and maintain. Those two elements don't necessarily cohere. Creating wealth and keeping it are not identical twins.

Although the details are irrelevant, and in any case no one's business, some of the principles I found useful apply not only in the financial realm but also more generally. Such capital notions which served me so well may be of some interest. In fact, among the titles I considered for this book was *Investing Secrets to Make You (and Your Dog) RICH and Thin for the Rest of Your Life and Afterlife*, which I believes covers all contingencies and needs. My publisher advised me that the book would sell far more copies with this title—money, dogs, slimness are all proven subjects to produce a best-seller—than with the one I chose. Combine these topics into one all-encompassing title and you have a block-buster. As it is, *A Play of Infinite Forms* is probably fated to be a worst-seller, a bust and a budget-buster rather than a block-buster. Later in the book I'll briefly discuss some principles which served me well in investing, a preview I now mention with the hope of keeping you reading. Please continue. My monkey mind remains filled with stuff which will interest you.

~ 6 ~

If the past is a foreign country, the future is even more so. The yesteryears predating our existence and all time to come after our time remain forever behind and beyond, hidden from view even more than is a ghost writer. We can grasp something of the past—of the future, nothing—vicariously but not what it was like to live before our time. We realize that past eras, like foreign lands, exist (or did exist) and that the future will, even if not for us. But they both remain removed from our immediate experience. It's hard to imagine now how the Christmas holiday operated back in the B.C. era—not Before Christ but Before Claus—when Santa Claus himself was a child. Back then, who flew in from the North Pole or from some remote foreign area to descend the chimney at the Claus house to deliver young Santa's presents? You also wonder about the future: Will the beloved Mr. Claus we know eventually opt to use Amazon for faster and more efficient distribution services? Will he be superseded by a soulless robot? Our very language makes us wonder about its ability to express concepts before Shakespeare lived, when early-day speakers, more tongue-tied then than later, couldn't spice the King's English with words and phrases the Bard originated. And how did we ever learn how dogs really behave and how people interact with them before Snoopy and the other Peanuts characters came on the scene? A strange country indeed is the past when you realize that at one time the present existed without so many familiar presences, ones which now seem normal and natural but in fact which then never existed and were completely contingent, just like you and me and all mortals and everything else. Everything which became a form or an event was absent for an eternity before emerging into existence. Even more foreign is the future, which ever remains empty of content and filled only with potential. At least the past once established is, North Pole-like, forever frozen in elapsed time. Times to come, especially those far removed from ours, would from our perspective seem even more strange.

In many ways the malleable future, empty like a desert, is more interesting than the unalterable past, a tangled jungle of happenings. Rich with potential, the future comprises anything, while in the past all possibilities have been reduced to a final fact. The whirligig of time spins an infinity of possibilities into a single outcome, transmuting what might be into what is in fact set in time and place. It could have all been different. At times I've imagined alternative scenarios for how my life might have evolved. I became especially contemplative when traveling in disadvantaged foreign lands where I saw the miserable conditions and situations suffered by inhabitants there. I'd gaze out the window of an air-conditioned train or deluxe bus and look down upon the unfortunates below sweating and swarming through the streets, a woeful scene. I saw my other self, the unlucky one, out there. But no matter how vivid my imagination, never could I get out of my own skin and into the bodily incarnation of another humanimal. I was only and ever a passing observer, not a fellow sufferer out there on the street. Viewing such scenes always made me appreciate my good fortune—how luck and circumstances happened to endow me with advantages—and also to lament the misfortune of those ill-fated earthlings which destiny somehow disfavored. The "why" of the random forces which produce such diverse life situations I never thought about, as no answer is possible. I only observed the what—that which actually existed, not the reason why. In any case, contemplating causes related to why things exist in a particular format set into time is as irrelevant as considering alternate scenarios for what might have been. It's unrewarding to imagine how anything or everything might have developed differently. Rather than speculating on the infinite number of other outcomes which might have occurred in one's life, it's more seemly to be grateful for the advantages you enjoy and the benefits fate has gifted you, while never forgetting that some or all of your good fortune, and life itself, can at any moment be taken from you. It could have been different for me, but better to put your energy into appreciation rather than speculation.

Wise historians appreciate the futility of counter-factuals. As G.M Trevelyan observes in *Stray Thoughts On History* (good thing he managed to wrangle and herd and corral those strays into a book), "speculating on the might-have-beens of history" is not a particularly useful exercise, for "if one thing had been different, everything would thenceforth have been different—and in what way we cannot tell. Out of a million, or rather out of an infinite number of possible lines that human affairs might have taken, only one materialized under the impulsion of chance." Whatever actually happened, Trevelyan continues, is an "arbitrary choice of Fate," and the dense reality of the past "wears forever the inviolate sacredness of the accomplished fact." By way of contrast, a wise perspective on the completely problematic future demands assessment of possible alternatives by scenario planning. With this procedure you consider various potential outcomes—based on extremely tenuous and opaque factors—which might result from different decisions, always subject to chance and luck which in the end determine outcomes.

When contemplating how a satisfactory earthly existence manages against all odds to develop, as was the case for me, it's both realistic and disconcerting to recognize the extreme contrast between the absolute plasticity of the future and the immutable fixity of the eternal past. You realize that, as Trevelyan observed, a slight variance in what occurred could have greatly altered what followed. As the present spins on, during each moment of incipient temporal duration almost anything is possible, even your very non-existence. Every next instant, time congeals into an unalterable past. A knife-edge line dividing past and future separates vast potential from unalterable reality. This sharp, acute and ever-moving line, the cutting edge of advancing time, shapes what happens to you, carving the content of your life from the vast range of might-be's which, after the fact, are discarded into the dust bin of mere might-have-been's. Time's waste basket overflows with all that's been discarded and can never be recycled.

Although chance always has the last word in influencing outcomes, believing that everything which happens arises only by an uncontrollable deterministic force such as randomness implies that personal agency plays no role and is merely a plaything of luck and fate. However, it's a mistake to abandon volition and put yourself completely in the hands of chance. While luck indeed greatly governs outcomes, personal will and intention can at times also exert some influence on what happens to occur. As influential as chance, fate and luck are, you can't leave everything to them. Although Sophocles-like "whirl is king" factors predominate, they're not exclusive elements. Even though His Royal Highness Whirl has the final word, whispering a few notions in the monarch's ear might influence him. To a certain but uncertain and unknowable extent, your own wishes, decisions and actions can at times in part shape your life, even if in the end other random factors govern. The earlier in life you activate your own mechanisms, the more likely it is that they can be effective. The longer you wait, the fewer the remaining options, for as time advances the avenues into future alternatives close until eventually, as for me now, a cul-de-sac confines you to a dead-end. The accumulated density of your elapsed past limits the range of future potential. In ways you can't understand, luck and happenstance in the future remain subject to the "sunk costs" incurred previously. What will happen depends in some measure on what already has transpired, but just how is unknowable.

While I always recognized the crucial roles of chance and luck, I also tried to nurture those capricious forces by purposeful proactive behavior. This involved both setting an agenda and pursuing it, all factors considered, as best I could. I sought not only positive results but also tried to avoid undesirable outcomes, especially those difficult to remedy. Both an offensive and defensive mind-set characterized my outlook. I sought to keep myself out of harm's ways and put myself in charm's way. Fate isn't a preordained destiny but a foot soldier to the general influence and commanding presence of luck. Verily, it's advisable to act as a handmaiden, or master, to try to tame in the minor way possible the wildness of chance and randomness which at a moment's non-notice can reshape for good or ill a humanimal's fortunes or misfortunes. Luck can turn on a dime, or even a penny. I hoped that my attempts to nudge luck my way would improve my chances to live in my own way. Of course, ultimately fate and luck have a way of slipping away from your touch, careening in unexpected and undesirable directions you can't anticipate or control. Nonetheless, fatalistic acceptance of determinism, and the corollary of remaining passive in the face of enigmatic destinies, reduces the potential for introducing at least a chance to influence results in a desirable way. Based on my belief in participatory behavior to importune luck to cooperate, I tried to make reasoned decisions within the framework of my situation and the prevailing knowns. In truth, almost every major decision and many minor ones essentially involve an experiment, just as each individual life represents an experiment by nature involving how any particular humanimal is created and experiences its existence. Although you can try to assess in great detail and predict how a specific choice might turn out, all the while hoping that the result will conform to your hopes, even a high degree of confidence in your decision remains vulnerable to the operation of chance. Humanimals and huwomanimals are random creatures, and their lives just as capricious.

Hope represents the human need to believe that good things might somehow happen, but it's a misplaced belief. Things just happen, regardless of how you hope they will. Hope is a useful illusion, but not one to be depended upon. It motivates earthlings to keep striving for desirable outcomes, even if no relationship exists between the wish and the results. In Chekhov's short story "The Lady With the Dog" Dmitri Gurov, a 40-year-old married man with three children, meets by chance Anna Sergeyevna when he pets her little Pomeranian dog after she sits next to him in the public gardens when on vacation in Yalta. The encounter leads to what the male gaze hopes to produce but seldom does, and Dmitri and Anna, also married, have a passing holiday affair, one in a series of Gurov's romantic adventures. Or so he presumed. But back in Moscow after they'd both returned to their spouses, Dmitri finds himself unable to forget Anna so he travels to her hometown and seeks her out to profess his love. She then starts visiting Moscow to continue the affair with Dmitri there. Each having finally found love outside of marriage, the two of them hoped eventually to live together, but a lot would have to change before they could at last realize that hope. How their story ends and the reward the dog received for its part in facilitating the chance meeting between Dmitri and Anna in Yalta Chekhov does not relate. Their hopes, the couple's and the dog's, remain suspended pending further developments beyond the limits of the short story. In the seaside promenade in Yalta stands a metal statue of the trio, a fixture there frozen in time and place with their hopes yet to be realized, if ever. Chekhov wrote "The Lady and the Dog," *Three Sisters*, *The Cherry Orchard* and other works at his house in Yalta not far from the statue. When at the house late July 2008 I sat on a bench in the garden where Chekhov and his fellow Russian writer Maxim Gorky used to sit and chat, and I meditated on the insubstantial but somehow persistent nature of hope, and also lamented that the dog failed to get its due as the go-between to enable Dmitri and Anna to meet. Every dog has its day, but not in the Chekhov tale.

Looking back, it greatly puzzles me how I for the most part managed to select options which enabled me to remain on course to live in my own way, while avoiding irrevocable setbacks. Although I'd like to claim that high intelligence, shrewd judgments and a keen perception of opportunities and risks comprised the operative factors, in truth I know that even the most informed, contemplative, meditated and thoughtful choices are subject to random forces and perverse outcomes. But somehow, inexplicably, instinct, intuition, awareness of nuances, accurate threat assessment and—most of all—an acute perception of reality served, even if only in a minor way, to help luck and fate favor me. My efforts in decision making, like everything else in life and life itself, were experiments which for me for the most part happened to work out in satisfactory ways. Any or all of those experiments could have proved otherwise. Circumstances alter cases, and so do luck, fate and chance.

Of course, I'm hardly the first person, nor will I be the last, to emphasize the role of those random factors in shaping a life. My purpose in repeated references to the influence of luck and chance and the other such random influences is not meant to be original but simply to highlight how those elements affected the life of one particular earthling, a humammal (the highest form of animal, so people believe) with whom I have a lifelong familiarity. In a sense, my book is an homage to the favorable luck that somehow, against great odds, I managed to enjoy and which brought me good fortune. While I believe that chance plays a definitive part in the existence of every earthly mortal, for me I'm sure that such was the case. This book presents many specific examples of how chance operated to influence my life. Somehow luck permeated my existence. My reminiscence is in large part an extended meditation on randomness, luck and chance as the operative and determinative factors in the rather odd cosmic experiment which is human life and the equally strange earthly experiment which was my life.

In attempting to cope with the hazards of luck, fate and chance and the uncertainties which shadow every significant decision, I found it useful to recognize how other peoples' agendas can affect your own. It's not a misanthropic view of human nature or an abnormal anxiety or any sort of "disorder" to realize that each humanimal has its own personal agenda. That agenda is in part open and transparent but in many cases also hidden and obscure. Those agendas often in some ways conflict with your own. Pursuing an agenda, explicit or implicit, represents the way people make their way through life. The agenda sets priorities, attempts to fill needs and wants, shapes outcomes to conform with hopes, expectations, aspirations, desires. For sure I had my own agenda, one item of which was to protect it from other peoples' want list. The degree to which you allow the agenda of others to impinge on your own affects how much your priorities might become subordinated to ones non-yous seek. For that reason it's important to monitor and to limit impositions so you can better remain able to pursue your own agenda, a "selfie" which is self-serving but not selfish. In living your own way you can of course take into account, to the degree you deem appropriate, the needs and wishes of other people. Indeed, for those in your circle you have to do that. The important element is to recognize that those needs and wishes exist and then to assess how they will affect your own agenda and how you might respond to what others seek from you. How you balance the two sets of priorities—yours and that of others—will determine both the quality and durability of your relationships and your ability to pursue your own agenda.

The most striking example of beings who relentlessly promote their own agenda is man's best friend. Doggies truly run the show, as shown by the little Pomeranian who introduced Anna to Dmitri in Yalta: no dog, no love match. Dogs induce you to feed, water, pet, walk, talk to, entertain, care for, cater to and spoil them. They live rent-free, paying no taxes or other expenses; they earn no income; they enjoy a fully paid (by you) medical plan and a 401(k)(9) retirement account (also funded by you). Hoping to be coddled and adored, they are. Dogs' hopes are realized more frequently than those of humanimals. Based on cunning, cajoling, intelligence, centuries of experience and innate genetic programming, canines know just how to realize their agenda. Wagging tails, pleading glances, soulful eyes, apparent love, at least at meal times, all enable your adored pet to mitigate fate and to eliminate chance, so ensuring canine contentment, courtesy of you. Your own agenda takes second place to that of your dog. It's not as easy for humanimals to accomplish their ends by wagging their tails and begging for treats.

One of the most intrusive forms of interference with an agenda arises from demands on your time. Time-abusers attempt to transmute their wishes into reality by using your limited temporal resources. Valuation of your time by another person differs greatly from what it means for you. To others, your time possesses no intrinsic value but represents only a commodity to be consumed to accomplish a goal which is most often not yours. For you, however, your time is a finite resource, its supply limited and your need for it far in excess of its availability, so making it scarce. I always tried to protect my time, using it to advance my own agenda rather than that of others. Of course, some demands extraneous to your agenda are inescapable but many, especially from other people, are quite escapable and can be reduced or even eliminated. You also need to protect your own use of time from self-abuse, misuse and wastage. For any undertaking of some duration or difficulty, such as writing a book, it's advisable to estimate realistically how much time you'll need and are willing to invest. This requires a well-judged evaluation of the possible return on investment, not a monetary assessment but one relating to how the proposed enterprise will serve your agenda. In some cases, of course, your agenda will mesh with that of someone else, and in almost all cases will be controlled by your dog's agenda, thanks or no thanks to a doggie's know-how to hound you into complying with his wishes.

It's of course perfectly reasonable to adopt some or all of another person's agenda as long as you've properly understood how that might affect yours. Among the most threatening situations are those which involve accepting another agenda with consequences which are difficult to remedy if need be. It's especially necessary in such cases carefully to consider the degree to which you amend your own agenda to accommodate another one. The eccentric but highly successful eighteenth century Yankee trader Timothy Dexter in Newburyport, Massachusetts viewed his wife Elizabeth as a shrew and a nag. Instead of divorcing her, Dexter decided to turn Elizabeth into a non-person, ignoring the woman and referring to his spooky spectral spouse as "Mrs. Dexter, the ghost that was my wife." Yet even relegated to a phantom presence in Dexter's life she no doubt continued to haunt and plague him, intruding on both his tranquility and his agenda. And in any case, both of them were essentially pre-ghosts, the same as all earthlings, even those who aren't ghost writers.

Unavoidable, of course, are the exigencies of earthly existence which impose themselves on you with many inescapable demands even more pressing than those of dogs. Whether you like it or not, such impositions interfere with your agenda. One obvious intrusion is the need to pay for remaining alive. The necessity to gain access to money to continue to live, for both you and your dog, requires that before you spend in order to survive you must spend a certain amount of time, effort and energy to secure financial resources. In a basic sense, everyone lives hand-to-mouth. The financial need arises not from someone else's agenda but is imposed by life itself. Various other intrinsic humanimal needs also present unavoidable demands which must be addressed. It's only the residual possible opportunities and choices which remain, after you fulfill the mandatory obligations and any others you accept, which offer the potential to live in your own way. Such debris represents the raw material available to build a life of your own.

To fashion a baseline agenda to live according to my own vision, I used a reductive but encompassing formula which for me summarized the components of life and how it might best be lived. I've challenged any number of people to suggest a significant element of human existence which fails to fit into any of the four general categories, and so far no one has managed to propose a valid additional factor (religion I exclude, as belief is based on faith which either exists or doesn't). By far the major factor which determines how well the other three can operate is health. Although a humanimal, by its habits and behavior, maintains some control over health, to a large and unsettling degree physical and mental well being depends, as usual, on luck, fate and chance. As a chemical broth with a few pinches of personality and individuality thrown in to add a little spice, an earthling is an organic creature subject to any number of disturbances, natural or man-made. No one cooks up his own specific genetic recipe; it's served to us unordered in a random á la carte way. That random menu to a large extent affects the likelihood of pathology and also influences for better or worse such characteristics as sociability, resilience, temperament, energy, intelligence and other personality traits. Health is an essential precondition for the other three categories described below to function well. Those three factors represent the elements which comprise the basic building blocks of a life. If all three work out well, your life rests on a solid foundation; if two, you suffer some imbalance; with only one, it's an unstable structure. Luck-deprived unfortunates lacking all three are just out of luck. The three categories are:

> **Personal life**—This refers to your ability to establish and maintain, how and to the extent you wish, a close and on-going relationship with another person (not including your dog). Although many who aspire to a rewarding personal relationship end up, like Timothy Dexter, with someone they opt to reduce to a ghostly presence, or absence, in some cases personal life proves less spooky and more substantial. Partly by chance, as for Dmitri and Anna, and in part by choice a lucky few manage to find a true soul-mate, maybe better described as a sole mate, the person we were somehow destined to meet and join with. But even an intimate connection short of that ideal "one and only" rarity can bring satisfaction. The ideal is not the only deal. Moderate and reasonable expectations enhance your chances for success in personal life, and life in general. Pleasant surprises are more likely with modest ambitions.
>
> **Social life**—This encompasses participating in a wide range of interests and activities "out doors" (as Andrew Grove expressed it) beyond your own private, personal and localized being. Hobbies, diversions, friendships, outside

connections, social relationships, out-going behavior, life enriching activities, a certain sense of curiosity and a participatory nature all contribute to facilitating a life with breadth and depth which connects you with the great wide world, even if just local, beyond yourself. I stretched my life to cover more ground by extending my interests, especially with outreach to people.

Professional life—This serves to give you a purpose, an identity, a place in the world beyond your private existence. Whatever your job or activity, not necessarily income producing, it connects you to the practical and productive work-a-day realm, so contributing to your sense of self-worth, maybe even your net worth, and providing you with some satisfaction and feeling of accomplishment. Professional activity also enables you to furnish your life with real-world experiences based on Jacob Bronowski's dictum: "The world can only be grasped by action, not by contemplation."

These four principles have guided me in two ways: first, in setting my own agenda and second, in observing other lives with the aim of learning from them by way of example the kinds of perils and pitfalls, the possibilities and potentials, which might apply to my own experiment in living. Never did I evaluate a major decision without considering how it might influence my health as well as my personal and social and professional conditions. Ideally, all four factors would be enhanced by a particular choice, but at times it's necessary to settle for trade-offs which in a workable way balance the various factors. To judge as well as possible, always under uncertainty, how a decision might play out you need to keep in mind and firm control of your carefully considered agenda, as it might from time to time be amended. Lack of clarity on what you want to accomplish will muddle the decision-making process.

The second way I used the four principles was to evaluate the lives of others with a view to gaining a better perspective on my own situation. Observing the successes or setbacks of many people I knew, I attempted to discern what factors may have contributed to those opposite outcomes. This exercise in reverse engineering enabled me to learn from the experiences of others, a method which gave me a multiplier effect by greatly expanding the sources of real-life evidence available to me. I pursued this not to form a judgmental opinion about other people as a voyeuristic Peeping Tom but only for the practical purpose of analyzing the mistakes and mishaps or the successes and accomplishments of other humanimals. Such valuable life lesson from other peoples' lives could well help me cope with my own aspirations and challenges. Nor was my interest in observing other lives based on curiosity. My motive was purely functional, an attempt to learn from other experiments in living what worked or what didn't.

In passing, I'd note that over a lifetime of scrutinizing and assessing other lives, both of people I knew and those I read or heard about, I never came across anyone past or present who I wished to trade places with. I always accepted and was satisfied with my own little life, obscure and modest as it was. My truly insignificant place in the world was adequate for my purposes, and I was quite content with my circumstances—both its limitations and its opportunities. I doubted if, realistically speaking, my situation could be better, and I knew for sure that it could be a lot worse. Many people enjoyed great success and highly respected accomplishments, but those lay far beyond my aspirations and capabilities. I never once imagined that I might be happier or more content in someone else's shoes. Mine suited me, and I preferred to keep what fit comfortably rather than risk exchanging my life, were it possible, for the hidden defects and disadvantages of others. I realized that often there was less than meets the eye, and that apparent advantages others enjoyed might be outweighed by less obvious problems.

Apart from choosing proactive moves to advance my agenda, my decisions also involved judging what to reject. What you choose to avoid is a factor quite as important as that which you opt to pursue, even if the consequences of a rejected choice are a counter-factual. It's impossible to know how things would work out for what doesn't happen. All you know for sure is how events actually happened to play out. Based on the random happenings which comprise the content of my life, I believe that for the most part any different choices I might have made would most likely have resulted in less favorable outcomes. For that reason, my retrospective perspective—now with little left to transpire before I expire—leaves me with few regrets or feelings of remorse. Apart from some relatively minor matters, neither omissions nor commissions vex me. You always wish some things had turned out differently, but taking a general and holistic view of the entire enterprise, the experiment of being alive, few complaints occur to me, and in any case there's no one else responsible for my choices. Moreover, had even one unfortunate experience or situation instead been positive, that single variance might have altered the future in undesirable ways. To tamper with the past, in fact an impossibility as it forever exists, would for me risk degrading my then future as it evolved into my past. It's all a done deal, one I accept.

My rather conservative and contemplative approach to weighing alternatives, my compulsively cerebral and realistic assessment of potential and risk, my instincts and intuition, the specifics of my experiences all served to guide me well enough to accomplish most of my goals. Even if not decisive, those factors most likely enabled me in some small way to moderate the stubborn and capricious nature of luck. At least it was worth a try, rather than allowing chance to engage in a rampage of completely undisciplined behavior. I tempered rather than tempted fate by recognizing its dangerous random force, and I attempted to mitigate, even if ever so slightly, that wild unpredictable power with a carefully curated assessment of the possible consequences of a particular decision. What little such efforts contributed to the results is impossible to judge, but for sure favorable outcomes for almost any enterprise derive more from luck than from sound judgment or other personal factors. We don't control luck; it controls us.

Most of what happens in life remains contingent and uncertain, another way of saying that luck and chance play dominant roles. The world swirls with potential, some of it favorable and some less so. Possibilities, offers, options, suggestions, attractions and distractions, diversions, enticements, poisoned chalices and ones containing ambrosia, opportunities clothed in rags and dangers dressed in finery all ceaselessly deluge you with a confusing chaotic jumble of widely ranging formats, each and every one freighted with uncertainty and ambiguity. What appears to offer benefits can lead to disaster, while seemingly adverse situations can result in unexpectedly favorable outcomes. A benign circumstance can mutate into a toxic one, a threat into a treat. You never know. A random confusion of choices confronts you, each possibility challenging you to discern on little evidence and with no certainty which of various options might prove to be the most appropriate. Long sought coveted appointments can turn out to be disappointments; unplanned chance encounters might develop into delightful experiences and a continuing connection. All is contingent, nothing certain. The whirligig of time, Shakespeare's spin on the dizzying whir of chance, twirls into view a blur of possibles, each of which will, if chosen, forge an endless chain of different outcomes. At every stage the play of infinite forms re-forms. The consequences of each scene lead to largely unscripted happenings. A right turn may not be right; a wrong turn to the left and you may be left on the path to a dead end. One slightly misguided choice will affect your entire life to come. In dealing with the confusion, I always bore in mind the four fundamental factors noted above, considerations which helped me to sort out some of the otherwise discordant elements inherent in the baffling chaos of one's existence.

Propositions, demands and the agendas of others come in many guises and disguises. Plots and counter-plots ever lurk. Situations involving personal relationships—romantic, family, professional or otherwise—entail especially problematic challenges, as each counter-party belongs to that capricious and unpredictable species, the humanimal. Although an intimate relationship can lead to a satisfying life-long emotional connection, it may also result in a corrosive and dysfunctional situation. Chekhov's Anna and Dmitri escaped unhappy marriages to find love elsewhere. Children can bring pleasure or woe, often both. A bow-wow may nip or lick, hound you for attention or patiently wait for it, serve as a loyal and faithful family member or behave as a black sheep.

Apart from personal relationships, more general agenda items can intrude. These include the needs, wants, wishes, hopes, desires, aspirations, impulses and other demands friends, acquaintances or even strangers try to impose on you. Then there are governmental regulations and exactions and also the constant intrusions of the commercial world. In a relentlessly consumer society, ceaseless efforts to induce people to spend, mostly on wants rather than needs, operate to degrade personal agendas into ones marred by the drive of organizations (non-profits as well as businesses) to sell goods and services. Only by will and effort is it possible in such a society to ignore an intrusive and constant campaign to sell, motivate, agitate, market, advertise, promote, induce people to buy stuff. Agendas remain vulnerable to distortion by commerce.

A commercial proposition often arrives as a wolf-in-sheep's clothing: beneath the soft woolly outer coat lies—truly lies—a rougher texture, abrasive and nettlesome for the fleeced buyer. One way to counter commercial or other intrusions is to limit the number of decisions you make. The fewer decisions, the less likelihood of poor ones. But some decisions are inescapable, mainly ones which relate to major matters with a profound effect on outcomes, both immediate and on into the future. The significance of a decision varies based on three factors: the degree to which it's irreversible or very difficult to amend; the degree to which and the length of time it's likely to affect your life; and the possible adverse consequences which might result from a wrong choice. These risks can be somewhat reduced by simplifying your agenda, by always keeping it well in mind, and by clarifying and focusing on your essential interests. One way to accomplish this is to lead a simple life, including one which avoids consumerism with all its choices. Long ago I withdrew from—in fact, never joined—the consumer society. I thus eliminated many distractions which for other people are attractions, and which entail the burden of making frequent choices.

By removing myself from the marketplace I almost entirely avoided buying decisions, restricting purchases to the necessities and, even with those, acquiring nothing fancy or superfluous. Those required little thought, expertise or expense. Other than buying the basics, I ignored almost all other acquisitive activity. Consumerism seemed to me akin to entertainment favored by people who lacked other interests. Although a large mall operates within walking distance of my small suburban house, the shoppers there at the nearby emporium and I lived in two different worlds, distanced by mutual incomprehension. I couldn't understand the attractions of malls, shopping centers, buying stuff and more stuff, nor—no doubt— could the shoppers comprehend my ascetic thing-less way of life. But lack of comprehension represents a chronic condition of human existence. As my reminiscence shows, not all that much of life did I ever truly understand, but that didn't bother me as no earthling truly comprehends what's going on here on the lonely planet. Some of what little I did manage to learn appears in this book.

I realized that the consumers, a category which included many of my friends, would never understand or endorse my decision in 1969 when in my mid-thirties to abandon from time to time my ordinary organized existence for the disorderly challenges of adventure travel, or my more recent decision to withdraw to produce for this reminiscence a bunch of words. But on second thought, maybe the conventional way of life typical of the frequent consumers and customers common in my setting represents a better format than my way and its byways. I took many planes and trains in my time, but perhaps in some ways I missed the boat. In any case, I was the one out of step with my surroundings, not the locals who enjoyed consumerism. At least they'd found a pleasing diversion, one less unsettling than what I'd chosen with my travels.

My homestead, pleasant and comfortable but without pretension, reflects my attitude toward acquiring stuff. In no way elegant, my residence is, in truth, somewhat tatty but in a genteel and charming way if I say so myself—which I should, as no one else would say it. Before long, after my time, the house will be sold "as is," which typifies my entire life. I lived an "as is" existence, always appreciating my lot in life, never attempting to analyze why the "is" for me was as it is, avoiding "why's" and not searching for the reasons why things worked out for me as they did. I shunned contemplation of how my personality or circumstances or fateful choices happened to be what they were. I just took it all "as is," soon to be "as was."

In one major respect, however, I opted to abandon the settled "as is" regularities of my life in favor of bringing to it a new and quite different dimension. At a certain point—after I was seasoned and had for some years participated in the regular work-a-day world and my life in the suburb where I clung to the format which represented my inherited provincial natural habitat—I decided to reshape the contours of my existence. Although I would in part retain that well-grounded home-based familiarity and security, I decided to pursue an additional agenda which would respond to that ceaseless heavenly but bedeviling trait which dominated my personality—curiosity.

In my early thirties I began to view my regular and routine existence—which seemingly stretched far ahead into the distant future along a straight and narrow route, as if on a train or trolley track—as too limited. This confined and constrained route would, barring the intrusions of chance and fate which might derail me, continue to carry me on comfortably without any twists or turns, onward to the distant horizon and beyond, on and on to the terminal at the end of the line. Such an uneventful ride, I reckoned, would provide a trip far too predictable and uninspiring, a one-way journey insufficient to quench at least some of my curiosity. I didn't have a one-way, one-track mind. I craved a more adventurous trip, one with some curves and detours, transfers to different lines with new directions and destinations. To accomplish this, I decided to plot my own route rather than simply following the conventional set path which stretched before me. It was time, I realized, to add to my life a new element which would enable me to proceed in accordance with Christopher Morley's maxim that the only success is to be able to spend your life in your own way. I decided that my own way would continue to include but not be restricted to the straight and narrow track into the future. To "Morley-ize" my life, I knew that a major change was necessary.

As something of a prologue, this Section I includes many of the main themes—perhaps too grandiose a word: less elegantly stated, call them threads or thoughts, cues or clues—contained in my reminiscence, as well as some of the beliefs, perceptions, practices and principles which served to guide me in formulating and executing my own way. You know the idea—a writer is supposed to foreshadow early what comes later. In the beginning is the end, and then near the end you back-shadow like an Indonesian *wayang* puppet play what happened to conclude the story. The teeming cast in the play of infinite forms casts lots of shadows, and in the end they all melt away into thin air like smoke.

To conclude the preliminaries, in the next Part, number 7 just below, I will now present what for me proved to be a startling and unusual early-day definitive episode, an unexpected happening which typifies the sort of new life and real world experiences I discovered with my new agenda. What occurred by chance so many years ago also exemplifies various verities I learned by experience over more than half a century of adventures. As I've noted above, they largely depended on chance, luck, fate, coincidence, serendipity and other wild cards characteristic of much of what happens as an earthling gropes its way through the brief strange experiences and experiment of being alive.

~ 7 ~

As much as you try to influence how to forge your own destiny, the way your life happens to develop results much more from uncontrollable factors than from your own choices and efforts. With the hope of exerting what little influence I could, I shaped my agenda based on certain interests, preferences and goals, always bearing in mind the four governing factors—health plus personal, social and professional life—noted in the previous Part. An agenda represents only a general aspirational outline, with the particular details which fill it in subject as ever to luck and to your ability to expose yourself to situations and potential experiences which might by chance produce what you hope to accomplish. To fill in the outline and fulfill my chosen agenda required me to go abroad., not just to foreign lands but, more generally, out into the world beyond the settled and comfortable perimeters of my existing circumstances. I aimed to enlarge the possibility of satisfying some of my curiosity, an agenda which would also produce some memorable experiences. But "memorable" might not necessarily mean pleasant. Each outcome remained unsure until after the fact. Out in the big wide world, where you're a hostage to fortune much more vulnerable than in your regular natural habitat at home, misfortune becomes more of a hazard. "Out doors" a traveler becomes subject to a chaos of random and unexpected events, happenings which at times threaten and challenge. Against those hazards you can arm yourself with a certain caution and optimism, with a positive but realistic outlook, and with common sense, all of which will hopefully serve to frame your reactions to facilitate positive results. In some cases it's not the random events but the way you respond to them which determines if they will fulfill your agenda. With other happenings you're at the mercy of events.

So it was that I decided to transcend my normal narrow conventional life to put myself out into the great spectacle of the big wide world, the *theatrum mundi*—the theater of the world—far beyond my comfort zone. It was a setting which in some ways became a discomfort zone. That pleased me. Out on that wider stage, where the play of infinite forms is the featured attraction, luck played the leading role and I was only a bit player subject to direction or misdirection by that dominant and dramatic actor and factor—chance.

In the summer of 1958, as a very young man, not as green as an unripe banana but not yet a full-blown fully grown fellow ripened by experience, I spent about a month traveling through Central America. This trip was not a foreign adventure activated after I restructured my life, for that came only a decade later. Nor was it my first trip abroad or to Latin America, as in 1944 my mother took me, then a young boy, to Mexico for an extended stay. At an impressionable age, this was my introduction to foreign lands and exotic ways, an early-day travel adventure I found pleasingly different from how I'd experienced life at home. My forced but nevertheless pleasant interlude living for a time in Mexico as a boy seemingly sparked my innate curiosity about the world, for over the years I continued to travel and eventually visited some 165 countries, including many return trips to Mexico. I always enjoyed that country for many reasons, not the least of which was that I in part credit that land with initiating and inspiring my life-long interest in seeing foreign places. With that extended youthful 1944 stay in Mexico, I began to suffer not from the common *gringo* ailment of "Montezuma's Revenge" but from a more chronic and incurable bug—"Marco Polo's Revenge." After Mexico I forevermore Polo-ed my way around the world. In addition to frequent domestic trips, I traveled on my own as a teenager to Europe in 1955 and then again two years later, covering the Continent with some grand touring even if not exactly The Grand Tour.

Those childhood, youthful and early-day travels, as I now see in retrospect, were precursors of my later more far-flung and more adventurous experiments in living which took me to the far corners of the world. The earlier journeys to easier and more accessible places foreshadowed the more difficult trips I later pursued in order to respond to my compulsive curiosity. Those travels before 1970, when my new agenda came into effect, were at the time ancillary to my regular home-centric working world. Only in 1970 did adventure travel around the world become a major element of my life, co-equal with my conventional provincial suburban existence. Up until then I followed the usual route common to my station in life, a track as fixed as the rails of a streetcar line which took me to the customary stops common in my day and age, place and station in life: education, profession, a club (inherited, not joined), the normal work-a-day routines of a typical member of my particular socioeconomic group. The context of my circumstances in my given place and time defined my existence. Although my curiosity originated quite early, during my younger years it remained subordinated to the practicalities of ordinary life and getting established. I dis-established myself only later. Meanwhile, I managed to indulge my interest in the outside world with occasional trips. Those for business often gave me an opportunity to divert a few days to non-professional activities. In other cases I contrived to travel on my own itinerary rather than as an add-on to a business trip. After travels around the U.S., Canada, and Europe, I initiated my first solo adventure to a less developed region—not exactly Third World, but perhaps an area better defined as Two-and-a-Half World. That trip took me in the summer of 1958, between school years, to Central America for about a month. I started in Guatemala, the northernmost country, then made my way through the other five on down to Panama (only later did I get to Belize, Central America's seventh country).

For my adventure travel I always preferred to go without advance hotel reservations or other prior arrangements. This allowed more flexibility, even if creating more uncertainty. With long and assiduous preparation for every trip I always embarked equipped with detailed information about where to stay, where to go, what to do, what to see, how to get around, and other practicalities. With me I carried my extensive notes along with copies of guidebook pages but rarely the books themselves, which were too heavy and bulky. City maps I studied carefully so that by the time I got to a place I already knew its geography. A need for efficiency to make the best use of limited time prompted my emphasis on preparation. I never arrived anywhere without detailed information about my destination. What remained for me was to see it—quite a different matter. Words can only take you so far. What transpired when I was actually at a destination couldn't be foretold by reading books. Any difficulties, challenges, plots, counter-plots and other happenings remained unknown until the tale was told. The uncertainty lent a sharp age to the enterprise.

From early days I tried to avoid upscale, fancy or chain hotels in favor of ones with a local ambiance offering historical, cultural or artistic elements typical of the area. In Guatemala City I chose the Pan American, a vintage colonial-style hotel with art deco trim, a patio-like dining area, a characteristic balcony wrapped around the middle of the three-floor building. I also always sought a convenient location, and the Pan American occupied a block in *Zona* 1, the *centro histórico* of Guatemala City.

History lingers at the hotel, which stands on the site of the 1914 Pension Amado, built to accommodate early-day tourists. After earthquakes in 1917 and 1918 destroyed the Amado, construction began in 1920 on the Hotel Astoria, which opened the next year. The architect was Rafael Pérez de Leon, who also designed the nearby Presidential Palace as well as the National Palace (Parliament). Back in the mid-1940s, when the Astoria changed its name to the Pan American, United Fruit Company and Pan American Airlines staffers stayed at the Pan American (for some reason it wasn't called Hotel United Fruit), the city's only luxury hotel until after World War II. Even years ago when I stayed there the Pan American was a period piece. By now I'm also a period piece, but back then I was a fresh-faced young fellow, even if not a complete travel neophyte or Twainian innocent abroad. But the summer of 1958 was early days on the long road over the long years, and compared to what came later was in large part uneventful—except for what happened in two episodes, one in Guatemala City and then later out in the countryside. At the time I was at the very front end of my life-long roamings and worldly explorations. Although I didn't and couldn't know what the future held, as things turned out for me thousands of miles, twilights and nights in foreign lands far from home were to endow my years with a rich stock of travel experiences and memories, and I somehow managed to live to tell the tale.

During the first few days of my stay at the Pan American I noticed at breakfast and dinner at the hotel restaurant a dapper well-dressed fellow about 35 or 40 eating by himself. After a time we started to acknowledge each other's presence by the nod of recognition common when a strange face transitions after a few days to a slightly familiar but still not well-known one. Before long the mystery man—and to him I was equally a mystery, a young American traveling alone, in Guatemala City for some purpose unknown to him—and I started to exchange a few words of greeting and then, in a rather normal progression, we began to chat, conversations no doubt provoked by mutual curiosity. When we were finally on speaking terms I asked the man what he did.

He replied, "I'm the secretary."

Not too macho or impressive, I thought. "You're a secretary?"

"No, I'm <u>the</u> secretary."

"Of the Pan American—of the hotel company?" I continued. That would at least sound impressive, more encompassing than just "secretary." He'd be a functionary of some kind at the venerable and well-known city-center hostelry. "Pan American" covers a lot of ground if those words figure in your job description. Maybe the hotel manager or perhaps even the owner was called "secretary" here in Guatemala. Things were different away from home. You never knew what curiosities you'd find.

"No, no," he insisted.

"Then whose secretary are you?"

"I'm the President's secretary. *El Presidente*."

Suddenly I understood. "So you're an important executive with the hotel company—the president's secretary."

"No, nothing to do with the Pan American. I live here—I don't work here."

"Then what?"

"*Señor*, I'm the secretary of the President of the *República*."

"The President's secretary?" I echoed.

"Yes—*El Presidente de la República*."

My new acquaintance's job description somehow sounded more impressive in Spanish. This was suddenly and surprisingly getting to be big time: this *hombre*'s domain was not the narrow precinct of the nominally grandiose Pan American but pan-Guatemala—the entire country, city and village and jungle.

"I see, I see," I semi-stammered. Maybe he took this as *Sí, sí*. I hoped so: I wanted to be agreeable—I was talking to <u>the</u> president's secretary. Already my thoughts were churning with possibility: how could I benefit from this entirely random encounter? Here I was, a very young *Norte Americano*, a newly arrived stray stranger in a banana republic and already by mere chance I'd encountered one of the country's top officials. It may have been a steamy, jungly *plátano república* but so what? Suddenly I had a potential in with this Banana Republic's top banana.

Later, much later, I learned more about Mauricio Rosal, my new local contact. Meeting him, I'd shot right up to the top of the pecking order. At the time I didn't ask him any questions about his background or life. In this case, potential benefits from the random encounter suppressed my curiosity, which Rosal might have found intrusive. I instead focused on the unanticipated possible opportunity which chance had brought my way.

"I've never met a president," I said. "Do you think you could introduce me to him?"

Contemplating my request, Rosal remained silent for a few moments. Finally he replied, "Come to the Presidential Palace tomorrow afternoon at 1:30. I'll take you in to see *El Presidente de la República* then."

Of course, at the time I suffered from a complete information deficit, not only about

Rosal but also about the President. I didn't even know his name or, in truth, that Guatemala had a *presidente*. For all I knew, the country's leader might have been called *líder* or *jefe*, *caudillo* or *cacique*, "premier" or "prime minister." There were a lot of ways to designate a top banana. Maybe he was even known as *La Cabeza Banana*. But *El Presidente* sounded rather more impressive. My detailed trip preparations had in any case had not included the prospect of meeting a president, let alone his secretary.

Ignorant of everything pertaining to the big man, I'd conduct my encounter with the chief executive the next day completely uninformed, quite unlike my usual hands-on thorough preparation for meetings, travels or anything else. I did know that in Guatemala grew lots of bananas, but was that enough information to start or conduct a conversation with the President? "Hey, *Señor Presidente*—tell me about all the bananas here in Guatemala." No, that wouldn't do it. It would be the shortest audience in history. I'd have to up my game—maybe talk about mangoes and papayas. I realized that I lacked enough information to engage in a meaningful dialogue. I wondered and worried about what we could talk about. On the other hand, I consoled myself with the fact that the President didn't know anything about me either, so in a way we were on equal terms—other than that he was *El Presidente* and I wasn't. I'd chat with the old boy man-to-man in a *mano*-to-*mano* match of wits, even if our lives hadn't been furnished with equivalent worldly experiences. The President was a seasoned general who'd somehow risen to the very top in the cut-throat machete-sharp local political jungle. I, on the other hand, was a wayward young *gringo* wanderer, a green bottom banana as low as you can get without dropping unripe into the jungle tangle below.

Only years later, long after my encounter with the President, did I learn the details about Miguel Ydígoras Fuentes when I read his account in *My War With Communism* (written with Mario Rosenthal and published in 1963). In his memoir, the general describes the many intrigues which swirled through the political scene and threatened him, including frequent plots and counter-plots and communist attempts to overthrown the government and grab control of Guatemala. The country seethed with conspiracies. The book's title would make it a sure-fire best seller at the Pentagon, the National Security Council, and at other Washington security agencies willing to pay big bucks (or *quetzals*) to keep *El Presidente* from becoming "ex" if the communists deposed him. It was the Yankees against the Reds, with Guatemala's incumbent politicians playing for the home team. As things turned out, what Ydígoras Fuentes calls his "involuntary retirement" occurred when he became "the victim of Fidel Castro's aggression." in late March 1963. The former president writes that on March 30 six tanks and 900 troops attacked his residence. He refused demands to resign and the next day flew out to Managua, Nicaragua to begin his life in exile.

Before meeting the president I imagined that he'd be the stereotypical Latino strongman, a general with gold teeth, a thick black mustache which bristled with menace, pockmarks, hairs growing from his nostrils and ears, and five o'clock shadow lending his face a sinister appearance. But Ydígoras Fuentes was in fact quite the opposite of the unkempt thug-like character I'd envisioned based on the stereotype I'd invented rather than on real-life experience. His teeth were polished politician white, no scars disfigured his clean-shaven slightly puffy face, every hair was under control and in its proper place. As I later learned from his book, the President's background and career had been equally civilized and well-groomed. He'd spent four years in Europe and the United States, attached to the Guatemalan embassies in Paris and in Washington, and had served as a member of his country's delegation to the Treaty of Versailles. Later, from 1944 to 1949, he was the Guatemalan ambassador to Great Britain. From there he returned home and "enrolled in the kindergarten of Guatemalan politics" as leader of an anti-communist political party.

In his book, Ydígoras Fuentes contrasted his life with the common conventional version of south-of-the-border political figures, noting that "Unlike the traditional Latin American military man who becomes a president without setting foot outside his own country...I was fortunate enough to be able to travel and to spend a good part of my adult life outside of Guatemala." This passage, in the book and in his life, of course quite appealed to me as it reflected my own values, which highlighted the merits of curiosity, travel, foreign experiences and gaining different perspectives by spending a good—very good—part of my life outside my home and homeland. I wish I'd known at least that one element of the President's mindset, his appreciation of travel, before meeting him. In that one way, even if no others, we were similar types—slightly soul-mates. But unlike Ydígoras Fuentes I was never president of anything, not even my kindergarden class or home owners association in my subdivision, where I was only residential and not presidential. I was unsuited to running anything except my own life, and I wasn't too sure I could manage that endeavor successfully: it was an experiment.

At the designated time specified by Mauricio Rosal for my audience with the President I arrived at the Presidential Palace, a short walk from the Pan American. The Palace wasn't particularly palatial. A low-rise colonial-style structure with no pretentions but electric with tensions, as I later learned from the President's—by then "ex"—book, the building occupied a large part of one of the squared-off blocks in the geometric center section of the city. The regular angles and shapes there seemed to me to be an attempt somehow to carve out a civilized and orderly enclave in contrast with and to keep at bay the natural vegetal chaos and entanglements in the jungly lands beyond the capital's cultivated man-made built environment. Within the civilized Presidential Palace the only jungle was the political one

When I arrived an attendant escorted me onward into the outer part of the inner sanctum where the President's office was located. There I met Mauricio Rosal, who proceeded to take me into the *sanctum sanctorum*, the chief executive's office, a process much as described by Guatemalan Nobel Prize winner Miguel Angel Asturias in his novel *The President*: his secretary ushers in a visitor as the President stands there, "his head held high, one arm hanging at his side in a natural position, the other behind his back." But contrary to Asturias's description of his fictional *Presidente*—"toothless gums, a grizzled mustache combed over the corners of his mouth [and] thin pendulous cheeks and pouched eyelids"—the real-life President was well-preserved and dapper, with no traces of pouched, pendulous or otherwise rogue skin, and his teeth (whether originals or surrogates) occupied their appropriate place in his gums.

I scrutinized the President closely, not his gums but much of the rest. This was my chance, most likely my one and only chance, to see at close range what a president actually looked and acted like. He had a ruddy, full, fleshy face—obviously well fed. Perhaps he'd enjoy discussing mangoes and papayas, and maybe even bananas: my conversational concerns would be solved. His manner was understated, not overbearing. He greeted me in a friendly, welcoming way. Maybe his pleasant demeanor arose from the fact that I wanted nothing from him, other than a few minutes of his time. I appreciated that he was willing to give up matters of state to spend a brief interlude with a young American traveler he knew nothing about. My visit with the President otherwise held no other agenda—I sought no job, no favors or favoritism, no commercial advantages, no hand-outs or hands-on help, no government contracts or subsidies, no string-pulling or rival-pushing, no concessions, no assassination of competitors. My agenda-free session made me a different, maybe even unique, kind of presidential visitor. In a context where normally everybody wanted something, I wanted nothing—except an experience, one I might recall with pleasure years later.

Gazing at the man, gums and all, I wondered what combination of chance, luck and fate, along with effort and shrewdness, had operated to lift him to the top, a height from which he was later toppled. But for now he was *El Presidente*. To gain a useful and often effective credential for success in other countries you can attend Harvard Law School, Oxford or Cambridge, *L'École Nationale d'Administration*, Tokyo University, Bocconi, work at Goldman Sachs or McKinsey, or study, remember and then apply Snoopy's wisdom. But what's the secret in Guatemala of rising from nothing to general and then on up to the *número uno*? What bayonet-sharp steely resolve and cunning plotting and counter-plotting operated to put

Ydígoras Fuentes into the top slot? Since I was not gifted to achieve any success in a leadership way, based on inducing and influencing other people to support an agenda focused on gaining a commanding position, I was always curious how someone managed that accomplishment. These sorts of traits or skills eluded me and, in other people who happened to possess such compelling characteristics, they remained mysterious to me. What methods or techniques would enable you to persuade followers to facilitate your own personal aspirations? Some aspirants to leadership or power seem to possess some hidden magnetic force which draws people to them and then allows the leader to pull his followers along in the direction he wants to go. All such traits—personality, authority, charisma, magnetism, an ingratiating manner, an emollient attitude—escaped me. Whatever the ingredients, Ydígoras Fuentes had them, while I possessed no leadership or for that matter followership resources whatsoever. For me, however, lacking those two traits was a win-win situation, as never did I aspire to be a leader and certainly not a follower. I only wanted to live my own life, and not in any way govern others or persuade them to endorse, promote and facilitate my agenda.

Although at the time I of course didn't know it, in July 1958 a senior CIA official described Ydígoras Fuentes as "known to be a moody individual" who "regularly disregards the advice of his Cabinet and other close associates." With me, the President displayed no moody behavior—he was cordial and friendly—nor did he disregard my advice, as I never gave him any (didn't even suggest that he'd look better if he lost a few kilos). I'm sure he appreciated my forbearance.

Mauricio Rosal lingered in the office off to the side as I engaged in chit-chat with the President, chat-chitting about who I was (I knew who he was), why I happened to be in Guatemala, how I'd met Rosal, where I planned to travel next, what I did back home. For once I was out-questioned. Perhaps the President wanted to make sure I wasn't an emissary with some sort of revolutionary mission. It felt strange being with the President there in his office, the very lair of political power in the country. A few days earlier I'd been simply a young traveler on my own, a transient tourist with no contacts and no commercial or substantive reason other than curiosity to visit Guatemala. Now, suddenly and unexpectedly, I was inside the Presidential palace in a get-together—I didn't dignify our session by calling it a meeting—with *El Presidente de la República* himself. My status had quickly risen from a nobody in a foreign land to someone with a providential presidential connection. Luck had somehow plucked me from the Pan American to become a lucky American confabulating with the President. For the moment at the top, my experiences in Guatemala could only go downhill.

After a time Ydígoras Fuentes said to me, "I want to show you a letter I recently received from a friend of yours." Now things were getting really interesting. The President and I knew someone in common? What a coincidence! The plot thickened. Which mutual friend could possibly connect the President with me? And what did the letter say? A president received all sorts of letter. Was it likely one came to him from a friend of mine?

As Asturias wrote in "The President's Post-Bag" (chapter 23 of the novel), the fictional president received a stream of mail, 16 letters in one group alone—petitions, requests, information, compliments, news, gossip, flattery, entreaties of a type which I, lacking an agenda, didn't bring to my real-life President. Nor did I discuss with him matters similar to letter number 16, the last one Asturias describes, in which one Tómas Jareli announces his marriage to *Señorita* Arquelina Suarez "and desires to dedicate it to the President of the Republic." Now that's what I call an exalted status: a marriage dedicated to you. A visiting Tómas, I didn't meet with the President to announce my marriage and dedicate it to him and, unfortunately, I didn't even know any local *señoritas*. I considered asking him to introduce me to a few, but thought better of it. I didn't feel assertive enough to turn the President into a matchmaker, and in any case I wanted to keep my session with him agenda-free. Besides, he'd no doubt prefer to keep any spare *señoritas* for himself.

Waiting for Ydígoras Fuentes to reveal to me which "friend" of mine had put a letter into his post-bag, I watched with anticipation as the President walked over to his desk, rummaged through some papers, and then retrieved the mysterious mail which he now handed over to me. Typed in English on crisp thick expensive paper (high rag content), courtesy of U.S. taxpayers, the letter, on fancy White House stationary, was from President Eisenhower to the Guatemalan President. The quality of the paper, if not its written content, was impressive. The letter spoke of the continuing good relations between the two countries and included other friendly and diplomatic anodyne "happy meal"-type State Department boilerplate. Eisenhower's emollient remarks didn't surprise me. After all, a former general in power in a Latin American land threatened with a communist take-over could serve as a reliable soldier in the war against Castro and revolutionary communism. With the Yanks you couldn't back then go wrong being an anti-Red and loyal Fidel foe. The *Norte Americanos* would pay up for those policies. Selling security services in the political and out in the natural jungle teeming with guerrillas could be a more profitable export than boatloads of bananas.

President Eisenhower's pen-pal told me that the letter had been personally delivered to him not long before my summer of 1958 visit by Milton Eisenhower, the president's brother, during his trip to some Central American countries. I was mildly disappointed to learn that I wasn't the first *yanqui* that year to meet with Ydígoras Fuentes in his Presidential Palace office. I was also bothered by my failure to arrive there with a gift, unlike the previous visitor. Milton had brought a pair of silver cuff links from the president for *El Presidente*. My host had thoughtfully worn them the day of our get-together so he could display the presidential present to me. I admired the cuff links, gratified that my president showed such good taste in jewelry. In his book, Ydígoras Fuentes briefly mentions Milton's visit: "We were honored with the visit of Dr. Milton Eisenhower in 1959 [sic: it was 1958] and we had several important meetings with him." Getting free silver jewelry is indeed an important occasion. The President's book for some reason fails to mention that my visit was "important"—I'd arrived empty-handed—and, in truth, didn't mention my visit at all. This is no doubt due to the fact that I didn't show up with a present for *El Presidente*. I'm happy now to try to make amends for my boorish behavior by presenting here my favorable report about how Ydígoras Fuentes received me. He couldn't have been more pleasant, even if perhaps disappointed at my presence without a present.

In his book *The Wine Is Bitter: The United States and Latin America* (published in 1963) Milton Eisenhower failed to mention Ydígoras at all. The emissary's account relates that "I had many talks with Guatemalan leaders," who claimed "a tremendous number of urgent needs." Maybe one of them was a need for cuff links. Milton found more interesting a visit with "the market women," practical and productive "small capitalists" who suggested that U.S. aid be more widely distributed to counter "the impression that our aid goes to a relatively few individuals, strengthens the rich, and does nothing for the poor"—and thus typical of much of our efforts to help Third World countries. Upon reading the emissary's comment long after my 1958 visit, I realized that I should have met with "the market women" as well as with the President to more fully experience and understand the country. In fact, over the years I learned to make a point of chatting with the locals in order to get an on-the-ground view of each country I visited. This conformed to Andrew Grove's practice of getting "out doors" more. Simply seeing the tourist sights gives you few insights into the realities of a place.

In Eisenhower's book I also found a useful formulation which to my mind partly explained the degraded state of society and the economy in various underdeveloped countries I saw around the world. "Guatemala suffers from under nationalism," he wrote, as "There is almost an absence of patriotism among the people." This illustrated for me one of the main reasons why so many places seemed dysfunctional and disabled by conflicting and incompatible agendas. Loyalty was not to the country but to clans, cliques, tribes, sects, factions, all of them competing for limited resources rather than cooperating to create new ones. Affiliation with one of those groups denationalizes patriotism and fragments national cohesion.

When I read my "friend" President Eisenhower's letter and saw the shiny cuff links there in *El Presidente*'s office I was truly impressed—two general-presidents in one day! That was two more than I'd expected when I'd arrived in Guatemala City a few days before. Now I was reading President Eisenhower's private, secret diplomatic correspondence with another president, written on fancy White House stationary no less. Its quality paper made me proud to be an American. You just never knew where life and luck and your travels might take you. What would be next?

After the President displayed his cuff links to me I sensed that our conversation was drawing to an end. My time with him was now short and before long I'd be back out on the street, just another face in the crowd, an ordinary *José* like every other Tom, Dick and Harry. I didn't want to give up my precious face-time with the chief executive without a few further questions which might enlighten me on what the life of a president is like. In that I'd never met a president before (or, for that matter, since) I wanted to seize the opportunity to get an authoritative inside view of the job. Various potential questions spun through my mind, and finally I asked, "What's it really like being president?"

It seemed that this surprised—maybe even somewhat startled—Ydígoras Fuentes. Not responding, he became contemplative for a time. "So glad you asked" (he didn't say) "because I'm going to give you an answer which will stay with you for years, an answer" (he did not continue) "which will be so thought-provoking and suggestive that it will ever haunt you and force you to contemplate all sorts of serious matters, and maybe one day," (he never said) "one day in the far, far future long after I'm dead and when you're old and about to become a newly dead, soon to be in your cemetery plot, maybe then you'll recall my answer and perhaps include it in an account of our long ago 1958 fleeting meeting here today." The President said nothing like this imaginary response to my question. In fact, for a good while he said nothing at all.

I waited patiently, maybe a little impatiently, for his answer. He deliberated, as if it were a great matter of state. He was obviously searching for a definitive response. I appreciated that he took my question seriously and did not simply give me a quick off-the-cuff link answer. Finally, in a low voice, he said, "There are many plots and counter-plots."

That was certainly not what I'd expected him to say. But years later, when I read Ydígoras Fuentes' book, I saw his unsettling answer reflected in the title of chapter 3: "Enemies to the left of me; Enemies to the right of me." Enemies, threats, hazards, conspiracies, rivals, plots and counter-plots—another typical day in the life of *El Presidente*. He wrote: "Like all presidents, and most people, I was confronted by problems that had no solution," adding (elsewhere in the book) that "while Guatemala enjoyed democracy, I suffered it."

Problems with no solutions, not only for presidents but also for almost everyone else: this was a thought-provoking observation which, in his book, echoed Ydígoras Fuentes' answer to my question during my long ago brief interlude with him, a response which over the years has continued to provoke my thoughts. As the President suggests in his book, there are indeed no ultimate solutions for any situation, or—to expand the notion—for existence itself, which is only an experiment without definitive answers. To each challenge and problem you can respond only by tentatively groping your way along to deal with the uncertainties which shadow every decision. Random and capricious factors, the plots and counter-plots intrinsic to being alive as a humanimal, abound. The President's answer reflected not only the title of his chapter 3 but also the comments—art imitating life or did life imitate art?—of Asturias's fictional *El Presidente*, who complained how "my enemies scheme [with] intrigues and malicious writings." But at least presumably Ydígoras Fuentes didn't suffer from rivals who, Asturias wrote, tried "to influence North American opinion, in the hope of discrediting me in Washington." To the contrary, the real President sported those silver cuff links, tangible evidence to prove his solid links with "the great colossus of the North," as he describes the U.S.A. in his book.

Although the President didn't last—evicted by force from his Palace—in my mind his reality-based comment about plots and counter-plots certainly did. Over the years I often recalled his ominous observation and took it to heart as a cautionary warning for my own simple life, less presidential but perhaps more providential than *El Presidente*'s. Unlike him, as time went by I at least managed to retain my un-palatial little house and had not (yet) been forced to flee for my life. In the typical banana republic way, a *golpe* deposed the President, the *coup d'état* overthrowing him in March 1963 after a rebel invasion of Guatemala promoted by the Cuban communist regime. Ydígoras Fuentes escaped by plane to Managua as a banana republic plot toppled the top banana, succeeded later by the counter-plot initiated by a new banana. In Latin America it's really hard for an incumbent to keep those cunning plot-driven ambitious banana types down. I wonder what became of those silver cuff links *El Presidente* received from President Eisenhower. Maybe after the coup a gold-toothed general added the silver adornments to his precious metal collection, or perhaps the gift from "the great colossus" was lost in the confusion of the plots and counter-plots which in the end brought a new president to the Palace.

As for my link to the President, the facilitator who arranged my visit, Mauricio Rosal, and I now shared an experience and, as such, we were on more than just polite chit-chat and nodding terms during the few additional days of my stay at the Pan American. We even took a few meals together. I reflected on the rather strange sequence of events my chance encounter with Rosal had activated. A week or so before I was at home in my own familiar surroundings, far from the Pan American and the Presidential Palace, and neither Rosal nor I knew the other existed. Then, suddenly, we happened to become joined in a fragment of time in a tiny corner of the world to briefly enter each other's lives. We experienced together some events, their centerpiece my brief session with the President, and afterwards held in common memories of those happenings. Of such random little vignettes, one by one by one, is a life formed, the individual components gradually accreting to add up to...a series of little vignettes. They remain such—disjointed, diverse, discrete—until, perchance, he who experienced it all might someday toward the end bestir himself to try to shape some sort of mini-meaning, artificial and arbitrary but better than nothing, by setting some of the disparate little pieces into some sort of framework. The authorial governance creates something of a plot, presiding (in this was I was presidential) over the available material accumulated at random by arranging some selected items into links, more rough than cuff, to create a sketchy impression of how the past looks from a late-in-life perspective. Reviews organize the original views into a pattern, one of many possible.

Not long after my presidential visit I left Guatemala City to travel around the country, and then continued on my trip through Central America. The strange interlude—me, a vagrant wayfarer of no standing, with *El Presidente*—soon seemed like a *la vida es sueño* episode, a dreamy and, in retrospect, quite improbable sort of happening.

I never saw or had any contact with Mauricio Rosal again, nor with the President, who never called me for advice about plots or counter-plots or any other matter unless, perchance, he dialed the wrong number and failed to reach me. Or maybe he called collect and I refused to accept the charges. As time went by, every so often I recalled with gratitude and pleasure Rosal's role in introducing me to the President. Only more than two years later did I hear of Rosal again. In early October 1960—by which time Ydígoras Fuentes had appointed him Guatemalan ambassador to Belgium, the Netherlands and Luxembourg—one Mauricio Rosal was arrested in New York City with 110 pounds of pure heroin he smuggled into the U.S. under cover of diplomatic immunity. As *El Presidente* had told me, "there are many plots and counter-plots." In his book, Ydígoras Fuentes for some reason failed to mention his trusty former secretary.

In the years to come, the President's reference to plots and counter-plots served as an extremely useful early-day lesson for a curious someone destined to roam the world seeking experiences and adventures.

Section II
THE FATAL CIRCLE

~ 1 ~

Many years—a good part and in fact the best part—of my allotted earthly quota have passed since my youthful long ago and far away chance encounter with *El Presidente de la República* in Guatemala City. Back then those years, almost an entire lifetime, all stretched ahead of me, an empty future filled with potential. All the then unknowable possibilities have by now occurred and congealed into my forever fixed past. As for every humanimal, the transition from the then to the now entailed a chaos of fateful chances and choices. Random happenings, uncertain situations, problematic decisions, unpredictable outcomes, twisty turning points, improbabilities which occurred and probables which didn't, unexpected challenges, chance and happenstance, plots and counter-plots all obscured my way. Not everything which happened derived from random and fateful forces. Existing conditions—inherited, acquired or willed—also influenced the evolution of my life, as did the cumulative constraints of all the past events which operated to shape the future. Many factors combine to determine what happens, but just how they will in turn influence future events is unknowable. The nature of a person's lived reality, as it happens to develop over time, is random and odd, but after the fact what took place and exists in the past soon seems settled and normal. Out of a nearly infinite number of possible outcomes, chance produces a single final result, so singular and specific it somehow seems inevitable. Once you have a certain son or daughter, once you get a particular Belgian Shepherd, Pomeranian, Chihuahua or mutt, once you chance a specific spouse, often a complete stranger at the time you randomly happen to meet him or her, then wild potential is tamed into reality. The tangible specifics of the being—dog, child, wife, husband—soon becomes a familiar matter of fact, one so settled you rarely if ever contemplate any of the myriad alternatives which in theory once existed before what actually happened finally occurred when elapsed time operated to reduce all possibilities into a single unalterable outcome. The outcomes gradually accumulate to form your life.

Ahead of events there's no telling either what will happen or how the cumulative events will, over time, operate to determine the content of your earthly existence. Only long after, as for me now, can you review everything and then produce a tell-tale account, telling some sort of story about what occurred. You can forget about all the what-if's and might-have-been's to concentrate on the was and the is. Different contexts produce different degrees of exposure to random factors. While remaining in your known and familiar surroundings, where many regularities prevail, fewer uncertainties exist. In theory, this reduces the likelihood of errant negative occurrences, but in fact almost anything can happen anywhere at any time. However, once you emerge from your normal narrow and habitual habitat to travel in the wild wide world you become vulnerable to many additional and unusual uncertainties and random happenings. The odds shift from somewhat predictable home-based routines to odd wild-card plays of infinite forms. Out there away from home, luck and all manner of surprises, countless unexpected events, unfamiliar settings and situations, alien practices, strange occurrences and many curious experiences await you. It's a free-for-all where plots and counter-plots, agendas hidden or obvious, abound.

As you grope your way through distant foreign lands, fate and chance somehow seem to conspire to push you now in one direction, now in another, ceaselessly plotting with no thematic plot to bring to you all sorts of unanticipated experiences, some favorable and others less so. Out there in the wide world so much happens by chance you never know if "perchance to dream" or perchance to nightmare will reflect what befalls you during the trip. Before

Section II: The Fatal Circle

you've flown it's all unknown; once you land luck deals its hand. Will you enjoy pleasant times comprised of such stuff as dreams are made of or a haunted incubus of a journey which trips up your desired trip? The contents which from moment to moment develop and come to comprise your life may unfold with some rhythm but without any rhyme or reason, other than what an arbitrary after-the-fact arrangement of the events happens to delineate.

The contrast between what might at any moment happen and what actually does—the vast variance inherent in myriad possibilities versus the one actuality—greatly influenced my view of the highly contingent nature of existence. After happenings become set in time they're so real it's difficult to accept that things could have been different. The extreme difference between the potential and the actual, the latter invariable but in no way inevitable, in truth entails a randomness which I found quite disconcerting. Such an uncontrollable, arbitrary and capricious force implies that how a life turns out depends for the most part on chance. My observations are hardly original, but that's not at all my intention. I mean only to describe my own view of the factors which most influenced my life, not to propose any new philosophical principles. As I've tried to show in this reminiscence, for me luck and chance ruled the day, the years. Other people can write their own account to show that things were different for them. All I know for sure is how my life happened to develop. In writing an account of a single humanimal's course through life I'm a one-trick pony, not a sage who means to present novel ideas for all humankind.

In any case, I don't view lack of originality as a flaw but only as an honest recognition of my inability to be original. My observations on what befell me are, admittedly, rather limited. Some, many, most or, in truth, probably virtually all of my comments in this reminiscence are quite unoriginal. My review of the world and of life as I experienced it record a few selections and observations I found of interest or noteworthy or useful to enable me to gain a perspective on what happened to me during my earthly existence. A work seeking to be original will be mighty short. What occurred in my life, or in anyone's life, is singular and unique. Many of my observations about those occurrences are commonplace and thus few, if any, original thoughts or perceptions endow my account with transcendent claims. I don't pretend to supplement the wisdom of the ancients or of the moderns or of any of the sages who lived in-between with Tommy-come-lately previously unnoticed revelations. By now, earlier generations have already thought of just about everything, and I'm incapable of knowing and am uninterested in presenting what little remains unsaid. My reminiscence includes only my own memories and impressions, based on my particular experiments in the experience of existence on the lonely planet where billions of earthlings have lived, are living, will live, all of them fated to cease to live, to disappear never really understanding the odd phenomenon of being alive. From the play of infinite forms I happened to experience I describe a few, reduced to the finite form of a single individual. None of my reactions to or observations about what happened are original except to me.

Like almost every life, mine was unique but not special. It all happened to me in one now after-the-fact set way rather than in any number of other possible one time pre-fact formats. Best as I could I tried, in the spirit of Christopher Morley's view of a successful life, to make it my way. Just as what occurred might have been entirely different, so too my reminiscence could have been arranged in any number of ways. Nothing is fixed or firm until it assumes its final form, at which time uncertainty ends. Until then all remains in flux. What the reader sees while reading this account, a minuscule and fleeting part of his or her life experience, represents just one possible version, only an artificial authorial imposition attempting to organize exact facts into some sort of order. Although the facts are true, their arrangement is arbitrary. The interplay of what's true—all the lived experiences, episodes, happenings—with the contrived order I imposed on the material reflects the juxtaposition of my real-life encounter with Guatemala's real President and Asturias's fictional *El Presidente*. The novel supplements my factual version of the episode with additional insights in a way similar to how my artificial selection and arrangement of the original confused life events serves to bring some clarity and meaning to what was otherwise just a chaos of dissimilar experiences. The events were contingent, mere potential until they happened and then became immutable. Their arrangement in a narrative remained completely mutable until—organized, reduced to print and produced in a final binding format—the account of what happened becomes a new fact. The two components form a kind of double-helix—a heltext?—comprising a factual-fictional construct intertwined into the final version.

The great existential uncertainty which confronts every earthling I took as a given. The reality that so much is unknown and unknowable didn't bother me: it was simply an unavoidable condition of being alive. I had no choice but to accept that a humanimal can engage in the experiment of existing only by gradually groping its way forward, at times suffering setbacks. You advance or regress along a shadowy path through a dark wood thick with mights, perhapses, maybes. Perhaps every now and then some sparks of illumination light your way as if by the flickering flames of a log fire burning into the obscurity a bit of brightness before dying into darkness. Looking back, I can only view with amazement how I managed to navigate through the perils and threats, the plots and counter-plots, now soon to emerge from the woods near the end relatively unscarred and unmarred by the thousands of threats, known and unknown, which lurked along the way. So fate-shaped and luck-dominated is every mortal's life, to this day I'm really not quite sure how my largely regret-free and for the most part satisfactory existence, soon to non-exist, happened to develop in the way I hoped in the face of so many contingencies. At any number of turning points, some with a finality which allowed no returning, I might have gone astray with unfavorable or even disastrous consequences. While managing to avoid those sorts of hazards, I also benefited from many unplanned favorable happenings (my thanks to, among others, *Señor El Presidente de la República*) which served to enable me to continue through the woods along my own path. Chance meetings with strangers who became life-long close friends, unplanned pleasures and delights, unanticipated opportunities, unexpected beneficial outcomes, quirky turns of fate which brought me serendipitous pleasant experiences, and many other random happenings operated to enrich my earthly existence, so making me a contented humammal who enjoyed creature comforts and relatively few discomforts while pursuing the creature's dual home and away allelomorphic-type double existence. Events often arrive in disguise. In some cases what at the time seemed freighted with great portent and potential fizzled into merely a transient and soon forgotten episode, while at other times what appeared to be a fleeting moment in time of no enduring interest by chance ripened into a highly gratifying and memorable happening. The older I became the more I attributed to such random factors as luck, fate, chance, destiny, serendipity, coincidence, happenstance the dominant role they play in shaping a life experience. By now, very near the end, it seems to me that those factors represent the controlling elements which determine how the experiment of living is conducted.

Although my good fortune, a residue of good luck, brought me great satisfaction I recognize that an account of contentment, rather than tales which detail travails, may be of little interest to contemporary readers. I was a happy camper, and a content ghost writer offers less haunting material than one spooked by life. But, on second thought, maybe when you finally close the book's cover after reading the entire text what I have to say might indeed haunt you. Read on through and you'll see what I mean. It's no doubt rather unfashionable these days to admit that one has enjoyed a satisfactory existence—unfashionable and thus quite in accordance with how I fashioned my life and how in fact it played out. I was largely indifferent to many customary and conventional practices and rituals. The security blanket of societal conformity I to a large degree tossed aside, just as I eventually rejected the pillowed comforts of home as the be-all and the end-all of my existence. I didn't rebel against the prevailing standards: I simply ignored many of them. Some I adopted, others I renounced; the few I disdained I refrained from criticizing. Willy-nilly, I conformed when it suited me or was otherwise convenient, while in other cases I silently dissented and went my own way. My protocol depended on how a decision, practice or attitude might advance or damage my agenda. Some of my behavior might have seemed slightly eccentric, but to me it always made sense in terms of my hidden agenda, hidden in the sense that I didn't make it manifest or explicit but no doubt otherwise obvious in how I behaved. I aimed to live in my own way, which in some ways failed to conform to conventional norms.

I certainly hope that a contented soul like me won't be seen as smug. I'm anything but, for I view my good fortune with amazement and great gratitude. Or, even worse, I hope I'm not viewed as suffering from a delusion based on some sort of strange psychological disturbance. A person who in this day and age actually professes his contentment, rather than lamenting a plague of sorrows and sufferings, perhaps seems to be possessed by a severe fantasy disorder evidencing mental illness. But I don't deem my mental or my sentimental attitude as being in any way sick—eccentric perhaps, but always quite logical and not at all pathological. My positive outlook is not a disorder but derives from an order, a life ordered and to a large degree structured, in so far as I could influence matters, to fit my chosen agenda. By and large, chance brought me a good fit for I was not a misfit, nor did what happened to comprise my life fail to fit my aspirations.

In order not to leave the wrong impression I should note that not everything went well for me. One way to look at existence is that an earthling is a magnet for problems—physical, mental, emotional, financial, situational, and other kinds. Corporeal existence by its very nature entails danger and vulnerability. The internal balance, homeostasis, is extremely delicate, and so are ambient conditions external to the humanimal itself. All too many plots and counter-plots, some natural and others man-made, threaten. To exist is to present a target. In my own case, like everyone else I suffered upsets and setbacks, defeats and disappointments, rejections, failures, losses. But why tell of or dwell on those negatives? With very few exceptions, these pages remain empty of such mishaps as they don't much interest me and are probably of little true interest, even if of some mild curiosity, to anyone else.

It's in any case desirable to bring a sense of perspective on long-past unfavorable events, bothersome at the time but which by now have faded into obscurity and no longer retain any enduring significance. Back when the incidents or situations intruded or existed I reacted to and then reflected on such negative developments, but over the years the bad times receded into the shadows of the long ago and here I am now, long removed from those old phantoms of ill-fated happenings which no longer haunt me. In fact, many were good for me. The hygiene hypothesis in medicine—pathogens at an early age strengthen your immune system—perhaps applies to psychological as well as to physical ailments. Dealing with life challenges early helps you cope with later ones—path-smoothing pathology.

Although quite a curious character myself, I don't wish to be a character in a book written to satisfy the curiosity of others. My reticence in discussing adversities arises not only from my sense of privacy and desire to omit self-revealing information, thus remaining a ghost writer rather than a character in the narrative. It's also because I thought readers might, perchance, find my cheery attitude refreshing. Perhaps they (you) will view my account as a welcome exception to the current practice of lamentation and confessional personal dramas. But maybe I'm wrong: if misery loves company, perhaps company loves misery. Traumas and dramas may appeal to readers more than does a reminiscence based on contentment and gratitude for favors fate granted.

Although my satisfaction originated in small part by my own attempts to shape my life, as I've noted by far the larger influences were my fatal circle and most of all luck. Chance operates within the givens of a fatal circle, one of the main factors which determines potential. A favorable fatal circle can facilitate but by no means guarantee safe passage through a tangled confusion of undergrowth consisting of accumulated facts and factors, plots and counter-plots, random possibilities. In this way an individual life evolves as an experiment similar to how the natural world developed, as infinite possibilities somehow mutated into a specific outcome. What comes to be operates to eliminates some possibilities but can influence even if not determine what happens next.

In *Chance and Necessity* (1970) Jacques Monod refers to "the general law: it is that of chance. To be more specific: these [protein] structures are 'random' in the sense that, even knowing the exact order of 199 residues in a protein containing 200, it would be impossible to formulate any rule, theoretical or empirical, enabling us to predict the nature of the residue not yet identified by analysis." From this extremely micro perspective Monod's macro conclusion holds that "Pure chance, absolutely free but blind, at the very root of the stupendous edifice of evolution: this central concept of modern biology...is today the sole conceivable hypothesis."

What might be called "Monod's Monad"—his unitary explanation of basic biological contingency, by which unruly chance rules—may well explain how organic evolution developed. In the realm of human affairs and how a person's life unfolds, as contrasted with how a protein folds, most everything can be attributed to chance and randomness. The initial conditions—a person's fatal circle, itself originated by chance—which set the opening terms can't be altered by volition. You're stuck with the limits and benefits decreed by fate. Pascal mused: why here and not there, why now and not then? When, where and into what sort of personal context you happen to appear arises from "pure chance," as Monod put it in a biological context. Those random beginnings greatly influence but don't determine what might later happen. Favorable and fortunate or negative and nasty, what you get for openers sets the controlling circumstances of how you begin your experiment in living. Moreover, what occurs after the randomly created initial conditions also depends largely on luck and chance.

Chance applies not only to individual organisms and to specific humanimals but also to the nature of things in general. In *The Logic of Life: A History of Heredity* François Jacob, who in 1965 won the Nobel Prize for Medicine with Jacques Monod, stated in the chapter on Time that only the random play of chance operates to create what exists "because no intention of any kind can be discerned in nature...There is no *a piori* necessity for the existence of the contemporary living world." Elsewhere he continues, "the idea of necessity in the living world" is refuted by "the contingency of living beings and their formation." There was no "preconceived plan," and "Had conditions been different, the living world would be different today or there might even be no living world at all." That in fact might have been a better fate for the lonely planet and for all its earthlings through the ages. No living world means no dying one.

For some years I pondered the mysterious interplay between chance and necessity, wondering how randomness and determinism operated in a synergistic way to affect and effect particular outcomes. By chance I finally came across a passage at the conclusion of Alexis de Tocqueville's *Democracy in America* which seemed to me to present the best working hypothesis on the matter: "Providence has not created mankind entirely independent or entirely free. It is true that around every man a fatal circle is traced, beyond which he cannot pass; but within the wide verge of that circle he is powerful and free." This analysis satisfied me: rigid constraints in a fixed and deterministic way, but within those limits volition, choice and discretion can operate to exert some control over events. Ultimately, however, it's not possible to know the precise demarcation and specific influence of the four factors—the fatal circle, choice, the weight of accumulated events, and luck. How the cards and discards of life are shuffled and played can perhaps best be suggested by viewing the fatal circle as an ace; the past as a king; choice, a jack; and luck a joker or knave who in the end trumps all. What William Goldman once said about the movie industry applies not only to celluloid but also to the cellular, for when it comes to how a particular life form assumes its experiential shape and content and then what happens to shape the form's passing existence, "Nobody knows anything." At least that much I know.

Because of all the uncertainties, irregularities and know-nothing elements of existence, being alive briefly as an earth-bound humanimal is truly a bewildering experience. You grope your way onward, maybe backward, through the shadowy wilderness. "Midway through life," in Dante's famous and mortal opening lines of *The Divine Comedy*—yes, mortal: the world's undivine devilish disorder and entropy will eventually destroy Dante's by no means immortal poetry, and everything else as well—"I strayed into a dark forest,/And the right path appeared not anywhere." Lost, the poet found himself in an oppressive "wood, so harsh, dismal and wild." Well, join the crowd: everyone is lost, lost in the baffling maze of existence and in the cosmic firmament, lost in eternity. No one knows anything—except that everything is nothing. In time all will be lost: Dante, Shakespeare, Goethe, Montaigne, Vermeer, Rembrandt, Charlie Brown and Lucy and Snoopy and all the "Peanuts" and popcorn and Cracker Jack, ball games and all games, all, all—even you, dear reader, *mon semblable*, and I as well, all will vanish and leave not a trace behind. But in the meantime what shall we do, how should we live? How do you arrange your life, to the degree possible given your fatal circle and the other factors, in light of the dark woods which represent the earthly landscape where earthlings dwell? Your guess is as good as mine. My reminiscence presents my guess and how it worked out. The book essentially gives my own answers to those questions, a solution I well realize few people could adopt and one which even fewer would wish to select. But it was available to and worked for me.

For me the fatal circle which circumscribed my life also endowed it with many advantages. My initial circumstances happened to be favorable. What I did with them and how things worked out for me is another story, one I've in part told in this reminiscence. A head-start in no way guarantees a strong finish. Some advantaged souls never manage to benefit from their good fortune. For them some sort of ill-fated combination of personal choice and impersonal forces—those random elements of luck, chance, fate—somehow operate to thwart a fortunate situation which exists at the outset. Even the most well-endowed souls can go astray. How it happens that some well-meaning, intelligent, diligent and financially secure aspirants to a productive, successful contented life fail to thrive is a mystery. It just inexplicably happens that some people make their way through the dark dismal forest to emerge at the end into sunlight, as dazzling as a desert's incandescent brightness, while others remain lost in the darkness. Maybe too many early benefits distort reality and otherwise contribute to later failures. In some cases an advantageous fatal circle can prove fatal.

My own mostly auspicious fatal circle happened to combine with good luck and, I'd like to think—but perhaps wrongly—some good judgment and a few other factors to produce a life lived largely in my own way. My wildly random and unexpected encounter with Guatemalan President Miguel Ydígoras Fuentes exemplifies how chance favored me. Never could I have planned such an experience. Unaccountably, little Miguelito grew up to become *El Presidente de la República*, while, equally unlikely, I was somehow incarnated into a curious humanimal, a creature who suddenly by chance found itself in the chief executive's office at the Presidential Palace far from my native habitat and beyond any expectations. Although our fatal circles and lives greatly differed, they somehow momentarily coincided. But that presidential encounter wasn't the only memorable episode I happened to experience during my summer 1958 stay in Guatemala. The other happening, a threatened mishap, almost resulted in my fatal circle ending in a prematurely fatal way. It was a lucky narrow escape, just before the circle narrowed to squeeze me lifeless.

~ 2 ~

When you embark on a new activity your hopes and expectations endow the project with a pleasurable aura. What then actually happens can tarnish the pre-image with blemishes as reality replaces possibility. In part emboldened by the random opportunity which permitted me to gain access to the President, I decided to keep my winning streak going by initiating a plan to visit Tiquisate, one of the United Fruit Company's Guatemala banana plantations. I suppose most people are quite content to live their entire life without ever seeing a banana tree but, as a curious lad, somehow I couldn't resist the opportunity to experience an interlude at a remote installation which seemed to symbolize the Banana Republic I was at the time visiting.

Two Guatemalas, each completely different, existed: the capital and the hinterlands. The neat square blocks of the curated, cultivated, civilized city—the country's economic, financial, media, cultural, governmental and political center—contrasted with the chaotic jungle beyond. Out there things were different. Primitive tribes lived in the wild, predatory animals lurked in the shadowy overgrowth, all manner of toxic insects and plants infested the tangled vegetal labyrinth to comprise truly a confusion of confusions environment. Structure contrasted with nature. In the city the built man-scape kept the jungle at bay, while in remote areas natural forces dominated. The city was settled, arranged, organized like an artificial construct. The jungle, thick with organic content, suggested the congeries of the dense past, once subject to the uncertainties of chance and luck until finally planted firmly and permanently in time. In contrast to the contrived capital, which seemed rational and purposeful, the jungle was a tangle of forms produced by happenstance rooted in chance.

Mankind's unrelenting enterprise in the context of his earthly existence, the fatal circle of the entire species, arises from the humanimal's struggles to temper the forces of nature and to mitigate the corrosion of time, also a natural phenomenon. Man desires to improve his given condition by protecting himself from both the ravages of nature and of time. The heritage of the past, the built environment and other artifacts from the mind and hand of man—culture, society, civilization, religion, beliefs, conventions and much else—can to a certain extent tame nature and delay decay and oblivion. But it's all just temporary Botox treatment. Everything's vulnerable. Just like Big Miguel—Miguel Ydígoras Fuentes, who fell from power—the Gros Michel (Big Mike) variety of banana was wiped out by Panama disease, black Sigatoka and other afflictions. Whatever is eventually isn't. In the end none of the earthling's works, whether built or imagined or otherwise, will withstand the forces of time and nature. The stubborn refusal of time and its offspring, history, to go into remission, to be refuted, defeats man's best efforts to overcome that inexorable execrable natural force.

It was curiosity—not philosophizing about structure and order versus the chaotic nature of the jungle's mesh of vegetation which mimicked the past's density—which motivated my interest in seeing another side of Guatemala. Milton Eisenhower had talked to the local market women to broaden his understanding of the country, and I wanted to diversify my experience by getting out of the capital to visit a banana plantation. Looking back, however, I can see that my desire to leave the comforts of the man-made city for the wilder Dante-like natural areas was perhaps an analogue to my interest ten years later in giving up my regular and settled familiar home life for the more problematic attractions out in the big wide world. But maybe that's reading too much into the excursion. Perhaps it was simply to see banana trees.

My emergence into the world of travel, exemplified by the excursion to Tiquisate, arose from two motivations: curiosity and to seek adventure. I abandoned the security of home for the desert-like emptiness of a problematic and uncertain future filled with many new and unknowable hazards. So it was that one day in Guatemala City I went to the United Fruit Company office to request a visit to Tiquisate. Although I'd lacked any preparation for my completely serendipitous session with the President, I did prepare for a possible visit to Tiquisate and was thus able to demonstrate to the local United Fruit manager some familiarity with the company as well as evidence of my serious interest in seeing the plantation. Just as for my interlude with the President, I requested the visit with no agenda, other than, as stated, curiosity and creating an adventure. I was perhaps the only stray character to ask to see Tiquisate with no business or professional mission in mind. I nonetheless managed to convince the manager to arrange the excursion, and he reserved a plane to fly me down to the plantation a day or two later. At the appointed time I arrived at the airport and boarded a frail looking and plain little single-engine plane. I imagined that out in the jungle swarmed and bit insects bigger than the diminutive flying machine. We soon took off, and behind Guatemala City faded into the distance as we transitioned from civilization to nature.

I rather doubt if it would be possible to arrange such a visit these days. In today's world, a stray traveler like me would surely not be allowed to fly down to Tiquisate and spend the day there. It can be really dangerous out there in the jungly hinterlands beyond the safety of the city. In the city you face pickpockets, aggressive beggars and such, but out in the jungle lurk innumerable hazards, not the least of which, for your hosts, is the risk of legal action if an accident happens to occur. Lawyers would swarm like insects in pursuit of suits: potential liability, damage claims, personal injury. Americans are quick to sue. As common as banana stalks at Tiquisate are Yankee lawyers back home who court clients and stalk cases to file. Out in the jungle anything might happen: you might slip on a banana peel and break a fingernail or worse; a clump of falling fruit may bean you; perhaps a territorial monkey might pelt you with bananas from the animal's perch by the creature's source of snacks. You could get a black eye or a cauliflower nose from the attack. Maybe a more serious cause of action, big damages possible, would be pain and suffering from a tarantula bite, or possibly insects would drill into your skin to inject poison, or a jaguar might attack or a Land Rover overturn. Nowadays any of these mishaps would mean big bucks in damages. Threats lurked everywhere out there in the rough teeming jungle. Insensate nature's plots and counter-plots could do you in at any moment. Any damage recovery would go to your heirs as you'd be beyond any and all damages—no more plots or counter-plots could disturb you. You'd be forever free from predators. Death activates the perfect immune system.

They say that money doesn't grow on trees, but at a place like Tiquisate that's not true. Bananas golden in hue and in value ripen on trees and yield money in the form of *Banana Gold*. So my friend Carleton Beals titled his 1932 book about the banana industry and about his trek through Central America. My predecessor—a footloose young Yankee roughing it in Central America, a character much more adventurous than me—I met through friends in Guilford, Connecticut. I occasionally visited Carleton and his wife Carolyn where they lived in a rural setting in Killingworth. His writing studio occupied a spacious book-cluttered one room schoolhouse moved to the site and emplaced near the main house. When I dropped by to visit Carleton and entered the schoolhouse unannounced I'd at times find him napping on the worn lumpy vintage couch. Disheveled as always, he'd languidly sit up and transition from deep sleep to chat mode. But Carleton was no layabout asleep on the job. He'd lived an active and productive life. There among the hundreds of books and piles of files and papers everywhere, works in progress or regress, Carleton dozed and wrote his days away in his own little outpost of the republic of letters far from the banana republics and other lands he'd frequented in his younger days.

Soft-spoken and, even apart from his frequent naps, laid-back Carleton was a modest and pleasant fellow, rather reticent and low-key, seldom hinting, unless asked, about his adventurous life and colorful career. I guess Carleton preferred to put his stories and thoughts into his books rather than into his mouth. In conversation with Carleton you had to draw him out to hear tales of his travels, word of his writings, and information about his various adventures. Many episodes of his colorful life you could read about in black and white second-hand from Carleton's writing hand. He wrote more than 45 books, including novels and, primarily, non-fiction works based on his travels and biographies.

Carleton's life exemplified the typical fate of an independent scholar and free-lance journalist. Unendowed with financial resources, he was a man of dependent means who lived catch-as-catch-can and hand-to-mouth as he put hand to pen and pen to paper. That exercise didn't pay much. The occasional book advance, a dribble of royalties and modest fees for articles and maybe some social security—for him more anti-social security as Carleton preferred to remain rather socially isolated there in his schoolhouse studio—represented his only sources of income. But this was his chosen path. He lived in his own way. Carleton received more stimulation and personal enrichment traveling to exotic places like the Ivory Coast (although he probably never went there) than sheltering himself in an ivory tower or holding a regular 9 to 5 job. The monetary element of Carleton's life was poor but in other enriching and gratifying ways he lived in a very prosperous manner and completely as he wished.

For his entire life Carleton enjoyed an interesting, stimulating, productive self-shaped existence and career, one which artfully combined intense activity out in the world with thoughtful responses to his many varied experiences, a way of life which recalls Wordsworth's view that "art is emotion recollected in tranquility." In his isolated little country schoolhouse Carleton engaged in his word-worthy activities, quietly recollecting and writing accounts of his vigorous worldly experiences over many years. Born in 1893 in Medicine Lodge, Kansas—perhaps the unusual place name foretold his outlier kind of life—at age three Carleton moved with his family to California where he grew up and worked his way through the University of California, graduating in 1916. His adventures began the following year when he arrived by wild burro in Mexico City where he founded the English Prepatory Institute and became principal of the American High School. But doing rather than teaching was his calling. Focusing on Latin America, Carleton began his writing career as a roving correspondent free-lance journalist for various publications. Never tethered to a payroll and reporting for newspapers and magazines but not to a boss, other than his nurturing wife Carolyn, Carleton remained independent and free-spirited his entire life. His first major writing assignment, for *The Nation*, took Carleton in 1928 to Nicaragua, which he reached partly by horseback hacking his way through the jungle, to cover General Augusto Sandino's revolt against the U.S. Marine occupation of the country. In another Banana Republic political disturbance years later another general-politician, one who I'd met, Miguel Ydígoras Fuentes, would flee from Guatemala City into exile into Nicaragua. Down in those lands the whirligig of time twirled a not-so-merry-go-round spinning blur of frequent revolutions.

Carleton managed to get a toe-hold, or a hand-hold, in journalism when his on-the-ground dispatches from Nicaragua were published around the world, reports which launched his career and over time led to far-flung travel adventures and articles, such as an account of Mussolini's rise to power in Italy. After a long career rich in experiences but poor in revenue Carleton died, age 85, in June 1979. Of a generation quite senior to mine, he was among the very first of my extended circle to disappear, in that way making him a precursor as an early-day harbinger of later lost friends who, one by one, eventually transitioned from pre-ghost to ghost. Not only did I lament each of those many losses, they also represented for me a frequent *memento mori* reminder which influenced me to pursue my own agenda without delay.

Although Carleton and I shared some similar interests and in a few ways pursued the same sort of activities, quite in contrast to him I never bothered much with politics or politicians, matters relating to social policy, economic theories, or the great public issues of the day. Carleton on the other hand was an outspoken progressive whose rather radical views extended well beyond the far left side of the political spectrum. Beyond liberal, he sympathized with the down-trodden and the disadvantaged, and endorsed the common description of the capitalist monster United Fruit as *El Pulpo*—"the Octopus," a designation which evoked the Company's tentacle-like stranglehold on the policies, politics and economies of the Banana Republics where the firm operated. In *Banana Gold* Carleton described how "Baron Banana" ruled the land: "The United Fruit Company and its lesser competitors can hardly keep pace [with the demands for the fruit]. Their engineers go slashing through the jungles; their railroads reach iron fingers out and out; their yellow clapboard buildings, hotels, bunk houses, stores, offices, hospitals, warehouses, spring up like mountainous fungi."

You'd think that taming the jungle and installing hospitals, shops and civilized living quarters and conditions for workers would represent some sort of progress, but for Carleton it was just an intrusive exploitive fungal growth. In contrast to his description, Miguel Ydígoras

Fuentes in his book depicted a United Fruit settlement at Bananera in less critical and more civilized terms: the enclave "corresponds exactly to the novelistic image of the collection of sickly yellow bungalows on stilts, rising over enormous smooth lawns in the shadow of palm trees silhouetted against an azure, cloudless sky of a tropical paradise." Two quite different perspectives, the one a progressive journalist-writer's earthy view which differs completely from the President's top-down perception: a slashed jungle marred by an infernal intrusion or a tropical paradise? These opposite views well exemplify what I came to understand as I traveled the world. The subjective ways people see the play of infinite forms and how the humanimal reacts to them and the conclusions reached are always only relative and tentative. How you perceive depends in large part what you believe.

As for the plantations around the settlements carved, or slashed, from the surrounding jungle, Carleton wrote: "The big green stems of fruit are cut from the tall, broad-leafed, plump-trunked trees, loaded onto" freight cars and sent to such points as Puerto Barrios where "Baron Banana rules supreme, and he rules roughshod with little regard for the beauty or happiness of his people." This sort of Beals-like focus on the socio-political dimensions of United Fruit's operations in Guatemala did not spark my desire to visit Tiquisate. Not the mind-set of an investigative journalist or social critic with a set mind but only curiosity and a sense of adventure motivated me. For my encounter with the President, luck and not a Baron Banana-like figure ruled supreme. However, for my plantation visit to the Baron's domain, choice rather than chance represented the operative factor as I myself activated the enterprise.

As the little plane winged its way over the jungle toward Tiquisate, skimming along as if some sort of giant flying insect, I gazed down at the landscape below, a seemingly serene green scene which from my distanced perspective above appeared pleasingly benign. But I knew that down below in the thick of the thicket seethed a Dantesque dark woods inferno. All sorts of creatures lurked there. In the jungle crawled, prowled, slithered, and creeped all sorts of creepy predators instinctively struggling in kill-or-be-killed, bite-or-be-bitten, eat-or-be-eaten combat. Fighting to survive, there swarmed countless rapine creatures armed with stingers and stabbers, biters, snappers, pincers, venom, poison, fangs and claws, toxic fluids and pathogenic potions, all surrounded by nettlesome poisonous plants, vines and spines and thorns to strangle and slash and rip and cut. Compared to the regularity and relative paradise of the banana plantation carved out of the jungle, diabolic nature created a dog-eat-dog world there among the hellish overgrowth where at any moment the struggle for life would end in death.

Seeking food and reproduction, the jungle competitors prowled and growled, snarled, screeched, roared, buzzed, hummed and hunted, driven by hunger and a deeply embedded atavistic natural lust to recreate themselves. Down there in the jungle blood flowed, pain seared, life ended. No creature was safe; each was both hunter and hunted. Animalistic plots and counter-plots, threats and dangers lurked at every moment in each place. Death lived everywhere. Which is which—victor or vanquished, diner or dinner—made no difference to Mother Nature, both a creator and a destroyer. As naturalist John Burroughs observed, "Nature does not care whether the hunter slay the beast or the beast the hunter. She will make good compost of them both, and her ends are prospered whichever succeeds."

What a whirl of a world existed down there in the untamed wilderness—nature red in tooth and claw, the cries of agonized demise a far cry from the civilized man-made artifact which was Guatemala City. But, I reflected high above it all there in the plane, were the two realms really all that different? Perhaps each was in its own way a kind of Dantesque thicket, with the Castro Reds who prowled the political jungle an analogue of the blood-red gashes slashed in the natural jungle by wild creatures struggling to survive. Lots of threats and plots and counter-plots played out in both the seemingly civilized world of man and the bloody natural world of beasts. From the far distant cosmic perspective of outer space, perhaps earthlings and their ways, their territorial cruelties and struggles, would resemble the teeming earthly jungle with its similar conditions. Whatever the view from above, nature doesn't care what happens below. The sub-lunar play of infinite forms were of no interest to the cosmos.

Above the dense jungle of vegetation below we flew on. During the flight I recalled some of United Fruit's audacious history in Central America, information I'd studied ahead of my Guatemala visit, both to educate myself and with a view to showing the firm's local manager my familiarity with and interest in the company. The operation originated in the 1870s when New Yorker Minor Keith Cooper initiated banana plantings alongside a railway line he was building in Costa Rica. Marrying the daughter of a former president, Cooper started the company's practice of deep entanglements with domestic politics, a connection which characterized United Fruit's method of doing business in the Banana Republics, better described as banana oligarchies. In 1899 Cooper and a Boston importer formed the United Fruit Company, and by the 1920s the firm owned great swaths of land in Latin America and operated the Great White Fleet of some 100 refrigerated ships, the world's largest private maritime enterprise. Earlier, in 1911, Sam "the Banana Man" Zemurray, later company president, had orchestrated the U.S. invasion of Honduras, one of *El Pulpo*'s strong-arm or tentacle efforts to grab local political and economic influence. Another such tentacular intervention occurred in 1954 with the company's role in the overthrow of the elected Guatemala government, yet another violent change of regime of the sort later suffered by Ydígoras Fuentes. Not only did United Fruit cultivate many plots to grow its bananas, the company also initiated and orchestrated many plots and counter-plots to cultivate its political influence and harvest the many benefits.

As we approached Tiquisate the little plane descended from our privileged high-sky position above the teeming agitations of the jungle to rejoin earth-bound terrestrial life. A clearing slashed through the surrounding greenery formed the short landing strip, a banana tree row rather than a Heathrow sort of airport. After we landed a jeep-like vehicle drove me to the plantation office where the local *jefe* greeted me warmly as if I were Jake the Flake himself. Maybe Chiquita Banana, the other member of the couple featured in United Fruit's famous advertising jingle, would soon join us. Perhaps preferring jungle to jingle, if Chiquita was around she never showed up.

Presently I started off in the jeep on a complete tour around the plantation, an extensive spread garnished with cultivated fruit-bearing trees, the bananas green but, with a certain alchemy, soon to turn into gold. I admit that if you've seen one banana tree you've seen them all, but nonetheless I enjoyed the tour, and in fact it was impressive to see the hundreds of plantings as a vast cultivated man-made unit. Created from the undisciplined landscape which previously dominated the area, "Baron Banana's" realm offered regular symmetrical rows of trees precisely placed in orderly alignments, nature tamed by man. Even if somewhat repetitious, the identical rows and rows intrigued me. Now, whenever I peeled a banana back home or heard Jake and Chiquita singing their ditty, I could recall all the effort which lay beneath the skin. A complicated infrastructure necessary to grow and an intricate supply-chain required to ship the fruit from the tropics to retailers in the far off U.S. underpinned the mundane commodity, less a baron than simply a commoner.

In truth, my effort to invest energy and scarce travel time to visit a remote banana plantation struck even me as somewhat eccentric. But the excursion typified how I arranged much of my travel life. In that bananas outsold even apples and oranges, I thought it would be of interest to see how the world's most popular fruit was grown—how banana gold was mined. While my curiosity took me to many conventional tourist sights—the Tower of London, the Eiffel Tower, the Leaning Tower of Pisa, Trump Tower—I also enjoyed off-beat off-the-beaten-track places like Tiquisate. It wasn't just bananas which drew me into the countryside. In almost every country I visited I made a special effort to leave the main city to see outlying areas. This was especially my goal where one major metropolis such as Guatemala City—a smaller but no less concentrated version of a dominant central city than other much larger all-encompassing centers—dominated the country. Places like London, Paris, Cairo, Bangkok,

Copenhagen, Amman, many if not most of the capitals in Latin America and Africa, held dominion over the economic, financial, media, cultural, governmental and political elements of a society. Those presiding places don't present a full or a nuanced view of the country, which is far more diverse than what a traveler finds in the sovereign city. As a provincial myself, I enjoyed seeing remote towns analogous to the kind of secondary sort of city where I originated, lived and will end.

Even though perhaps rather eccentric, my plantation visit by no means proved a fruitless enterprise as it served to satisfy my interest in seeing areas beyond the capital city and in learning about one of the country's main economic activities. I thought that a visit to a Banana Republic would be incomplete without seeing any bananas growing. I went to all the trouble to get to Guatemala, so I wanted to experience as much as possible there. As memorable as my visit with the country's top banana was, an encounter with the President hardly offered a complete picture of the country. Wherever I went over the years I always followed the same practice, seeking from a visit to each place the maximum possible within the available time. To help to accomplish this I prepared extensively for each trip and knew in advance most of what I wanted to see and do. Of course, I never knew what unexpected problems or opportunities might arise, but at least I was familiar with the main sights and activities. Efficiency represents an essential component of serious travel. Time while away, and especially in remote countries, is always limited and must be spent wisely and efficiently to yield the most desirable results.

Tiquisate also interested me as an example of man's struggle to shape nature in ways which addressed the needs of humanimals, ever subject to systemic conditions and impositions. Confected out of the original unruly landscape, Tiquisate was in effect press-ganged into existence to serve humankind by feeding its appetite both for nourishment and for order. The civilized little settlement contrasted with uncultivated nearby areas beyond the plantation's manicured enclave. The town arose not as an organic community which grew bit by bit from the ground up but as an artificial construct imposed by man from the outside. In *Empire in Green and Gold: The Story of the American Banana Trade*, Charles Morgan Wilson describes how in the 1930s "Tiquisate became the first completely mechanized banana farm. Its lands were latticed by 400 miles of roadways, and tractors and tractor-drawn carts replaced long caravans of banana mules." There in the jungle the United Fruit Company built "a new banana division with a work force of some 4000 local Indians, many of whom had never seen a banana. Company planes flew in men and supplies. A hospital and a clinic were completed, and sanitation squads went to work. Railroads, homes, clubs, stores, roads, farms, irrigation systems, schoolhouses, and playing fields appeared and multiplied." It all sounds to me like a lot of work just to enable Chiquita and Jake to sing the promotional ditty, "I'm Chiquita Banana and I've come to say/Bananas have to ripen in a certain way." It would have been a lot easier just to let the *Norte Americanos* and other hungry nationalities stick to apples and oranges, with occasional ice cream and M&M's thrown into the mix. Dairy Queen, White Castle and Burger King rather than Baron Banana could rule the menus and satisfy appetites. But as things turned out, a new town carved neatly out of the wilderness brought to the world banana gold, a treasure for which Chiquita and Jake sang the praises. I never understood why Carleton Beals and observers with similar attitudes didn't recognize and appreciate at least some of the benefits banana settlements carved out of the jungle brought to provincial folk otherwise stranded in poverty.

My Tiquisate visit represented what to me is the essence of curiosity—an admittedly somewhat useless and gratuitous activity likely to provide an interesting experience or redundant knowledge with no practical benefit. No functional purpose motivated my desire to see a banana farm. Curiosity and a sense of adventure pure and simple inspired me to arrange the visit. I gained no advantage or profit from the enterprise, but simply just one tiny fragment of all the many which formed the eventual mosaic I've arranged in this reminiscence to bring an imposed pattern to my variegated disconnected adventures. Of the millions, billions even, of possibilities which might have comprised my life components only a very tiny few eventually actually happened to occur. All the others failed to attain existence. I only wish that I'd had the time and opportunity to create even more sensory impressions to add to the content of my earthly existence. In any case, I'm quite pleased that a banana plantation visit numbers among my travel accomplishments. I believe that I am the only resident on my little cul-de-sac street back home who has seen a banana plantation.

A keenly curious character constantly faces the Shakespeareian dilemma that "The will is infinite but the execution confined." Although you desire all manner of experiences, finite time and other limitations restrict you to relatively few of the potential experiments which comprise a life. Some, like the plantation visit, I activated while others, such as my session with the President, happened by chance. Either way, I aspired to be someone on whom, as Henry James put it, nothing was wasted. From my experiment in living I aimed to extract as much as possible, banana peel and all, as my fatal circle and other factors would permit. Every opportunity to gain experiences should be exploited to the maximum. At Tiquisate, however, I recalled not Henry James but O. Henry, who in one of his stories invented the expression "banana republic." Both Henrys provided me with useful notions.

It so happened that my fatal circle and the other relevant factors—including my own efforts and, primarily, luck—enabled me to create some experiences which in the spirit of James I tried not to waste. As best I could, I tried to remember, reflect upon and draw some sort of meaningful interpretations from much that I was privileged to see and experience. In fact, this reminiscence represents a final attempt to create from the welter of all that happened over many years something of a coherent account meant to mitigate the wasteful humanimal practice of disappearing with all your memories in mind but not on paper. Without my book, everything which formed the content of my earthly existence would be scrapped, not a very mentally or environmentally friendly practice. I greatly treasure my good fortune—soon to end—and before it all vanishes am attempting in this reminiscence to salvage a few stray shreds and shards from the debris left by time. One goal was to put myself in field position to score a goal by producing this account before leaving the field. In that way I could purpose my place in the play of infinite forms. Living and now recording some of my earthly adventures brought me an advantaged existence of a kind unavailable to most earthlings. Unlike me, all those sweating machete-wielding plantation workers I saw in passing lacked access to banana-free life experiences. Those jungle-limited laborers were condemned to hack their way through the dark woods of existence confined to Tiquisate, while I only passed through on a fleeting flying day visit before escaping to wing my way on to other adventures.

As things turned out that day, my departure was sudden and forced. After the tour I returned to the office for lunch. I wondered if the menu would feature a house specialty such as banana pudding, banana jam, banana juice, banana cookies or cake, Bananas Foster. But none of those, nor Chiquita Banana and Jake the Flake, made an appearance for lunch. It proved to be a "Yes, we have no bananas" meal. Halfway through the lunch, however, the pilot made a quite unexpected appearance. Interrupting us at the table with great urgency, he shouted, "We have to leave now—immediately."

Oops—there goes dessert. His agitated behavior disrupted the leisurely tropical lunch and the languid almost soporific atmosphere which verged on a siesta-like torpor. The scene suddenly crackled with an electric urgency. Although I asked the reason for such an abrupt departure the pilot didn't pause to explain. He insisted on leaving without delay and without explanation. Maybe Red revolutionaries were about to invade the plantation or a pack of wild beasts approached, or perhaps an infestation of aggressive stinging insects whirred toward us. For whatever reason, according to the discomposed pilot immediate departure was necessary. My brief dalliance in the kingdom of Baron Banana was coming to an unexpectedly abrupt end: I faced sudden exile from the Banana Gold realm.

Jolted from the pleasantly slow-paced only half-finished meal, with the pilot I scurried out to the jeep and we then sped over to the landing strip, there to take off without delay. Once airborne, the pilot told me that not far south a huge tropical storm was developing and unless we left now we'd be grounded and unable to return to Guatemala City that afternoon. Stranded in bananaland—hardly a fate worse than death. But would the risky flight tempt fate and end in death? As we gained altitude I could see in the far distance behind us an ominous black sky. Thunder growled, lightning flashed. Sensing the storm, the wildlife in the jungle below would now be taking shelter as best the beasts could. Up in the plane, however, we remained exposed to the raw elements of nature's force. In the sky there was no place to hide.

Down below the cultivated plantation with its neat precise rows of banana trees, each aligned in its allotted plot, gradually gave way to the unruly counter-plots of the surrounding tangled jungle. After we reached flight altitude and leveled out into cruising speed I gained a more extensive and disconcerting view of the storm. Behind us in the far distance—but was it far enough for the plane to out-fly the tempest?—the ugly black sky drenched the earth with great sheets of rain. I didn't know how fast we could fly, but whatever our speed it seemed that the storm was advancing even more quickly. A strong sense of mortality—death by storm—rather than mere curiosity provoked me to wonder if we could manage to fly faster than the unruly hurricane-force elements. We were playing tag with death. If the turbulent winds and rains caught up with the flimsy little plane carrying two even more fragile earthlings we'd be engulfed in the torrential tropical storm and washed out of the sky to crash back to earth in the beastly jungle. As a party in interest, the pilot himself had been willing to take a chance on beating the storm, but was his risk-assessment well-calculated or poorly judged? We'd soon find out. I looked back: the storm was gaining on us. Maybe the pilot was driven in a flight of fancy to take a foolish chance to enable him to reach a waiting *señorita* back in Guatemala City. If that was indeed the case, his possible pleasure became my peril.

We were losing the race. The storm drew nearer. Even here, high in the sky above the dangerous jungle seething with predators and their victims, we couldn't escape the forces of nature, random wildness which threatened the lives of man and beast alike. As we hurtled ahead the slashing sheets of rain moved closer and the black sky began to close over us, darkening the day and our prospects. It was going to be a close race to the very end, either touchdown in the capital or crash in the jungle. Rain now started to pelt the plane, and before long the intense downpour battered the little capsule, the storm as stinging and perhaps as fatal as the bite of a toxic jungle insect.

Call no man happy until he's dead said the ancient Greeks. At any moment now my life, perhaps soon to end, might qualify for that final judgment. It would be an unhappy death of a happy life. The plane began to buck and shake as the agitated winds, precursors to the storm's full force, battered us from behind. My hold on life suddenly seemed as shaky as the plane. Perhaps now, at the premature age of my early twenties, I'd crash to an early death. This would at least seal my earthly happiness and fix it forever in immutable finality. I'd no longer be a hostage to fortune and luck. One major misfortune would eliminate those random forces. An early demise would serve to bring forward the eventual inevitable, eliminating all uncertainties and making certain that I'd indeed die happy. This would avoid all manner of possible future miseries and misfortunes. Why delay the end when you can exit existence sure that you've lived a happy life? Quit while you're ahead. Get it over with—become a ghost without further ado or adon't or any other earthly concerns. Once safely installed in your cemetery plot you're immune from any other plots or counter-plots. My transmutation from a living humanimal to a dead carcass would endow me the the blessed fixity of a completed life, one which satisfied me. On the other side of death, when for you it exists—after you don't—to define your final state, no further risks exist. Protected from any future unhappiness or misfortune, I could then finally and irrevocably be called happy. Non-existence, either never to be at all or removed from being, is not a serious matter. For millions of millennia I was nobody; now I'm somebody; soon I'll revert to my original natural state by ceasing to exist, once again being a non-being nobody. Surely the certainty of dying at any age, even in your twenties, after a happy life was worth something, But was it worth losing your life that young, before your experiment in living had run its course? All things considered, crashing to death is a pretty heavy trip just to be able to say that you died happy. Better, perhaps, to risk for a while longer future uncertainties and possible unhappiness rather than to gain the deadly security of premature oblivion.

At the time of my flight from Tiquisate toward Guatemala City I was a very young man with very nearly an entire lifetime ahead of me—if I survived. I of course could not know back then what the years would bring. If I managed to reach old age would I still be able to say that I died after a happy life? Looking back with a very late-in-life perspective, I now know that few regrets vex me and that, so far, no serious reversals of good fortune or ill-fated happenings prevented me from living in my own way in a satisfactory manner to enjoy something of charmed life. I don't mean to tempt fate with such a statement, as I realize that things can change in an instant. What's clear to me, however, is how my life has so far played out for me as the future became the past. Back when I faced the real possibility of a fatal plane crash, one great regret occurred to me—I mean, other than no more ice cream or M&M's. If we went down my life-span would have been much too short even to begin to make much sense out of what it means to be a humanimal and more specifically what my life was about, if anything. Never would I enjoy sufficient time to arrange a certain coherence, even from the relatively few events which at the time comprised my brief earthly existence. But by way of compensation, a short life brings fewer disparate experiences, and those more limited components are easier to shape into a cohesive whole. As a life grows longer it becomes ever more difficult to fit the irregular pieces into a meaningful pattern. This became obvious to me in the course of writing this book, a summing up by which I can offer only a gist but in no way a complete picture of the whole. There's just too much to cram into the story, which as I've told it is a summary serving only to suggest the entirety.

My fleeting flying regret—a life too short to understand it—which came to me on the wing as we headed to a landing hard or soft might have been quite valid. It may in fact be more advantageous to die completely disregarding wisdom from the ancients. Although the age old opposite proposition—an unexamined life is not worth living—appears to be wiser, the reverse seems quite as sensible as the original concept. Why even bother to examine your life? Better just to live it. Does anyone's existence, or existence in general, really bear analysis? What difference does it make to attempt to puzzle out any sense from one's brief earthly existence? Most people seem to get along quite well without examining their life. They take it as it comes, and goes. They may even flunk the examination. A few mortal souls may from time to time give a passing thought to what it all means, but then the events of daily life crowd in to evict such transient notions. A chronic detailed examination is too intrusive and in any case deals with a chimera, for who can say that an examined life is worth living any more than an unexamined one? The plantation workers who for years on end wield their machetes to chop banana stalks from the trees may seldom or never examine their lives. Those laborers no doubt examine their blades to keep them sharp, a much more practical exercise than wasting time sharpening their thoughts about life. Maybe that's the right way to live, and to die—leave well enough, or bad enough, alone and avoid any introspection, just crash without regrets. That demise would be a crash course in how to exit smiling.

Back then in the wind-blown plane, lightning rather than enlightening existential conclusions represented the only illuminating moments, a split-second, split-sky brightness surrounded by darkness. As the storm's maelstrom neared us from behind I gazed ahead into the distance to try to gain a glimpse of Guatemala City's lights, tiny dots of brightness signaling man's presence in the ever darker cloud-clogged day. The capital city stubbornly refused to materialize. No city, Guatemala or otherwise, evidenced signs of civilization. We remained lost in nature, whose winds and rains toyed with the tiny plane. Santayana said that everything in nature is lyrical in its essence, comic in its existence and tragic in its fate. Lyrical or comic nature may in fact be, but any individual or collective fate is merely a fateful fact, not a tragedy. If a living growing thing like a banana is cut down in the prime of life from its mother stem and orphaned by being sent far from home and then devoured, is that a tragedy? If our fragile little plane with which the tempest toyed crashes us to death that would for sure be a messy final fate but not a tragic one. The pilot's passing would simply leave his *señorita*, if there was one, waiting for him there in the big city doomed to wait forever for her no-show and never-will-show boy friend. As for my own late existence, an early death would eliminate an eventual and perhaps much later demise, a blessed youthful end releasing me from many possible miseries yet to occur. Ending the story much sooner, at a time when my life could be called happy, might be better than prolonging my earthly experiment. Surely dying young is no tragedy, nor is it comical or lyrical. It just is—or isn't. With such a premature disappearance I'd benefit by avoiding any and all future threats, hazards, damage, disappointment, plots and counter-plots. I could rest in peace. This sort of advantaged non-existence hardly seems to be a fateful tragedy. One good hard crash and I could get the whole thing over with. The plane descends to the ground, a let-down after a happy life, and then—my life grounded by a sleep—I descend underground with a sort of poetic symmetry which would define my demise. I could hardly imagine a better way to go—no feeding tubes, chemotherapy, surgery, injections or infections, no pathology, just a quick plain vanilla disappearance. What's not to like about a sudden crash?

It's of course obvious that my account of the flight and my existential contemplations lack suspense because the outcome is evident in the telling.. Had I perished in Guatemala back in the summer of 1958, these lines would not have been written by the defunct ghost writer I would have become. Only a real live ghost writer, as I briefly remain—not merely texting posthumously from the grave—could produce this text. For now, but not much longer, I'm a living ghost writer, still able to bring my spectral presence to these pages. But I'm keenly aware that at this late date I will soon reach a terminal point more fatal than the airport terminal in Guatemala City the plane finally reached just as the storm's full force hit. Escaping from death by crash simply postponed death at a later time by other means, There's no doubt about our fate. The only suspense relates to what happens during our brief earthly suspension in time, that fleeting interval between our pre birth non-existence and our post-death forever after. The ultimate outcome is a given, or a taking, with the only element of uncertainty pertaining to the means to the end. How do we get from nothing to nothing? From the first wail of the new-born to the last death rattle of the expiring imminent ghost, the outcome is clear. What happens during the years from start to finish becomes known only after the content which—with the ceaseless play of infinite forms— informs and forms or deforms a life as it takes shape.

Implacable, irresolute, insensate nature, of which each humanimal for a brief time forms a part, should be viewed for what it is: a creator of death. The truth is that from day one a traveler's inescapable destination is the terminal. As Mr. Hemingway put it, "all stories, if continued far enough, end in death, and he is no true-story teller who would keep that from you." Never let that be said about me. Much have I omitted from this book but, as will become apparent, I haven't refrained from being a true-story teller. Such truth may spook the reader but that's one of the functions of a ghost writer. Whether "Death in the Afternoon" or morning, noon or night, in the Guatemalan jungle or in a hospice or hospital, whether far from home among strangers or comforted by family and friends as you ebb away in your own pillow-cushioned bed, he who would keep that end from you should not be telling stories. However and wherever and whenever the moment comes, the story of each and every humanimal reaches the same conclusion, as do I in the conclusion to this book.

Just after we landed at the terminal in Guatemala City the storm, a tremendous force of nature, broke over us. The pilot and I remained sheltered in the plane until the tempest passed. We then emerged from the metal carapace, he perhaps to meet a waiting *señorita* and me, lacking such a local *amiga*, to return to the welcome haven of the Hotel Pan American. In the empty rain-swept but not broom-swept littered streets on the way to the hotel lingered in my mind phantom images of the world, so unlike my own, as described by Miguel Asturias in *The President*. The conditions outside there on the streets contrasted with the comforts at the Presidential Palace where the other Miguel, Ydígoras Fuentes, sheltered from the storm. Asturias's street people populated a pageant of misery. Beggars shuffle along and at night they sleep on a church porch, a "confraternity of the dunghill [which] had never known pillows or mutual trust"; "a pregnant deaf-mute"; a dog "vomiting against the grille of a shrine"; an idiot with a broken leg "whining...like a wounded dog"; groups of boys catching insects "submitted to a series of tortures"—a grotesque macabre nightmare of appalling urban vignettes. "The street is an unstable, dangerous, adventurous world, false as a looking-glass—the public laundry of all the dirty linen in the neighborhood."

Presiding over it all, the street jungle and the natural one beyond the city, was *El Presidente*, his police "keeping their nightly watch over the President of the Republic." Although fictional, Asturias's description in fact depicts the world as I saw it in many places, an odd sort of lonely planet whose dark corners and sad settings and situations lingered in my memory all through the years. My stay-at-home family and friends would find it difficult to envision such sordid sights and scenes, but even now I can recall many because they continue to haunt me.

~ 3 ~

In the course of my world travels over more than half a century I spent a total of some three or four years in underdeveloped areas. The more undeveloped the region the more I was able to develop my agenda for adventure travel. In those Third World lands, some of them more Fourth or even Fifth World, I saw many wonders and unforgettable places along with street scenes and conditions similar to those described by Miguel Angel Asturias in his novel. Quite striking were the extreme contrasts between ruler and ruled, the ins sheltered in their palatial mansion and the outs out on the street, the have-a-lots and the have-nots, those with things and others with nothings, the pampered and the paupers. Those blessed with a favorable fatal circle and good fortune, or at least a fortune, could enjoy their lives far more than ill-fated souls doomed to suffer bad karma or bad luck. As I made my way around the world, observing in so many places human suffering in close proximity to great wealth, I meditated on the fatalistic roles of luck, chance, happenstance and random events and how they influence individual lives. "There's a destiny that shapes our ends,/Rough-hew them how we will," says Hamlet. Out in the rough-edged Third World a harsh destiny shaped most of the means and the ends in a rather ragged, rugged and thread-bare way.

From my privileged and protected fleeting vantage point as a traveler passing through impoverished areas, often did I wonder how chance happened to tease out from all the possible outcomes the particular ones which define a life. Just as *de gustibus non est disputandum*, so there's no disputing about luck. It's just there, for good or for ill. Like the one-time boast of Citibank, luck never sleeps, not even a quick siesta. Like the Pinkerton eye, luck always keeps watch, and often intervenes. Luck comprises the ganglia of the nerve-wracking, capricious and contingent system which operates to shape lives. Although it's possible to exert some influence on what happens, many poor souls remain stranded in a no-man's land where personal choice doesn't exist. Out at the banana plantation you cut banana stalks—that's it. For the fortunate few life proposes possibilities, but life is imposed on the millions of earthlings who lack the means, financial or otherwise, to escape the tight constrictions of their limited fatal circle. But even those of us with some freedom and flexibility to choose our way ultimately remain subject to outside forces which operate beyond personal agendas and individual volition. You can't beat the system, but only try to live as well as you can within its rules. There's really no explanation for why a fortunate few happen to luxuriate inside the palace while other humanimals are fated to remain part of the "confraternity of the dunghill" outside. Fatal circles are arbitrary and, once drawn, impossible to erase.

Why some souls somehow seem destined to enjoy good fortune while others are doomed to dwell in dung is a mystery. Dung is largely an equal opportunity employer, but a lucky few humanimals somehow manage to avoid the excrement. One puzzling oddity of the random play of luck and chance pertains to how people of apparently average ability, ordinary intelligence (social and intellectual), uninspiring personalities and with seemingly mediocre prospects, happen to prosper in many ways, not just economically, while other much better endowed earthlings will at best lack success or, worse, suffer setbacks and failures. One's lot depends a lot on a lottery ruled by luck. In truth, no operative or definitive explanation exists to account for "top of the hill" or "dunghill" variances. They for the most part depend on capricious forces essentially determined by chance. Both on the upside and the down side the wheel of fortune spins unequally. Not only does chance favor few mortals, it also often operates to reduce those few to less fortunate circumstances. It's more likely that those at the top will decline than that unfortunates at the bottom will rise. To whom much is given, much can be taken away. Status and standing and good fortune are as slippery as a banana peel.

However the chips may fall, good or bad, uncontrollable random factors dominate all life outcomes. While the unfavored may curse their bad luck, many of the fortunates seem not to appreciate their good luck. Speaking only for myself, this entire reminiscence is in a way an extended appreciation of the winning leading role Lady Luck played in my story. Because of the uncertain and arbitrary nature of outcomes, I've always been thankful for the fateful forces which somehow favorably shaped both my fatal circle and the events which by chance conspired to bring me good fortune. From early days I realized that my fatal circle endowed me with many advantages. I saw both at home and during my travels how the tight confines of a constricted fatal circle greatly delimited the space in which a person could operate. For me the circle was more liberating than confining, as it gave me a privileged situation which enabled me to attempt to spend my life in my own way. But even with that head start, success in fulfilling my agenda remained uncertain, even improbable. I realized from the beginning that living one's own life represented an exceedingly rare phenomenon. Very few mortals enjoyed the opportunity, let alone the actual realization of the goal, to aspire to an independent way of life. To accomplish my aspiration, not only the givens of my fortunate fatal circle lay beyond my control but luck also remained a wild card. My chances for success thus depended on elusive random and unwilled will-o'-the-wisp factors I couldn't influence.

A *fata morgana* of sorts, luck is a kind of shadowy mirage you can never seize and guide. It seizes you. An apparently lucky circumstance may appear out on the horizon, only to disappear or change as you draw closer and attempt to grasp the phantom opportunity. Like the erratic winds of a tropical storm, plots and counter-plots can suddenly shift and in an instant blow luck away. No matter how well meditated, reasoned and analyzed and how deliberately and artfully curated a decision might be, never can you be sure what will happen as you try to pluck good luck from the pot-luck mix churned by the whirligig of chance happenings.

The uncertainty of random events is both disconcerting and exhilarating. Not knowing what lies ahead as you grope your way through the Dantesque dark wood adds drama and suspense to the journey. The extent to which your life offers some apparent security conforms to the degree to which your existence remains confined. At times that sort of predictable probable narrowing of possible outcomes offers great comfort. I enjoyed and valued the relative regularities, repetitions and limitations of my normal home-based routines. I especially appreciated them after the extreme contingencies, uncertainties and rigors which typify adventure travel. My intention was to balance the two ways of life to benefit from and enjoy both my settled home base and the unsettling but stimulating challenges of the open road. For me it would have been a far too limited existence to pass my entire life restricted to the comforts and confinements of a conventional Midwestern suburb. But ceaseless globe-trotting would represent an unwelcome vagabondage equally limited and unbalanced.

My goal was to retain my roots while stretching my wings and experiences. I sought a balance between home and away, roots and routes, settled and footloose. This in turn reflected my interest in gaining a larger perspective on existence, a viewpoint which facilitated an ability to see matters from both an intensely personal and a worldly angle. I realized that focusing only on my own extremely narrow and limited field of vision wouldn't provide an accurate or realistic experience in the experiment of being alive. Nor would viewing the experiment only from a more general perspective. It was the synthesis of the specific near and narrow plus the play of outside infinite forms I happened to encounter which brought me a wide-ranging way of looking at life I found satisfying.

The contrast between home and away represented a subset of my belief that the interface of the individual with the ineluctable general characteristics of human existence comprised the essence of the life experiment. Neither a purely bourgeois comfortable conventional provincial suburban existence nor a constant peripatetic roaming life filled with a wide variety of experiences would prove satisfactory. I needed both, and luckily could access the two quite different modes of being. By complete chance I'd been given a unique one-time opportunity, if being alive is such, to exist. To make the best of it I tried to interlace my experiment with cross-connections to tie together like a shoelace two separate but related elements of existence—the finite one, consisting of my fatal circle and circumstances and individuality, and the wider play of infinite forms around the world.

In the course of my journeys I often attempted in a small and temporary way to demote myself from privileged traveler to a level closer to that of the local unfortunates. This temporary relegation of comfort enabled me to expand my travel experiences. I rather relished lousy hotels, hard-class transportation, simple but hopefully hygienic meals. Given my rather pampered home-based existence—nothing special, just a regular suburban American bourgeois way of life—a sub-basic standard of living while away was something new to me. Among other things, that low-level life conformed to my frugal instincts. In Katmandu I stayed for a week in a cubicle at a primitive but serviceable hotel for a total of $14, alas not including meals. Nor, back then, did I qualify for a senior discount. Although I at times briefly lowered myself toward the bottom to experience how the locals lived, I of course realized that it's quite impossible to walk in the shoes of people stuck with a bare-bones existence, and anyway some of them, especially children, were shoe-less and barefoot. I never really existed as they did but only flitted into and out of their way of life. I sampled a few privations, ones I could quickly abandon, while for the unfortunates there was no escape. Those down-and-outers remained stuck at the bottom.

All along the way all around the world I passed by the down-trodden but didn't really see them. The humanimals I glimpsed in passing remain a blur of anonymous faces in the crowd with no individuality I can recall. Of all the thousands and thousands, I vaguely remember only a handful—or a faceful. By a perverse process, some of the most wretched with severe deformities I can still recall, a macabre low-caste cast of outcasts, benighted characters in the infinite play of forms. Glancing at the time- and life-worn faces of the street people gave me only a fleeting superficial, or superfacial, impression. The rough reality of their daily life and inner being remained out of view. You can only barely read between the lines of their wrinkled brow or furrowed face. The ultimate material and existential degradation the unfortunates suffer you can't begin to imagine. My sightings of those miserable conditions and luckless earthlings profoundly deepened my appreciation for my chance place in the scheme of things. I observed the street people *de haut en bas,* looking down at—but not on—them, while they saw me from the ground up. Although I knew how I perceived them, I wondered how those at the bottom, looking up, saw me as I flashed by, a blur to them as they were to me, to continue on my way. I treasured being a privileged passer-by, by-passing the impoverished masses, a transient ever on the move able to abandon an unsavory scene festering with dung, disease, filth, misery, hopelessness. I could continue on to other places and new experiences, onward from where locals remained forever entrapped in their unfortunate circumstances. While fate snared millions, it spared me. I doubt if the impoverished much noted me. Forced to devote all their energy and efforts to scratching, scrounging, scraping and scrambling simply to survive, the poor souls no doubt had no interest in or curiosity about a stray stranger blurring by. To them I was only a momentary sensory impression never to be seen again, while to me the street people were only a quickly passing "see" of faces. I remember not the individuals but only their collective miserable conditions.

While traveling I observed not only the disadvantaged but also many other characters. I was a people-watcher, one who hoped, as does a ghost writer, to be present but unnoticed by others. I enjoyed studying and sometimes eavesdropping on both locals and fellow travelers, and on occasion briefly chatting with a passing stranger. I was curious about other characters in the play of forms, even people I'd never again see. Every so often I tried to imagine the contours of their lives—the realities of their daily existence—what they did, how they lived, what they were about. It's of course impossible to know such details merely by looking and guessing, but nonetheless it was entertaining to attempt to envision, based on very limited evidence, a few of the characteristics and circumstances which might define the person. Whatever version I fabricated was simply one way to interpret what I saw. Even for people you know well it's sometimes difficult to understand their essential inner being and granular reality. But whoever and whatever a stranger might in fact be, we always had at least one thing in common: we both happened briefly to exist as live humanimals in the same place at the same time. That much was certain. What by chance brought the other person to where I happened to be just then was no doubt usually quite different than my reason, but in any case we were both there simultaneously. Perhaps other personal similarities also existed, but such coincidental resemblances would forever remain unknown to me. However, a more general existential reality always united us, as for the moment, until death did us part from life, both the stranger and I existed, concurrent members of the human race. I knew my place in the race but not theirs. I had no way of telling how in the long run the passing strangers were running the race. Were they prize-winners or only also-rans? Phantom-like, strangers come and go like ghosts momentarily haunting the scene before disappearing. In the play of infinite forms the passing strangers had only speechless walk-on parts. Of the thousands of earthlings I encountered over a lifetime, only very few did I get to know. All the others vanished, for me, without a trace, as did I for them.

On one occasion in a far-off land I engaged in the exercise of imagining into existence the life of a complete stranger who I'd never laid eyes on but whose name I knew—namely, my own. On my second visit to Buenos Aires, in early 2001 on my way to Ushuaia in far southern Argentina to begin a trip by ship through the Drake Passage to Antarctica, I indulged my curiosity by thumbing through the local phone book. When in another city I often looked through the directories, both the white and the yellow pages. I found that the ads, some of the listings, local historical and cultural sections and other content offered enlightening insights and information about the area. Much of my paging was random, but almost always I checked to see if any "Weil" locals were listed. In B.A. I struck pay-dirt, finding not only a "Weil" but one "Tómas Weil." So there was my name, mine but not mine, in its Spanish version in the phone directory of a remote city far from home and foreign to my natural habitat. Who was he, that *Porteño* Tom Weil? He was doubly my double: nominally, as we shared the same names and also because, unless Tom had already died, we were both living sentient humanimals. But beyond our names and existential characteristics did Tómas and I share any other traits? Where and how did Tom live? Was he a Tómas Weil I'd find compatible or one who in my view would tarnish the name which since my birth had designated me? I fervently hoped he wouldn't degrade the brand. To satisfy my curiosity, I thought about phoning Tom Weil.

Maybe he'd turn out—like the call I contemplated— to be a phony, an unpleasant character who would make me regret that we shared the same name. If he had any apparent hang-ups I could always hang up. I imagined how the conversation might go:

"*Señor, usted se llama Tómas Weil?*" I'd ask.
"*Sí, quien llama?*"
"Tom Weil."
"*Sí, me llamo Tom Weil. Quien habla?*"
"Tómas Weil," I'd respond.
"*Soy yo Tómas Weil—quien usted?*"
"*Yo soy tambien Tómas Weil.*"

Would Tom Weil hang up on me, Tom Weil—no doubt a crank caller, maybe even a madman—before I had time to explain. How would Tómas respond to the Abbott and Costello-like "Who's on first?" dialogue? Who was first? Was Tómas older or younger than me? What would he think about another Tom Weil now on his turf in B.A.? Would Tómas be as curious about his name-twin as Tom was, or only irritated that Tom Weil had disturbed Tómas Weil, maybe interrupted his siesta or a fiesta. Perhaps he lived an incurious life, perfectly content to Weil away his days there in B.A. without the slightest inquisitive instinct. As vast as the Argentine capital is, Tómas might nonetheless deem the metropolis too small for two Tom Weils. Buenos Aires was not so *bueno* with a foreign interloper. One Tom Weil was quite enough; two would be one too many. Apart from my brief presence in B.A., Tómas enjoyed a local monopoly on the name, exclusive to him unless, perchance, an unlisted Tómas Weil also existed, in which case we were a trio—and three's a crowd. A friend once told me I was so private I had an unlisted life. Maybe a third B.A. Tom Weil was similarly private and unlisted. I hope he wasn't listless—that would present a bad image for our name-brand and embarrass me.

Possibly the listed Tómas would in fact be delighted to learn of another Tom Weil. He might well relish the chance to compare genealogical backgrounds to see if we were somehow related. Maybe he'd organize a big family fiesta, with Weils galore and platters of savory juicy Argentine beef. Should I press my luck with Tom Weil and dial his number? All sorts of scenarios and possibilities tangoed through my mind. In the end, however, I sat this one out, bringing to an end all the animated imaginings which danced into and then out of my consciousness. They remained mere potential, never to be converted into the reality which would then represent part of my past. I decided not to intrude on Tom, so allowing him the pleasure of remaining oblivious to my existence and continuing to suppose that he represented the one and only Tom Weil in B.A.

The Tom-Tom call, then, never drummed into being. Maybe the entire course of world history—or, on a more modest scale, his-story (Tómas Weil's) or my story—would have evolved to a different beat had I phoned Tómas. One tiny variant can result in vastly altered outcomes. Not only what actually happens affects the future but also what doesn't occur. We know only the actual past, not the inchoate dimension which might have materialized but didn't. That part remains forevermore unknown, completely dominated by the unalterable reality and finality of what happened to happen. In the play of infinite forms that which took place upstages what didn't. Although myriad nothings which never happened far outnumber the one and only reality which occurred for each situation, we ignore the might-have-beens. What has taken place, forever settled in time and our past, seems so real and concrete all the other unrealized possibilities which in theory once existed are only ghostly remnants of faded moments fated not to produce a different outcome. Far too numerous to contemplate, those phantom variants, too unformulated and spectral to create vivid impressions, remain relegated to an obscurity overshadowed by the one and only outcome which happened. Still-born, potential alternatives exist after the moment has passed only as theoretical realities forever dead.

Given the great contingency of any outcome, it struck me as strange that any particular person, such a highly contingent inconceivable creature before its conception, happens to exist. Tómas Weil may not know it but he's an oddity, just as is Tom Weil and all the other Toms, Dicks and Harrys and everyone else, however named. A vast cast of characters in the play of infinite forms exists, but it's odd both that people in general and each individual happened to come into being. Where on earth did they all come from? Each humanimal represents an unrequested and unexpected opportunity to experience some weird stuff—that is to say, the particular play of infinite forms which comprise the content of each individual earthling's life. Those forms, many of them insensate and most of them random, combine, contort, mutate, reshape, re-mix in a ceaseless dance of colors, shapes, formats. It's a great spectacle, for which each humanimal is both a participant and an observer and then—after the creature's time—neither. With fondness I remember some of the sensory impressions that happened to reach me. It was all a dazzling display, an improbable and incredible play of infinite forms, from moonglow to moonshine, afros to aphrodisiacs, plots and counter-plots, shadows and light, substance and smoke, lives charmed and hexed, desert and desserts, empires and pyres, awake to wake, composing to decomposing, being and nothingness, born and borne away, carry-on to carrion, remain to remains, here today and gone tomorrow, passing fancies and passing on, spectacles and specters, the infinite forms playing on and on until now, for me, the play is almost over—from overture to final curtain.

Without a herd of humanimals populating the world it wouldn't exist—not necessarily a bad thing—in the form as known by humankind. The play of infinite forms would play on without being observed. Just as people in general aren't necessary, no particular person in and of himself is essential to the scheme of things. The shenanigans and henanigans of any one person add almost nothing to the play of infinite forms. The macro system as it exists encompasses a mass of people, but specific specimens of the species don't figure in to the larger scheme of things, which depends on bulk. The macro doesn't depend on any individual micro. Individual beings are usually desired, at least by their parents, and some are even desirable: it would have been a much poorer world had Shakespeare or Vermeer or Snoopy never lived. But in general things would go along quite well, or ill or whatever, without a Tómas Weil, a Tom Weil, or anyone else as a component of the entirety. Those two characters represent simply random finite units in the play of infinite forms. I could have been that other Tómas Weil, he could have been me, or the molecular and genetic combinations which happened to comprise either of us might by chance have been thrown together to create a completely different creature, like Lassie or Rin-Tin-Tin or Snoopy. And after our time, when our T.W. forms have become de-formed when we give up the ghost, the released molecules might re-form to produce other living things, like a banana tree, a rose, a turnip, an elephant, an anything. Created by chance as temporary sensory receptors, the two Toms are only toys of nature and one day the impressions they absorbed will vanish as if the receptor and the received never existed. The fact that either Tom does exist is passing strange even if, once alive, they come to seem un-strange, ordinary, unremarkable, a typical humanimal, one of billions. Although it seems normal that—once thrown into life—a Tómas or a Tom Weil exists, their being is a brief aberration from the truly normal state of non-being. When a person or an event becomes real they seem so vivid their fundamental contingency and natural state, non-existence, is often forgotten. It's hard to re-imagine reality to perceive it as lacking what you know exists.

Although I briefly entertained myself in that Buenos Aires hotel room by inventing an imaginary dialogue with my local nominal double, my thoughts didn't linger on the might-have-beens of a possible encounter with Tómas Weil. Such theoretical scenarios were for me always passing fancies. I preferred to concentrate on the more granular realities of what I actually saw and what really happened rather than alternatives which might have occurred. Those realities, especially in the Third World, exerted a strong influence on my views regarding the operation of luck. Clearly, not everyone can be lucky. If luck favored each and every humanimal the world would be quite a different sort of place. As it is, in general luck disfavors most people. Human life is far too complex and chance too capricious for luck to endow more than just a few fortunate souls with good fortune. The world operates with too many moving parts to produce smooth processes with friction-free outcomes. Good luck lubricates only a small part of the lonely planet's machinery and only too rarely serves as a balm to sooth and smooth one's way. Most earthlings are embalmed or burned—vanishing in puffs of smoke—at the end without that emollient balm.

With so many competing needs and wants—so many different agendas, plots and counter-plots aplenty, humanimals struggling to survive all manner and womanner of hazards imposed by random forces, natural and man-made influences and intrusions, chronic confusion and disorder—luck faces many barriers for it to show its hand as good rather than bad. Good luck is severely rationed, and there seems to be no rational way to explain how it's distributed. Is there one large lump of luck which, by some sort of cosmic process unknown to man, a presiding agency doles out unequally? Or is luck randomly dribbled into human affairs higgledy-piggledy just by pure chance, for some reason favoring only a few fortunate individuals? How does the system work? Is there a system, or only a play of insensate fateful formats and infinite mutating forms?

What I saw in the Third World persuaded me that the benefits of luck enjoyed by the favored few are distributed in ways which seem for the most part arbitrary and unfair. Providence endows some souls with enough luck to live especially fortunate lives, while most humanimals suffer the ill effects resulting from lack of luck. The distribution of such different outcomes seems out of kilter. And of course even the most fortunate remain ever vulnerable to a reversal of fortune. Luck can turn on a dime or a nickle or even a penny to mint within an instant a completely different kind of fortune or misfortune based on a random flip of the coin. Heads or tails? Chance determines if you get ahead or if you're doomed to trail behind at the tail. In a more just world, not so many individuals would remain low caste outcasts, down-and-out outsiders rather than up-and-in, like those few blessed with luck. But it's just a world, not a just world. For all I knew, perhaps some of those unlucky souls consoled themselves with the hope that good luck would one day favor them so they could then eventually gain purchase on a more humane way of life. Better yet for the unfortunates, perhaps they believed that in after-life they'd enjoy permanent celestial bliss, a heavenly state of being which would justify the hell on earth they suffered. Maybe the more extreme the earthly suffering, the greater the eventual postmortem reward. Bad luck would serve as a prelude to endless good fortune. The fleeting earthly existence of the luckless would then represent only an unfortunate transient prelude to eternal happiness. One of the Spanish words for "beggar" expresses the optimistic view of a better life after life—*pordiosero*: "for God I'll be." In those heavenly after-life realms, the infernal conditions back on earth would no longer define the oppressive fatal circle of an unfortunate. Maybe divine justice would even operate to turn the tables with a topsy-turvy effect by lifting the previously luckless up while those once favored when above would in the end be relegated to life at the bottom of the heap. Speaking for myself, I hope that role reversal does not in fact represent what's in store for mortals who have enjoyed good luck.

As I traveled far and wide, the contrast between the millions in deep poverty and the few at the top favored by fate seemed to result from random distributional factors. Of course, I had no way of knowing the back stories, but the humanimals who happened to live in the palaces and those who languished on the sidewalks outside might well have interchanged their places had luck operated for them in a different way. A few lived well while many dwelled in conditions unfit for man or beast. From what I observed over the years, a fine line—capriciously drawn by the hand of fate—often separated those who enjoyed relatively favorable lives from the great majority of earthlings who suffered oppressive conditions or otherwise lacked any elements of a charmed life.

I can still recall the vast masses swarming the streets, a seething stew of benighted beings in a boiling cauldron of hellish conditions. There were millions out there, pullulating and pulsating, perspiring and respiring and aspiring and conspiring and expiring, eating and excreting, expectorating, womenstrating and demonstrating and remonstrating, defecating and micturating, scheming and dreaming, plotting and counter-plotting, agitating and agglomerating, copulating and propagating, sweating and begetting, coping and syncoping, scrambling and scampering and struggling to survive—a jungle of street fauna existing, just barely, on the lonely planet lost in the cosmic vastness of space and time. Where on earth did all those humanimals come from, and what were they all doing clinging to life as the planet spun through the emptiness, all of them stranded upon the crust of a mite of dust in the firmament? It all presented a spectacular spectacle, churning with forms, but one shadowed by sorrow and misery

The unfortunates made a great impression on me, but only as types. The play of infinite forms populated the scene with a vast variety of human characters: beggars, the *pordioseros*, urchins, street vendors, slum-dwellers, maimed and mutilated, the decayed and deformed, panhandlers and garbage handlers, barefoot paupers and bone-thin skeletal near cadavers, rag-clad ragamuffins and unruly ruffians, scruffy scroungers and scavengers, stranded refugees and furtive illegals, street people, outcasts, homeless, bums and tramps and vagrants, the ten-to-a-room families, the alcohol or drug-dazed zombies, harlots and varlets, pick-pockets aplenty (but just which ones among the mob were they?).

Then there were those ill-fated captives sentenced to a lifetime of hard labor in menial, oppressive or degrading jobs: sewer workers, packing house laborers (their blue collars stained blood red), booth-confined toll-road attendants (endless hands handing over payment, receiving change), pedicure ladies (an endless parade of feet), errand boys (some of them old men), busboys and dishwashers, grease trap cleaners, door-to-door peddlers, coal and lead miners and minors carrying bricks or dung and maybe diseases, enema technicians, banana plantation peons, laundry room and other sweat shop toilers, cruise ship kitchen staff, garbage collectors, corpse handlers (also garbage collectors), nursing home attendants, shoeshine boys, maids in no-star hotels, septic tank crews, ditch diggers, subway train drivers ever underground in life and then in death forever so.

And there were those partial people, fragments of a humanimal, some with a screw loose ranting and raving and babbling incoherently, others missing body parts, maimed souls who earned a living (some life!) from their deformities, like the lepers who displayed their half-limbs or nose-less faces or other hideous features or lack of them to coax alms from passers-by—curiosity seekers like me, transient observers who could at will escape the miserable menagerie of grotesque creatures and move on to less gruesome new sights. All those unfortunates were out of luck, never had any to begin with, completely luckless and stuck in their oppressive fatal circle, chained to place and locked in a relentless struggle to survive. Such servitude and desperation made me ponder how it happened that they suffered such an ill-fated destiny while I was a stranger to such misery, a passing outsider looking in rather than trapped inside looking out. Yet there I was—always only passing through to see the spectacle without being forced to stay and play a role in the production. Free and flexible, I remained in motion removed from the commotion. If my sightings became too unpleasant or unbearable out there in the big wide world I could always retreat to my privileged existence in my secluded dead-end street back in mid-America, a haven far removed from lepers, beggars, filth, dung and poverty and all the rest of such raw humanimalistic scenes. The scenery was less jarring at home. As Graham Greene once noted about his adventures, like him I traveled with a return ticket home. Flimsy as it was, back in the days when a thin paper voucher represented your flight from misery that ticket made all the difference in the world, the Third World I often roamed. By means of that little piece of paper I could extricate myself from unpleasant upsetting settings and return to my cozy little home, closing the door to leave the outside world and its sordid scenes and difficulties behind.

I don't mean to exaggerate the dangers, challenges or difficulties of what I experienced with my travels. Although I faced my fair, or unfair, share of problems, risks and discomforts during my adventures, any such threats or unfortunate episodes never resulted in lasting or ultimate troubles. They were all transient, as was I. I never sought macho challenges but did my best to avoid them. I enjoyed seeing Mount Everest, but didn't want to climb it. I also tried to avoid placing myself into potentially dangerous situations. At times risks were inevitable, inherent in a situation or setting, but by instinct, intuition, reasoning and experience I did my best to modulate and curate exposure to possible dangerous exposures. I was a curiously hybrid sort of traveler, both bold and cautious, a chancer while ever careful, adventurous and at the same time timid. It was an odd admixture of the conservative and the bold, an analogue to my home and my away lives.

With me as I traveled I carried the enablers which would allow me to withdraw from undesirable, uncomfortable or unpleasant places or situations. Those flimsy, light, small little pieces of paper and plastic were worth much more than their weight in gold. The credit cards, travelers' checks, cash (no soft stuff but only hard currency coveted everywhere: U.S. dollars) and the return air ticket gave me freedom. My controlled-risk and protected mode of travel contrasted greatly with such true derring-do adventurers—I was more daring-don't—as Edmund Hillary in the high Himalayas (or Hermalays these days), Thor Heyerdahl on the high seas and, to mention only a few others, Thesiger, Burton and Doughty, so renowned their last names serve to evoke their lasting reputations, and for good measure an American, the hyperactive but less famous William Augustus Bowles.

Back in earlier times the going was much tougher and the privations suffered by adventurers much more oppressive. Compared to the sufferings early-day travelers bore my own difficulties, while concerning at the time, were quite mild and much more easily resolved. One of Hakluyt's most remarkable accounts of early-day travels and explorations is Job Hortop's tale of his capture and 23 years of imprisonment by the Spanish. Hortop begins his story by referring to the Biblical Job, noting "that man born of a woman, living a short time, is replenished with many miseries: which some know by reading of histories, many by the view of others' calamities, and I by experience of myself, as this present treatise ensuing shall show." Of the three sources cited by Hortop I accessed written accounts and engaged in many observations, but compared to him my travails were merely minor passing inconveniences

In October 1567, Hortop sailed as a gunner with John Hawkins' third expedition to the Caribbean. A venturesome sea captain—who in 1588 helped command the ships that defended England against the Spanish Armada—the daredevil Hawkins and his crew suffered hellish experiences. Hortop relates episodes of encounters with "wild Indians" and their poisoned arrows; "monstrous" sharks; a monster sea-horse, "his teeth very great"; "a monstrous venomous worm with two heads"; "tigers, monstrous and furious beasts"; "a monstrous *largato* or crocodile"; a monstrous adder. Facing all those monsters was the easy part. Hortop's real problems were yet to begin. As President Ydígoras Fuentes had observed to me when I was a very young man, plots and counter-plots abound, both out in the wild and in the civilized if not always civil jungle where humanimals compete. Alexander Hamilton complained in *The Federalist*, Number 29, how his political opponents depicted his argument with "frightful and distorted shapes—Gorgons Hydras and Chimeras dire—discoloring and disfiguring whatever it represents and transforming every thing it touches into a monster." Hortop and Hamilton both faced monsters, the Englishman natural ones and the American the man-made political variety.

Far worse than all the beasts Hortop encountered were the Spanish, human monsters, and how they treated Hortop after capturing him following the September 1568 fight between Hawkins' ships and the Spanish fleet at the Battle of San Juan Ulua (at Vera Cruz) in Mexico. As the captured victim summarizes in "The Computation of My Imprisonment" at the end of his account, Hortop spent a total of 23 years as a captive in prison, in an Inquisition

House, as a galley slave (12 years), and in other detention confinements, occasionally being threatened with death by hanging or burning. Hortop faced inconveniences such as gallows and galleys with a gallant galliard attitude. Finally back home in England, Hortop concludes his nightmarish story—a remarkable tale of suffering, human endurance and survival, all told in a matter-of-fact manner—by noting simply, "Thus, having truly set down on to you my travels, miseries and dangers, the space of 23 years, I end."

Job Hortop was for sure no armchair traveler, a cushioned stay-at-home who only read about or observed the travels, travails, trials and tribulations of adventurers. My travels represented a mere very pale copy of Hortop's truly horrendous experiences. Compared to him, only slight risks tinctured my adventures with danger. I was willing to expose myself to the problems, some known and others unknown, of far-flung adventure travel but the potential threats and discomforts were more in the nature of inconveniences than dangers of the kind true adventurers faced. I gave up not 23 years, as did Hortop, but only my cozy shelter in the protected precincts of my coddled and comfortable natural habitat. I willingly renounced those benefits to venture abroad and to brave whatever hazards and uncertainties might lurk out there on the open road . Monstrous challenges such as Hortop encountered didn't endanger or concern me. But there were quite enough lesser lessons and minor vexations to educate and to challenge me. There is no such thing as adventure travel without exhilarating and sometimes perturbing adventures.

For me, traveling the great wide world proved to be an enthralling and memorable enterprise, one which enabled me to satisfy much of my curiosity and my wish for a diversified existence. I could have more easily and conveniently just stayed at home, back in the pleasant provincial subdivision where I hang out (not, like Hortop, almost hanged), back where the only poison is ivy (not poisoned arrows), and only wild weeds (not wild tribes) and monstrous cars (not beasts—although pampered doggies and the squirrels get a little fat) threaten. Compared to Job Hortop and to other true adventurers and explorers my travels were benign, relatively problem-free and for the most part safe, even if fraught with some anxious moments. At least that's how my far-ranging trips worked out for me, although it could have all been different. A microscopic stray pathogen or errant vehicle, a violent tropical storm while airborne or many other random threats could have done me in. For sure I tempted fate and teased luck by undertaking such adventures, as all motion and almost everything else comes with risk of one type or another, risk all too often without reward. But most of the time good luck, often a fickle companion but for me usually a faithful one, happened to accompany me.

Although I was keenly aware of my favored situation as I traveled through the Third World, I also realized that in an existential way my life was no different than the downtrodden humanimals I saw and still so vividly remember, at least in mass. Even with all the advantages my fatal circle and good luck had given me—means, mentality, attitude, independence, interest, the return air ticket—I faced the same mortal reality as did every *presidente*, prince, pirate, pauper, peon, potentate, prisoner. With my air ticket home I could always escape from a place, but no passage allowed me to avoid my final destination. Like everyone else I was stuck in a finite existence while surrounded by infinite forms. Along with all the millions I saw churning through the overrun streets I too would soon be gone, succeeded by millions of other humammals (and lower forms of life) equally mortal, and so on and on and on. In *Tamburlaine the Great* Marlowe noted how even majestic royal figures only pass through and, in time, pass away: "Is it not passing brave to be a king and ride in triumph through Persepolis." Such line did I remember when climbing the Apadana Staircase in Persepolis, and at many other times. Emperors either lose their empires or are eventually lost to them. Advantaged as a regal or a favored life may be, even the most fortunate soul enjoying great luck possesses only passing benefits which soon disappear with the person. In polite parlance humanmals don't die, they pass away as if passing on to a new and better state. Lots of luck on that one.

In addition to mortality a second commonality existed between the favored few, like me, and the unfortunate many, for we all faced the same vulnerability to fate and chance. No one escapes their influence. Because luck and fortune by definition operate in a capricious manner, those random factors remain subject to sudden change. Any particular existing status not only depends largely on chance but is also highly provisional. The ragtag rag-clad emaciated barefoot rickshaw runner in Calcutta, bone-thin as the poles on the cart he pulled, may once have been a nattily-dressed prosperous passenger. Fate or bad karma, however, had somehow decreed that he was now a sweaty pull-man dragging the wagon through Calcutta's mean streets. While rolling along in a Calcutta rickshaw I meditated on the possibility that a nasty twist of fate might have operated, or at any time still could, to mutate my life by reducing it to the level of the lowly laborer who like a dray animal pulled me along. But somehow I was able to remain up there in the seat while others were condemned to languish down in the street. Shaw and rickshaw—plays playwright George Bernard Shaw's vehicle, a cart, the low-caste street person's vehicle, each humanimal a random character in the play of infinite forms cast in his part by Chance, an inscrutable dramaturg, the whirl-a-world spinner of fates.

The rickshaw driver I hired in Calcutta for one short pull—a new experience for me, as in my suburb back home taxis had replaced the fleet of rickshaws—dragged me at my order to the Park Street Cemetery where long defunct English functionaries and their families from the Raj era repose. I remember wondering as I paid the driver, a fare so low as to be unfair, if he envied the dead there in the cemetery. So miserable was his life, trotting barefoot through Calcutta pulling passengers for a pittance, maybe the poor fellow would prefer to be released from his present existence to try for better karma in the next version of being. But once thrown into life, no matter how unfavorable the fatal circle and other circumstances, humanimals seem to prefer to cling to life, to continue on in misery rather than to risk oblivion or, even worse, perverse postmortem conditions unimaginable to the living. As bad as things might be here, perhaps they were better than there—wherever and whatever the "there" was. Some deeply-embedded instinct caused earthlings to favor chance-riddled life, even if luckless, than the alternative.

Calcutta provided a memorable introduction to India. I arrived there in late November 1973 from Rangoon, Burma, as the first stop on the first of my three trips to India. For travels in the East the name "Calcutta" evokes a nightmarish metropolis teeming with humanimals reduced to the most abject and miserable conditions, the very avatar of poverty, an accursed hellish inferno which Vishnu "the Preserver" had somehow failed to bless. In fact, the oppressive conditions in Calcutta at bottom seemed little different from those I found elsewhere in the Third World, for at the very lowest level poverty the world over entails the same sort of deprivation and sense of hopelessness. The cry of the down-trodden, "Gemme outta here," falls on deaf ears if, indeed, anyone's listening. There's no escape. Although Calcutta serves as a symbol of an extreme state of misery and human degradation, such conditions no matter where encountered reveal an unfortunate underclass which exists, barely, way down in the pecking order, the lowest of the low bottom dwellers who occupy a Dantesque "dark forest," one "so harsh, dismal and wild" that no route out could ever offer an escape from the oppressive entrapment. In such places a passing visitor encounters not simply nightmare scenes—at least bad dreams, as unpleasant as they are, end—but daymare realities, an irremedial wretchedness from which waking brings no relief. There in Calcutta Mother Teresa worked in her own merciful hell-hole, comforting a few of the lucky unfortunates who wasted away to their death with compassionate care while the unlucky ones died unnoticed in the streets. The standard ghastly image of human horrors which describe life, and death, in Calcutta invariably includes people dying on the streets. Although I did notice many near-ghost skeletal people living—barely—on the sidewalks, I never saw anyone dying there. Maybe they're removed before the last gasp.

Section II: The Fatal Circle

As poor and as misery-filled as Calcutta was, the city seemed safe, even at night. After dark, small kerosene lanterns cast eldritch shadows on the bed-less human lumps bedding down for the night on the hard sidewalks. The luckless homeless lacked shelter, privacy, privy, running water (except when it rained), kitchen, solitude, quiet, books, beauty, greenery, refinements or finery and most of all dignity, all of which amenities comforted and coddled me back home. In Calcutta things were different, other than for those of us just passing through and for a few fortunate locals. Once I emerged into the world, and especially the Third World, I quickly realized how fortunate are those of us who are privileged to live in the First World. Many of us there enjoyed a fatal circle completely different from humanimals in underdeveloped lands. I no longer took my home-based comforts and conveniences as a given, for I knew that those benefits represented exceptional conditions unknown and unavailable to millions all around the lonely planet. I lamented how so many souls suffered such limited luck-less lives, impervious to improvement, and I always greatly appreciated my own fortunate situation which I knew depended on chance, happenstance and luck.

There were in general two possible responses to the poverty and misery I saw in India and elsewhere in underdeveloped lands. I could have emulated Mother Teresa and become Father Tereso, working among the deprived to try to alleviate their suffering. Of the millions who needed help, Mother or Father T could minister only to very few, but at least that humanistic effort would do some good and set an example. But—as explained below—I chose the second alternative, which was for the most part to ignore the misery while ever bearing such degradation in mind so as to avoid wasteful practices or frivolous consumption, and never to claim a sense of entitlement or to ignore the role of fate, chance and luck in my favorable circumstances. I didn't view the unfortunate as essentially different from me but as me had I been deprived of luck and a positive fatal circle.

Experiencing a place like Calcutta brings more than just culture shock. It's an existential shock. It was quite sobering to see how people lived, or tried to live, in the conditions which existed in Calcutta back when I was there in November 1973. Maybe things are better now. Back then the scenes I saw there and in other such purgatories around the world exerted a profound and life-long effect on me, one which greatly reinforced my already existing tendency to live simply and frugally and to shun possessions. People who had nothing reinforced by way of example my own practice of living with as little as possible, possessing only the necessities without any surplusage. From the day, now so long ago, when I watched women and children in Calcutta pawing through mounds of fetid, putrid, smoldering garbage rummaging for cans, bottles, some scraps of paper or cloth the scavengers could sell for a few rupees, I've always recycled as much as possible long before the environmental movement made that practice fashionable. Most of my "post-consumer waste," as garbage is called in the First World, my world, came not from unbridled consumerism but only from necessities. Even though not much of a consumer, what waste I produced I did not allow to go to waste. With a want-not waste-not attitude I refused to toss refuse away into the trash, for I found myself unable to scrap materials which on the other side of the world represented bare survival for the unfortunates scrambling and scrapping for a living. Visions of Calcutta garbage heaps and the scavengers there were for me out of sight but never out of mind.

Just as I didn't waste waste, so never did I waste energy on gratuitous physical activity. It seemed to me perverse to spend energy needlessly at home when in the Third World so many people lacked energy to spare. All their energy was devoted to bare survival. In many places they didn't even have enough electrical energy. Back home life was different. People made playful use of their surplus energy, working out to burn calories, to trim fat and to get fit, while at the same time in far-off impoverished lands locals had no extra energy or fat for such frivolous pursuits. Visualizing how those Third World calorie-poor souls struggled their entire lives just to survive deterred me from dissipating my energy in unnecessary self-oriented work-out exercises. I was unable to engage in that sort of self-indulgent exercise, knowing that all around the world so many down-trodden earthlings exerted every ounce of energy just to continue to exist.

Upon arriving in India for the first time, it didn't take me long to suffer both the shocks—cultural and existential—which unsettle newcomers. After landing at Dum Dum airport in Calcutta I took a taxi to the hotel I'd selected but had not booked ahead of time. I seldom reserved a room in advance. My reservations related not to booked accommodations but to the limitations of a fixed arrival date. I didn't want the trip to be governed by the need to be in a specific place at a set time. Based on my research for each anticipated stop, I picked out a few possible hotels and arrived at random on a day of my choosing. At times I had to tramp the streets of a strange city, lugging my luggage, to find an available room, but I preferred that inconvenience to the restrictions of being locked in ahead of time with a reservation.

In travel there's almost nothing like a first arrival in India. As the taxi advanced through Calcutta I observed the passing scenes—or, rather, the permanent and for the locals inescapable ones, for I was the one passing through. For the throngs I saw, this was their daily and life-long reality. I enjoyed mobility and the ability to leave it all behind. Urban pathology festered: on the sidewalks people lay inert (dead? dying? sleeping? dreaming of living like a maharajah, of paradise, of dying?); beggars, some maimed, worked their way among the swarming crowds; garbage-strewn streets and littered byways disfigured the city-scape. Cows ambled along, blocking the chaotic traffic which spewed noxious exhaust fumes. Feces dotted the street, a dung deal. Incessant noise, a cacophony of discordant sounds, and harshly colored garish billboards coarsened the teeming scene. Seeing the misery, I fleetingly contemplated a "stop the presses" decision—to abandon my agenda, my home life, and everything else in favor of converting to Father T (Thomas or Tereso) by joining Mother T in her angelic work. But I immediately realized that the kinds of hellish problems she addressed were intractable and I doubted that I could have the slightest effect in dealing with them. It was all I could do to try to solve limited and tractable problems. The misery was so widespread even the most well-meaning good Samaritan couldn't help much. Never would I become a St. Thomas but always was I a Doubting Thomas, and for sure I doubted my capability in dealing effectively with any of life's transcendent and immutable problems, my own as well as those of the masses of unfortunates around the world. Leaving the oppressed behind, I soon pressed on with the agenda I'd established a few years earlier. Mother Teresa would have to soldier on without me.

My hotel of choice happened to have a free room. While checking in to the establishment, a welcome refuge from the noisy kinetic chaotic world outside, I chatted briefly with the desk clerk, who'd moved to Calcutta from the adjacent state of Orissa. Curious, I asked what motivated him to leave his family and homeland for a place like Calcutta. "Things are really bad in Orissa," he replied.

It's of course very nearly impossible to experience how Third World inhabitants actually live. A few truly intrepid travelers manage to lower themselves to the rock bottom local standards, pursuing low-life to gain a perspective on how things look from the bottom up. Although I didn't want to descend to the very bottom, occasionally I opted to reduce my standard of living to a level below its otherwise fairly modest format in order to see the world from a new perspective. I figured that since I was traveling in disadvantaged places I might as well include in my adventures a few experiences which somewhat—but not entirely—typified how the locals actually lived. I went to all the trouble to get to those distant areas, so why not extract as much as possible from the experience by renouncing the comfortable and coddled life I enjoyed at home for a different, even if passing, way of being. Where I came from, comfort ranked as a high priority. Americans crave comfort and expect it. Social conditioning and air conditioning and all the rest was meant to ensure that home folks were comfortable. People of means are described as being "comfortable," but even mean-less folks coveted comfort. Americans don't tolerate discomfort or inconveniences well. Flip a light switch and the electricity better come on N-O-W. That's the world where I came from. Where I traveled to, comfort often remained out of reach, with only endless discomforts marring how the local humanimals lived. Life out there in the big wide world, especially the Third, was for many people, and even in some ways for visitors, chronically uncomfortable. That attracted me—as long as I had my return ticket home.

It's sometimes said that life is an adventure, but that applies only to a fortunate few. For many people, life presents a great challenge and imposes a continuing struggle to survive. Only a lucky minority enjoys the opportunity—the means, temperament, will, fatal circle, circumstances—to pursue an adventurous existence. The road to adventure is a dead-end for most people, blocked by the demands of everyday exigencies. Life in general, even for the fortunate such as those who ride through Persepolis passing brave, is not an adventure but only an experiment, one with two elements. The first involves a specific chemical-genetic mix which by chance creates an individual, a unique organic combination but with essentially the same components as for every other humanimal. The other pertains to the reaction of the sensory receptor unit created by the chemical stew to the stimulations produced by the play of infinite forms. Although luck and fate play a role in those two experiments, for most people adventure seldom adds an enhanced dimension to their life experience. Other than for a favored few, life is more of a struggle than an adventure.

My passing experiments as a well-heeled First World traveler living at times somewhat in the style-less style of a Third World indigent gave me only a slight taste of the low-life, not a full course meal and for sure not a permanent diet. I only snacked on the pitiful buffet, the meager offerings available to the under-nourished down-and-out masses. Moreover, my buying power and my return ticket home always buffered me from the ultimate realities of Third World life. I had a way out and could at any time revert to a more pleasant standard of living, even if not a lavish one, and escape from zero or sub-zero-star hotels, hard-class train accommodations, dilapidated third class Third World buses, and all the rest of the hand-to-mouth inconveniences which subjected the entrapped locals to a life sentence without parole. My own sentence would be only in a book describing an experience, one I read about or had myself encountered and wrote about. No matter how hard the going, the soft life always remained available to me. I could revert to comforts at any time. The poor disadvantaged would permanently remain stuck at the bottom, as stasis rather than upward mobility governed their lives. Perhaps some of them hoped to escape their fate, but more than mere hope comforted me. Hard dollars guaranteed me a soft landing whenever I opted to abandon the low class discomforts to resume my normal way of life.

My endowments, financial and otherwise, far transcended the suffocating fatal circle in which the Calcutta street people were confined. As I strolled through the crowds I realized that I carried on my person—in cash, travelers' checks and credit card buying power—more than the total life-long earnings, by sweat or by alms, available to any of the unfortunates languishing on the sidewalks all around the city. Even the collective income of the thousands of street people over their entire lives, admittedly probably short, wouldn't match the financial resources I carried with me for just this one trip. You can't describe those miserable souls as "dispossessed" because they never really possessed anything to begin with, and they never would. In a perverse sort of way, paupers enjoy the ultimate security as they have nothing to lose. The impoverished conditions differed starkly from life back home where stuff clogged closets, garages, basements, attics, rooms. Those spaces brimming over with possessions presented an extreme contrast with the possession-less beggars and rag pickers and rickshaw pullers and millions of others I saw over the years in Calcutta and in many other places all around the lonely planet.

At home I chose to live in a dispossessed way. Partially influenced by the conditions I saw in the Third World and in part by my temperament, I followed a have-not format rather than the "have" way of life typical of a consumer society. This pared-down existence I selected not for want of buying power but for lack of want. From my earliest adult years I had no interest in using money to acquire stuff. Material objects never interested me. I thought that they'd only distract me from my true interests. I instead preferred to spend on the intangible but enriching good of good experiences. The return on that investment greatly multiplied my riches, as I defined them, by endowing my existence not with collections but with recollections I could profit from for years to come. My experiment in living entailed accumulation of memories, not things. My most treasured possessions are the least tangible invisible ones—ghostly but still vivid original experiences, and also my relationships with family, friends, colleagues, acquaintances. To me those are worth far more than their weight in gold, which anyway would amount to zero as my true treasures are all weightless. I possess them without fear of loss by outside forces, but will before long lose them all by the ineluctable force of nature. No insurrection, revolution, confiscation, appropriation, war, civil disorder, thief, burglar or any other intrusive agency can take from me those sweet golden memories. Only time—time that thief who loves to put sweets into its purse, a bag which contains alms for oblivion—can, and will, rob me of my fond memories. My experiences and relationships remain burglar-proof while I live and retain my recollections. Thereafter all will vanish, like smoke into thin air. Losing your treasured past and the rich after-glow of its gilt-edge moments in time is the price you pay for having lived them.

Contrary to many of my friends and neighbors in the comfortable Midwestern suburb back home where I lived, I viewed possessions as a burden rather than a benefit. Out in the world where I saw so many unfortunates who had next to nothing but still managed to survive, even if only barely, without things, I realized that less was more. I acquired only necessary basic things with no frills or extras. When a Good Will truck appeared on my street the neighbors didn't know if it was a pick-up or a delivery. FedEx and UPS never delivered parcels to my house, My only on-line connection was to hang my laundry, not to buy wares or wears from internet retailers. I've never purchased one thing off the web. For me, Amazon was a virago, not a source of stuff. Function rather than fashion determined what I bought. I viewed things as boring. They didn't change much, other than breaking, gathering dust or—the worst fate of all—going out of style. Many of my contemporaries wouldn't want yesterday's vacuum cleaner, last season's unfashionable fashions, plain merchandise without a status logo, brand-less generic products lacking prestige. Relationships and experiences—more dynamic than static stuff and with the potential to bring new, stimulating and enriching experiences—appealed to me more than things. Things don't change but I could, and so I revised my life to enlarge the intangible dimensions, which for me were more satisfying than objects. I endorsed William James's view that "lives based on having are less free than lives based either on doing or on being." I believed—and lived by my belief—that primacy of possessions and possessing interferes with the highest and best use of the brief opportunity afforded by existing as an earthling. For me, the fleeting experiment of living pertained to the essence of the exercise, characterized by impermanence as reflected in experiences and relationships all of which, unlike stuff, eventually end. If you happen to exist, a chance occurrence, and, also unlikely, by chance happen to be in a position to afford whatever stuff you want, it's better to avoid it. Things divert you from more gratifying pursuits.

In *The First Time Around: Some Irreverent Recollections* (mine are all reverent), Joseph Wechsberg recalls how in March 1939, by which time he was safely in New York, his mother back in Czechoslovakia "waited too long" to escape. After the Germans had invaded the country, she delayed her departure to obtain a special permit to take out the family's paintings, rugs, china, silver. Instead, she herself was taken away, forcibly removed from her cherished heirlooms and sent to a concentration camp. Wechsberg observes: "Ever since I've disliked possessions. Albert Einstein, a fellow refugee, called them 'ein Stein am Bein' "— "a stone around the leg." For Einstein *ein Stein* was relatively burdensome.

As already noted (in Section I, Part 4 above), Virginia Woolf was "exhilarated" when she lost her household possessions in a World War II bombing raid. This freed her up to restart her life, "free to go anywhere." In my own case, what I'd most miss are the personal objects with a meaning beyond their decorative or functional value. Those things consist mainly of inexpensive travel mementos I acquired during the years I was free to go anywhere and in fact did so, pursuing my wide-ranging trips to the far corners of the globe. Each object evokes a particular time and place, a memory of where I happened to find and acquire the item. Such possessions have no monetary worth—only sentimental value. Taken as a whole, those worthless but precious mementos represent tangible evidences of many pleasant intangible experiences, long gone but not forgotten as each item offers an *aide-memoire*. In *Thieves in the Night*, Arthur Koestler notes how "the feeling of possessiveness towards an object is the recollection of memories it represents; its value is crystallized memory." Should a bomb, earthquake, fire or other disaster, natural or man-inflicted, destroy my house I'd regret that those glass menagerie-like crystallized memories would be fragmented by loss of the objects. I'd also miss some familiar functional familial household objects which furnish not only my house but also some memories. But even without freedom of the kind Virginia Woolf's total loss of possessions gave her, I decided to restart life with a new agenda, one which made me "free to go anywhere," and over the years I very nearly did go everywhere.

~ 4 ~

All around the world over many years my journeys took me to a crazy-quilt of places, each interesting in its own way, each different in its glory and diversity, and all of them populated with the same fauna—the humanimal. Whether Buenos Aires in Argentina, the sometimes *malos aires* of an impoverished Third World place like Calcutta (in the race to the bottom Haiti tops places like Calcutta), whether the *fríos aires* of icy Antarctica (a few earthlings—or icelings—at times live there) or in the *calientes aires* of a sweltering tropical banana plantation in Guatemala, each corner of the world offered something of interest and then memorable experiences. By now long past, my experiences and my more recent memories of them form the content of my reminiscence. Its text puts them in context. Based on happenings widely separated in place and time, I might have imposed on my account any one of many arrangements, each equally arbitrary and artificial, to bring some order to the unruly materials. The original events, many of them random, might also have been completely different. Going forward, what happened to form the contents of my life was contingent, until in the course of time it all became irrevocable.

Just as my life was tentative until definitive by lapse of time, so this reminiscence remained mere potential until arranged into a certain order. But unlike my nearly completed life, my observations and opinions in this book remain tentative. In no way prescriptive, they're based on and relate only to the life events of a particular person who experienced existence in his own way. My Buenos Aires namesake double, one Tómas Weil, has perhaps written his own reminiscence—same author, nominally at least, but entirely different content and point of view. Even one's very name, as familiar and as close to us as our own skin, represents only an arbitrary construct. Were my account in Spanish, I'd be writing as Tómas; in Italian, as Tommaso. The book itself presents an order imposed on a disparate hodgepodge, an arrangement as artificial and arbitrary as a name. In any case, whatever the author's name, once the work exists he's irrelevant, and in my case I prefer to remain irrelevant. Although I've tried to be a ghost writer removed from the story, my narrative hopefully retains sufficient substance, even if I'm absent from the tale, so that the account offers more than simply a bare skeleton of a book as if seen in a disembodied X-ray.

Looking back, maybe I should have phoned that Tómas Weil during my pre-Antarctica B.A. visit in February 2001. Perhaps Tómas was well connected, like Mauricio Rosal in Guatemala City back in the summer of 1958. Through *Señor* Tom I might have been able to gain access to the Casa Rosada for a second Latino presidential palace visit. Not impossible:

those Tom Weils really get around. As things turned out, no Rosal Rosada figure graced my stay, and so I saw there no President of the Republic. Instead I found myself in the Republic of Letters, that imaginary literary land which transcends national boundaries to form a virtual international intellectual community populated by citizens devoted to the realm of words. I'd come a long way since Guatemala 1958, from a Banana Republic to the Republic of Letters. That new Republic I entered without a visa, except my Visa and American Express cards, in late December 1971 during my five-month-long trip all through the vast incontinent South American continent, the first of three such extended trips around the world. The other two, in the early 1970s, later took me to the Middle East and then to the Far East, all far removed from my comfortable U.S. Midwest base.

It was by my own volition—rather than, as in Guatemala City, how I by chance happened to meet the president—that I met in Buenos Aires Argentina's renowned and distinguished literary figure, Jorge Luis Borges, who presided not over an entire country but only over the country's National Library. No Mauricio Rosal-type intermediary facilitated my access to Borges, as I sought him out on my own. As library director, Borges had a local habitation with a name, and so I knew just where to find him. Not a presidential palace but a fanciful palatial pile then housed the library, since moved elsewhere (an undistinguished eye-sore-style structure I visited in February 2001). Arriving there at the Republic of Letters outpost in Argentina's capital, I threaded my way through the labyrinthine and slightly lugubrious (all those dead books buried in the stacks) building on Calle Mexico, that name evoking memories of my first foreign trip imposed on me as a young boy nearly 30 years earlier, to arrive unannounced at the director's office. That's how I did it with hotels—just show up. Maybe the same technique would work and Borges would be accommodating. I asked to see the world-famous fableist, realizing that the blind Borges would not be able to see me. But hopefully he'd be willing to receive me. Borges in any case still retained his ears and tongue, so he could listen and speak.

Borges did receive his intrusive Yankee stranger cordially and without delay or ado, just as had *El Presidente* in Guatemala City thirteen years before. Such welcoming behavior, I must admit, wasn't due to my importance or my reputation for a sparkling personality. For one thing, my personality didn't sparkle—it was unbubbly—and for another, neither of my exalted Latino hosts had ever heard of me. But perhaps Borges knew Tómas Weil and supposed I was related to him, in which case my name (or Tómas's) served as an "Open Sesame." Or did the erudite Borges imagine that I was somehow part of Marcel Proust's clan? His mother was a Weil, one Jeanne Weil, daughter of Nathée Weil, a stockbroker whose grave Marcel visited every year until his health failed. Borges would know all that. Or possibly the writer perceived no supportive family connections on my part but was simply curious about a presumptuous young *Yanqui* who with no prior arrangement requested an audience. Or maybe he thought the *Norte Americano* might be good for a substantial contribution to the library to help fund the local Republic of Letters installation. Or, per chance, he could be looking for some curious stray stranger to inspire one of his strange stories—in which case Borges had found the right guy.

For whatever reason, the renowned author promptly saw me, or at least invited me, into his spacious office, saying, "You found me in my lair." The appropriate precise word speaks volumes about Borges' fluency in English. As we chatted in English, I contemplated the metaphysical question whether, in using that language, Borges had become "George Louis" rather than "Jorge Luis." In fact, he often called himself Georgie, as did others. Clothes make the man (I dressed in Good Will styles, a wardrobe which did not create good will for me among my friends) but does language? Whatever the case, Borges' English brimmed over with a rich and exact vocabulary, spoken with a slight accent. Endowed to me by my fatal circle, my vocabulary was the equal of Borges, even if my choice of words less measured and refined, and my own accent very pronounced. I spoke with a strong U.S. Midwest accent, which in fact is quite neutral and not very accented. But no matter how natural and native my tongue sounded to me and to my fellow Midwesterners, for many other English speakers around the world I had an accent. In India, Pakistan, Belize, Guyana, New Zealand, Australia, Canada, Ireland, Wales, Scotland, England I was the one with the accent. Even some of the words were different. Even one's native tongue—so intrinsic to your very being and which seems so normal and natural—is subject, as is everything, to many variants. You needn't travel very far to be viewed as someone who speaks with an accent. Below the Mason-Dixon Line, down in Texas, or up in New England I spoke differently, even though using the same words. I came to realize that all around the world, including places not all that far from home, I was an outsider, a stranger, and in some places probably even strange (and maybe in some ways so viewed even at home) in the way the locals saw me.

Travel taught me to try to perceive myself as others did. Their view of me in many ways didn't conform to my own limited perceptions of myself. One especially pronounced and obvious novelty was the way I spoke. Of course, in a number of places speaking was useless as the language was foreign to me. But seldom did I encounter any great problems even if, as in many areas, I was ignorant of the local lingo. Among the many various enlightening new self-perceptions gained from my travels, I came to realize that listeners in most regions heard from me a rather strong accent. This was news to me—which pleased me. Travel served to revise and renew me by bringing novel views on practices so common they never gave me a second thought at home. Out in the world, second and many more thoughts came to me. Travel endowed my entire way of life with different and refreshing perspectives, angles of vision and revision of a type little accessible to stay-at-homes unexposed to the outside world. My travels educated me, more than did classes or books. My journeys informed me not only about places but also in more general ways. By now perhaps it's clear that although my account is based on my travel adventures, my reminiscence is more in the nature of a description of a journey through life—a trip through time as well as place—rather than only roamings around the lonely planet. I do realize that it's something of a cliché to equate one's progress through the years to a trip, but the two—an individual life and travel—have much in common. Although the reminiscence indeed includes a travelogue, my account is in fact more of a life-log. At my very advanced age, that "log" has burned almost to ashes, the smoke vanished into air, into thin air, and by now nearly all gone. My trips brought me a new life, a second one with second and more thoughts. The Third World, especially, inspired many third thoughts. Out on the open road I became more open minded, and my sensory perceptions seemed more acute than when at home where daily life was rather too regular and mundane. A change of scene and of seen proved stimulating. Away from home base, I perceived and learned many things and enjoyed many experiences which never occurred to me in my normal everyday existence. Circumstances alter cases; places alter perceptions; travel altered me.

There in Borges' office at the National Library surrounded by thousands of books, the author was in his element, or "lair" as he called it. During the Perón era the writer-librarian didn't enjoy such a Republic of Letters setting. In 1946 Perón demoted Borges from his post at a neighborhood library to work as a chicken inspector. After the *caudillo* was overthrown Borges was appointed director of the National Library in 1955. What the chickens lost in the way of a highly intellectual and no doubt conscientious inspector, the nation gained in the way of a highly qualified librarian.

On learning where I was from, Borges asked me if I'd ever met "Tom." Was he confused? I was Tom. That mysterious third Tom, along with me and the local Tómas Weil, now completed the trinity of B.A. Toms. I failed to recognize the Tom who Borges referred to so I remained silent. Was it Tómas Weil? "Tom...Tom," Borges drummed, as if the reference was obvious. Upon hearing my silence—do you hear silence?—he elaborated: "Tom Eliot—did you ever meet him back at home?" With Borges I may have been briefly in the Republic of Letters but I was only a mere visitor in that rarefied land. I'd met the president of Guatemala, but the likes of Tom Eliot remained well beyond my reach. After moving from his U.S. Midwest home town to England, Tom Eliot transmuted his name to the more formal and poetic "T.S.

Eliot." This designation no doubt seemed to him more fitting for a London literary figure than the casual "Tom" or "Tommy" Eliot appropriate for his backwater native city. For my part, as a pretender—but not a citizen—in the Republic of Letters, I was content with the simple succinct "Tom," a direct single syllable that conformed with my natural habitat, the basic elemental Midwest, a provincial region without any airs, *buenos, malos* or otherwise. I was quite flattered by the idea that Borges supposed I, plain old Tom, might have met T.S. But perhaps my genial host was simply being polite by suggesting that I traveled in cultural circles exalted enough to encounter the famous poet. In fact, I traveled in no cultural circles. My fatal circle failed to include such refined rings, to which I lacked access. My group was more a circlet, and my activities designed more for adventure travel than for flights of fancy like Borgesian fables or the poetry of T.S.

Borges himself traveled mainly in his B.A.-based imagination. Preferring roots rather than flights, with few exceptions he stayed home. In 1914 when a young man he moved with his family to Europe where they lived during World War I. Following the war, Borges traveled through southern France and in 1919 in Spain. In 1921 at age 25 he returned to Argentina to live in B.A. After publication in 1923 of his first book, *Fervor de Buenos Aires*, Borges returned to Europe, staying there almost a year. After visiting England, Switzerland and Spain he finally returned to Buenos Aires in July 1924 to establish himself definitively in his home town where he remained some 37 years until September 1961 when Borges left home to spend a semester as a visiting professor at the University of Texas in Austin. Borges as a "Longhorn"—hard to imagine, but there he was deep in the heart of Texas. His "fervor" for and attachment to Buenos Aires Borges described in that first book: "This city that I thought my past/is my future, my present; the years that I have lived in Europe are illusory,/I was always (and will be) in Buenos Aires."

The past, and much else, indeed seems illusory and elusive. What you saw and experienced becomes ever more remote, a will-o'-the-wisp gossamer kind of ghostly shadow you can't quite grasp. Proust (perhaps my very distant relative through his mother) captures the uncapturable nature of the insubstantial past in the concluding words of *Swann's Way*: "remembrance of a particular form is but regret for a particular moment; and houses, roads, avenues are as fugitive, alas, as the years." Even a character who remembers everything still perceives the past with "much of the quality of a dream," as was the case for Ireneo Funes, one "Funes the Memorious" in Borges' short story of that name, a chapter in *Labyrinths*. Thrown by a horse, Funes was left paralyzed and lived out his years "as one in a dream" who noticed and remembered everything: "his perceptions and his memory were infallible." Funes boasted: "I alone have more memories than all mankind has probably had since the world has been the world." He could grasp every errant evanescent event and nuance in his "vertigenous world," much as normal people perceive such simple and eternal regularities as a circle or a triangle or a comic strip featuring Snoopy. Snoopy may not survive forever, but his wit and wisdom will.

Funes was of course a freak of nature, but the larger cosmic oddity is that all nature as it manifests itself since the world has been the world is a complete freak. It's doubtful that there exists any place anywhere else like the world. The natural conditions which allowed the lonely planet as humanimals know it to exist are most likely unique. How nature on earth originated and evolved, the planet's fatal circle, developed largely by trial and error, by trials and tribulations, by random experiments, by chance. It's hard to say if the advent and continuing existence of the world represents a lucky or an unlucky outcome of all the experimentation. The benefits the Big Bang and its play of infinite forms brought into being may be outweighed by the disadvantages. Whatever the cosmic experiment might have wrought, for ill or for good or for nothing at all, the chances that nature similarly developed elsewhere are virtually nil. Not only the macro of the cosmos but also the particular micro of the lonely planet evolved with the same sort of arbitrary contingent pedigree. Any specific humanimal—and, once existing, the creature's lived experience—as well as all earthly lives, individually and collectively, the entire world and all its natural characteristics, the universe, the cosmos all derive from a random play of infinite forces. The earth and life on it is an unprecedented, unrepeatable and unique freak of nature. Similarly, a Funes, a Borges, a Tómas or a Tom Weil or Eliot are each simply examples of the singularities which characterize all that exists. The strange world and the earthlings in residence on that sub-lunar pebble are oddities, curiosities of no ultimate significance but which offer irresistible attractions for an acutely curious character. Like everyone and everything else, such a character is strange. It was the exotic nature of the world which so attracted me. I thought it would be a shame to experience the experiment of being alive without getting to know something of the world as it presented itself to earthlings with so many infinite forms in all their spectacular variety. A spectacle attracts spectators, and the world thus drew me into it. I became a spectator at the play of infinite forms.

There I was, then, in the office of George Louis or Jorge Luis Borges, one Tom Weil or Tómas Weil, an earthly form in the person of a lone American who'd wandered far from home. Like the imposed arrangement of my reminiscence, the names essentially meant nothing. They were just arbitrary labels without any intrinsic significance. I could just as well have been called Snoopy (a curious person is snoopy), Spot, Fido, Bowser or Bow-wow, Elvis, Snowflake. Any Tom, Dick or Harry could be renamed with no change in his essential characteristics. Similarly, calling a dog "Tom" (or a feline a Tom cat) is as appropriate as calling a Tom "Fido." Both man and beast have the same morphology, the only essential difference between the two creatures being that the biped hears time's wingèd chariot drawing near while the quadruped lives its entire life oblivious to time's passage, a passage as fixed as a trolley's rails, which channels mortals to the terminal. Although "Tom" is a better name for humanimals than Fido or Snowflake, dogs can be named "Tom" to good effect as it humanizes them, not that they need it. Naming a dog "Tom" would flatter me, if the not the dog, and might help raise my standing among my friends, which my Good Will wardrobe certainly can't do. I would be proud to endow a dog with my name—unless he declined the honor and nipped me for the insult of being called "Tom."

Borges' name wasn't at all unique, even if he was. All around Argentina, thousands—every Tómas, Ricardo y Enrique—could be called Jorge or Luis, and in many lands George or Louis. Even a city could be San Luis or St. Louis, Buenos Aires or Good Airs. As for everyone, Borges' individual oddity was the person beneath the skin, the inner being—as seen with X-ray vision—and not the outer indications. Like Everyman, he was both an archetype and an individual—a never before and never-again-to-exist humanimal. He was unique, both generically and individually, but also just another one of the boys, one of billions to be born, live, vanish. As a particular person, Borges was a one-off sum of his inherited characteristics and of all his earthly experiences up to the very moment I arrived unannounced at his lair there at the National Library. Whatever his name, Borges might have lived an entirely different life at a different place and at a different time and created another illusory past. His fervor could have been for Tirana or Ougadougou or Nome. Some of the locals in those places no doubt have fervor for them. Whatever the essentials of our personal histories and our being, for a brief interlude Borges and I faced one another in the fluid momentary present, soon to dissolve into the unalterable past.

Borges cherished his past, as indistinct and dream-like as his and everyone's might seem. The fourth generation of native-born Argentinians, Borges refers in some of his writings to his ancestors. A Borges sort of coincidence accompanied his first job, as the Avenida de la Plata library where he worked was named after Juan Crisostomo Lafinur, a forebearer who died "in exile" in 1824. Borges dedicated his essay "A New Refutation of Time" to Lafinur.

Borges was not all lit-chat chit-chat. Although I preferred to talk to the author about his writings and literary matters, he mentioned some topics relating to Argentine politics. When I asked Borges why that subject interested him he replied, "This is my country. I feel very

Argentinian." To my response that his published works seemed worldly and cosmopolitan Borges countered by emphasizing his Argentinian roots. He mentioned his grandfather, a cavalry colonel who died in 1874, the subject of Borges' short poem, "Allusion to the Death of Colonel Francisco Borges." Later in the conversation, when we discussed foreign travel, Borges underlined his contented settled existence in B.A. I understood how he felt. I, too, was a fourth generation local legacy on my native turf and I valued both that long continuity and my own comfortable and familiar deeply rooted home life. But I also wanted to add another dimension, world travels, to my settled existence while, unlike me, Borges had been content to stay at home. For him the wordy rather than the worldly—the Republic of Letters and not the republics around the globe—sufficed. Stating that he had no desire or plans for a trip abroad, Borges (born in August 1899 and so age 72 when I met him) said, "I'm an old man. It's too hard for me to travel." Ouch! I'm now older than was that old man back then but I still occasionally travel, even though it's gotten harder as I've grown older. Is it the travel or me?

Oddly, Borges' fervor for Buenos Aires didn't extend to his end or to his permanent resting place. Borges wished to die and be buried (June 1986) in Geneva, where as a young man he'd lived with his family, and in that Swiss city he lies in foreign soil far from his beloved homeland. For the long-run, Borges' preference for a distant land superseded his fervor for Buenos Aires. Peoples' enthusiasms come and go, just like the people themselves.

Many years after I met Borges at the National Library in B.A. I encountered him again when he gave a talk at a university near my house. Our lives once again momentarily joined. He was traveling with María Kodama Schweizer, his young and shapely companion, who—Braille incarnate—he could at least feel even if not see. He later married her, some seven weeks before he died. The then elderly and soon to be dead groom obviously rejected a better-never-than-late attitude. After Borges' lecture I went back stage to see him. When I reminded the author, now a traveling man, of our by then long ago encounter in Buenos Aires he hesitated a moment and then, sightless but insightful, offered an artful response: "Of course." This nondescript but perfect comment neither confirmed that he remembered our meeting nor did it imply, most likely, that he recalled nothing of our brief get-together. Borges' answer represented one of those functional and non-committal succinct responses reminiscent of the generic reply H.L. Mencken used to stamp at the bottom of critical, complaining or otherwise objectionable letters he received: "You may be right." The famously cantankerous critic could have also used a reverse rubber stamp legend with the message, "You may be wrong." On receiving such a comment from Mencken, Borges in turn might have replied, "Of course." Andrew Grove of Intel favored a more earthy legend on one of the short-cut reply rubber stamps he used: "BULLSHIT." The irascible Mencken might have adopted that pungent message, but not the more fastidious Borges. My own preference would be for the Grove version. Hemingway said that writers need a built-in shit detector. It occurred to me that if highly successful professionals in two such different fields endorsed recognition of bullshit, this must be a useful practice and I tried to follow it. I've always been concerned that spare parts may not be available for my much-used detector device.

Although fervorish about Buenos Aires, his life-long lair, not only Borges' death and decay arrangements but also his life offer evidence which in part contradicts his stated deep connection with his home town. "I was always (and will be) in Buenos Aires," he wrote, an invented poetic sentiment. In reality, things were somewhat different, as exemplified by his brief wife and future long-term widow. Daughter of a Japanese father (Kodama) and a German-Spanish mother (Schweizer), María met her future husband in Argentina at a Borges B.A. Lecture on Icelandic literature. They married (by proxy April 26, 1986) in Paraguay, they traveled, she buried him in Switzerland. Quite a cosmopolitan mix. Borges was less B.A. and more away than he professed. What authors invent by abstraction—rather than imagine based on lived experience—doesn't necessarily reflect reality. As an infrequent traveler, Borges had to invent his dreamy fantasized writings as contrasted with authors who imagine them as derived from immersion in the real world.

Before visiting Borges I bought a few of his books at the Ateneo, a large vintage bookstore on Calle Florída, the pedestrian shopping street in downtown B.A. In December 2000, 29 years to the month after I'd bought some of Borges books at the store and met him, the Ateneo converted the former Grand Splendid Theater into a showplace of a bookstore and now also occupies that splendid facility. The old theater offers a grand stage for the play of infinite forms as found in the pages of the thousands of books on sale there. When with Borges I asked him to sign one of the books I'd bought and brought with me. With his greatly limited vision Borges could barely write my name and his signature and so, as a ghost writer, I helped guide his hand as he scratched the inscription on the title page. Only later did I notice that when Borges added the date he'd entered the wrong year. Although I didn't want to read too much into this slip of the pen and of time, maybe the mistake represented a playful Borges-like "refutation of time."

At the end of "A New Refutation of Time" Borges concludes that our destiny "is not frightful by being unreal; it is frightful because it is irreversible and iron-clad. Time is the substance I am made of." It's a "fire which consumes me, but I am the fire." The substance is insubstantial, like an ashed log of a campfire which flames something into nothing, a thing which once existed burned into nothing except smoke which disappears into the air. Years before, as a boy at Camp Ironwood in Maine, I'd seen campfire logs flame into nothing. Even if hazy and dreamy, the past is iron-clad, unlike the evanescent Ironwood wood which vanished right before my eyes.

You can refute seemingly reasonable and logical propositions and beliefs or opinions, but there's no way to refute time. You can't argue with time or counter its authority. Perhaps writers believe they're able to refute time by binding duration onto and into a printed page, manipulating time with alternate fictional scenarios such as Borges did in his *Ficciones*. You really don't even have to write fiction to attempt to refute time. You need only bring an authorial mindset to the exercise. Anyone with half a mind for thought experiments can, even without revision of time in written form, envision any number of temporal refutations. Written works, however, offer a briefly durable effort to refute time. In a novel you can fictionalize the past, recasting it or, as I've done in this reminiscence, it's possible to describe what actually happened but structure the events in a different order. This rearrangement of time adds some meaning by fracturing chronology to connect disparate happenings in order to create an artificial structure which in a way revises time, even if not truly refuting it.

But I may be wrong. A reminiscence such as mine may not at all refute time but only distort it. Chronological order would have presented events as they came to me, but in fact that was a quite disorderly jumble. Although that temporal order would have been their genuine arrangement, such straightaway "time's arrow" sequence would offer less meaning than the one I imposed. What I contrived—an imagined order to order the narrative as well as the observations and conclusions I've included in this book, all based on real life actual happenings—may be a version less informative than others I might have chosen. My interpretation of what befell me over the years represents only one view, the one I happened upon, of how the various events should be arranged and what they meant. My doubts about the format arise not because my explanations may be incorrect but because I recognize that all of the opinions, observations, commentary and conjectural speculations which serve as the sinews and connective tissue of my account represent only subjective views, not certainties based on transcendental principles. My reminiscence could well be titled *The Weil Conjectures,* as Karen Olsson named her 2019 account of philosopher and Christian mystic Simone Weil and her brother André, a mathematician. I propose no irrefutable generalities or finalities but only tentative observations based on what happened, events specific to my own limited experiences.

I describe what occurred and, in a suggestive rather than definitive way, my reactions. It was all an experiment, one not peer-reviewed or subject to proof.

If the belief of Simone Weil (perhaps related to that *Señor* Tómas Weil in Buenos Aires or to the family of Proust's mother, Jeanne Weil, or maybe even distant kin to me) is correct—that "doubt is a virtue of intelligence"—then I am indeed an intelligent fellow. But, in truth, my doubts arise because I question my intelligence. For me IQ designates "I Question." My realization how tentative and provisional my thoughts actually are originates from a fundamental existential attitude that views most everything as contingent and subject to revision. This perception, which came to me as a young man, was for me both frightening and liberating. Doubt and skepticism, but not cynicism, served me as intellectual tools to establish a realistic view of the world. Long ago I realized that although time can't be refuted, its yield is random and not authoritative, for it might have developed in any number of different ways. What actually happened to occur became definitive but was in no way inevitable, nor was it meaningful. I imposed some meaning, but only one out of many possibilities.

Writing this reminiscence is the best I can do to review and recast time's residue—my experiences and memories of them—into an artificial format which in a small way interprets time rather than refuting it. This book, almost any book, may be an attempt to deal with time's chronic depletion, its corrosive effects, its inexorable alchemy in transmuting vivid lived events into memory. Through writing it's possible to imagine and preserve a contrived meaningful order, so making somewhat permanent and integral what otherwise constantly melts and mutates into the past, as burning logs become smoke. But then again, maybe the opposite is true, for in truth almost everything I say in these pages by way of interpretation—in fact, much of what anyone says about anything—could be reversed into an equally and maybe even more meaningful and reasonable statement. Perhaps, say, time is a refutation of writing rather than, as proposed above, the reverse. Because every last word, sentence, paragraph and page faces a death sentence and is doomed to oblivion, the operation of time in effect counters, or refutes, all attempts to preserve the past in print. Everything is smoke, existence like a log vaporizing into nothing. Libraries are just tome tombs where books interred on dusty shelves decay away day by day, finally to end in disintegration. The dis-aggregated words become scattered letters, the letters only litter and in the end no longer literate.

The fate of books mirrors that of humanimals. Millions upon millions of hard-won published thoughts fixed in words in print languish forgotten, and even more millions lie moribund and ignored as they were never read in the first place. All those volumes, so neatly organized, printed and bound, were in fact bound to repose in obscurity, stranded and forgotten in collections never recollected. Like their authors and their readers, the volumes are all eventually dust. Of course, my book too will suffer the same fate—the volume you right now momentarily hold in your hands is destined to crumble into nothing, just as will all the earthlings who ever wrote and all who ever paged through the doomed books. Sorry about that, but that's just the way things are (don't blame me)—or aren't. In Eugène Ionesco's play *Exit the King* one of the characters refers to 400-year-old King Derenger I: "He will be a page in a book of ten thousand pages in one of a million libraries which has a million books… until the day comes when the paper has turned to dust." Meanwhile, pending the inevitable fate for my books and for myself as well, if my statements anywhere in my account strike a reader as illogical, incorrect, obtuse, ridiculous or otherwise objectionable he or she need only look elsewhere, maybe resurrecting a few long-dead volumes, to find more compatible countervailing views. Before doing so, however, please finish this Tom-tome, even if you find some, many, most or all of my observations unacceptable or (like my reminder just above that you're dust) unsettling. It's good for you to get some views which disagree with yours or which provoke you to consider your Everyman or Everywoman pre-ghost status as a transient earthling. In any case, my observations are not meant to be definitive but are based only on my own particular individual experiences and how I reacted to them.

In truth, I've never been particularly attached to or certain of some of the principles which helped me to function in life. A few tentative certainties (perhaps a moron's oxymoron, but there it is) have endured to guide my way, but others I've adjusted, revised or even discarded. Like everyone, I groped my way forward, and at times, when I suffered reverses, backwards. Some of the main factors I relied on were such intangible elements as intuition, instinct and guesses, bolstered by logic, rationality and reasoning, best as I could apply them. These all combined to help me assess risks and opportunities and to judge how decisions might affect my agenda. But in each situation, uncertainty and doubt shadowed my views and choices. I was never sure if they would lead to favorable results. To the mix I also brought practical experience which gave me lessons in the real world. To learn such lessons I paid a high tuition, but the curriculum proved useful. As Bronowski observed, you understand the world by action, not by contemplation. But even the most enriching and varied experiences, the keenest intellect, the most proven principles, and the most acute instincts—all of which describe not me but the ideal— bring no certainty. Little conviction, only shots in the dark, guided me through the dark woods, for the world is a shadow play and life a fleeting phantom-like experience haunted with ambiguity and chance. Those realities don't spook me—it's just the nature of things. By chance, those intangible factors—reasoning, instinct, experiences—combined with luck to produce for me the life Morley described, one I spent in my own way. But it could have all been different, and for most people is.

Some of the favorable outcomes in my experiment in living I ascribe to a flexibility based on my doubting, questioning mindset. This sort of outlook may typify the mentality of a curious person, and also represent a writerly sort of attitude. A curious sort sees the world in a quizzical, tentative way, while an author considers different scenarios and ways to organize the material. This habit of contemplating alternatives serves to enlarge your vision, an amplification which brings you different perspectives. This mindset also persuaded me that reversing one version of a proposition can better enable you to comprehend the subtleties and nuances of a situation, decision or opinion. Diverse perspectives broaden perceptions. If you examine matters both backward and forward, upside down and downside up, inside out and outside in you improve your chances to perceive reality correctly and to avoid mistakes based on misapprehensions. In this way, pluralistic perspectives and reverse reasoning resemble travel, much of which offers eye- and mind opening perspectives based on unfamiliar practices and scenarios. As with thought experiments, travel enables you to see things in new ways, with the important difference that journeys take place in the real world of hard knocks rather than in the soft and malleable invented realm of the mind. Out in the real world—where you need to keep your wits about you, not let them dash off into speculative frolics—you had to live with the consequences of your decisions.

The most creative type of contemplation derives not from invention but from imagination. A dreamer like Borges could make up all sorts of stories, invented notions which toyed with the real world. But out in life it's imagination, based on realities, which counts. Invention is pretend and often pretentious and too assuredly tendentious as to a large degree inventing lacks sufficient grounding in reality. On the road I had to face facts, not *ficciones*. My judgments were in large part based on various real world factors which facilitated my ability to imagine viable solutions. Inventing them based on airy trial balloons unanchored to earth would leave me adrift. Admittedly, it's a lonely planet but its terrain offers enough company and much firmer support to structure a life than do fantasized thought games.

Living with the consequences of a decision endows it with a momentous significance. Contrariwise, exemption from the consequences removes the incentive for carefully weighing all the factors needed to make the best judgment under the circumstances. This represents a key element when considering observations, advice and agendas offered by other people. Much of what others propose or attempt to impose is at best simply an opinion, often a self-serving one, and as such not to be taken seriously or as in any way authoritative, definitive or actionable. Those who offer and, especially, sell advice seldom have a stake in the outcome. Advice is easy to give; consequences can be hard. Sir John Cowperthwaite, financial secretary of Hong Kong in the 1960s, observed (as quoted in *Architect of Prosperity: Sir John Cowperthwaite and the Making of Hong Kong* by Neil Monnery) "We suffer a great deal today from the bogus certainties and precisions of the pseudo-sciences which include all the social sciences including economics. I myself tend to mistrust the judgment of anyone not involved in the actual process of risk-taking."

An entire universe of scholarship and purported expertise—the social sciences including economics and many other claimed authoritative doctrines and disciplines—remained alien to my mentality and were largely ignored by me. Rather early I decided that what many authorities proposed as definitive is merely opinion, often self-serving and many contradicting other strongly-held views, and offered by operators who have no stake in the outcome. Those who purvey opinions in the guise of certainties include not only social scientists but also many other types, such as financial advisers, wealth managers, politicians, bureaucrats (exemplars of the play of infinite forms), soft-subject (non-scientific) academics, newspaper columnists and editorial writers, reviewers and critics, forecasters, political and investment and psycho-analysts, preachy religious professionals, consultants in many areas, and various other such characters. Even astrologers, witch doctors, fortunetellers, soothsayers and similar occasionally hit upon useful perceptions and opinions, so I don't mean to suggest that the professionals noted above never manage to profess some valid or actionable notions, and in fact I've mentioned a few in this reminiscence. Although eliminating all those categories both saved me time and greatly simplified my sources of information, those omissions no doubt also prevented me from accessing some useful learning. But ignoring it reduced what influenced me and left me free to focus on authoritative figures I deemed more credible, such as Elvis A. Presley and Snoopy from my era and other ever-relevant sages from bygone ages. In truth, my failure to find very many reliable sources in the listed disciplines says more about my defect—too much a Doubting Thomas and too obtuse to perceive the benefits—than it does about the supposed experts, most of them no doubt competent and well-meaning specialists who are true believers in their area of competency. If you've benefited in the way of solace or guidance from views I mostly ignored consider yourself lucky. In a way I envy you for having found some certitude in the way of reliable perceptions and advice from the belief systems I largely eschewed. I for the most part relied on my own opinions and suffered or enjoyed their consquences.

Pseudo doctrines comfort those seeking certainty, but assurance which in fact is opinion based on dubious principles serves only to deceive. The Carvaca, followers of the seventh century B.C. anti-religious doctrine called Lokâyata, held that only the material world exists. A commentary on the views of the sect stated that doctrines of religious believers were created only "as the livelihood of those lacking knowledge." One element of so-called experts which particularly bothered me is that many of them lacked a Simone Weil-like doubt—"a virtue of intelligence." I was at times even skeptical of my own views.

My Doubting Thomas mentality tinctured my judgment calls with uncertainty. Even more did I doubt outside opinions or recommendations or pronouncements. Rarely have I ever adopted, and never without close scrutiny, advice in soft-subject matters—those not derived from deductive, empirical, evidence-based sources— from someone immune to the consequences of their opinion. Most such opinions I ignore, and some I also disdain. Lest it be claimed that I lack self-perception, as I've already explained all my comments and observations in this book, but not the fixed realities of the facts, represent only provisional opinions, views based on how I perceived what happened and with no claim to more general validity. I always remained open to revising my views, and I was eager to find new ways of looking at things. In his *Memoirs* novelist Kingsley Amis refers to someone who commented on Margaret Thatcher's "readiness to be interrupted." This extremely useful and all too rare trait conforms exactly to my own attitude. I've always greatly valued having my settled notions interrupted. In fact, one reason I sought travel adventures was to experience new and unusual places and situations which might serve to unsettle and to reset my perceptions and opinions. With travel I made myself ready and eager for my regular way of life to be interrupted.

Because of my Doubting Thomas attitude, I sought and readily accepted exceptions and variations which contradicted some of my long and strongly held views. I believed they should always be subject to revision and even complete reversal. I occasionally convened in my mind a kind of "States of Deliberation," as the island of Guernsey's Parliament is eccentrically called, to consider new views and ratify them when appropriate. Although variances, reversed perspectives, counter-factuals and alternate scenarios often seem rather amorphous and even frivolous, I must admit that every once in a while I amuse myself by pondering such mutations and "what if" possibilities. This flexible and mind-flexing exercise reflects my tentative view on many matters and my recognition that anything and everything might have been different.

By now many raw materials based on experience exist to enable me to re-imagine some of the past. I'm now older than was Borges at the time I met him in late December 1971. In fact, I'm getting uncomfortably close to the age at which he died. In my own mind-games, playing with a kind of refutation of time, I occasionally think back to various "thens." Those memorable way-stations and decision points along my passage through the dark Dantesque woods induce me to imagine different paths I might have taken. What if I could revert to a long-ago pre-Borges era, back to the early twenties youth I was in Guatemala City the summer of 1958 when I met the President. This sort of thought experiment refutes time, in theory at least, by deleting all that came after. In my mind I'd rewind the years as if setting back a clock at the end of daylight saving time. What happens to that autumnal hour you lose when you turn the clock back? You've lived those 60 minutes but somehow, poof, they disappear. If you eliminate thousands of hours to revert to a years-earlier time, then all the then future you actually lived and know, the certainty of the past, would be erased. The years to come would lie before you, as it did back at the time, as a great blank slate of possibilities awaiting actuality. But I seriously doubt if an alternative version of my life's content would improve what actually occurred.

Years ago in London I saw and heard the musical *Blood Brothers* which featured the song "Roll Back the Years." In my memory still echoes the repeated refrain, "Roll back the years... roll back the years." It's an alluring idea to imagine turning the clock back to revise time, refuting it. But if your life by chance has in general turned out as you wished to have lived it—conforming for the most part to your hopes, dreams, wishes, aspirations, agendas—it's much too risky to regain time by rolling back the years, for you'd lose far too much in exchange for great uncertainty. In fact, time doesn't represent the essential value but only a necessary one as it's the medium in which we exist and in which happenings occur. Water enables but does not define the essence of a fish's existence. Time is of the essence, it's said, but it's not the essence. That essential existential quality derives from time's content.

For anyone whose existence has been endowed with good fortune, it would be a far too risky roll of the dice to roll back the years, even were that possible. As my earthly tenure draws to a close, I wouldn't want to exchange what actually happened for an alternative scenario. Had I been a Calcutta rickshaw puller, rolling the carriage along day after day, I would have been quite willing to roll back the years for another chance. But for me, recovering past time to live it all again, deleting what I already experienced for another try, would surely produce a less satisfactory result. Given my fatal circle, any reshaping of it would only shrink its circumference, making it more restrictive and tightening it from loose to noose.

It may represent a failure of imagination, but I can visualize very few enhancements for what might have in general improved my life. It's highly unlikely any alternative scenarios realistically available would have benefited me. How the once completely contingent and random events which now comprise my life happened to evolve satisfies me. By now they're safely set in elapsed time, other than the few stragglers yet to be determined. From the "then" of Guatemala City in 1958 on through to Buenos Aires in late 1971 and then onward to all the subsequent "thens" and up to the "now" as I write these words I appreciate and treasure most of what happened. I'm a happy camper. For me, rolling back would likely bring a roiling going forward, a more turbulent journey than the relatively smooth sailing and flying and busing and, in all senses of the word, training I in fact experienced. As a character in Italo Svevo's *Further Confessions of Zeno* says, at a certain point you have "too much experience of life to become a real young man again." I would not wish to refute time in order to start over as a young man. Being old with my particular life experiences safely behind me is far better than being young with many open years ahead which might or might not prove satisfactory. It's not youth which is the prime value but elapsed time filled with the kind of content which represents having lived in your own way.

To be young again, or even to live backwards in a magical Merlin-like reversal of one's existence, would require the erasure of everything and not just the bad things which transpired since your youthful days, and even your initial fatal circle would be wiped out. Some unfortunate earthlings like the Calcutta rickshaw driver might favor reversion to younger times for a chance at revision of his existence, and even a return to pre-birth with the hope of better karma which might produce a more desirable fatal circle. But for me that would tempt fate. My second time around would almost surely confine me to a less favorable fatal circle and result in more unfavorable outcomes. It would of course be appealing to eliminate some of the negatives which marred my way, but you really can't pluck out as fleas from a dog the irritating elements while retaining all the beneficial ones. If I rolled back the years, not only the bothersome biting bugs would be eliminated but also would be extinguished all the glowing enlightening lightning bugs, summer delights where I live, which helped to illuminate my way so well.

Even eliminating adversities wouldn't be an unmixed blessing. For one thing, setbacks serve to educate you. Sometimes the curriculum is tough, but if you pass through some hard times you learn valuable lessons. Suppressing difficulties might also produce even more undesirable collateral effects, as removing adversity of one sort could perversely bring about other more difficult problems. You'd be trading bad for worse. Unpleasant transient problems could be replaced by the much more serious disadvantages intrinsic to structural, irremediable and irrevocable happenings. Over time, most of the negative situations I faced lost their force, and even those unpleasant passing random events left me with some fond memories. Compared to the permanent privations and conditions of Third World unfortunates, my problems in such lands were relatively momentary and minor, even if upsetting when they occurred. The immediacy of challenging situations makes them seem all too vivid, but in retrospect they become ghostly memories from long ago which lose their power to haunt.

An example of such an annoying but passing minor miserable episode I suffered occurred in New Delhi on Christmas Eve 1973, a brief interlude which by now survives as merely a nearly half-century old remembered remnant of my long gone travel days. I returned from Jaipur to the capital that holiday evening (a holiday at home but not in India: places alter circumstances) after abandoning a thwarted excursion to points beyond Rajasthan, my advance blocked by a general strike. No rooms were available in New Delhi, even in low-star or no-star hotels. I faced becoming a street person that night. Forlorn, I dragged my bag through the inhospitable city, seeking shelter as darkness shrouded the metropolis. India has the image of being a hot country, but in the dead of winter cold winds sweep through the northern regions bringing cutting currents which bit into me as I trudged through the night, all the while recalling the Christmas Eves I'd enjoyed back home.

Finally I happened upon a free room—free but very expensive considering that I paid to stay in a five-star flophouse (minus five stars)—and there I flopped for the night. The unkempt and filthy room featured a broken window which allowed cold currents to pour over me—air conditioning at its best. At least I couldn't complain that the room was stuffy. To keep somewhat warm and to avoid contact with the unwashed dirt-streaked sheet, I slept fully dressed. The intrusive wind wasn't all bad, as it scoured from the room some of the odors which lingered there from unknown emissions or residues. I can't remember the name of the establishment. Probably something like the Royal Waldorf-Astoria Ritz Palace Hotel. As I learned over my years of travel, pretentious and grandiose names often designated the most modest places. A name represents only an arbitrary label, not a reality.

As I lay there alone on the stained sheet, shivering the night away under miserable conditions, I felt sorry for myself. This was a do-it-yourself enterprise, as no one else was present to perform that humanitarian service of feeling sorry for me. I wondered what on earth I was doing there, stranded in a two-bit, actually a one-bit at best, crummy hotel in a foreign land, a far country strange and distant from my own. That solitary Christmas Eve I thought only about myself, lamenting my own condition, with not a thought spared for the millions of impoverished Indians who existed, barely, out there beyond the jagged broken window pane. Those unfortunates suffered a lot more and a lot longer than did I, miserable only that one Christmas Eve night. If I could refute time by reverting to earlier days would I eliminate that benighted night-time experience? No, for it's now a wonderful memory, a pleasing component of the mix and matrix of the play of infinite forms which forms and informs my past. The next morning Christmas Day dawned clear and cool and fresh, reviving my spirits. With renewed enthusiasm I continued my travel adventure, the first order of business being to find another hotel. My vagrant mood lamenting my overnight fate was as transient as my travels, which unfolded as an ever-changing kinetic jumble from the ceaseless play of infinite forms.

Although below low-class, those not very accommodating New Delhi accommodations were hardly the worst I ever experienced. I over-nighted in many other ultra low-level joints, not out of frugality but often because of late arrival which required me to find immediate shelter. No place was a roach hotel, the kind of flophouse you can check into but never leave, but in some cases roaches did share my room, with no additional charge to me for that extra, nor did the roaches have to pay for their stay. However disgusting my quarters, I knew that by the next night I could upgrade my living conditions, an improvement unavailable to the impoverished locals.

In my experience the world's worst hotel was the Sultan Mohammed in Ghazni, south Afghanistan. If you ever happen to visit Ghazni try to avoid that hotel. For Edith and me at the time (autumn 1977) this wasn't possible because the Sultan Mohammed was the only place to stay in Ghazni. A monopoly supplier can get away with a lot. Sleeping out in the desert would have been an upgrade, and even the shabby New Delhi hotel offered better accommodations. It seemed as if the filthy Sultan Mohammed had never been cleaned, and it probably never would be. Why waste money on frills? The sheets were in a way colorful—no fancy Neiman-Marcus designs, but at least the random patterns of black streaks and various other residual emissions of unknown origins lent a certain panache to the bed clothes. They must have been purchased from the same supplier which sold linens to the New Delhi hotel.

The rest was even worse. The communal toilet facilities were so vile (unwilling to re-imagine them, I'll leave the description to the reader's imagination) I immediately retreated from the putrid chamber. Returning to the room I stood on the bed and, thankfully able to avoid the disgusting toilet for my necessary function, I improvised my own outhouse-like outlet by voiding out the window onto the void of an empty alley beside the hotel. This spray perhaps represented the first irrigation the sandy strip below had received in months. My poor traveling companion, Edith, suffered from a certain structural anatomical disadvantage which prevented her from adopting my window toilet solution. Before she proceeded to the fearsome facility I warned her in graphic detail about the conditions she'd encounter there. On returning to the room she claimed that my description didn't really do justice to the actual repulsive nausea-inducing facility. I agreed, as the chamber was indeed indescribable. Bad as the hotel was, neither of us would opt to give up our respective memories of the Sultan Mohammed. Those oddly patterned sheets, the execrable excrement-littered toilet area, the dirt, the odors and all the rest—a play of infinite forms less savory than most others—brought us a great experience, one which added to the experiment of travel in an exotic place like Afghanistan.

Although the Sultan Mohammed was somewhat below my usual standards, I've never been particular about the quality of accommodations. I've frequented a number of five-star hotels, but usually one star at a time. I always favored a central setting convenient to transportation and some of the main sights. Apart from location I tried to stay at places like the Pan American in Guatemala City which offered some characteristic local flavor or history. Although in some ways quite colorful, the Sultan Mohammed was, however, somewhat too extreme in the way of typical local sensory impressions. After one night at the hotel I certainly hoped that the management, if there was any, didn't intend to develop a chain based on the quite distinctive original version in Ghazni. I preferred not to face the possibility of staying in another such place elsewhere with the same format. The Sultan Mohammed seemed to me inappropriate for expansion as a franchise operation, but private equity and venture capital investors may disagree. Many other mildly misbegotten and still unforgotten unfortunate experiences bedeviled me during my travels around the Third World. Such discomforts and hardships which befell me I greatly value and would never want to delete any of them. I took to heart, and mind, Aeneas's suggestion in the First Book of the *Aeneid* to his followers, about to be shipwrecked: "Maybe someday you will rejoice to recall even this." Although I don't rejoice in recalling the Sultan Mohammed toilet, the memory does alert me to the advisability of using my frequent guest points at the hotel during my next visit to Ghazni to upgrade to a better room.

From Ghazni Edith and I continued west along the desert road which loops through desolate southern Afghanistan to Kandahar where, on an isolated side street, boys hurled rocks at us. More than a decade later, in 1989 in the marketplace at Ta'izz in southern Yemen, Edith and I were similarly bombarded by rocks, but unlike the Kandahar attack we couldn't see who our assailants were as they remained hidden in the crowd. Such rocky episodes represent harder experiences and create less pleasing memories than do other passing travel hardships. At travel moments like those in Kandahar and Ta'izz I longed for home where a target was a store, not me, and where I could peacefully occupy a rocking chair with gentle rocks rather than hard ones hurled at me.

Air travel in the Third World often presented problems. Flying in late 1987 from Abidjan to Ouagadougou in Upper Volta, now Burkina Faso, Air Afrique lost my suitcase, so stranding me without any of my meager but essential accouterments. I was left with nothing except the clothes on my back plus a few items in my small shoulder bag. I travel light but not that light. The bag's contents had very little monetary value but they were worth a lot to me: money doesn't measure utility. Although I refrained from wearing any of Edith's clothes—which anyway didn't fit or suit me—I did make use of her toothpaste, comb and a few other minor amenities. After my few days of deprivation Air Afrique finally managed to find my bag in Dakar, Senegal (to which the plane had flown from Ouagadougou) and returned the stray suitcase to me. It was a very happy reunion.

Trouble of a more serious nature arose when one evening we had dinner in Ouagadougou at l'Eau Vive. This supposedly reputable restaurant was one of various, including a branch in Rome, operated under that name by nuns to raise funds for good works. For me the nuns' works were not so good. I drank some bottled water which proved to be rather too *eau vive* as it was alive with pathogens straight from the tap rather than purified mineral water. From experience I recommend that on your next visit to Upper Volta you do not drink tap water. That evening after dinner I become violently sick, and for the next few days couldn't eat. That's one travel experience I'd gladly give up, although even that unpleasant episode came with a beneficial collateral effect. I found over the years that the best and most certain weight-loss diet is Third World travel. Because I ate so cautiously and exerted myself so vigorously traveling in underdeveloped areas, I always lost pounds as I gained experiences. Most of my other travel discomforts—including in a way the polluted water which sickened me—I value even if at the time they were annoying, dispiriting, disgusting, distressing or worse. So to the question of which events, experiences, happenings I'd wish to delete from my past the answer is, with very few exceptions, "none of the above." For me, rolling back the years would require not only eliminating all the hassles and problems but also un-spooling the reel to wipe out many favorable happenings and impressions which enriched my experiment in living.

It's of course possible to imagine refuting time not only for what occurred during the specific time-frame of your earthly existence but also by envisioning living in an entirely different era, past or future. Such extreme temporal displacement is unsettling. Like the "vertiginous" multi-memory chaos in the mind of Borges' Funes the Memorious, it's head-spinning enough to contemplate reliving your existence just during your own particular time without speculating about being alive in more remote past or future periods. Recalling your early days represents both an exercise in nostalgia and a process of discovery from a long-gone era. In old age the youthful years seem quite remote, almost as if a stranger had lived them. Another person, little known to you, seems to have existed in those vanished yesteryears. Backward have run my thoughts, as I near the end of my time, to remember some of the remote happenings, back from the now to Ghazni in 1977 and then farther back to Borges in 1971, then to President Ydígoras Fuentes in the time-distant summer of 1958 and, a reversion to the truly ancient very early days, rolling back the years, way back to my time in the summer of 1945 when I was a boy at camp in Maine.

Back then, a lifetime ago, almost everything remained in the future for me. Nearly my entire life was shapeless, mere potential and unformed. The play of infinite forms which over time would form my life lay before me like a featureless desert empty of content. Only as time elapsed would the jungle-like thicket of random events take root to comprise my past. As a boy I knew the general conditions of my fatal circle and how it had so far influenced my mini-life, at the time so short compared to my present extended longevity, the early-day brevity of my youthful existence comparable with my now very limited old age life expectancy. Both in the early days and now in the closing years—or maybe only months, weeks or perhaps just days—my life-span was and is defined by short tenures. Now, near the end, the fixed past in all its plenitude comprises almost my entire earthly existence. Little remains to complete the picture. The play of infinite forms, very nearly played out, has for me almost reached its final curtain. It could have all been different, but what became my life now remains forever without refutation. Back at summer camp in Maine, the lightness of being—my brief existence—left the future free. Because so little of my life had by then accrued, I hardly felt the weight of the past. A vast future rich with potential and rife with hazards lay open before me. Plots and counter-plots lurked ahead, but what, when, where and how I couldn't at all imagine. The yet to be written story would over time develop for good or for ill. The unknown future then outweighed the short known past, while at this late date the balance has shifted from what will be to what has been. Time has nearly tolled and the tale is almost told.

Camp Ironwood nestled in the woods in the Sebago Lake area of Maine near the town of Harrison. The landscape around the camp is better described as a lake-scape, for Sebago and Phantom Lakes and other waters delimited the dark woods which dominated the setting. Every so often campers would sit around the campfire at Ironwood singing a rather mournful song which intoned the sentiment, "I want to go back to Phantom with the wood smoke in the air." Still today, decades later as I record this memory with pen and paper, a method as archaic as my youthful past, I can hear in my mind's ear that slow dirge-like refrain, "I want to go back to Phantom with the wood smoke in the air." Sparks, bright bits of fleeting finite light, danced in the campfire's flames, then vanished. The camp was Ironwood but the logs weren't. They soon disappeared, cremated into nothingness as languid swirls of smoke gradually rose into the air until finally a phantom fire glowed with a few remnants of warmth before darkening and cooling. All gone. I suddenly realized then, as a boy, that burning logs would be my life—Everyman's life—and that my life-long life-log would in the end all disappear into smoke.

I wondered how it could be that those Ironwood logs, at one moment so solid and firm and iron-hard, soon burned into nothing and were gone. Back then, a kid, I didn't think of Donne's lines as I'd never heard of the English poet, didn't in fact know anything about poetry, not much about England, some remote land where, presumably, the Engs lived far away somewhere across the sea, didn't know much about anything. By now, however, I've been to England many times and I've heard of John Donne and maybe even understand something of what he was getting at with his lines, "Go and catch a falling star, .../Tell me where all past years are." They're gone, lost in oblivion—that's where they are, dead and gone forevermore, vanished like the campfire logs or a fallen star, one you couldn't catch and hold. The past, stars, time, the play of infinite forms—can't grasp them.

Decades later, now, I recall how the boy camper, the once upon a time me, would sit there by the camp fire in Maine in 1945 singing, "I want to go back to Phantom with the wood smoke in the air" as the flames burned into smoke which writhed and rose into the air, then vanished into nothing, all gone. Although rather less spritely and upbeat than United Fruit's advertising jingle featuring Chiquita Banana and Jake the Flake, the campfire song was no less promotional. The refrain sung around the ashing campfire was intended to instill warm feelings for Ironwood. The boys were supposed to tell their parents to send us back to Phantom, there where the wood smoke hazed the air. But the song's sentiments left me cold. I didn't particularly like the camp—too much discipline and hazing and that hazy evanescent acrid smoke as the logs melted away—and I really didn't want to go back to Phantom. It was all just a phantom shadow-show. Let the campfires burn without me. I preferred to let it all—Phantom and Sebago Lakes, Camp Ironwood, the fires—remain phantoms, moribund memories not to be revived during another summer. Let them rest in peace as ghosts of the past. That single summer I attended camp soon passed into the past, just as has everything else in my life up to this very moment as I resurrect memories of those long ago long extinct campfires. My entire past, no matter how recent or vivid, has gone up in smoke, vanished to leave behind only an insubstantial vaporous residue. Kindling now as a ghost writer a few phantom misty memories from my boyhood, the long vanished past seems more ethereal than real, as evanescent as the kindling of the campfires which soon burned away to nothing there by Phantom Lake a lifetime ago. All gone.

Years after my Camp Ironwood summer I once visited Rye in England, a land which during my camp days seemed remote but a country by now familiar to me. On the cathedral steeple clock at Rye appears the inscription, "Time is a shadow which soon passeth away." Looking back, I can now see how passing time operated to fill in the then emptiness of my future, now a past which for me teems with shadowy memories haunting my thoughts. I remember just where and when I suddenly saw the Rye inscription momentarily come to life, in the Afghan desert back in 1977. Etched in my mind is the memory of a line of camels advancing atop the ridge of a dune, their shadows casting silent moving forms on the sand slopes below as the caravan, silhouetted against a twilight sky, moved on. Momentarily there passed a phantom-like pattern of dark-light-dark upon the desert's empty sands, a soon-to-vanish play of forms forming a visual echo of earthlings' never before, the now, and the nevermore. And then, suddenly, *Ecco è fuggito*—all gone, as Leopardi wrote at the end of his poem "The Evening of the Day of the Festival": "...and fiercely tightens my heart/To think how everything in the world vanishes/And leaves hardly a trace. *Ecco è fuggito*." Heine echoes this fugitive sense of life in the first lines of his poem "Die Lorelei" ("Die" a German "The" and not the English "die"): "I do not know what it might mean/That I am so sad." But Heine did know, just as Leopardi did, and just as I do: *Ecco è fuggito*.

After Camp Ironwood the years brought me some familiarity with a few poets, and I also came to know that once remote foreign land, England, across the ocean, as well as many other countries around the globe. One by one by one accreted all the diverse random experiments, experiences, fragments of information, knowledge, events and happenings which during the now fugitive years combined to comprise my life. Verily, many strange and wonderful sights did I see. Not quite done with Donne, I now add a few of his other "catch a falling star" lines which, better than can I, express my eye enterprise of global sightseeing:

> If thou be'st born to strange sights,...
> Thou, when thou return'st, will tell me,
> All strange wonders that befell thee

My reminiscences in these pages, now that my adventures are done, represent my attempt to recount some of the strange sights and wonders I sighted, experiences that befell and beguiled me over many years. Those varied views furnished me with enough delightful memories to last a lifetime, and by now they very nearly have. What more could I want? Never would I wish to risk rolling back the years, and never did I long to go back to long ago Phantom with the wood smoke in the air. Time and luck, chance and fate gave me too much to flame it all into ashes and return to the past for another chance.

~ 5 ~

My travel experiences which gave me a post-graduate course in life enabled me to proceed with my learning by degrees but not academic ones. My studies took place far beyond the confines of a classroom. Not practice exams but practical learning out in the big wide world educated me, to the degree that I was educable. No tests or thesis marked my progress and no professors supervised my work. My studies were not academic or bookish but endemic and lookish. I traveled to places and experienced them. After my formal education, I realized that I was uneducated. I gave myself an academic appointment as an autodidact and, a hard grader, flunked or passed myself based on realistic assessments of my performance in various fields of study. Some, like travel, I passed with flying colors; some remained stuck at half-mast; a few matched the debased quality of Ghazni's Sultan Mohammed Hotel. No grade inflation distorted the report card, and when in doubt I deflated my marks, all the better to be able to improve. Under-performance, lowered expectations, moderate aspirations are more likely to lead to positive surprises than are exaggerated hopes or unrealistic self-deceptions.

One of my strongest post-graduate courses over the years and around the world dealt with economic matters, but not in the way professors teach economics in the ivory tower halls of ivy. My version of the subject entailed day-by-day applied economics as systems operated out in the real world. Academic versions invented by professors who profess to understand economic realities based on econometrics, models, mathematical formulas, algorithms, abstruse theories, number-clogged papers and other such arcane studies and theories formed no part of my curriculum. I never understood all that anyway. My training derived from real life functioning, or dysfunctional, economies I experienced in operation and needed to deal with. You really don't want to rely on an economist or a professor to tell you about the real-world economy. Observing how things actually operate in economics and many other fields and then being involved in them yourself offers lessons far more valuable than those you learn in a classroom or from "experts" who profess knowledge. In Jules Verne's *20,000 Leagues Under the Sea*, Captain Nemo showed excellent judgment by filling the library of the submarine *Nautilus* only with "books in science, ethics and literature...[but not] a single work on economics." It's uneconomical to waste scarce space on economics, a discipline which even an economist like Professor Hutchison believes seems in many ways to lack validation and practical utility.

Economics professes to understand and be competent to modulate in some ways an intricate human system resulting from the often unpredictable decisions of millions of participants. This purported competency contradicted my belief that in the convoluted affairs of mankind little was certain and much depended on random and incomprehensible forces. As explained in his 1977 book *Knowledge and Ignorance in Economics*, English economics professor T.W. Hutchison at the University of Birmingham (U.K.) well understood the limitations of his specialty. Noting the "significant currents of increasing unpredictability gathering strength in the economic and social world," he criticizes the "naive intellectual pretentions" of economics and quotes a fellow academic who faults mathematical economics for "the vacuity or irrelevance of most of its theory and the patent unreliability of its statistical information," and its tendency to "be a gigantic confidance game." Often based on "clairvoyent hindsight," economics, Hutchison states, is "full of exaggerated theoretical claims put forward in order to sell particular professed policies of one political strip or another." Even though on economics, the professor's book certainly merits a place in the *Nautilus* library.

After my time in Buenos Aires when I met Borges in late December 1971, I flew to western Argentina to visit Córdoba and Mendoza before continuing on across the Andes to Chile. As I noted earlier, you can't really take the measure of a country if you remain only in the capital or the main city. Restricting your trip to the dominant metropolis limits your experience by preventing you from gaining a rounded picture, especially in countries such as Britain, France, Belgium, Thailand, Mexico, Egypt, Guatemala, Burkina Faso, Molvania and many others where the seat of government is also the main commercial, cultural, educational, financial, media, economic, political and governmental center. This concentration of power, wealth, intellectual resources and elitism doesn't completely represent the entire land but is only partially representative of the country. Out in the hinterland, beyond the confines of the main metropolis, a traveler finds many other elements which help to define a country's nature. In some places where an arbitrary and unruly rule of law prevails, venturing out to the provinces often makes you an object of suspicion. This jeopardy, too, is part of experiencing how a country functions—or dysfunctions. I would rather have experienced my adventures without such disconcerting episodes, but at times an errant traveler far from the usual tourist areas unavoidably attracts unwelcome attention and intensive intrusive scrutiny. More than once a driver or a guide I engaged doubled as an observer, watching where I went, who I talked to, what I said and what I did. This makes you feel uncomfortable, let alone unwelcome. It's far preferable to be a passing observer in a country—there today, gone tomorrow— rather than a lone stranger in a far land who is under observation and at risk of detention. In one Middle Eastern police-state country my escort asked me if I was a "spyman." Obviously assigned to keep an eye on me, he was the spyman. I tried to behave so as not to give the agent any cause for suspicion.

During a group tour in Saudi Arabia in February 2000 we flew up to the little-visited town of Al Jauf in the remote northern desert, supposedly the first foreigners there in many years. During our visit a somewhat mysterious small band of armed escorts accompanied us everywhere. I asked if their function was to protect us or to observe and control the group. "Both," said our tour leader. Similar armed escorts accompanied our group on a later tour in Algeria. In Harar, eastern Ethiopia, a plainclothesman shadowing us everywhere followed closely behind as Edith and I made the rounds of the tourist sights. His presence was so obvious it hardly seemed shadowy. The agent's mission was as plain as his clothes—continuous observation of our whereabouts and whatabouts. In Russia the summer of 1985 our Intourist escort, without the slightest attempt to conceal his main function, admitted he was a KGB informant as well as our tour leader. At the Kosmos hotel in Moscow Edith and I addressed some of our conversation to the hidden microphone in our room by loudly discussing our laundry—underwear rather than undercover matters featured in our conversation—and other such top secret topics. More than thirteen years earlier, in late December 1971 after my stay in

Buenos Aires, out in Córdoba in western Argentina a challenge more menacing than merely being under observation confronted me.

Although unexpected happenings can add spice to travel adventures, at times an incident can prove a little too hot for comfort. While sauntering around the center of Córdoba, I was suddenly detained by a uniformed character of unknown status. Based on his nondescript appearance I couldn't tell by what authority he operated: soldier, sentry, policeman, militia, secret service, non-secret service, private guard, paramilitary, bandit, shake-down artist, hold-up man. I had no idea what his mission was and why I'd suddenly become a part of it. His garb, bellicose but bland, failed to reveal which force he represented. Or perhaps he prowled the streets only as a lone wolf, a free-lance operative working for himself, his compelling authority arising from his weapon rather than from his rank. He now raised his rifle and pointed it directly at me. "*No se mueve*," he growled, ordering me to raise my hands. Otherwise, I didn't make a move.

Even though it was hot that December I froze. But my thoughts became hyperactive and churned through my mind. I couldn't imagine why a maverick gadabout like me happened to attract his attention. Maybe he preyed on stray tourists. Or perhaps he somehow took me for a guerrilla. If so I couldn't decide if I should be flattered or anxious, or both. It seemed that suddenly some guy with a gun viewed me as some sort of a threatening character, a new role for me. Maybe my captor thought I was a revolutionary, truly a revolutionary change in how I was perceived. Back home friends might have viewed me as a character but not a threatening one. No one ever viewed me as a guerrilla. Maybe they saw me as something of a gorilla or a baboon or as an eccentric with a monkey mind. Admittedly, I was in a way a rebel, dissenting from a number of norms, but in no way a revolutionary. I didn't want to change the system but only to ignore some of it. At my house only the blades on my porch ceiling fan were revolutionary. Perhaps the rifleman saw me as a Che Guevera type, a threat to the established order. After all, I was now visiting the general area where the revolutionary had originated, in the same neck of the woods or pampas over in Rosario some 230 miles southeast of Córdoba. Guevera was even in Guatemala during a United Fruit-orchestrated CIA-backed coup which overthrew communist president Jacobo Arbenz, four years before I met with the *Presidente* in 1958. Did some sort of deep-state plot or counter-plot dossier somehow link me with Guatemalan revolutionary politics and Che? Although far more fanciful than even Borges' far-fetched *ficciones*, when you face the barrel of a rifle all kinds of strange notions shoot through your mind.

The gunman stared at me, drilling his gaze into me. If he had an itchy trigger finger I hoped he wouldn't scratch it. In his homeland Che might be a folk hero, but here in a land far from home I was seemingly perceived as a desperado, a stray lone foreigner prowling Córdoba's streets on some sort of nefarious mission. The alert guard had now reeled in a big fish and headquarters would be pleased. The authorities would extract from me an admission of mischief, so enabling the diligent agent to gain promotion and a bonus, at my expense. Or maybe he was just bored and pleased to come across by chance a lone stranger, a mark to bring some excitement to the *hombre*'s otherwise routine day. I hoped I wasn't a marked man.

"*Papeles*," he demanded. They always ask for papers, as if I were a newsboy. One morning in 1985 when I was wandering around Moscow on my own minding my own business a uniformed soldier or policeman or militiaman stopped me and demanded *documenti*. I played dumb, an easy role as it was in character, and walked on. The guard was less persistent than the pesky pest in Córdoba.

Papers? In Córdoba I had no papers on me except some Argentine and U.S. banknotes. The sentry was probably putting the bite on me for a payoff. But if I handed some bills over to the rifleman as a *mordida* he might take the bribe as an admission of guilt. If so, guilty of what? My "papers"—passport, return air ticket home, other identification—were back at the hotel. The guard extended his hand, palm up, awaiting my *papeles* or perhaps my cash. I now faced his arms and his arm, a double-barreled threat. Suddenly I remembered that although I lacked papers I did carry with me some plastic, and from my money belt I fished out my American Express card. Pointing to the "American" as I displayed the card, in an authoritative voice I said, "*Embajada Americana—diplomatico.*" Down fell the rifle, away went the outstretched hand. I'd given him no money but instead something to think about. Did he want to detain a diplomat from the American embassy, which would be rather undiplomatic behavior on the gunman's part.

The rifleman took the card and carefully scrutinized it. He took his time examining the document, and then with an inscrutable expression he scrutinized me, then again studied the plastic piece as carefully as if it were his bonus check or pay stub. The card certainly looked quite official, or so I hoped. He kept staring at it. I wonder what he saw in that plastic rectangle. I certainly hoped he'd see an official U.S. Government credential. I could almost hear the guard's mind clicking away: the document was not exactly a *documenti* item but it did say "American" and the *Yanqui* did claim some sort of connection with the embassy in B.A. Perhaps he was even the ambassador himself, or another big cheese official functionary. Could be dangerous to tamper with an oversized *queso* like that. From suspected terrorist to diplomatic status, from extinguished to distinguished in a few moments, quite a hey-presto sudden change in perceptions. And anyway—his thoughts clicked on as he processed his developing revised views—he doesn't seem like a revolutionary. Looks something like a gorilla, perhaps, but not really like a guerrilla. Don't want to mess with a *diplomatico*, he'd be thinking. Too risky. I'll have to get my bonus another way. The whatever-he-was character finally abandoned his interest in me and with a curt flip of his tan hand, that demanding hand which had awaited my *papeles*, waved me away. My American Express card to the rescue, a use of the card not even the company's marketing department or advertising agency had envisioned. Maybe they'd pay me for the idea: when detained in a foreign country it pays to play your cards right.

Following the incident, I contemplated the image and power of a brand, just as back in B.A. I'd pondered the worth of the "Tom Weil"/"Tómas Weil" brand, which—only a name and not an intrinsic identifier—may in fact be valueless. Thanks to the impressive official looking piece of plastic I'd managed to buy my freedom. Impressed by the name "American," the rifleman had decided to let me go. Like American Express, "America" is a brand, one often useful out in the world. The nationality serves as a passport which helps surmount many barriers along the way. Americans aren't universally popular, but in general they're allowed to circulate around the world without ado. Back in my day, some traveling Yankees used to wear on their backpacks a Canadian flag patch to avoid being identified as Americans. I didn't hide my origins and always highly valued the considerable privileges my nationality brought. Because a U.S. passport often immunizes you from travel hassles, you're usually waived through borders and barriers, inspections, confrontations and control points while other nationalities face closer scrutiny and delays. Only rarely, but memorably, was I viewed as a "spyman" or otherwise as a suspicious character.

Back in 1958, when the American Express card was first issued, my father signed me up, and to this day my card bears the legend "charter member." He told me back then that this kind of charge and payment system represented the future and that holding the card would make make me a very early adopter of a cutting edge format. On the major anniversaries of the card's issuance, about every ten years, American Express sends out some sort of memento to recognize charter members. In 2018, however, the sixtieth anniversary went unnoticed by the company. On the rare occasions when I happen to speak with a phone representative they're amazed to hear from a card-holder dating back to the original customers more than 60 years ago. I imagine that few charter members still exist, most of them having by now gone to their reward, and not one based on points earned by spending. Before long, all the original members will become discards as they enter uncharted territory and the "charter" designation will no longer appear on any card.

Section II: The Fatal Circle

The unanticipated versatility of my American Express credential there in Córdoba so many years ago suggested to me later how financial instruments used in commerce could be extended to unexpected ancillary functions. I realized that economics and financial dealings—buying, selling, financial instruments, money matters, business endeavors, commercial activities, markets—encompassed far more than simply the narrow aspects which pertained to those specific categories. In reality, a rather broad range of considerations and elements pertain to what might appear as only an economic matter. Failure to consider forces beyond the immediate and obvious ones can result in misbegotten financial decisions. Economists and advisers all too often focus on what they know, or think they know, rather than the many additional factors which theorists or "experts" should know but don't. It's a bother and time-consuming to educate yourself about all the matters which matter beyond your own specialty. What's measurable with charts, graphs, tables, equations, algorithms, numbers and similar professed profundities omits the essentials and the on-the-ground realities. You learn those only by being out there—out doors, where the cold breezes of the real world blow away windy hot air theories.

Some of the most useful economic insights I absorbed on my travels came to me after I passed across the Andes from Mendoza, my last stop in Argentina after Córdoba, to reach Santiago in Chile. On New Year's Day 1972 I traveled by shared taxi through the mountains, passing the Christ of the Andes statue, and on to the Pacific to Chile's capital city. There I began the second part of my five-month-long South American trip, finally having crossed the continent from sea to sea and headed North toward home. At last I was facing North, a very welcome new direction leading to my own little corner of the lonely planet, still distant but at least now ahead of me rather than at my back. With every long trip you reach a turning point at which you start to return home. Few such pivot points seem as geographically striking as for my South America trip when finally, in Chile, I had transitioned from the Atlantic to the Pacific and, changing direction, turned North. My home no longer behind me and ever more distant as I descended the east coast, in Santiago I looked forward to the day when I'd leave the open road behind me and settle back into my cozy little house. Heading South during the first part of the trip, at my back had been not only time's wingèd chariot but also the one and only place in the world where I truly belonged. As much as I enjoyed the stimulations and even the tribulations of the adventure, I was elated to be on the road back to my natural habitat. Being a wandering stranger in alien lands can only take you so far.

Home I treasured even more than travel. As much as I enjoyed seeing the world, I was essentially a provincial homebody. My own sentiments toward home—a long-familiar setting where I began, grew up and still remain, and where my remains will forever remain—differed from Willie Morris's view of his hometown, which he abandoned to become editor of *Harper's* and dwell "in the middle of New York letters and publishing," he wrote in *North Toward Home*. Morris gave up Yazoo City in the Mississippi Delta country for New York City, the Big Cave he called it, to become part of the literary establishment. The litterateur occasionally returned to his roots, the place where his dead ancestors reposed, where he once again, as when a boy, ate his mother's fried chicken and casseroles and pie, a setting where he "still knew where every tree was, the angles on the roofs of every house, the hidden alleys and paths and streams ... every street corner and side street had meaning for me." But no longer was it home, for Morris returned to his Big Apple-Big Cave literary life, the departing plane—he ends his book—circling "in widening arcs over the city, over the landmarks of my past, and my people's. Then, slowly, with a lifting as heavy as steel, it circled once more, and turned north toward home."

My North toward home from the Santiago pivot point in South America took me not to a new place like New York but back to my original homeland, the town where I knew the graves, the trees, the roof angles, the hidden alleys and paths and streams (one flow behind my back yard) and where everything had meaning for me. On the lonely planet meaning is difficult to find and I didn't want to give up my home-based meaning. All those familiar and mundane localized examples of the vast worldly infinite forms I valued and was willing to leave behind only temporarily to adventure abroad. But to home I always returned. I was too much of a settled provincial to give up permanently the comfortable and comforting forms which comprised the only place where I was truly at home in an otherwise foreign, alien and inhospitable universe. Whatever direction the whims or winds of wanderlust took me, my iron-clad attachment to my personal magnetic north invariably drew me back to my home habitat.

At the time I arrived in Santiago, Chile's political and economic situation was quite unsettled. The incumbent regime was under threat and the economy in tatters. Even an outsider could feel the tremors shaking the land and observe evidences of the convulsions which rattled the country. The thin crust of a once-settled society and civilization was trembling, as were many of the inhabitants. Through the streets of Santiago surged crowds demonstrating against Salvador Allende's communist government, soon to be overthrown. This was an occupational hazard of Latin American presidents. Less than a year after I visited Chile, the many plots and counter-plots and counter-counter-plots there operated to eject Allende from office. The president suffered his own 9/11 when on September 11, 1973, about to be captured at the presidential palace by insurgents, Allende shot himself with an automatic rifle, a gift from Fidel Castro. Allende inflicted upon himself the fate I managed to avoid when the Córdoba rifleman had detained me. The president might have fared better with an American Express card handy. Being a *presidente* in Latino lands seems a high-risk profession. Are the *jefes* down there insurable?

Turmoil, more than Allende, ruled the city. While watching from the balcony of my hotel room a raucous *desfile de ollas*, the marchers banging on pots and pans, I became a target of their rage. A hail of stones rained down near me, but fortunately the attackers had bad aim. The hit parade passed on by as I retreated to my room. The unrest had greatly destabilized the financial and economic situation in Chile. Political instability produced many negative collateral effects. The country's currency was rapidly losing value, so making the black market exchange rate extremely favorable for someone like me fortunate enough to possess U.S. dollars. Not only an American Express card and an American passport but also American banknotes endowed a Yankee traveler with many benefits. Although the black market rate was ultra-tempting, like many possible high-return propositions the illegal transaction came with great risks. I feared that I might be entrapped by an undercover agent—you never knew who was a spyman— and arrested for violation of exchange regulations. This time my American Express card wouldn't serve as an express get-out-of-jail credential. I assessed as best I could whether such a lucrative exchange transaction was worth the risk, a cost-benefit analysis based on very incomplete information. Given the compelling economics of the exchange ratio, I decided to chance a deal. I pondered how I could best reduce the risk. As I'd learned during my professional career, effective businessmen didn't try to avoid risk; they embraced, assessed and tried to manage it.

It happened that at my hotel one day a bellboy took me aside and nervously whispered an offer to change money. Was his nervousness an act? Was the attendant in fact a plant to entrap me? You can never be sure until after the fact, and by then it might be too late. Cautionary alarms invaded my thoughts. Through my mind swirled all sorts of imagined plots and counter-plots, analogues to the real ones which were bringing turmoil to beleaguered Santiago. Was I about to become a victim of a plot? Who you're dealing with when you encounter and chat with a random stranger in a foreign land always involves uncertainty and in some cases risk. Is he a president's secretary, later to be revealed as a narcotics smuggler, like Mauricio Rosal in Guatemala City, or a president's henchman trying to entrap a foreigner? The current economic chaos in Chile produced a currency exchange opportunity, but was it to benefit me or to enable a government agent to euchre me? The fellow seemed genuine but maybe he was just a good actor posing as a *botones*, as Spanish calls bellboys. He buttonholed me, motioning me into the elevator, and as it rose we concluded the exchange, my hard dollars for his soft pesos. I was reassured when, even in that isolated cubicle with no chance of being observed, *botones* instinctively glanced over his shoulder to ensure that no police spy was scrutinizing the illegal deal. The bellboy's caution evidenced his concern, an example of what lawyers call *res gestae*—-things done indicate truths. His offhanded behavior while handing the bills off to me—a spontaneous probative gesture—evidenced to me that he was not an agent but a principal working for his own gain.

The great rate for the exchange yielded not only an enormous foreign currency arbitrage profit but also some valuable insights. I realized that the then current currency situation and various other unsettled factors in Chile represented a vivid and practical case, one I personally experienced, of how political turmoil could damage a country's economic system, greatly diminishing the value of the paper means of exchange and rupturing social norms. Thanks in part to this experience, I became acutely aware of the need to consider many factors when assessing decisions regarding investments and other money matters. Unlike operators who focus only on the narrow specifics of a transaction or on what's conveniently quantifiable, I learned that it's necessary to contemplate a wide range of possible influences before making an economic decision or—for that matter—almost any significant decision. This perception accorded with and reinforced my more general view that when addressing choices you need to take into account more factors than the obvious narrow parameters of the matter. The wider and more holistic your perspective, the greater the chance to avoid or at least mitigate mistakes. Extraneous elements, not just the more obvious intrinsic ones, can greatly bear upon what might appear to be a limited situation. The difficulties inherent in this sort of expanded view prevent many people from considering factors other than those only readily available which can be easily observed or quantified. Factors less accessible and more problematic often comprise the very elements which in fact determine success or failure.

The considerably elevated buying power resulting from my elevator-protected black market exchange allowed me to buy a round-trip air ticket on LAN-Chile to fly out to Easter island and back to Santiago (I always traveled with a return ticket), a total distance of some 5,000 miles, for a very nominal price. The trip was almost as cheap on a per-mile basis as hard-class train travel, but not a dirt-cheap rickshaw ride, in a country like India. Air service all the way to Easter and back for an ultra-fair fare was perhaps the best travel bargain I ever found, my good fortune due not to my diligence but only to Chile's unfortunate political and economic disarray. My long flight out to Easter and back to Santiago represented one of my most far-flung excursions involving leaving a country's political, commercial and cultural metropolis to visit a provincial area. It was a very long trip for a short stay of a few days, but the journey showed me an area of Chile far from the capital and remote from any other settled areas. Easter—perhaps the settlement furthest from any other inhabited point—may be the lonely planet's most lonely spot. The approximately 2,500 mile distance out was about the same as from New York City to Los Angeles (2,448 miles), but with the flight over the unpopulated waters of the Pacific, all a natural seascape rather than over developed landscapes cultivated and occupied by humanimals. Because of the degraded socio-politico-economic conditions in the country just then which greatly damaged how the Chileans lived, I benefited from a great bargain fare while the locals paid the cost. The problems Chileans then suffered gave me the opportunity to travel almost for nothing, a free ride—an economic rarity. Although some business transactions manage to strike a fair win-win bargain favorable to both sides, in other cases one party gains and the other loses. The varying outcomes often result from forces beyond the control of the participants or from factors which influence only the specifics of a particular transaction. Circumstances can distort the normal economics of a deal.

Back when I visited Easter, almost no commercial tourist facilities existed. The residents on the remote island provided only informal but perfectly adequate services. As in many other formerly undeveloped tourist areas, by now things on Easter have become more sophisticated, maybe even too much so as perhaps the vintage unspoiled and delightful original ambiance no longer exists. On Easter these days the upscale Explora, a fancy low long Frank Lloyd Wright-like structure, now caters to visitors. Located near Hanga Roa, the island's main settlement, the establishment offers all meals ("cuisine," always an expensive designation for grub), alcohol, activities (tours), all at prices which "start" (always an ominous word for buyers) in the thousands in hard dollars, not in depreciated currency. Things on Easter used to be much simpler and more basic. There was no "cuisine" but just tasty home-cooked family meals at the private house where visitors stayed. Prices started and stopped at the same amount.

In my day, the workings of Easter's tourist industry was not really an industry but just a down-home *ad hoc* system. It exemplified how an informal economy operates, casual but functional like a black market. No hotels and not much of anything else existed back in early 1972 to serve the needs of travelers in a regular commercial way. When you landed, locals waiting at the primitive airport invited you to stay in their simple houses. No thousand dollar tariffs applied but only a modest payment for shelter and meals and jeep rides. I encountered around the island no islander who spoke English. Travelers lacking Spanish were out of luck, but luckily I knew the language. ("*No se mueve*," ordered in Córdoba by a man pointing a weapon at me, spoke volumes.) Out on the remote self-contained island the informal economy operated at its best. It was delightful to live with islanders in a friendly familial and hospitable setting. For a few days you enjoyed the opportunity to experience life as the locals lived. But unlike the inhabitants I was only a bird of passage, soon to fly away and resume my wanderings, a privilege few if any of the islanders could claim. From what I could tell, almost none of them had ever left that tiny speck of land lost in the Pacific Ocean to reach *El Continente*, as the locals called the mainland. Confined to Easter, the islanders remained out of the mainstream. The typical inhabitant was born, lived and died within the space of a few miles, never to know the great wide world beyond the little island, so much of which over the years I managed to see.

One day I conversed at length with an old fellow, but then younger than I am now, about his life on the island. "El Viejo" had passed all his earthly days, amounting to some 75 or more years, out there on the claustrophobic speck of land. His basic existence on the remote island entailed few needs and even fewer wants. The nearest *supermercado* or convenience store was some 2,500 miles away, not too convenient. In a way, El Viejo as well as the other islanders lived in a timeless mini-world, removed from seasonal change and from the dynamic and revolutionary changes which typified areas in the maelstrom of current events. Not much happened out on Easter—no resurrections or insurrections or other such noteworthy developments. The days just drifted by like the errant clouds which occasionally passed across the open sky before disappearing. El Viejo and the others also lived outside the main-line mainland economic system, far removed from a *Homo Economicus* existence, as Rolf Dahrendorf titled his book on man's place in modern society. El Viejo was deeply embedded in the tiny confined social and economic structures which prevailed on the minuscule island. Unable to extricate himself from those conditions, El Viejo couldn't function in the way Dahrendorf advocated: "it is crucial that the individual be capable of holding his own against the claims of society." So circumscribed was society on the island, no resident could effectively dissent from the prevailing "claims" which mandated conformance to the local value system. How vastly different, I meditated as I chatted with the old man, were our respective fatal circles. For me the world was my oyster, but for him the place was only a tiny hermetic shell of land. Of course, before you start to believe that the world is your oyster you better be sure that oysters are in season—otherwise you can end up with a bellyache.

"No man is an island," said John Donne, now a Donne deal as I won't again refer to him. But someone sentenced to a lifelong existence on a dot of land far distant from the mainstream is island-bound, imprisoned in a kind of no-man's-land, an Alcatraz of isolation, a Robben of remoteness. I found it difficult to imagine how someone like El Viejo—untraveled, uneducated, unsophisticated—perceived the world and life. Not withstanding his limited existence, the old man did have uncommon common sense. He told me that he was quite content to pass his entire life on Easter. He of course had no choice, so his attitude simply reflected an acceptance of the conditions which defined his greatly constricted fatal circle. El Viejo was incurious about what lay beyond the island. His constrained mindset obviously contrasted with my own, which was anything but insular. Stranded there on Easter, El Viejo possessed no curiosity about the remote wide world and which in any case was inaccessible to him. Not only did isolated islanders lacked curiosity about the world, but so did many mainlanders. Over the years I encountered people who enjoyed wide opportunities to enlarge their lives but who continued to lead an insular existence. Although their fatal circle would allow more expansive experiences, the temperament or other factors of such characters somehow limited their life to one as confined as that of the Easter Islanders. Some people are quite content to remain within the confines of their comfort zone. Such a narrow-minded mentality in people with options facilitated by a capacious fatal circle puzzled me. I couldn't understand why advantaged people remained indifferent to the great Technicolor panorama of spectacles out in the wide world, but somehow the play of infinite forms failed to exert their magical effect as they did on me. On the other hand, those self-limited listless characters lacking a broad and abroad agenda no doubt never comprehended why worldly adventures so motivated me to abandon from time to time home comforts for the rocky road of travel. Each humanimal's experiment depends on different premises and perceptions as well as other personal considerations, including emotional, mental and temperamental, psychological and logical, location and motivation, and various other factors. They combine to shape your subjective objectives based on what you value and on how you view and respond to the random sensory impressions which happen to reach you.

Even with all my real world graduate education by far-flung travels, my many peripatetic pleasures, my reading, my range of friendships near and far, my exposure to many different places and experiences the world over, I accepted that El Viejo out on Easter might possess a greater grasp of the verities and existential realities of life as humanimals know it than did I. Instinct and intuition rather than worldly adventures perhaps brought to the land-locked old man lessons far more meaningful than the ones I happened to learn over many years with book-learning and all my peregrinations to many nations. Stranded life-long out on a place like Easter, a person has a lot of time to contemplate such existential matters. While I thought about them on the fly, El Viejo considered such matters in his well-grounded island native habitat. Nature overwhelms the nearly earth-less earthlings out there on the tiny dot of land where the islanders enjoy very little solid ground. The sky is vast, the sea endless, the years long on a tiny remote island with no attractions or distractions. Lacking external stimulation you turn inward. The "longest journey...[is] the journey inwards," claimed Dag Hammerskjold. But since I wasn't introspective, for me the longest journey was out to Easter Island and back. Perhaps El Viejo was a philosopher in the rough like longshoreman Eric Hoffer, a pensive sage whose highly confined and repetitious insular existence with few new sights brought him great insights. Unlike *botones* back at the hotel in Santiago, El Viejo knew nothing of exchange rates or the fast-moving current changes on *El Continente* which swirled through Chile. Easter not only confined the old man but also happily removed him from the turmoil back on the distant mainland far beyond his natural habitat.

In his isolated setting the old man might well have gained understandings which had escaped my notice in spite of my familiarity with a variety of infinite forms experienced around the world. The islander was perhaps far wiser than implied by his narrow and uninformed existence. His horizons and perspectives might be broader and more solid than the invariable watery panoramas viewed from Easter. Simplicity and repetition characterized his life, while disjointed and ever-changing randomness defined my existence during my travel ventures and adventures. His elemental existence which furnished him with only limited happenings and few impressions to serve as food for thought perhaps gave El Viejo an uncluttered mind to enable him to see with a clarity more acute than could I, even though my perceptions were based on a wide range of the play of infinite forms out in the world. More likely, however, was that despite his advanced years they had not brought him enough input to facilitate in-depth contemplations, so depriving his thoughts of texture, substance and granular content. To fuel illumination, logs and not just a few twigs are necessary for a bright campfire.

My thought-provoking—for me, if not for him—conversation with El Viejo concluded with his comment that not only did he not wish to leave the island but that he didn't at all dread departing from life. After death, he observed, *"no hay gastos"*—no expenses once you're in the grave. Indeed, mortality can help you balance your budget. Using zero-based budgeting, you eliminate all the line items. Deficits, economic ones and those of ill luck or a disadvantaged fatal circle, no longer matter. Once gone, you're written off as a discontinued operation and not subject to any further depreciation. Although death comes with a few disadvantages, its many benefits should not be overlooked.

Back on the mainland after my Easter Island visit, I found further practical instruction in how unsettled and dysfunctional economic conditions can damage an entire country. Browsing one afternoon in an antique store in the center of Santiago, I learned in real time in an actual transaction a useful lesson in real-world applied economics. Paintings, sculpture, silver pieces, cases stuffed with jewelry, antique furniture, a miscellany of household items and other sorts of once treasured and now orphaned possessions crammed every corner of the shop. Seeking a new owner, each item, like doggies in the pound, begged "Take me home." Mostly elegant and tasteful, the objects were offered at exceptionally low prices which reflected not their intrinsic worth but a greatly stressed economic situation. Forced sellers desperate to convert valuable but illiquid tangibles into devaluing but spendable money had been compelled to sell family heirlooms for next to nothing. Even without the benefit of my very favorable *botones* elevator black market exchange rate I could have bought most anything in the store for a few dollars. Huge supply completely overwhelmed nearly non-existent demand. Due to economic realities in Chile, great distortions contorted the market. The proprietor explained to me that under the radical policies of the Allende regime, once better-off classes had greatly suffered and were driven to sell their possessions to survive. The supply-demand imbalance in objects mirrored the same distressed factors which influenced the currency's poor exchange rate. The societal pathology was contagious: a collapsing economy, no longer virile, is viral.

As often happens in markets, random elements affected prices. Values remained about the same—the pieces on sale retained their intrinsic worth—but the price a buyer was willing to pay in the marketplace had plummeted. So vast was the supply of discarded items and so small the demand for them, Adam Smith's invisible hand had become strikingly visible. The ultra-low prices evidenced that the material wealth of the nation had greatly diminished. Even at those rock-bottom bargain prices few purchasers existed. Scarce buyers plus numerous desperate sellers in need of cash equals an unusually depressed market. The prices dropped and continued on down and down in an attempt to reach an equilibrium balancing supply and demand. But the bottom had dropped out of available demand. Even give-away prices failed to tempt buyers, as virtually none existed. My buying power in fact represented a rare phenomenon there in unsettled Santiago early 1972. Although my dollars weren't burning a hole in my pocket—ever frugal, I always lined my pockets with fire-proof material—I nonetheless at least possessed the means, if not the will, to buy. This time I could use my American Express card for its intended purpose. One swipe of the card would enable me to acquire the entire stock of quality outcast objects for almost nothing, ship it all home, and then sell the goods at a high mark-up. The very same items which were nearly worthless in Chile would in another setting be transmuted into expensive valuables. Uncertain, however, were such additional costs and factors as shipping, export permits, bureaucratic hassles, pay-offs to officials, theft, breakage and other unquantifiable expenses which would lower my profit margin. Arbitraging the very low cost of goods with the high sales prices available at home could more than pay for my entire five-month-long South American trip, but in the real world such a textbook high-profit transaction came with many uncertainties and risks. In any case, the purpose of my travel adventures was to deal in experiences, not stuff. As always, material things didn't interest me. My agenda didn't include becoming an importer and dealer in distressed merchandise, an activity which would distress me more than it was potentially worth.

One item at the store, however, attracted my interest. I came across a large-format more dining room table than coffee table size thick slick-paper book published around the turn of the century containing period photos of Chile. My library back home featured a collection of old guidebooks, early-day works on travel, long-ago museum catalogs, yesteryear picture books, and other such vintage antiquated materials which presented the world as it had been before my time. As the illustrated volume on old Chile fit right in with my collection, the book greatly tempted me. But the large format and weighty tome presented a problem. Still ahead for my trip stretched a long hard road—the entire Western region of South America: Bolivia, Peru, Colombia, Ecuador. Even though I was now finally headed North toward home, many miles and days and unknowns separated me from my own cozy corner of the world. Lugging the heavy bulky book so far so long presented an unappealing prospect. My small suitcase offered little extra space and, book-less, was relatively light. The volume weighed too much and would take up too much volume. When deciding what to take on a trip, weight and bulk represented major factors. I carefully selected each item to assemble a highly curated and limited but quite functional travel kit. I considered mailing the book home, but the unsettled situation in Chile made that option uncertain. It was a cash-and-carry deal or nothing.

Noting my interest in the book, the proprietor offered the antique item to me for a modest price. I demurred, not from resistance to the asking price but from a disinclination to add the heavy bulky object to my gear. The owner then reduced the price, but still I hesitated. I wasn't bargaining, only procrastinating. After he again lowered the price to a bargain basement level, I continued to refrain from buying. Finally the shopkeeper said, "Just take the book for nothing. At least it'll have a good home in someone's library." For me this was a poignant moment. Wishing for someone to adopt an orphaned book, the owner would give it away. The book was literally priceless. I explained to the dealer—under the circumstances, no wheeler-dealer— that my reluctance to buy didn't represent a bargaining technique but arose only because of my hesitation to take on the burden of carrying the weighty item. I objected to the object, not its price. But the tome did indeed need a new home, and the contents conformed well with my interests and my collection of travel-related vintage volumes. Reader, I bought the book. The market-clearing equilibrium price of zero was unfair, and much to the owner's surprise I offered him the original asking price. We then engaged in the odd process of a reverse auction, with him insisting that I take the book for nothing or very little while I forced on him the full price. This bargaining was truly a perversion of the usual economic regularities. But such occurs in distressed situations, as in Chile at the time when great distortions contorted normal financial, political and social behavior into uneconomic formats based on extreme conditions.

Not only place but also time can cause prices to deteriorate, especially for perishable services rather than for goods. In late 1879 Robert Louis Stevenson boarded a dilapidated coach attached to the end of a freight train to head west from Omaha. An attendant offered the components for a bed—a plank, three straw-filled cotton bags for mattress and pillows— for $2.50. Just before the train left, the price was reduced to $1.50, and a few stations onward the outfit was on offer for 45 cents. "The Scotsman recorded these details as his contribution to the economy of future travelers," wrote Oscar Lewis in *The Big Four*. Perhaps my description of the economic realities in Chile at the time I was there, when the country was suffering great turmoil, will serve as my minor contribution on such matters for future travelers visiting unsettled lands.

Still today, after all these years, the book I bought in Santiago occupies a place in my library and continues to remind me of the circumstances which prevailed when I acquired along with the item a memorable travel experience. Up the long South American west coast I carried the heavy volume, lugging it over miles and weeks until finally the book reached its adoptive home. The early twentieth century scenes pictured in the volume serve to remind me of how old Chile once looked and also conditions there in more recent times when the country suffered from turmoil under Allende. My experience in Chile back then taught me some useful lessons about practical real-life economics in ways theoretical or bookish notions lack. The process of acquiring the book, not its contents, served to educate me. The episode enabled me to see how things actually worked in an economy subject to the stressful circumstances which then existed. Contemplating what I'd experienced in Chile, a ruinous time for the country, I drew a number of conclusions which helped me over the years. My lessons included: the relativity of value in different contexts and under varying conditions; the vulnerability of entire classes to misbegotten socio-politico-economic forces; the ways that unmanageable influences affected transactions; how supply and demand function in distressed situations; the different ways dysfunctional economic conditions affect people and commercial operations; how possessions viewed as valuable don't in fact necessarily represent long-term stores of value; the need to be ever aware of and to analyze risks; the perverse influence of political extremism, rigid and radical ideology, governmental incompetence and mismanagement; how developments beyond your control can enhance your economic well-bring or operate to damage or destroy not only your own situation but also society in general and even an entire country; the need to apply cost-benefit factors for any decision, economic or otherwise; the difference between wealth and cash; how assets deemed valuable can become liabilities, and are ever subject to confiscation or devaluation in one way or another; the effects of a currency's collapse, and the resulting increased worth of safe and reliable money such as, for me, dollars in Chile; and much else.

The *in extremis* situation I encountered in Chile prompted me to recognize that such devastating and damaging intrusions could happen anywhere at any time. No social, political or economic system, no seemingly settled arrangements or governance structure, no actual structure or civilization or culture is immune from shock and radical change. The more seemingly secure, the greater the shock when revolutionary changes shake the foundations and tumble the constructs. Like the Camp Ironwood logs in Maine long ago—no iron wood but in the end only smoke and ashes—all which exists remains vulnerable to becoming nothing. Everything forming the play of infinite forms created by both earthlings and nature is subject to random forces, whether natural or from the hand of man. Unexpected tectonic movements can upset what seems to be fixed, permanent and relatively immune to change. Destructive instincts typify both nature and mankind. Each influence imposes plots and counter-plots on the play of infinite forms. In *The Denial of Death*, Ernest Becker described the pervasive ever-present fragility, "the rumble of panic underneath everything," which at all times threatens whatever has by chance happened to assume an earthly form. In an economic rather than a more existential context, as Becker addressed, Joseph Schumpeter (my apologies for quoting an economist, but every so often such types happen upon a useful idea) observed that successful businesses in fact stand on shaky ground that is "crumbling beneath their feet." And back to Santiago January 1972, to set the fragility of things into a Chilean context, Charles Darwin described in his *Beagle* voyage account February 20, 1835 a severe earthquake in Valdivia, southern Chile, which lasted two minutes but seemed to the naturalist much longer. Darwin felt "giddy": "A bad earthquake at once destroys our oldest associations; the earth, the very emblem of solidity, has moved beneath our feet like a thin crust over a fluid; one second of time has created in the mind a strange idea of insecurity which hours of reflection would not have produced." Reflection doesn't reflect the real world as much as does actual experiences of the place. Thought, contemplation and reflection—and their progeny in the form of books and other artistic creations—are only derivative, removed from the granular substance of life as it's actually experienced on the lonely planet. Like the earth-shaking tremors in Chile which rattled area residents into his-panics and her-panics, the uncertainty and insecurity faced by earthlings embeds in them a Becker-like existential pan-panic.

Everything inanimate which exists remains ever vulnerable to damage, decline, destruction, disappearance, and all living things also face disease, decay, deterioration, decline and death. At any passing moment, that which exists can radically change form, even cease to exist. The incessant play of infinite forms toys with the forms. From one minute to the next, anything and everything might suddenly mutate or disappear, be it a Guatemalan president, Che Guevara's life, my own life, Snoopy, El Viejo on Easter, Pompeii, or an inflamed London or Chicago burning into smoking nothings. Fragile and vulnerable and rare are the conditions necessary to create and sustain a civilized and orderly society. Such an unusual combination of necessary factors is truly an exception, a situation unlikely to last. On a more individual scale, to enjoy a charmed life of one's own choosing also depends on an unlikely set of conditions. For both a country and a person, and for all civilized life, many improbable elements must at the same time exist and then persist.

Such conjunction of necessary elements rarely cohere. Referring to ancient Athens and Rome, Marguerite Yourcenar in *Memoirs of Hadrian* notes: "These subtle and complex forms of life, these civilizations comfortably installed in their refinements of ease and art, the very freedom of mind to seek and to judge, all this depended upon countless rare chances, upon conditions almost impossible to bring about, and none of which could be expected to endure." An intricate and unlikely web of institutions, conventions, beliefs, practices and values form the fragile gossamer strands which knit together a functional society, economy and system. The provisional and somewhat arbitrary conditions which facilitate civilization depend on far too many intangible and delicate factors to be common or enduring. In a *Harper's* magazine review (quoted in *Peter Taylor: A Writer's Life* by Hubert H. McAlexander) by Jack Thompson of Peter Taylor's short stories, the reviewer observes that "all culture, all social contrivance, country club or Army, family or corporation, knife and fork and automobile, all those are only the fragile and necessary contrivances that stand between us and the essential horror of our animal condition."

With its sometimes benign and beneficial and sometimes beastly behavior, the humanimal creates and destroys all manner of earthly wonders and beauties as well as bringing into existence and destroying less desirable phenomena, often perverse and ugly. Nature joins in with its beauty and destruction. The works of both man and nature create a play of infinite forms, all of which represent raw materials for destruction. That which by chance appears as a form represents what will before long disappear. Whether Santa Claus or Santa Fe, or Dracula, Hitler, Hadrian or Heliogabalus, Snoopy, Dairy Queen, King Kong or kings and queens passing brave or timid, Al Camus and Albania and Alice in Wonderland, the Promised Land, all unpromised and unpromising lands, a rare day in June or a tropical storm, the entire zoos who of humanimals, it all slips, slides and slithers away to disappear—time, status, power, fame, prestige, wealth and health and every other thing and attribute valued, revered or feared by earthlings vanishes. In Chile, money lost its value; it can also lose its very existence. It's said that when one-time Abu Dhabi ruler Sheikh Shakbut bin Sultan al Nahayan found some 50,000 British pounds worth of oil credit vouchers eaten by mice he asked the United Kingdom's Political Agent to print future credits on decay-resistant paper infused with rodent poison. Whether nibbled or dribbled away, all wealth, tangible or intangible, remains subject to loss, as exemplified by the orphaned objects in the Santiago antique store and by the country's greatly devalued currency.

Whether grounded on the lonely planet in Santiago, Chile or Santiago, Cuba or Santiago, Spain, all remains at risk. Conditions in the Chile and Cuba Santiagos prompted me to consider how economies are subject to abuse, while the Spanish Santiago, with its dominant ecclesiastical dimension, inspired me to think about the institutions humanimals create to cope with the mysteries and uncertainties of existence. Everything in the human realm depends on scarce and fragile factors which remain ever vulnerable to the vicissitudes of chance, fate and luck. Leaving aside the sacred as represented by Spain's Santiago for more profane considerations prompted by the Cuban and Chilean economic situations I encountered, my visits to those two countries made me reflect on the financial element of travel. It's obvious that wherever you go one constant is the need for spendable funds in order to function. No one cares where your money came from—they just want it. Food, shelter, transportation and whatever else you need or want requires buying power. Without financial resources, you won't be able to fly out to Easter Island or, once there, unable to return to *El Continente*. Like El Viejo, you could be stranded on the island for the rest of your life. A return ticket in your money belt is desirable, especially when visiting remote dead-end places.

Non-monetary assets may be enormously valuable but, unspendable, they won't help you in daily life, at home or away. At times liquid money—cash, credit cards, travelers' checks—tends not to stick to your fingers. It dribbles away. Unlike valuable fixed assets, which can't be directly spent and thus afford a certain store of value, cash—unless confiscated, stolen, destroyed, devalued by Chile-like turmoil or other damaging intrusions by mice or men—lubricates economic exchanges. You can of course pledge some categories of capital assets like cars or houses to borrow spendable cash, and some sorts of stuff can be pawned for money or sold to an antique store. To avoid delays and the complications of borrowing or selling in unsettled and uncertain social, political and economic conditions, cash becomes more valuable than its face value. In Chile when I was there a U.S. dollar went a long way, with very few dollars equivalent in local buying power to otherwise expensive long trips out to and back from Easter Island. People endowed with fixed assets, no matter how valuable, remain highly vulnerable to loss of the properties. Liquidities and small portable valuables can usually more easily be channeled out of harm's way.

My experiences around the lonely planet observing and coping with many different situations the world over enhanced my understanding of applied economics. I never understood theories but only the practical elements. Much of what I learned was rather obvious, even if not to me before my real-life education. I was something of a slow learner and needed real-world experiences to absorb the lessons. Not that I'm a halfwit or a nitwit. Although I may be wrong (I hope so), I don't think I'm a complete idiot—just an incomplete one. It's a work in progress: please be patient and you'll eventually see the completed version. As my observations about economic matters suggest, they're mostly just common sense. I didn't sense them very vividly until I experienced them. Living those experiences enabled me to demystify the subject and to ignore the abstruse models, complex equations and other academic constructs professorial or professional economists formulate. Jargon and gobbledygook proposed by experts and guilds conceal what practical experience can clarify. Empirical realities rather than imperious experts served to educate me. Early eighteenth century London coffee houses were known as penny universities, as the cost of a cup of coffee represented the tuition a customer paid to listen to and participate in conversations. I imagine it was possible to learn a lot more about life in a coffeehouse university than at an academic one. So it proved for me as I journeyed around the world, a vast penny university which instructed me about many practical matters. There's always more to learn, so my graduate course in applied economics which I learned slowly by gradual graduated experiences continues. My life won't last long enough for me to graduate.

In most personal accounts, the book usually ignores books of account. Like a ghost writer, money matters remain a ghostly invisible but influential element. Authors often suppress the topic, leaving it an unnoticed phantom factor. Apart from books relating to finance, business, businessmen and similar commercial topics, monetary considerations are frequently omitted. Works about lives focus on literary, philosophical, psychological, familial, personal and similar such matters to the exclusion of how the characters described pay their expenses. Money's treated as a given, like air— it's just there. Spending power is taken for granted, with no mention of where the money comes from. The memoir writer discusses at length how he flits here and there, from East Hampton to West Egg, Southampton to South Kensington, from Bridgehampton to Knightsbridge. You read about cruises on private destroyer-size yachts, exclusive parties lubricated by streams of fine wines and rivers of single-malt scotch, weekends at palatial country houses, swanky *soirées* with elegant guests, refined *haute cuisine* at restaurants with menus in French, luxury travel, trips in high-style on private jets, but nary a word about the source of funds to support such a lavish way of life. Even more modest living levels remain unexplained from a monetary standpoint. Tell-tale memoirs and other such life-story accounts contain more information about sex than about bucks. Talking about filthy lucre is apparently too pornographic. Mind you, I'm not implying that the sex sections are without interest. But the money passages are also interesting, and too often they're omitted.

Although I mentioned above, in passing, Santiago de Campostella I didn't discuss my visit there. There's no shortage of descriptions or commentaries about a place like Santiago, and I can think of nothing to add. But seldom have I come across in a personal account much about the nitty-gritty grubby topic of money and its role in travel and in life. Every so often a writer will complain in print about book advances, commonly so small they're more like retreats, or being shorted on royalties, or the ruinous costs of his (seldom her) most recent divorce. Otherwise, mum's the word (or for heirs, mom's the word) when it comes to the financial aspects of trips, possessions or everyday life. You can live without ever visiting or writing about Santiago, but it's hard to live without money. Even a Calcutta rickshaw driver needs a few rupees to survive.

Reluctance to discuss a subject so vulgar as money conceals how noteworthy people actually manage to pursue their active and sometimes costly ways of life. Reading such accounts, I've often wondered where the bucks come from. The disinclination to deal with money matters arises from any number of considerations: modesty, privacy, disinterest in interest or dividends, ignorance (somehow, one's bills get paid: cash flow gushes, the plumbing's unimportant), a fear of appearing uncouth (sex, no sweat—let it all hang out; money is gross), a desire not to attract unwanted attention to wealth, or for other reasons. In accordance with my intention to remain aloof from my narrative as a ghost writer, facilitating but not intruding upon the text, I won't include in the account any information about my bank account, so the book contains nothing about my financial books. That element of the story remains beyond the scope of my agenda for the reminiscence. Unlike many authors of personal accounts, however, I've at least called attention to the topic and will offer a few general generic thoughts on some of the financial aspects of an active life.

Back in the 1970s and 1980s when I undertook many of my wide-ranging travels, such trips cost considerably less on an inflation-adjusted basis than in more recent times. My three five-month-long journeys in the first half of the 1970s to South America, the Far East and the Middle East were relatively inexpensive, among other reasons because I traveled on the cheap. Just as at home, on the road I lived frugally. This constant in my life, wherever it happened to unfold, represents a deeply embedded attitude. Like curiosity, it was inherent in my personality. While away, however, I faced duplicate costs, a kind of double-entry bookkeeping. Expenses accrued back home at the same time that I was paying for travel. While out and about, my out-go increased as I had to pay for both of my separate lives. Logging miles along with logs to keep the home fires burning required money. Home operated whether I was there or not. I was out of sight but the bills didn't disappear when I did.

Back at home lived and ate rather regularly such critters and creatures as dogs, some cats (another always on the way), a few hamsters, a pair of parrots, a goldfish (the only gold, other than fillings), wife, some kids (while away I couldn't always remember how many) or (a) all of the above, (b) some of the above, (c) none of the above. But that information isn't at all relevant to my account. The only salient factor is that—unlike the late El Viejo buried out there on Easter Island and forever beyond *gastos* and with no risk of an Easter-type resurrection—when away I incurred double expenses. For the most part I anticipated and budgeted for the simultaneous costs, and in any case travel expenses represent the best money I ever spent, which may not be saying much because I really never spent a lot on anything else. But what little I wanted in the way of things and the lot I sought in the way of experiences I acquired without hesitation. Based on my values, I spent willingly. Based on my means, I spent rather little. Based on how my friends and neighbors lived, I existed as if on welfare, faring well enough for me.

Obviously, an essential element of being alive is access to sufficient means to remain alive, with the additional factor that if you wish to live in your own way money plays an even larger role. Financial resources help improve your life in this regard by mitigating, even if not eliminating, some of the adverse effects of chance and unlucky happenings and situations. Wealth buffers you from the consequences of unanticipated risks which happen to intrude. Money also facilitates choice. How you use financial resources to enable choice depends on various factors. In my particular case, the configuration of my fatal circle endowed me not only with something of a head-start financially but also with some other less tangible favorable characteristics. My temperament and my curiosity played major roles in how I wanted to live and in how I did live. Money served as a catalyst to help me carry out the experiment of being alive. From early days I subscribed to Ruskin's view that "There is no wealth but life." At the same time, I remained realistic and practical enough to realize that without wealth there is no life, or at least not much of one. As I observed all around the world, although a humanimal can exist and subsist with extremely modest financial means, the way I wished to shape my life required some spending power, even if I didn't seek a high-powered existence.

Even with non-grandiose goals I realized that the resources my fatal circle had by chance provided me would probably prove insufficient, if unaugmented, to support the full range of activities I aspired to pursue. I also understood that wealth is only a facilitator. Like time, money is a necessary but not a sufficient component to produce well-being and even being itself. The possession of resources is a necessity, but how you use them represents the key element. My agenda excluded many things and wants many more ambitious souls strove to attain. Possessions, luxury, prestige, image, standing, enhanced standard of living, fancy objects, stuff, and other such values and wares I ignored in favor of my own "wheres"—all the places everywhere I hoped to see. To conspicuous consumption I preferred inconspicuous non-consumption. My journeys brought me experiences invisible to anyone else but invaluable to me. They enriched me not with tangibles my contemporaries could see but with a treasure trove of travel memories as insubstantial but as haunting as phantoms. In that way, a good part of my life seems to have been a spectral sort of existence, rather like that of a ghost writer. I lived with a view to creating the delightful nothings of vanished but at times recalled memories created from passing, fleeting moments which remain in my consciousness, at least until it ceases to exist.

Unlike some trust fund babies who trust that their fortune will always yield independent means, I tried to nurture what good luck had by chance given me. Many of those well-endowed characters who originated in a coddled cradle rich with a financial inheritance never lifted a pinkie to manage their money or to generate additional funds. On the other hand, I used all my digits to try to enhance my good fortune, good but not great, which happened to come with my fatal circle. Never did I take for granted what was by chance granted to me, nor did I simply bank my inheritance under the assumption it would support my agenda. I actively managed the account in an attempt to produce an amount likely to meet my needs, which were modest for possessions but ambitious for experiences. Whether one is a self-made millionaire or a millionheir or a combination of both—an inheritor who greatly expands his initial stake—the roles of fate, chance and luck always influence how you fare financially. Those random elements apply both to what you received and the results you manage to produce on your own. Although you may inherit wealth, it should never be considered as a given, even though it is. Without constant care and diligence, money given to or earned by you will one way or another be taken from you.

There's of course really no way to explain why fate favors with financial resources some lucky souls while ignoring other earthlings who live in the same setting. Be they Rockefellers or rock-bottom dwellers, each humanimal owes its initial economic situation to chance. Once aware of the monetary elements of my fatal circle, as I began to devote myself to dealing with financial matters I spent little money but lots of time, effort and energy to acquaint myself with the realities and practicalities of the investment world. Exactly what I did in that lonely sub-planet of a world is irrelevant to my reminiscence. I'll note only that by wide reading, extensive research, working and networking, connecting with many sources and resources, I managed over time to advance my financial interests to gain sufficient resources for my enterprise—to live in my own way. All the decisions were mine and none depended on "advisers." My investing activity comprised part of my post-graduate real-world course in applied economics. Although many people dread and avoid as much as possible any financial and investing dealings, I quite enjoyed the challenge. An autodidact, I pursued a curriculum which over the years brought me in contact with many interesting, intelligent and accomplished figures beyond those in the pages of financial reports. Although I listened keenly to what the professionals I respected professed, in the end I always made my own decisions. The views of such insiders represented only opinions, not actionable information. I was interested in how the operators thought more than in what they thought. From early days in my enterprise of learning about investing I realized that only very rarely did "advisers" follow their own advice. What they did with their money bore no resemblance to what they told other people. Wall Street sings off of two different song sheets, one for the customers and the other for the insiders. For the clients the music is discordant.

In the course of my course in investing I came to realize that if you don't take care of your money it won't take care of you. Neglect of wealth doesn't last long for the simple reason that if you ignore your assets they'll soon disappear, leaving nothing to neglect any more. Lacking wealth once owned but lost will in a way simplify your life, as poverty relieves you from worries about managing money or losing it. For some people that may be a relief, for then you no longer need be concerned about the Dow Jones or keeping up with the Joneses. To preserve or enhance wealth, the mercurial nature of money demands that its holder exercise constant care and vigilance. Many plots and counter-plots out in the financial jungle aim to gain access to your wealth. Predators lurk everywhere. With its "advisers," salesmen, brokers, wealth "managers" and other such types, the financial industry can be dangerous for an investor. Those characters relentlessly attempt to transmute your wealth into theirs. In *Entrepreneur Extraordinary: The Biography of Tomas Bata*, Anthony Cekota describes how the up-from-nothing founder of the famous shoe company viewed money:

> This thing—money, so powerful and secure in one's hands, all of a sudden appeared to him to be highly elusive and highly volatile. It is here one moment and gone the next if one forgets to watch it, to observe where it comes from and where it is going, or to direct it where it should go before it left...
> Tomas learned more about time and money than most men learn in a lifetime. It taught him that nothing was as subject to tempting dissipation than time and money when one was free to do as one pleased. Time and money disappear quickly, and there is never enough of either.

Although from completely different worlds and generations, the Czech Tomas and the U.S. Middle West Thomas writing these words held identical views about money and its relationship with time. The two are inextricable linked. It's said that time is money, but it's also true that money is time. Money represents monetized time, for only in the course of time can wealth accrete. Ancient Athens or Rome as described by Margurite Yourcenor in *Memoirs of Hadrian* weren't built in a day, nor was the Rockefeller fortune. Not only does wealth require time, money frees up time. Financial resources can buy unstructured time. Money purchases freedom. T=M and M=T, hardly Einstein-quality equations but formulas just as valid and useful in terrestrial terms as his. Money brings access to something more substantial than tangible matter. What matters are freed-up intangible time and the means to live in your own way.

To nurture, produce and preserve wealth requires its manager "to direct it where it should go," as Tomas Bata believed. Intelligent investing entails such factors as analysis, diligence, logic and rationality, contemplation, planning, patience, reasonable expectations, a common-sense understanding of basic real-world applied economics, awareness of incentives and motivations of financial world actors, and a keen sense of reality. But in the end those elements don't represent the essential ingredients for investment success. You might recall (but most likely won't) that at the end of Section I, Part 5 I foreshadowed a discussion of investing "secrets." Some of my views were shaped by what I observed in places remote from the investing world, such as the economic conditions at the two Santiagos, *de* Chile *y de* Cuba, both far from *la Calle de Muro,* which in Spanish somehow sounds less predatory than does "Wall Street." The apparent disjunction of trying to gain insights from a context seemingly remote to an enterprise typifies my practice of attempting to draw lessons from situations and sources beyond those usually related to a particular activity or project. Such a heterodox process in fact represents the way discordant factors, especially those experienced in travel, randomly reach a humanimal as the animate finite form engages in the experiment of being alive. The poor creature receives an endless jumble of chaotic, confusing, jerky, unrelated sensory impressions which make a scintilla of sense, if at all, not at the time they happened to strike the temporary earthling's consciousness but only later with the perspective of time. With its unpredictability, uncertainties, confusions and randomness, both travel and investing resemble life in general. Mistakes in each endeavor can cost you a lot.

Much could be said, and far too much has been said and day-by-day is said, about investing. I'll briefly add to the vast and mostly useless corpus a few hopefully useful comments to address a topic usually ignored in personal accounts. Because money represents a vital component of a viable life, that seldom discussed subject merits at least a brief mention. Savants, academics, wealth "advisers," market operators, commentators and other observers and participants deal at length with such concepts as regression to the mean, diversification, "beta," supply and demand and risk and reward—those two pairs the Mutt and Jeff and Abbott and Costello of economics as they encapsulate contrasting but complementary factors operative in markets. In fact, all of these economic principles apply not only to investing but also to much else in life, such as adventure travel.

The title of the world's earliest book on the stock market, the 1688 *Confusion de Confusions* by Joseph Penso de la Vega, summarizes in those three words the existential reality of both investing and living. The author's observations were inspired by trading in Amsterdam of Dutch East India Company shares. One of de la Vega's purposes in writing the book, he says, was to describe stock trading and reveal its "trickery." He advises market participants to take profits before they slither away like an eel in the slippery way Tomas Bata described. Better, says de la Vega, to grasp gains while you can without hoping for "the persistence of good luck." This comment offers a great clue to success in the financial markets or in life.

It's useful for investors to recognize that markets essentially operate in "confusions," ones so unpredictable and confusing that in fact no "adviser" or portfolio manager, pundit, computer or program can really help you much. Analyses of markets which purport to explain their actions are presented only after the fact, when what happened seems to make more sense. It's the same when you're writing a reminiscence near the end of your time: clarity sharpens with a retrospective perspective. So-called "advisers" don't really know any more than do you, by definition an astute, intelligent and discerning person as proved by the fact that you're reading (still reading, I hope) my book. From what I observed and heard from other investors and from informed financial professionals I respected, the essential enterprise of the financial advisory industry is to give investors little or nothing for something—your fees. As Wall Street puts it, the operators turn your money and their experience into their money and your experience. As much as I liked experiences, that's one kind I did my best to avoid. To Wall Street your wealth is simply a *piñata* full of treats, a treasure trove of goodies a broker and other financial types want to break open to grab some of the sweets inside. You can keep more of the treats for yourself by avoiding "advisers." What those characters tell you is mere opinion. You take the risks while the adviser harvests his fees, regardless of performance. Exceptions to such self-serving behavior surely exist, but I didn't seek to find those outliers as I preferred to make and suffer consequences of my own decisions rather than to pay advisers for ill-advised choices. Because I avoided advice, I didn't need to adopt a *caveat emptor* caution as I wasn't a buyer of what Wall Street sold. Our economic system has become far too exploited by the financial industry, which drains resources from the productive sector, exemplified by an activity such as writing a reminiscence. Over my lifetime it seems that the United States has ever more acted to make the world safe for hedge funds, high frequency traders, portfolio managers, wealth advisers and other such financial operations and operatives of questionable social utility.

Boiled down to its essential reality, like almost everything else in life investment success depends on luck. No matter how diligent and astute an investor is, his or her fortune always remain a hostage to fortune. To keep the tune in fortune, noteworthy fortunate happenings help. As de la Vega says in *Confusion de Confusions*, it's ill-advised to expect "the persistence of good luck." You can only hope that from time to time luck will happen to reward your efforts. This perception may be disappointing, but—based on my own experience—that's really about all I can say for sure about investing. These comments no doubt comprise just one of any number of disappointing passages in this book. Unlike other authors, however, I readily admit that my account includes, inadvertently, such misbegotten material. Every reader, like every humanimal ever alive, is bound at times to face disappointments. I only hope that yours—in reading this book and otherwise—are as few and far between as in general mine have been so far, which by now is about as far as I'll be going.

How to induce luck for investing and, more crucial, for life in general I'm unable to say. How luck happens to operate, for good or ill, is a complete mystery. Although it somehow happened to favor me, I always realized that Lady or Laddie Luck is ever capricious, a protean force now beneficial and now harmful. A delicate, unpredictable and shifting balance with no comprehensible rationale determines good fortune or misfortune. At times a poor decision results in a self-inflicted disadvantage, which is the case for my comments describing my disdain for the financial industry. With those observations I've no doubt alienated many potential buyers of this book, a large and prosperous—far too prosperous—group now unlikely to purchase my reminiscence. Although my candid comments about Wall Street make good sense, they represent bad economics. By reducing demand for the book I took a risk of damaging sales, so kindly repay my refreshing candor by buying multiple copies of my work.*

By now my late-in-life perspective on human existence based on my own experiment should be evident and, if not, *mea culpa*, an apology which sounds somewhat less damning, and with more gravitas, in Latin rather than in plain English. My impression of the situation of earthlings on the lonely planet is that life as humanimals know it spins on and then soon— sooner than many people imagine—winds down and wears out, the energy dissipated by the dizzying twists and turns brought by fate, luck. chance, destiny, serendipity, coincidence and other such random factors. The whirligig of time determines how each humanimal happens to fare, be the outcome fair or unfair. Those intangible capricious factors originate at the very beginning with each individual's conception, and then chance draws the design of the person's

---

*Editor's note: I agree. Publisher: a splendid idea. Barnes & Noble: a noble sentiment. Amazon: operators, or robots, are standing by. Borders: you're too late.

fatal circle which sets the when/where/what of his or her beginning circumstances, and those in turn influence what occurs to comprise the person's life. As for everyone, my fatal circle set the terms of engagement for how my experiment in living began. But what you get for openers doesn't necessarily lead to what you might hope for in the way of closures. Too many other factors intervene. In my case, family circumstances, temperament, inherited characteristics and resources, the times and the society in which I happened to live, the prevailing political and geopolitical contexts, and much else endowed my life chances with many advantages. Those factors, combined with good luck, served to enable me to live in my own way, a coveted but—given all the variables—unlikely outcome. It would be churlish to fail to appreciate how things turned out for me, foolish to suppose that only my own efforts produced such favorable results, and folly to wish to roll back the by now phantom-like years in order to start over by returning to Phantom Lake and the wood smoke in the air at Camp Ironwood in Maine.

My emphasis on luck as the determining factor in life arises only from my own view of how things developed for me. Other earthlings may attribute what happened to them to other factors, but I believe that in my case luck represents the operative force which shaped my fate. Luck itself is not fair or unfair, but to a large degree its outcomes are. From what I saw, for many and perhaps even for most people their experiment in living is shadowed by a certain cosmic unfairness. Lots of luck to anyone who believes that luck isn't a major factor for outcomes, incomes and other comings and goings. Lady and Laddie Luck companion our path through the dark woods whether we want those escorts or not. In his *Autobiography*, Mark Twain fulminated that "persons who think that there is no such thing as luck—good or bad—are entitled to their opinion, although I think they ought to be shot for it." Such fatal punishment is quite unnecessary, as fateful consequences await the unfortunate earthlings who fail to recognize the role of luck.

My own theoretical solution to the problem of unfairness would be to rearrange the distribution of luck. With my plan, the phrase "some people have all the luck" would become obsolete. My scenario would more equally distribute to more people the benefits of luck while reducing them for earthlings overly endowed with good fortune. Such a plan would have greatly diminished the fortunate fatal circle and the luck I enjoyed. But by now, for me, the end is near and so it's for future generations after my time to arrange a way to redistribute luck as I've suggested.

At this late date in my earthly existence, virtually all of my past, that random product of luck and chance, is fixed in time and very little future remains for me. I will soon repose forever beyond the hazards of fortune and the random play of luck and chance. For me now luck has almost run out, as soon I'll be ghosted into a luckless lifeless ex-humanimal. Writing this reminiscence sparked many warm memories of those long gone cold old campfires at Ironwood and all that came after. By now the fleeting fugitive years and what they brought me as time unfolded are like the Phantom campfires burnt into ashes and smoke a long, long lifetime ago.

Section III
CHARMED LIVES

~ 1 ~

Starting in my thirties I diversified my life into two different and quite distinct parts. The division wasn't temporal to establish a clearly demarked break between a before and an after, or geographical as with the obvious directional change in Santiago, Chile when in January 1972 I pivoted to head North toward home. Instead, the new format involved a functional reorganization which offered both continuity and change. Both parts of my revised life continued to exist, one along with the other in a balanced and mutually satisfying way which kept the best of the old while adding to it some potentially beneficial new elements. The dichotomy didn't arise specifically from an attempt to feature any of the four factors (discussed in Section I, Part 6) which determine the quality of a person's life—health plus personal and professional and social activities. Rather, I restructured my existence in a general way to add a second major dimension to supplement but not eliminate my life at home. My circumstances, including my fatal circle and other considerations, allowed me in my early thirties to revise my agenda to suit my interests, temperament and personality. How matters would actually work out with my new structure remained to be seen and lived.

My regular home life was known and somewhat predictable. World travel represented a completely unknown and uncertain way of life. The two modes of being both contrasted with and complemented one another. My settled home life unfolded in an upscale Midwestern suburb, but even there chance, luck and random happenings could upset my routine existence. The most seemingly stable and regular situations remain vulnerable to hazards from unexpected quirks of fate or mischievous tricks of capricious forces, even if less threatening at home than during adventures out in the big wide world. Although the regularity and constancy of my comfortable settled domestic routines seemed somewhat certain, that sort of way of being suffered from being rather too settled. I valued and enjoyed the laid-back security of my natural habitat, a provincial corner of the world bequeathed to me as part of my fatal circle. In my thirties, however, I decided that such a routine way of life, as pleasant as it was, should not comprise the entirety of my earthly existence. To the repetitious home format I added activities so different and diverse that they in effect represented a second parallel life.

My travel adventures involved a completely different world, not only by place but also in a substantive way. The trips brought constant change, surprises galore, excitement, exotic scenes and happenings and, in short, a kinetic existence quite the opposite of my stable life back in the suburbs. This second life added a welcome and refreshing dimension which pleasantly supplemented my original set-piece, which in turn made the new one more meaningful. Either format without the other would have left my life incomplete. I wanted and needed both for a full experience in the experiment of living. Although I appreciated the relatively predictable regularities at home, I also greatly enjoyed the stimulating uncertainties of travel. I never knew what the morrow, my own private little tomorrow, would bring while I was away. Events and situations were malleable. So different were these two modes of being they barely overlapped, so serving to afford me two fairly complete separate lives rather than just partial elements of one. Indeed, each half added up for me to more than just a single integral whole. My away adventures and my at-home interludes produced a net result which yielded a benefit greater than the sum of the parts.

With this bifurcated structure I didn't become a split personality but remained balanced comfortably between the two alternating and alternate ways of living. They remained juxtaposed while I switched back and forth, now at home and now away. I found the contrast satisfying and it made me appreciate all the more each mode. One foot I put into the outside world while the other remained firmly planted in my homeland. This bi-pod posture gave me a solid footing in the best—at least the best available for me— of two possible worlds. For a complete experience in the experiment of living, the suburbs weren't fit for purpose. My little home was only a halfway house. But without that settled half my life would not have been as full and fulfilling.

To preserve my ability to lead this sort of privileged bi-functional existence I remained ever alert to conditions or situations which might upset the balance. Plots and counter-plots can at any time threaten an agenda. Even the most benign activity, such as sitting in an easy chair reading a book, might result in an unanticipated and unwelcome happening. In *La Reine Margot* Alexander Dumas describes an episode relating to how the enemies of Charles IX coated the pages of a book on stag hunting with arsenic so that as the king kept wetting his finger to turn the pages he gradually poisoned himself. I realize that some of the material in my account may be hard to digest and in some places even strike a reader as toxic. Any side-effects, however, are not the responsibility of the author or the publisher, but—a trigger warning—just to be on the safe side you should get a robot to turn the pages, or at least wear disposable gloves (as should the robot).

In an effort to preserve my agenda, and especially to keep on course when engaging in adventure travel, I subscribed to the formula in tattoo artist Ed Hardy's memoir *Wear Your Dreams: My Life in Tattoos*: "Lay low, stay anonymous, keep your eyes open, and watch out. It's all about control." Unlike those tattooed characters who wore their dreams, I didn't wear my heart or anything else (except buttons) on my sleeve and no decorations on my skin. I kept my dreams to myself, hoping at least some of them would ripen into reality. Only time would tell. I couldn't know what its story would be until nearing "The End" line. By now that once distant finish line is coming into view.

Only in retrospect can you perceive how any particular decision has reshaped your fate, for better or worse. After the fact, all the wild cards in the play of infinite forms for you have been played and the outcome reduced to an end-game with a single reality. Enterprises involving "sunk costs" made me especially cautious. These consisted of a decision which would be difficult and maybe even impossible to revise or revoke. Even less binding decisions depend in part on guesses, opinions—based on hopefully informed and not deformed views—and on perceptions, probabilities, subjective perspectives and other such instinctive or intuitive factors. You thread your way through a labyrinth, diverted by many false starts, wrong turns, dead-ends, one-way traps, points of no return and disappoints producing no returns. Threats, hazards, risks, unknowns, President Ydígoras Fuentes-type plots and counter-plots spin a web of dangers which can entangle the unlucky or the unwary or even very wary wayfarer who may fare poorly. Tattooed into my memory, out of sight but not out of mind, is the slogan "Lay low, stay anonymous, keep your eyes open, and watch out."

Decisions in the face of uncertainty, a redundant statement as every major decision is fraught with uncertainty, is an art, but not one you can learn at the École des Beaux Arts. As usual I tried to educate myself not only by observing but also by participating in real-life experiences. I watched how some of my accomplished friends addressed important decisions. A few people I knew managed to live as least in part a charmed life so they must be doing something right. I tried to discern some of the principles and techniques my successful friends used to run their lives. One common element was the role of non-rational but not irrational considerations. I noticed that many people, no matter how hard-bitten or hard-headed, often decided on a basis as much emotional as promotional—that is, not only a calculated choice which seemed best from a bottom line standpoint to advance the decider's interests. Instinct, intuition or emotions often over-ruled logic and rational thought. The experiment of being a humanimal in an uncertain world is an enterprise which entails far too many random factors for pure reason and rationality to assess choices. So baffling and puzzling is the lonely planet and so vulnerable are earthlings to the forces of chance and luck, a purely cerebral approach fails to provide a sufficiently complete perspective on problematic matters. As I saw with my accomplished friends, logic can only take you so far, beyond which a gut feeling takes over. Ideally, all major decisions depend on a synthesis of thought and feeling. One without the other falls short of a relatively complete reality check. My successful friends who enjoyed something of a charmed life employed both factors to guide decisions, and in making my own way I tried to follow that example.

The luxury of choice for a matter as existential as how to live endows you with a rare and valuable privilege, one which is also quite challenging. Faced with limited discretion, most people lack the opportunity to choose. If by chance you're lucky enough to be able to select your own way of life you then face the fundamental question of deciding how to exercise that freedom of choice. At the Abbey of Thélème, says Rabelais, rules the mantra (and womantra) *Fay ce que vouldras*—do as you wish. This is a tall order with a wide set of possibilities. What, in fact, should you wish to do? As Rabelais understood the concept, it's not an Open Sesame to allow anything. Rather, the idea involves a more measured Renaissance humanistic creed to encourage and facilitate a free and yet purposeful and productive life unfettered by medieval beliefs and constraints. In an analogous but more limited personal way, I de-constrained my life by removing barriers which impeded my ability to add adventure travel to supplement my life at home. Once the new and refreshing balance was struck, I enjoyed and greatly valued both lives.

On its face this sort of binary structure, one person operating simultaneously in two major dimensions, seemed to me fairly unremarkable. Other people sometimes successfully pursued two or more quite different kinds of endeavors while retaining both or all of them. Multiple lives aren't all that unusual, and such modes of being help to create charmed lives. It eventually occurred to me that my sort of duality represented a much more profound transcendental and existential dichotomy than simply engaging in two quite different major activities. At home I dwelled in a conventional Middle West suburban setting surrounded in my subdivision by home-bodies whose lives in large part revolved around *Kinder, Kirche und Küche*. Children, church, cooking, metaphors for traditional bourgeois concerns, typified the homey ways which motivated my friends and defined their existence. My contemporaries at home found pleasure and value in cars, cards, clubs, diamonds, spades (gardening), social life, shopping, parties, child-raising and child-schooling, and all the other usual ways of life common in the provincial setting where I lived. For my coevals those interests and activities represented the good life, whereas I deemed that many of those practices comprised elements of the bad life. My attitude wasn't meant to criticize or demean my friends or neighbors—I understood the merits of their value system—but served simply to evaluate and delineate how I wanted to live my life. My intended new agenda differed from but was no better than any which others pursued. They followed their ways, I followed mine. We lived in the same suburb but in many ways occupied different worlds. The difference was that I retained my footing in the home-based world but also lived part of the time in the quite different realm of far-flung adventure travel, while my contemporaries seldom abandoned their home comforts for the rigors of independent adventures in the outside world. My friends were comfortable with their lives, and I was comfortable with the discomforts of travels to far lands.

As I eventually realized, I was a closet revolutionary, someone who revolutionized his own life but had no interest in changing the system. Mine was a private revolt. Politics never interested me, not even as a spectator sport, and in contrast to Che Guevara I didn't conspire or aspire to overthrow governments. By personality and temperament I had no interest in changing the world. Let it be. I wanted only to change my life. I rebelled internally but never by agitation or any overt behavior against the provincial suburban mentality and practices. At home I continued to live something of a bourgeois life, which in a way seemed to me an odd paradox as I rejected much of what typifies that sort of life. Hemingway (supposedly) described life in Oak Park, his family's Midwestern suburban Chicago hometown, as a place where the lawns were wide and the minds narrow. My friends' interests weren't narrow; they were simply different than some of mine. In reality I lived in precisely the same sort of wide-lawn context, one from which I from time to time wanted to escape. This paradox puzzled but didn't at all concern me. I readily accepted the fact that although I shared few of the beliefs, interests, values, mores, priorities, aspirations or characteristics of my fellow suburbanites, nonetheless I clung to my native turf, wide lawns and all. Why not just give it all up? Never have to cut the grass again; no weeds; leave the oaks in an Oak Park-like Midwestern suburb to shed their leaves without me raking them. I viewed how others around me behaved (for certain matters) as object lessons on how not to live, and yet at times and in some ways I lived just as did my cohorts at home. A Doubting Thomas regarding the merits of some of their practices, I nonetheless in part duplicated the ways of the very people I sought to distance myself from both by behavior and by travel.

Although I considered myself a fish out of water in the suburbs, in truth I really swam in the same little aquarium as all the other minnows and goldfish. All of us, myself included, lived in a way I found fishy and to some degree inauthentic, a mode of life I sought in part to shun and yet to which I in part clung. My homestead and what it represented anchored me in a context I was determined to exit, but only temporarily, in favor of a free-form existence far from home where the play of infinite forms offered a spectacle unavailable in the suburbs. Overseas the "sees" expanded my horizon beyond the narrow limits of the fishbowl. No one at home would ever recognize how their mild-mannered suburban friend and neighbor lived on the road. My adventure travels entailed an irregular unsettled sometimes risky and on occasion dangerous catch-as-catch-can-or-can't, on-the-go stop-and-go, here-today-gone-tomorrow, skelter-helter, burly-hurly, turvy-topsy, zag-zig confusion of confusions teeming with all manner of new and different exotic, strange, weird and wonderful foreign experiences, a great ceaseless play of infinite forms all around the great wide world.

For ever so long every so often I puzzled over the seeming contradiction between my two greatly contrasting ways of life. I couldn't reconcile how two such extremely different modes of being could so compatibly exist in one person. In fact, when in each mode I often longed for the other. Wherever I roamed in foreign lands I carried with me a photo of my little suburban ranch house to remind me of home, a haven to which I could always return using my return ticket if the outside world became too challenging. Back at the ranch I could resume my normal mundane suburban existence, sheltered there from the hazards of fortune intrinsic in adventure travel. Still, the disjunction between my two radically different lives continued to perplex me, but not in a bothersome way. It was just something I occasionally pondered. I could understand how shades of difference in one's daily routines might have made some sense. Variations on a theme, modulated adjustments to regular practices, fine-tuning and such similar minor gradations on a continuum were logical. But two utterly different formats embedded and both functionally workable in one person struck me as odd. Both modes greatly contented me. I was neither fish, in or out of water, nor fowl. How did I fit in to the human menagerie? I seemed to be something of a twin ("Thomas" in Hebrew, or Shebrew), two personalities incarnated in one corporeal container. Clearly I wasn't identical twins as my two individual components were quite separate and unidentical. Finally, one day in a far land, I by chance happened to come across a succinct explanation for what had seemed to me the baffling matter of how I lived two such dissimilar but completely compatible lives. Suddenly, it all made sense.

~ 2 ~

Although I've read in my time thousands of pages, experiences always appealed to me more than reading about them. The benefits of books are in a way highly overrated, my highly beneficial reminiscence being an obvious exception. H.L. Mencken observed that "There are people who read too much: bibliobibuli." Such print fanatics, "Constantly drunk on books," are chronic readers who "wander through this most diverting and stimulating of worlds in a haze, seeing nothing and hearing nothing." A tenth-century Persian grand vizier was accompanied on his travels by his 117,000 volume library carried by 400 camels, trained to walk in alphabetical order so the librarian could easily find books. My journeys were less elaborate and bookish: only 26 camels, one for each letter of the alphabet, accompanied me to carry my favorite volumes, and eventually I left the camels at home to graze on grass so I could cancel my lawn service. I traveled only as a lowly passing visitor, not as a grand vizier, and both away and at home I was rather nondescript, just another earthling, one of billions, briefly aboard the lonely planet before the final trip when I departed forever from my terrestrial home and its dazzling play of infinite forms.

Among those forms were the many books I read. The millions of words which passed through my mind didn't furnish my life with its most important component, and anyway I forgot most of what I read. Reading was mostly just a time-filler, and when I put the book aside after finishing it the time had been filled but not my mind. Books represented for me not life but only a reflection of it. Some of what I read enabled me better to profit from experience, but as much as I enjoyed reading I never confused the two activities—living life and consulting its word version. Books were a side-show ancillary to the main attraction, which for me was the play of outside infinite forms. Writings often passingly stimulated, informed, entertained or amused me, but never did I read anything which could match the vivid immediacy of actual experience.

Word-immersed bookish types often attribute to printed materials a status beyond their merit, a mistake I tried to avoid. In *Avid Reader*, Robert Gottlieb, former president, publisher and editor-in-chief of Alfred A. Knopf and editor of *The New Yorker*, observed more than once that "nothing is real until I read about it." For me, such a confined view was far too limiting, as books didn't conform to or confirm the outside world—they distanced me from it. Words on paper could never replace in my mind the actual play of infinite forms which churned and spun into the sensory impressions forming my real-world experiences. In *Ferocious Alphabets*, Denis Donoghue warned that "people who deal in language are easily tempted to believe that the entire world is linguistic or at least susceptible to linguistic processes." M&M's and ice cream tempted me, but not the notion that nothing is real until I read about it. And in any case I doubted that the world is susceptible to any processes—linguistic or otherwise—to order it. The place seemed rather unruly and disordered.

Books can be dangerous, not only by their ideas but by other contents. As Charles IX learned from reading the arsenic tainted pages of the stag hunting book, reading can even be fatal. The king would have been safer hunting the stags than paging through a book describing the hunt. I realized that reading, an inactive activity and a passive pastime, doesn't offer a substitute for living. Books represent only derivatives of real life. But just as occasionally an economist or social scientist or humanities professor offers a noteworthy thought, every so often a passage you happen upon in a volume speaks volumes by illuminating a matter which applies to your life. I hope my reminiscence offers readers a few such enlightening passages. The words I happened to find in a stray book far from home served to crystallize what had previously remained only unformed shadowy shards of thought, stray notions which resisted formulation into an integrated concept. Thus did print imprint upon my mind a new and enduring perception.

In *If on a winter's night a traveler* (to quote a book to refer to books), Italo Calvino lists among the contents of a book shop "Books That Fill You With Sudden, Inexplicable Curiosity." It thus so happened that one day in April 1973 I by chance found in a Tel Aviv used book store a vintage volume published in Paris in 1915 titled *Pensées de Gustave Flaubert*. I wondered how Flaubert's "thoughts" had happened to migrate from France to Israel sometime in the previous 58 years, and who before me had read the passage (and what they thought of Flaubert's "thought") which finally so suddenly served to explain to me my seemingly contradictory bifurcated life. Pages 54-5 presented Flaubert's thought that one should live in two parts—as a bourgeois, while at the same time thinking in a radical way. Although your daily life might be conventional, within that framework you can also believe in and pursue a completely different sort of existence. Bourgeois comforts and stability don't contradict but supplement and enhance your more radical approach to the world, an approach which represents your authentic being. The duality entailed no anomaly. How you lived didn't necessarily signify your inner being. Like an X-ray's image which reveals the skeleton which structures the animal, a person's essential nature can differ from what appears on the surface. I now understood how the two so radically different ways of life managed to co-exist so compatibly. Each served a different purpose, and together they facilitated a larger integrated diversity.

Prompted by Flaubert's thought, I came to realize that my conventional home life was in fact the very factor which inspired and informed my other way of life out among the worldly play of infinite forms. I owed a debt of gratitude to my suburban bourgeois existence. My provincial setting both inspirited me to escape while also providing me with a comfortable settled home base to which I could always return. That peaceful little corner of the world, the lonely planet's only turf—with its wide lawn but, so I hoped, not a narrow-minded resident—which belonged to me, provided me with a place of my own. The more bourgeois the better, for after a long and grueling trip I longed for the dull routines, familiar sights, mundane daily activities and certainties of my provincial life at home. While away I appreciated ever more my bourgeois home life, and when there I often wished that I was once again in far lands pursuing exciting adventures. I was at heart a radical in bourgeois clothing—more authentic, so I believed, than characters who are pretenders in radical dress but cling to their bourgeois comforts—and I valued both aspects of my existence. The two modes of existence which represented such extremes both suited me. I shed my suburban outfit to travel and then back at home willingly and eagerly redressed the lack of comforts and conveniences of my completely un-bourgeois worldly adventures. I was at heart a provincial, but in my mind worldly.

Well did I realize that my format didn't represent an arrangement which would appeal to or even made sense to my friends at home. My chosen modality was as foreign to them as my trips. My cohorts seemed to view my world-wide adventures as rather eccentric behavior. People toiled away their days at 9-to-5 jobs to earn enough for a way of life I was willing to abandon, at least temporarily. It seemed to puzzle others why someone in their world would choose to forsake the comforts of an advantaged suburban existence for the rigors and uncertainties of the open road. I must admit that on occasion when I found myself in far-off lands, and especially when I faced particularly challenging or unpleasant circumstances, I asked myself the same question. But it was only a passing thought. I enjoyed and treasured most of my travel experiences, even the less pleasurable ones. Similarly, back home I valued even the most bourgeois routines, which greatly comforted me after the chaos and confusions of my adventures in the outside world. Although Flaubert described this sort of synthesis as including a radical element, my new agenda was only evolutionary and not revolutionary. I didn't seek to change any system except my own. I realized that it was extremely unusual for a person to enjoy the opportunity to live life in his own way. While such a potentially charmed life remains out of reach for most people, somehow it was available to me—possible but for sure not inevitable. It was up to me to make something out of the opportunity at hand. You can wear your dreams in the form of a tattoo, but to accomplish them in real life is quite another matter. An extremely successful and widely admired friend of mine, a public figure who at age 87 became fatally ill, told me a week before he died, "Don't feel sorry for me—no sympathy necessary—I have no regrets or complaints." He paused, then concluded, "I had a charmed life, and now it's over." It's precisely this I hoped I could say during my last week of life. This was the ultimate be-all and end-all of my new way of life.

An opportunity to try to establish a charmed life depends on a very unlikely concatenation of various factors. Such wild cards include a favorable fatal circle, fairly consistent good luck, an ability to recognize and to deal with plots and counter-plots, and a realistic agenda whose proper execution might enable you to live in your own way. In any case, "a charmed life" defines only a relative desideratum, not an absolute one. As generally satisfactory as a person's existence might be, many exceptions always intrude to de-charm his or her life. In a more existential way, no life is charmed because in the end you lose it. So it is that the more charmed your existence, the greater your loss—not very charming. Fated to expire, a charmed life—ever haunted by mortality—is bewitched. In the end all charm, hexed to death, passes away. This is all the more reason to appreciate a charmed life while it lasts. Its inevitable destruction lends such a life a fleeting intensely bittersweet tone.

The ability to establish a charmed life represents only an opportunity. The unlikely potential you've been given is a long-shot, one which often misfires. It's in fact as much of a challenge as an opportunity to face a blank slate of possibilities with the responsibility of filling in the empty space with appropriate content. Designing and producing a satisfactory scenario presents many problems. It would have been far easier for me to leave well enough alone, forget the charms, and continue to exist in my pleasant bourgeois mode at home. That comfortable format provided enough charmed benefits, and as a stay-at-home I would've avoided many discomforts and challenges in the outside world. But at the same time I would have missed many of the delights which served to give me wonderful experiences and to furnish me with many fond memories, heart-warming and mind-warming recollections which now companion me in old age.

To prepare my agenda I cast my thoughts into the far future to visualize how my life might look after most of it had been lived, which is my present situation. To activate this theoretical backward-looking view I tried to imagine how my eventual retrospective perspective might look so that near the end it would conform to my interests, hopes, aspirations and agenda. Back at the time of this thought experiment I still retained some degree of influence on how my life might develop. By peering ahead I tried, going forward, to curate rapidly disappearing time. I didn't want the years to vanish without trying to shape them to my ends before I met my end. By now, however, it's too late to chance further attempts to infuse additional charm elements. Of course, no matter how I chose to proceed back at the beginning the ever controlling factors of luck and chance would in the end determine the outcome of my experiment in living. My scenario planning always remained subject to the capricious intervention of random events. I could only produce a rough outline, a directional guide which pointed me to the way I hoped to find through the dark woods. But no plan ever survives contact with the real world.

Because of all the random factors which might impinge on plans and goals, a charmed life always remains vulnerable to forces which at any time might deprive a favored life of its charm. Just before losing his head to Macduff's sword, Macbeth at the end of his saga claims invulnerability: "I bear a charmed life, which must not yield/To one of woman born," to which Macduff replies, "Despair thy charm/[for] Macduff was from his mother's womb/ Untimely ripp'd." You just never know when a charmed life is heading for a beheading. Charm is as vulnerable as sword-slashed heads. I tried to avoid the plots and counter-plots of the Macduffs. I aimed to keep my wits and my head about me. By chance I was given a head-start in life; I didn't want to lose my advantage.

One basic attitude I found useful to meet the challenge of envisioning a new way of life,

Section III: Charmed Lives

as well as for many other matters, is "TCB"—not "Tom's Cautious Behavior" but Taking Care of Business, a formula I owe to the acclaimed poet-philosopher and mover and shaker Elvis Presley. The now defunct Tower Records rather than an ivory tower characterized the pop star's musical musings and amusings. For my purposes, "business" referred not to a career or commercial goals but to the business of life. Presley's phrase encapsulated for me an entire way of looking at the world. TCB meant always to keep my agenda in mind so that in so far as possible I could make decisions on how they might affect my chosen path. I sought to retain sufficient flexibility and discretion to be able to realize the potential my fortunate situation afforded. I tried to shape my life experiment as it was being formed rather than leaving it all to chance, even if that factor represented the dominant influence. In economics a "price taker" is a seller who must accept without any bargaining power the prevailing price being offered. I didn't want to be a life taker, accepting what befell me, without trying to negotiate as best I could better terms. While time was on the wing, flying away day by day, I aimed to guide its trajectory. Many people look back on their lives with remorse and regret. I opted to look forward, well aware that before long the now will become the then, and by then it's too late. At this point in my existence, all my thens have coalesced into very nearly my entire life, which will soon be deprived of any more nows.

Presley's useful phrase served to remind me always to bear in mind the business at hand, which was to live life in my own way. However one's life business may be defined, practicing TCB will help you to carry out that enterprise. In the mid to late nineteenth century in the Japanese commercial city of Osaka residents commonly greeted one another with *mokari makkan* ("how's business?") rather than the traditional *konnichi* ("good day"). From time to time I asked myself, "How's business?" This periodic self-audit kept me focused on my agenda.

Although a reductive formula such as TCB or "How's business?" might over-simplify matters, I found those extremely terse notations useful to clarify a wide variety of situations. For this purpose it's necessary to select realistic and presumably valid aphorisms rather than clever but misleading ones. These condensed formulations of wisdom I in turn abbreviated as "BST"—Bumper Sticker Theory. Far briefer even than a Tweet, such micro summaries reduced complicated categories into simple and more workable rubrics. These extremely basic and brief slogans enabled me to "Take it down to the physics," a mantra favored by Elon Musk, who liked to start with a problem's underlying first principles.

For me the ultimate BST would be a single simple symbol which completely reflects my puzzlement about the phenomena of earthly existence and the play of infinite forms—a "?" In truth, little of how the cosmos happened to be arranged makes any sense to me. Apart from one obvious existential certainty, few if any others exist. The one certainty I don't doubt is the absence of certainties. I view with skepticism "isms" pertaining to transcendental matters. To look at the world through the prism of an "ism" —capitalism, Marxism, altruism, mysticism, egotism, optimism or pessimism, Snoopyism, whatever—distorts reality. Doctrines, ideologies, cults, religions (oops—there goes the church-going book-buying readers: lost a lot of sales there) and similar systems claim an authority unsupported by evidence. To the contrary, the more certain and didactic the dogma the less believable the professed belief is.

Although a question mark most accurately represents my state of mind, an exclamation point—"Wow! What a trip!"—could be used to punctuate my final impression of the play of infinite forms which happened to comprise my life. In the end, however, a question mark truly designates my last word on my experiment in living. In a cypress grove outside Vienna the grave marker of physicist Ludwig Boltzmann, who committed suicide, bears the inscription "$s = k \log w$," a BST-size formula for his famous theory of entropy. This may be the perfect epitaph, as it summarizes in a few letters the fate of every humanimal, every other living thing, and in the end the destiny of the entire world. Boltzmann's theory holds that every activity results in an irreversible loss of energy which gradually forces the world to run down. For my own burial plot, a counter-plot to existing, I'd favor a gravestone in the shape of a question mark to represent my life-long Doubting Thomas mentality. Given my age, I'd suggest that the stone masons get to work on my monument soon.

Elvis Presley's TCB BST seemed to me more practical and useful than many pronouncements professed by academics, intellectuals and other bookish types. When Borges asked me in B.A. (a reductive BST-size miniature) if I knew T.S. (also BST-type initials) Eliot I should have responded with a reference to E.A. Presley, a lyrical poet quite as worthy and maybe more so of a Nobel Prize for Literature as fellow musician Bob Dylan, Tom E. and other Thomas writers such as Nobelist Thomas Mann, Dylan Thomas, Thomas Pynchon (like me, a ghost writer type and typer), Thomas Hardy, Thomas Browne, Thomas Wolfe, Thomas Merton, Thomas Aquinas. My fondness for Presley and his TCB motto may degrade and dumb down my narrative, but such a pop culture reference hopefully might on the other hand add to the gravitas of my text. In some respects Elvis's oeuvre is more authoritative than Eliot's. TCB trumps TSE for practicality and real-life guidance. While Elvis favored a whole lot of shakin', Eliot did a whole lot of thinkin'. The wisdom of Elvis was plain and simple. Tom's writings, confected by exquisitely filigreed literary artifacts laced with intricate thought patterns, you had to read carefully to tease out practical wisdom. As if feeling my way through the thicket of Dante's dark woods, I once groped through a collection of Eliot's essays entitled *T.S. Eliot: Selected Prose*, edited by John Hayward. (Unfortunately, a copy of E.A. Presley's essays continues to elude me.) In his July 4, 1950 "A Talk on Dante" lecture at the Italian Institute in London, Eliot observed in reference to Dante: "The explorer beyond the frontiers of ordinary consciousness will only be able to return and report to his fellow-citizens, if he has all the time a firm grasp upon the realities with which they are already acquainted." Although not condensed to Bumper Sticker Theory concision, Eliot's comment matched Elvis's useful TCB doctrine. Like Flaubert's *pensée*, which essentially attempts to reconcile how a creative person should relate to ordinary everyday life, Eliot's observation addresses the relationship between a poet like Dante, the "explorer," and his intended audience. The analogue to Flaubert's notion that a creative person should live like a bourgeois while thinking radically is Eliot's view that an "explorer" of extraordinary creative ability like Dante must understand and sympathize with ordinary people if his work is to have any meaning for and impact on them. In both cases, an artful amalgam of quite different components produces an alloy richer than either element on its own could.

My own interest in being an "explorer" didn't include Eliot's sense of a creative type exploring great thoughts, like Dante. For me, thinkers like him represented sources, not role models. I sought only to experience the world while retaining my place at home. I sought not to pass "the frontiers of ordinary consciousness" but only to pass across earthly frontiers which divided the lonely planet into separate sections. The world wasn't beyond me—some of it I could access—but great thoughts were. I never had an original great thought—and not all that many lesser ones—in my life. Unlike pundits who professed wisdom, I realized that I was incapable of sage observations of general application. I could only modulate my own little life, not pontificate on how my fellow earthlings should carry out their own experiments in living. I accepted my limitations and stayed within my range of capabilities.

Oscillating between the two extremes, comfortably bourgeois at home and adventurous explorations abroad, never disoriented me or required much in the way of adjustments. I easily transitioned from one mode to the other. My equilibrium didn't depend on Librium but was based on my rather impassive personality. My mood almost always remained level, like the heart chart of a just defunct humanimal. Ghosts no longer circulate. Seldom did I rise to an elated state, nor did I descend into dejection. My steady-state and somewhat phlegmatic personality saved me from both euphoria and moroseness, the modulated mood blocking me from the delights of deep emotional highs and also from the depths of despair. A personal account meant to detail my gyrations between highs and lows and other psychological topics pertaining to my life would contain mostly blank pages. Just as my two ways of life—here today, gone tomorrow, then back home again—offered a balanced and measured format, so too did my personality. My equanimity enhanced my ability to adhere to a Taking Care of Business focus on my agenda. Anyone who aspires to something of a charmed life has to retain control of his own agenda. Other than the degree to which you're willing to accept other peoples' needs and wants, all too often intrusive, you need to protect your agenda from outside demands and requests. Such importunities can damage opportunities related to your agenda by degrading your ability to fulfill your own intentions. In *The Poorhouse Fair* by John Updike a character says " 'No'— The word hung high in the living room, the 'o' a hole that let in the cold of the void." A "no" will stop cold those who try to get from you a "yes" which would interfere with your agenda. My patron saint—were I blessed with one—would be Jean-Claude Richard de Saint-Non (later changed to "Nom"), an amateur painter and engraver whose family homestead was the Château de Nom in north-central France. "Monsieur No" was a patron of Fragonard, who in 1759 accompanied Saint-Non (who said "oui" to his traveling companion) on a prolonged tour of Italy and who later painted portraits (one in the Louvre, the other in Barcelona at the Catalonia National Museum of Art) of his benefactor. In matters pertaining to protecting my agenda from outside demands, "Non" could have been my "Nom."

My agenda included not only adventure travel but also remaining involved in many aspects of regular suburban life, with all its advantages and conveniences but without completely enmeshing myself in the bourgeois setting where I contentedly remained—some of the time. My goal was to remain attached in a detached manner. Many of the local rituals and practices didn't interest me, but I always enjoyed my home-base comforts, especially after a long trip. In *A Matter of Passion*, Bernard Berenson stated that it was preferable to "live the life of a looker-on, nothing but a looker-on, without getting mixed up, committed, tangle-footed like a fly-paper with the life of others." For the most part I preferred the freedom to fly away rather than remaining stuck in sticky no-fly fly-paper. But even though I treasured my freedom to diversify my experiment in living, at the same time I sensed a danger in cutting my ties with regular conventional life. I felt that living an exclusively on-looking life would be rather charmless. To be only a looker-on as Berenson favored removes some of the substance of life and reduces you to an outsider. I didn't want to be a drop-out but sought only an occasional time out. My trips were interludes which would distance but not sever me from my familiar suburban surroundings and the ordinary work-a-day world. Cut off, like Macbeth's severed head, from my life as a home body would have deprived me of an essential component of my life. Participating in some chosen normal societal procedures and practices kept me tethered to the routines and rituals of my natural habitat. This helped me from becoming alienated from the everyday bourgeois provincial realm—real life, as millions of suburbanites experienced it—I inhabited. In *The Future of Industrial Man*, Peter Drucker noted that for "the individual without function and status, society is irrational, incalculable, and shapeless. The 'rootless' individual, the outcast—for absence of social function and status casts a man from the society of his fellows—sees no society." In my inner being I was to a large degree an outlier, but to retain at least some status at home I tried to remain an "inlier" connected to but not caught in the web of societal conventions. For that reason, among others, I cultivated my roots, even while occasionally cutting loose from them at both the native ground and the societal setting where I'd originated.

Not only did place, habit and tradition exert a strong hold on me, so too did my basic personality. It's obviously possible to change where you live, and perhaps even to adjust if not completely to alter your personality to adapt to new circumstances or for other reasons. But I sought no such changes. Among other reasons, it was easier just to accept the hand— and the elbows, arms, legs and all the rest, physical and mental—fate had dealt me. Like every humanimal, I was endowed with a certain kind of temperament, another way of saying that much of a person's personality arises from his genetic fatal circle. For better or worse I tended to be stable, and this proved convenient for me. Mood swings never see-sawed my balance. I had no interest in exploring the reason for my steady-state trait or any other personal attributes. I sought only to determine my strong and weak points so I could concentrate on the former and try to work around the less favorable ones. How or why I originated and developed was irrelevant to me, and how my inherited traits happened to shape my personality made no difference to me. I didn't care about the unanswerable "whys." The essential matter was to comprehend in a realistic way what I was about and then, within that framework, to try to operate effectively. I needed to "know thyself" (myself) only in a functional way which would help me cope with the challenges of being alive. Both accurate self-perceptions as well as realistic perceptions in general were preconditions for me to live an authentic and hopefully charmed life.

One element of my self-knowledge was my perception that my temperament led me to doubt the likelihood of being able to live a charmed life. I wasn't a skeptic but a realist. I believed that to live in the way I hoped would require a strict focus on TCB, which was largely up to me, and lots of good luck, which was out of my hands. My practical and sober temperament tempered my expectations, leaving me with moderate ambitions, modest aspirations and muted hopes. I realized that it was a tall order to shape nearly infinite potentials fraught with hazards into a life which even in a small way could be deemed charmed. I tried to arrange my agenda based on what seemed reasonably attainable results rather than unlikely successes based on fanciful visions of an ideal outcome. Although some ambitious strivers believe "the sky's the limit," I avoided pie in the sky and refrained from reaching for the heavens. I wasn't moonstruck, sun-stroked or otherwise influenced by remote trans-earthly or transcendental factors. I was content to remain earthbound and to deal with the world as it and as I were.

In addition to moderating my expectations, my equilibrated character also operated to limit my desire for accomplishments. In truth, never did I have much ambition. I never aspired to any sort of over-achievement, and in any case doubted if I was capable of such excellence in my life. I decided that it wasn't worth going to a lot of effort to seek excess success. I had absolutely no desire to make a dent in the universe, as Steve Jobs wished for himself. The world's best body repair shop couldn't remove the dent in Jobs life caused by his collision with mortality. I didn't see the point of banging away to dent anything. I was quite satisfied simply to keep the universe from making a dent on me. W.H. Auden realized that humans "hold a mere chance/Of significance/And call the scratch on the surface of things/Wisdom of ages." Denting and scratching marks that would evidence my significance didn't appeal to me. I had my hands full simply coping with my own immediate situation. My values and goals were challenging enough without tempting fate with overweening ambition or unrealistic expectations. Just attempting to live in my own way to create something of a charmed life was quite enough of an agenda for me. If I managed to accomplish those goals, that would represent success as I judged it.

I found it hard to imagine why anyone would want to pursue great accomplishments in order to induce other humanimals to believe that he or she was a larger-than-life character. I deemed it too much trouble to try to expand my earthly identity to a scale beyond its fundamentally diminutive dimension. I was content to remain smaller than life. As a realist, I recognized that my modest potential would limit the chances I could in the eyes of the world, even in my own tiny corner of it, become noteworthy for anything. In any case, I was indifferent to how other people viewed me. Ever private, I sought no public notice, other than the little necessary in connection with my writing. I subscribed to tattooist Ed Hardy's BST to stay low and anonymous. Before long I'll be really low—underground—and enjoy the anonymity of the obscure dead.

My general agenda lacked ambition but not activity. It included such goals as remaining somewhat productive; travels to new places; creating memorable adventures and experiences; keeping and gaining friends; nurturing personal relationships; staying connected with my local world; and satisfying as best I could some of my compulsive curiosity. Those were all ambitions enough for me. Moreover, I felt that those goals were in reach, and I pursued them with an optimistic can-do frame of mind. I simplified my attempts to fulfill my agenda by limiting my goals. I eschewed those which would have distracted me as well as ones which seemed out of reach. Not only demands or requests by other people but also my own mission-drift could interfere with my agenda. My lack of ambition also deterred me from seeking many of the accomplishments of the kind which often motivated people with higher expectations. I didn't want to waste my limited time, energy and effort attempting to achieve what some other humanimals deemed worthy of pursuit. Their values weren't mine. Absent from my agenda were fame, prestige, power, social approval, glory, acclaim, standing, influence, business success, public office, prominence, or any other status ambitious earthlings commonly covet. My agenda in fact lacked the trappings and the traps of success, which in general related to external notice and acclaim. My goals were internal and largely invisible to casual observers.

As for wealth, it interested me only to the degree that I needed resources to enable me to live in my own way. In my early years I perceived the ultimate futility of spending my life on the accumulation of assets, especially material ones. Earthlings who greatly enjoy and succeed at the process of amassing great wealth eventually face the problem of disposal—of themselves and their estates. What no doubt brought over the years treasured delights in creating assets becomes toward the end an incubus as the burden of providing for successorship looms. How to hand off the wealth presents a major challenge. It made little sense to me to devote my brief life to pursuing an accomplishment, in the way of building an estate, which in the end would prove to be a problem. For wealth and in other matters which the world applauded as worth-while, success mostly struck me as requiring unrewarding pursuits and as such they failed to motivate me. I didn't want to spend my life pursuing will-o'-the-wisp endeavors which held no importance or meaning for me. I was happy to strive to achieve only a few limited accomplishments sufficient to tint but not to taint my life with a scintilla of success as I defined it. I participated in the hustle-bustle and hurly-burly of human affairs only to the extent necessary to find for myself a small place so that I could remain connected. I wanted to retain a link to the practical and functional every-day world without being chained to it, but I had no ambition to excel.

Because over-achievement represented for me overkill, I deemed that being too productive would be counter-productive. Before undertaking an enterprise I tried to assess its difficulty and how well it fit into my agenda. The only time I failed to pursue such an evaluation was for the challenge of carrying out my lifelong experiment in being alive. I somehow lacked the opportunity to ascertain ahead of time if being thrown into existence was worth its many obvious disadvantages. I was by chance formed into an earthly form by a procreative operation which lacked my informed consent, so I was unable to select the non-being option. I'm still not sure how I'd judge the matter. My cost-benefit analysis for possible activities attempted to determine how much return on investment—of time, energy and effort—my exertions would yield. Contrary to common practice, I counter-intuitively decided that a yield too high might well bring diminishing returns. Over-achieving would both complicate my life and saddle me with that much more to lose, possibly while alive but for sure at the end. If you aspire and perspire to build an empire and by chance happen to succeed, you're doomed eventually to lose your domain one way or another. It's only a passing fancy to proceed through life with a regally fancy quickly passing journey. No matter how elaborate an earthling's life course, time's wingèd chariot carries every passenger to the desert of vast eternity.

Accretion of fortunes, worldly goods or accomplishments make their possessors a hostage to fortune. In *The House of Rothschild* Niall Ferguson quotes the comment of French author Prosper Merimée when James Rothschild, "one of the richest men in history," died in 1868: "It seems to me that it must be more disagreeable to die when one has so many millions." This is an especially meaningful comment, as it derives from someone whose first name was

"Prosper." Even as a boy I realized that the more prosperous and successful you are the more disagreeable it is to lose it all at the end. I aimed to limit my earthly accomplishments so that my disappearance would not operate to sever me from hard-earned but transient and, for me, meaningless successes. This attitude didn't arise from despair, nor did it lead to inaction. I wasn't a nihilist but simply a denialist as I denied the merits of what by common consensus among conventional circles passes for success. For me, however, the only true success was to be able to live in my own way.

I felt that if you run your life based on what others deem important you no longer control your agenda. Living according to societal values which don't represent your own value system produces an inauthentic life. I accepted some of the prevailing conventional values, but I selected only the elements that suited me. Casting my thoughts forward to a much later time, which is now, I didn't want to find myself at the end with meaningless accomplishments based on strenuous life-devouring efforts. And in any case I might be wasting my time and my life trying but failing to produce successes with nothing at all to show for my efforts. I decided not even to try. It was a poor bargain to invest time with the hope of obtaining possible but most likely improbable success. For sure you lose the time, while the success you may never win. The ultimate consideration which influenced my lack of ambition arose from the reality that when you reach the end all accomplishments of any sort are overtaken by oblivion. Even as a young man, at the time remote from a normal terminal age, I felt it made no sense to devote my brief earthly existence to attain notice and maybe acclaim only to have everything vanish with the person. To spend limited years gaining what would inevitably be lost represented a bad trade. It was quite enough to lose my life; I didn't want to lose a lot more as well.

Somewhere between certain permanent oblivion and very brief uncertain existence lies some sort of balance to make living worthwhile. My existential agenda was to find that balance and live in accordance with it, even in light of the dark reality that ultimately all is lost. Worldly experiences can as a general category be replicated, but not the specific experience of an individual life. It's an odd one-off phenomenon. Even for the few of us favored with a charmed life, human existence seemed some sort of strange fleeting shadow play of little substance, a phantom-like experience with no enduring value. The experiment in living was nothing more than a ghost dance like the Lakota Sioux tribal rituals during the 1890-91 Ghost Dance War when at the Wounded Knee Massacre the U.S. Seventh Cavalry converted 300 native braves and their women and children into ghosts. During my pre-ghost earthly span of time I didn't want to devote my years to insubstantial phantom activities which at the end would haunt me with regrets and remorse. What was viewed as success in my circles, both fatal and situational, I largely ignored. My challenge was to create another sort of life, one with some validity for me in light of my belief that my accidental earthly existence was only a brief ghost dance without an encore. My reminiscence describes my response to that challenge.

~ 3 ~

My refusal to adopt many of the values which prevailed in the provincial bourgeois society where I'd originated and remained left me dissenting not only from some contemporary practices but also from how mankind has functioned over the ages. Ancient Biblical wisdom in "Ecclesiastes" advocated forceful efforts to achieve. The argument was exactly the opposite of mine. The Biblical text urged: produce results now as post-death your chance is over. The Scripture advises: "Whatsoever thy hand findeth to do, do it with thy might; for there is no work, nor devise, nor knowledge, nor wisdom, in the grave, whither thou goest." Like most non-science based pronouncements, the Good Book's view is only an opinion, no better than the gospel according to Elvis regarding Taking Care of Business, wisdom I readily adopted, or the philosophical musings of Snoopy and the Peanuts gang, whose views I read religiously.

Plain ole Snoopy was in some ways a role model for me—a laid-back guy who didn't take things too seriously and didn't suffer from dogged ambition to achieve high accomplishments. For him "the sweetest sound of a summer day" was the crunch of a can of dog food being opened, a limited and practical pleasure in line with my own modest aspirations. But they weren't so diminished that I was reduced to eating dog food, even if some of my cooking tasted like it. I of course recognized the indisputable truth in "Ecclesiastes" that once in the grave your chances are over, but I drew different conclusions from that reality. I questioned why I should exert all my might to produce what would before long disappear. For me the opposite was true: worldly success was so impermanent it wasn't worth the effort. Why bother? Now at the end I'm glad I didn't. Losing a charmed life is quite enough without the added loss of high status and its components. My agenda was to exert myself just enough to keep connected but not so much that—like the ultra-successful—I'd lose a lot more than my life when eventually disconnected. I attempted to modulate between the now and the never. All the success-blessed undertakings of the over-achievers will be over once the undertaker takes over. After their time, the biggest names will shrink into less than small nothings. Bill Gates will before long arrive at the gates of eternity and the cold winds of mortality will soon buffet Warren Buffett. Billions of admirers and dollars count for nothing in the trillions and trillions of years out in the desert of vast eternity where nothing lives. Sorry guys—that's just the way things happened to turn out here on the lonely planet. Better luck elsewhere. Music to their ears, undertakers savor the wingèd chariot's bone rattling racket and are standing by for Gates and Buffett, for me, for you, for Everyman, Everywoman, Everydog.

    No matter how long my earthly tenure lasted, it would be quite short and oblivion far longer. The play of infinite forms would be followed by infinite formless dead time when the play for me would be over. It made no sense to waste time to enhance the by far shorter duration, my existence, when in fact eternity represented my permanent state and fate. Being alive spanned only the briefest interlude, a moment in time—a passing glimmer of life flanked by forever darkness—which interrupted my eternal non-being. Both my prenatal and posthumous absence from life lasted infinitely longer than my momentary passing earthly existence. A realistic sense of proportion led me to de-emphasize the brief interval of being which was insignificant compared to all past and future time in which I didn't exist. Given the circumstances—the fatal circle, truly fatal, which applied to all humanimals—I tried to make the best of things, an attempt to create something out of nothing, and to shape as best I could a charmed life, but without bringing to my experiment too much to lose. I realized that my demise would annul all efforts and accomplishments, as well as all memory of me. I saw no need to excel when I'd soon be extinct.

    My stoical outlook represented one of my personality's set pieces, given to my by chance as part of my fatal circle. As time went by, my observations and experiences reinforced my endowed fundamental stoicism, a trait which characterized the attitude of a kindred spirit, one of the least known outstanding figures in American history. Yes, he was a Confederate, but in him I saw a confederate with whom I was allied by temperament, even if not by accomplishment, and as a fellow humanimal from whom I could learn how to look at life. Rather forgotten these days notwithstanding his remarkable accomplishments, which I didn't view either as exemplary or at all attainable for me, Judah P. Benjamin represented a great role model, not for what he did but for his level-headed and positive temperament. I completely ignored his burning ambition and his political career and views in favor of focusing on how he looked at life—his stoicism. Impressive as they were, Benjamin's professional activities and successes were relevant for me only to illustrate how he viewed what chanced to befall him. His reaction to defeats, not his victories, served as the best object lessons.

    Benjamin was born in 1811 on St. Croix, then an English island. Just after the War of 1812 his parents took the child to Charleston where the family settled. At Yale College, which Benjamin entered at age 14, he was an accomplished student and a popular figure, but financial difficulties forced him to drop out before completing his junior year. In 1828 at age 18 Benjamin moved to New Orleans where he learned French and Spanish while studying law and working in clerical jobs. After being admitted to the Louisiana bar at age 21 he enjoyed a distinguished law career, specializing in appeals, and in 1842 was elected to the Louisiana House of Representatives. In 1849 he became involved in railroad enterprises, including construction of an inter-oceanic line across the Isthmus of Tehuantepec in Mexico. After his 1852 election to the United States Senate, Benjamin was offered and declined an appointment to the Supreme Court and the post of Minister to Spain. Not long before the Civil War he was reelected to the Senate where his colleagues viewed Benjamin as the best debater and "the ablest and best equipped Senator" in the last 40 years. But Benjamin chose to abandon that august body, and on New Year's Eve he delivered a farewell speech to his fellow Senators before resigning to join the new Confederate government: "And now, senators, within a few weeks we part to meet as Senators in one common council chamber of the nation no more forever. We desire, we beseech you, let this parting be in peace...[I]ndulge in no vain delusion that duty of conscience, interest or honor, imposes upon you the necessity of invading our States or shedding the blood of our people. You have not possible justification for it." As events would show, in this case Benjamin's renowned powers of persuasion failed.

Joining the Confederate government in early 1861, Benjamin was appointed Attorney General and later Secretary of War, then in early 1862 Secretary of State, a position he held until the Confederacy's defeat. After Lee's surrender at Appomatox in April 1865, Benjamin fled under disguise to Florida from where he proceeded to England, starting over there at age 54 working as a law clerk. Five months later he was admitted to the English bar, in 1872 was appointed a Queen's Council, and by 1874 Benjamin had established the largest appeals practice at the English bar. In 1883 at age 73 he retired from the practice of law to live in the mansion he built at 41 Avenue d'Iéna in Paris. "I long for repose. I seek rest and quiet after the exhausting labours of 68 years of a somewhat turbulent or rather adventurous life," he said. Benjamin died there at his Paris home on May 6, 1884, his turbulent adventures at an end.

Benjamin's life, career and accomplishments were outstanding. He somehow possessed that mysterious and elusive quality which enabled him to mold men's minds just as he wished to shape them. Articulate, persuasive, charismatic, Benjamin dominated every activity and enterprise in which he played a role. Endowed by his fatal circle, by his native ability, and enhanced by his own efforts and the operation of luck, Benjamin managed to live a charmed life, even if one filled with much turmoil and many setbacks. The accomplishments so many seek and so few find endowed his life with success: admiration, respect, prestige, power, influence, fame, wealth, standing. Although the Confederate rebellion proved to be a "lost cause," Benjamin recovered from the defeat to find a sterling career in a foreign land. Benjamin's accomplishments I greatly admired, but in no way did I aspire to emulate them on a much smaller scale, nor was I at all capable of doing so. His "rather adventurous life" seemed to me too adventurous in ways which didn't interest me. What appealed to me was Benjamin's exemplary temperament. I found compelling how he responded to what luck, fate and circumstances happened to bring him. In January 1880, toward the end of his life, Benjamin wrote the wife of E.A. Bradford, his New Orleans law partner years earlier:

> You say I look on things through the *couleur de rose* of success. Not so, but through the *couleur de rose* of a temperament, blessed birth gift, that under all circumstances and every trial, took to the bright side and absolutely rebels against distrust of the future. I was complimented on my "pluck"...it was simply elasticity of natural temperament: a total absence of...despondence and brooding over adverse circumstances.*

---

*Quoted in *Judah P. Benjamin: The Jewish Confederate* by Eli N. Evans. This biography gives a fascinating full account of the life of a truly remarkable man.

Pluck plus luck, an inherited plasticity of temperament, an immunity to worry over the misfortunes of adversity and other random elements which happened to define Benjamin's character served to make him one of the great winners in the lottery which is the chance-fated experiment in living.

Benjamin's temperament represented for me the ideal attitude, one I aspired to emulate even if not altogether successfully. His elasticity was far too flexible and his equanimity was too serene for imitators to match. Benjamin's outlook reflected C.P. Snow's comments in *Corridors of Power* regarding "a man of action's optimism. The optimism which makes a gulf between men of action and purely reflective men." Benjamin's type of temperament exemplified one of the keys to living a charmed life, which is to view the experiment with a positive attitude. I felt that I wouldn't get very far, either with my travels or my winding path through life, focusing only on the risks and brooding about the plots and counter-plots which might arise. Although I hoped to duplicate Benjamin's serenity in the face of adverse circumstances as best I could, I was more wary of the future than he was. I feared that in un-elapsed time lurked fateful events which would upset my agenda. A temperament like Benjamin's, ever optimistic and trusting the future, is in fact rather rare and—with my somewhat skeptical and "Doubting Thomas" attitude—was for me unachievable. Nonetheless, I tried my best to follow his example.

The ancient stoics, precursors of a Benjamin-type attitude, advocated calm acceptance of misfortune. But did the stoics actually live that way, as did Benjamin, or just philosophize about it? As a practicioner, Benjamin offered an exemplary real-life illustration of the principle that what happens to befall you, much of which lies beyond your control, is less meaningful than how you respond. It's not difficult to understand the age-old wisdom of the ancient sages or to admire Benjamin's beliefs and behavior. The challenge is to apply the doctrine in your own life to cope with random mishaps. But when misfortune intrudes, it's difficult to react to the bad luck in a stoical way. Such a devil-may-care response no doubt comes easier to someone who, like Benjamin, managed to attain the rose colored tincture of success. This rosy outlook with its "total absence of...despondence and brooding" represents the beau ideal which few other humanimals can match.

Less ideal for my own purposes was Benjamin's career, which represented ambition far beyond my interest or capabilities. In that my aspirations differed completely by type and degree from those of Benjamin, I arranged my agenda in a way quite contrary to how he operated. In contrast to the turbulence of his adventurous combative encounter with life—sedulous with legal, public, political activities typified by contentious, argumentative, bellicose conflicts—I restricted my endeavors to personal and private pursuits. It baffled me why anyone would seek power as a politician, fat cat businessman, or otherwise. Those were the last goals I'd covet. I was content to remain a thin cat, if not an underdog, and I had absolutely no desire to exert any dominion over others. It was all I could do to control my own life in order to live it in my own way. I held no doctrines or beliefs I sought to impose on other humanimals. There were already enough characters and wise guys (and all too few wise men) telling earthlings what to do. Politicians, bureaucrats, advisers, social scientists, consultants, preachers, Holy Scriptures of one sect or another and much else offered all manner of advice, much of it contradictory. In any case, I really couldn't say what was best for my fellow earth dwellers, or which policies should be advocated to improve society at large. All that was far beyond my interest and my competency.

Dealing with the challenges of my little earthly experiment in living presented me with quite enough to occupy all my time, energy and attention without complicating my life trying to solve problems relating to the outside world. In any event, I had no solutions for mankind's state or fate, and only a few tentative experiments to apply to my own circumstances and agenda. Only Benjamin's attitude, not his achievements, appealed to me. Unlike his turbulent life, my own seemed rather sedate, even with my adventure travel. While he was ambitious and daring, I remained satisfied with my circumstances as they were and cautiously tried to preserve them. I attempted to eliminate a frequent need for stoicism by avoiding pursuits likely to require such a reaction. Risking passing excitement which might lead to permanent regret didn't tempt me. Turmoil and thrills of the kind Benjamin experienced in his hyperactive public life contrasted with how I preferred to remain in the shadows, a private person—a ghost writer in the making—whose risks related mainly to those connected with my far-flung travels. My adventures in foreign lands gave me quite enough excitement. I shunned controversy, conflict, arguments, public life, high-wire high tension pursuits, politics, Mexican railroad promotions, civil or uncivil wars, court appeals, and other unappealing activities of the kind which occupied Benjamin. Contentious and abrasive situations would divert me from my agenda. Why argue? I endorsed H.L. Mencken's attitude of responding to critical or provocative statements with the rubber-stamped legend, "You may be right." Disputing the views of others—"You may be wrong"— was too much trouble. I had enough trouble reconciling my own conflicting viewpoints so I could act. To Mencken's attitude, a sentence from *The Dhammapada* added a more existential dimension: "The world does not know that we must all come to an end here; but those who know it, their quarrels cease at once."

My attempted avoidance of situations likely to lead to conflict matched my renunciation of possessions. With only a limited amount of time and energy, both on a day-to-day and on an existential basis, I preferred to concentrate on experiences and also on relationships which promised to be friction-free and, if they kept their promise, rewarding. Not "Ecclesiastes" and other Biblical passages but Buddhist teachings in *The Dhammmapada* and elsewhere represented the scripts and Scriptures I adopted. Buddhism featured the merits of a quarrel-free existence and an escape from *tanha*—the thirst to satisfy one's ego. All around me swirled the energetic quests of humanimals seeking fame, fortune, power, possessions, prestige and all the rest of what society valued. I stood aside to focus on my efforts to make my way through the Dantesque woods without getting lost in the thicket. My narrowed path both limited my life and liberated it. Benjamin's career put him in harm's way; my path tried to reduce the need for a stoical response to unfortunate happenings. Like everyone else, I remained a hostage to fortune, but I hoped to reduce the ransom payable by limiting my vulnerability. The ultimate vulnerability, mortality, no one could escape, but pending that fate I tried to shape the means to that end. I ignored consumption and ambition as irrelevant to my agenda, preferring instead experiences and relationships.

Meeting the individuals who will suitably populate your circle represents one of the life experiment's most important challenges. Expanding the cast of characters in your own little play of finite forms involves a lot of trial and error, hopefully with a minimum of trials and tribulations and a maximum of satisfying long-term relationships. Although associating with other people offers great potential, such human connections also come with the hazards of complicating your life. Discerning the benefits and the burdens presents a challenge. As things worked out for me, for the most part my family, friends, acquaintances, colleagues, associates and casual connections served to enrich my life experience. But like everything, those relationships were time-limited, as by now all too many of them have expired. The departed left me behind with only completed final memories to evidence what was once a living, on-going and ever refreshed relationship.

Out of the billions of humanimals who represent the totality of those you might befriend you're destined to meet only a few. A few you inherit from your family, others from classmates, some from the social circle included in your fatal circle, and some by similar proximity or connections. Those characters are in a certain way preordained, as they derive from the setting you happen to inhabit. They include such types as your Sister Carrie and Uncle Sam and Auntie Mame; your school buddies, the Hardy Boys (Laurel Hardy flunked third grade, his brother skipped it; one of the boys was study-hardy, the one who failed, frail); the Andrew Sisters (one with bangs, the other with braids, the third home-schooled so hairdo unknown to me);

two doors down, the Bobbsey Twins (I had a crush on one of them, but which was which? I could never be sure); Snoopy the dog across the street; up the street near the corner, those two rascals—Tom Finn with freckles and Huck Sawyer with pimples (or was Finn pimpled and Sawyer freckled?). Looming over the whole scene were Father Time and Mother Nature, ever present to spawn the cast of characters and the entire play of infinite forms. All those people by chance present in your natural habitat existed there in place (yours) and time (yours) as if they somehow belonged there and then. You didn't choose to be there at the same time and place as all the rest of them. It just happened that way. For openers you take what you get. Perhaps a few of those set-piece contacts will ripen into a friendship, but most of them will wither and disappear from your life. The people who surrounded you in the early days and who once seemed such familiar fixtures eventually fade into phantoms, long ago presences now ghostly absences like a campfire's smoke which has vanished into the air. Where are they now, those boyhood humanimals who a lifetime ago represented what appeared to be a fixed and enduring cast of characters? They did not endure.

As you get older, your relationships originate less by inheritance and more by your own "line-extention"-type efforts. Such widening hopefully connects you with new people to form a network which operates to hand you on, moving you along to yet other relationships. Apart from referrals, you can also seek people out on your own initiative, a process which requires you to determine who you want to meet and then how to gain access to them. Yet another way which determines how you expand your relationships depends on those ever influential intangible factors which ultimately rule our lives—luck, chance, fate, destiny, happenstance, coincidence, serendipity. In a way, meeting someone as a result of those elements represents the most satisfying method because capricious and unexpected encounters which lead to enduring relationships are especially delightful. Some of my best friends I happened to meet merely by chance. Both of us were somehow fated to be in the same place at the same time, an overlap which automatically evidences commonality and, especially in remote corners of the world, potential. Unplanned affinity of time and place can create possible opportunity. Lady Luck has suddenly stepped in to make an introduction. Admittedly, such an unexpected encounter usually leads nowhere, but every so often a fortuitous meeting happens to develop into a lasting friendship.

The elusive role of chance in meeting people prompted me to ponder the opposite case—the hundreds or thousands of potential candidates who you might have befriended but to whom the Lady failed to introduce you. Who are they, all those unlucky souls who never got to meet me? Other than that Tómas Weil fellow in Buenos Aires I didn't phone, I'll never know. All the possible acquaintances I might have met disappeared before they appeared in my life. They're ghosts who, for me, were never alive—phantoms of nothing. Those spectral figures who never materialized are like scientist Dr. E.E. Denziger, erased from time at the ending of Jack Finney's novel *Time and Again*. The story reverts to the past to retell the tale, a Borges-like refutation of time, by revising what had the first time around occurred to enable Denziger to exist. In the new scenario, his parents never meet, their chance encounter failing by a few seconds and thus eliminating from future existence their child, E.E. Existence of a years-long relationship can depend on a fateful encounter occurring within just a few seconds. Out there somewhere exist thousands of kindred souls you never happen to meet, potential friends who might have greatly enriched your life, and hopefully you theirs, who forever remain unknown to you, lost in the limbo of unrealized potential.

Possibles which never mature into reality of course applies to everything in life. Henry James said he wanted to be a person on whom nothing was wasted, but in truth almost everything which could be meaningful to you is wasted. Potential vastly exceeds the actual, but even the residue of the possible—that which actually assumes existence—represents far too much for earthlings to experience or comprehend. No one can even begin to absorb it all. The play of infinite forms offers too many sensory impressions for any one life, and many of the few forms you happen to experience go to waste. Of all the multitude of possible friendships, of all the specifics of your experiment in living, of the huge number of events, happenings, situations, spectacles, impressions you might experience, only an infinitesimally tiny few ever materialize, and much waste occurs with the ones which actually happen. It's all too much—sensory overload results in great waste. No matter how curious you may be, how active, how much of a self-starter and a lucky finisher, how adventurous, how ardently you seek experiences, how much you try to absorb the world and attempt to understand it, no matter how resourceful, diligent and intensely observant you are, how avid for information, knowledge and even a little wisdom, you're doomed to failure. So multifarious is the vast cosmic spectacle and so countless the possibilities, one finite consciousness possesses no chance to be someone on whom nothing is wasted. A humanimal is essentially a creature for which virtually everything is wasted.

Given the constraints, I did the best I could to salvage from the infinite possibilities and the disparate actualities a few which would serve to help me achieve something of a charmed life. One of its most essential components was the range and type of relations I managed to establish. Among the best and most meaningful examples were four in New York City, that metropolis teeming with infinite forms and possibilities. In two cases only happenstance and in the other two necessity, not any intrinsic interest in New York, determined that the Big Apple would prove fruitful as the setting for four of my most delightful and memorable friendships. The two people I sought out were accomplished figures in two contrasting professions, investing and publishing, while the other two happened to be family friends. All four of them belonged to the generation senior to mine, and by now they're all gone.

Although New York City represented an unlikely place for a provincial like me to visit so frequently, that's where the quartet happened to live and so I went to Manhattan a few times a year even though it wasn't my favorite destination. My personality was unsuitable and my regular suburban life back home too unsophisticated for me to adapt to the Big City. Whenever I found myself in New York I felt lost. The provincial in me surfaced and I became disoriented. For me the City seemed in some ways like a foreign country—even if in my homeland—as it differed so greatly from my own simple and relatively uncomplicated natural habitat out in the Midwest. I was more small time than Big Time, a seedling rather than a Big Apple type.

The place was too convoluted and too busy for me. I didn't particularly enjoy the metropolis, as the City seemed antithetical to my slow-paced Midwestern suburban way of life. I wouldn't go so far to say, as did Henry Miller in *Tropic of Capricorn*, that New York City was "a piece of the highest insanity, everything about it, servers, elevated lines, slot-machines, newspapers, telephones, cops, doorknobs, flophouses, screens, toilet paper, everything. Everything could just as well not be and not only nothing lost but a whole universe gained." Surely Miller is wrong regarding the toilet paper, for I can speak from experience that it's quite as useful in New York City as anywhere else. I mean, in a country like India toilet paper can be scarce (and it's not soft but abrasive), so you appreciate it that much more when the item is readily available in New York. I certainly respected but did not envy my friends who lived in that demanding metropolis. Although I enjoyed some of the restaurants and museums, the theater, book stores and other attractions, for me the distractions and disadvantages outweighed the benefits. Although I tried to keep my life in balance in many ways, being in the center of the action represented an uncomfortable equilibrium. I preferred to remain off to the side, away from the Grand Central complex which seemed truly a terminal dead-end for my own purposes. In New York people seemed to confuse motion with action. There was a whole lot of shakin', non-Presley kind, but I questioned how much productive output the City generated with the many inputs utilized. From what I saw, much of the highly charged live-wire nervous energy there was really just insensate kinetic motion with a lot of short-circuits and burnout. And yet even my pace and pulse rate accelerated during my visits.

No question about it, the place was stimulating, and so were my four metropolitan friends who over the years so enriched my life. Unknown outside of their own circles—family, friends, professional—each of the four represents an example of the millions of capable and productive unsung individuals who remain obscure to the world at large but meaningful to those who know them. I had the good fortune to meet and befriend the four New Yorkers, who—in the rest of this Section—I'll describe to try to show their personalities and attitudes, their generosity to me, how they enhanced and influenced my life, and how they exemplify the best qualities of capable and effective admirable earthlings making their way through the dark woods. Each of the four symbolized for me an ideal relationship of the kind which is essential for anyone who wants to live a charmed life.

~ 4 ~

There's a lot to be said for living an adventurous life. That's one of the reasons I decided in my thirties to augment my sedate and settled Midwestern suburban existence with a large quantum of adventure travel. New York wasn't exactly what I had in mind but, as fate would have it, I nonetheless often traveled to that urban jumble and jungle even though I preferred real jungles and deserts and other natural landscapes. I went there mainly to see my friends, two of whom were at the beginning only professional contacts I'd met on my own. Although New York is a great city, toilet paper and all (with apologies to Henry Miller), it's a place incompatible with my personality, interests and preferred way of life. The City is a lively and exciting metropolis, but mostly in ways I didn't enjoy. For me, coming from my sleepy suburban subdivision, the BIG City was far too crowded, too noisy, too fast, too raw, too urban, too competitive, too sophisticated, too tough, too claustrophobic, the streets too lit and many too bright like a Great White Way Broadway of a place, a too big Big Apple with too few apple or other trees, and all too many other "too's" to note. Unlike many provincials who longed to escape the suburban for the urban, to leave behind laid-back villages for keyed-up metro areas, or to abandon small towns in favor of the big time, I preferred to remain rooted in the small-time Midwestern setting where I'd originated, and where I lived contentedly, occasionally upsetting my quiet existence to see the world beyond but always returning home. For sure I didn't have a "bridge and tunnel" mentality which attracted me to those connectors giving Manhattan access to exiles stranded in the outlying areas around the island. Those routes into the City were not the kind of connections I looked for there. Never did I desire to live in New York, a curiosity of a place I visited but didn't want to befriend. For me the Big City served only as a utility, like the telephone: a system to access the networks and then, when finished, you hung up. I was far too unambitious to survive on the fast track and compete in the rat race. I preferred the more leisurely mouse race back home.

Forced to choose between living in New York City or somewhere in the provinces, I'd opt for a smallish Midwestern or Southern college town, a place on a human scale with some intellectual activities and stimulation combined with youthful exuberance and energy brought by the presence of the university and its students. There in mid-continent, far from the coastal extremes, I'd contentedly live out my time as a librarian or as a prof teaching something easy and of little consequence like writing, or as the town banker, editor of the weekly newspaper (less work than a semi-weekly) or, most desirable, as owner and operator of the local Dairy Queen franchise. By contrast, no position at all in New York City could have induced me to live there. I realize this represents a great lack of ambition and a failure of imagination on my part, but I never claimed to be ambitious and my imagination I used for purposes other than dreaming about the Manhattan version of success.

Only once did I ever venture to stay in New York City more than just in passing, and that episode didn't originate with me. The two month interlude served to satisfy almost all my curiosity about the place. But perhaps I'm too easily satisfied. Although I enjoyed the extended visit, my positive attitude was in part facilitated by the fact that my stay was limited, and so from the beginning the end was in sight. At Edith's request for professional reasons on her part, in March and April 1987 I rented a small apartment at the Park Vendome, a genteel vintage building—"pre-war," the Manhattanites called such older places—on 57th Street a few blocks west of Carnegie Hall. A pre-war model myself, I found the old place suitable for our stay. As is typical in New York buildings, a low ceiling and small rooms made the apartment seem closed-in and confined, a contrast with the Midwestern space and openness I was accustomed to. In fact, I found the entire city claustrophobic—the skyscrapers which dwarfed the humanimals swarming through the streets, the crowds which jostled along the sidewalks, the massive built (over-built) environment and, most of all, the warren of subway and train tunnels roaring with hurtling rush-hour cells jammed with passengers. I'd be underground quite long enough after my time, so I didn't much enjoy going below before my temporary option to stay over the underworld expired. The clattering subterranean trains made me long for my rush-less quiet suburban life back home where I could remain above ground and take time to smell the flowers.

New York seemed to me like some sort of artificial giant Lego construction, a bewildering artifact made from thousands upon thousands of components which all somehow fit together to form a vast unnatural human construct. The laundry room in the Park Vendome basement, subterranean like the subway but less noisy and oppressive, offered a retreat, even if a confined and gloomy one. I was usually alone down there, just me and the clothes to be washed. As the machines whirled and whirred away, spinning and agitated but less unnerving than the city above, I studied the fittings, fixtures and furnishings in the laundry room. Jammed into the cramped space was a chaotic maze of crisscrossed pipes, tubes, wires, lines, connectors, along with valves and vents, metal boxes, various switches and knobs and handles and other assorted mechanisms, equipment and gadgets, many of mysterious purpose and the ensemble of which evoked the complexity of the metropolis and the tremendously intricate infrastructure required to sustain the lives of the millions of humanimals clustered together on the crowded island. Only X-ray vision could reveal the convoluted underlying systems which enabled the City to function. But they were child's play compared to the hidden inner workings of what lies beneath the skin of humanimals—a nature-designed organic network vastly more complicated than anything man has created. The chaotic web of lines and connectors and fixtures there in the Park Vendome basement represented a mechanical analogue to the innards of the humanimals who prowled the urban jungle above, as described by Nathanael West in *Miss Lonelyhearts*: "Under the skin of man is a wondrous jungle where veins like lush tropical growths hang along over-ripe organs, and weed-like entrails writhe on squirming tangles of red and yellow. In this jungle, flitting from rock-gray lungs to golden intestines, from liver to lights and back to liver again, lives a bird called the soul."

The body's internal twists and tangles and the laundry room's convoluted systems were brothers under the skin—one natural, the other an artifact—except that New York City didn't have a soul. Whether or not a soul inhabited the body, most likely a soul-less container, I couldn't tell, nor was the city truly a jungle but only a man-made version of the real thing. But in its own way New York seemed to me as forbidding and as threatening as did a natural jungle, the City's streets teeming with life and with plots and counter-plots galore like the steaming vegetal tangles I'd seen in Guatemala and elsewhere. Just as in the City, out in the jungle swarmed all manner of life, mindless insensate beings driven by instinct struggling to survive. In the tangled web lived and swarmed insects, bugs, beasts, snakes and grew vines, trees, plants, flowers armed variously with thorns, burrs, stabbers, fangs, claws, stings, bites, venom, poison and other weapons. New York wasn't like that on the surface, but I sensed a certain suppressed menace in the place, something I could feel but not see. I was too much of a provincial to understand the inner workings of the City or to feel completely comfortable there.

Section III: Charmed Lives   131

I was glad to be only a bird of passage in New York City. Irredeemably a Midwesterner, I clung to my roots, and they to me. Like the natural features of a jungle, New York seemed overgrown and, for a provincial like me, bewildering and daunting. Over time I began to see the City as a remarkable example of the humanimal's efforts to keep nature at bay by creating artifacts—buildings, culture, society, civilization—to counter the inhuman cosmic forces which prevailed on earth as in the heavens and everywhere else. The City's convoluted infrastructure and complexity seemed to me highly vulnerable and subject to complete collapse someday. Unlike small town America, the Big City depended for its survival on big systems, a wonderland of connections and facilitators, as evidenced in a small way by the intricate pipes and lines and boxes in the Park Vendome laundry room. Just as for a human body, easily disrupted physical networks had to function seamlessly to allow the millions to continue their functioning First World standard of living. They expected a lot, those demanding New Yorkers. A confusing chaos of machines, equipment, logistics, transportation, communication, computer banks and investment banks, maintenance workers, technicians, high frequency trading and high frequency dating and much else had to mesh to ensure that a crowded fast-paced city could continue to operate effectively. New Yorkers demanded speedy and efficient responses to demands. Everything must function all the time without delay. All American were pretty much like that, but it seemed to me more complicated to respond to needs and wants in a jam-packed complex metropolis like the Big City. Out in the Third World, the locals were happy if anything worked just some of the time. Even low expectations were often unmet. A developed society and economy, which to operate properly requires an intricate and delicate support system, wouldn't tolerate the typical dysfunctional Third World conditions I encountered the world over.

The efficiency and functionality of a place like New York depends on a highly unlikely and ever vulnerable combination of factors, an interface which balances on a knife edge above the abyss. It's really not far-fetched to imagine the whole thing crashing down, all of New York City collapsing into rubble. Perhaps a plane would crash into a Manhattan skyscraper, the start of the demolition process. As Richard P. Brickner imaginatively wrote in *My Second Twenty Years: An Unexpected Life*: "Nothing, no one, can stop a plane from crashing into the Empire State Building." The book was published in 1976.

I didn't remain completely immune to the fascinations and charms of the Big City, but for me its primary attraction related to my friends there and to a few professional contacts who soon became friends. Little else motivated me to visit, and now that my social and professional circle there has shrunk I seldom go to the City. My New York nexus, which included the four especially meaningful relationships described in this Section, serves to illustrate how fortunate I was in finding such friends and how they contributed to my attempts to live something of a charmed life. Each of the four in his own way in some respects also enjoyed a charmed life. I'll now try to explain why each of them is noteworthy and what they meant to me.

William K. Jacobs, Jr., who I sought out, approaching him with no introduction, ran a New York Stock Exchange-listed investment company whose record I admired. I by chance discovered the firm in the course of my reading and research on business, financial and stock market matters. Most of my reading was more general, to satisfy my too wide-ranging curiosity. My interests were less limited than was my time. A reading machine, I read anything and everything from ticket stubs to entire encyclopedia volumes. A good—very good—part of what I read consisted of delightfully useless information. My mind was more a hotel than a residence. The words checked in, stayed briefly, then checked out never to return. All sorts of writings passed through my brain but, alas, all too few settled there. I read a nearly ceaseless flow of books, magazines, newspapers, journals, reports, articles, product labels, candy bar wrappers, fine print and fine printing, but relatively little did I retain. A scattering of words and thoughts stuck but more just transited through. My mind was a way-station, like a small town train depot, rather than a big city Grand Central terminal stop. All that I read which I fail to remember could fit into hundreds of volumes, and in fact does. The part of my Taking Care of Business reading which focused not on the business of life in general but on business matters happened to lead me to Bill Jacobs. This proved in various ways one of the most enriching results from my compulsive habit of reading.

By way of contrast, to the best of my knowledge Bill never cracked a book. Bill didn't depend on reading to enhance his wealth. Among other things, he showed me that you could largely ignore reading altogether and still find great financial success. This was a very informative lesson for me, one I didn't emulate but which I certainly found surprising. His disinterest in printed resources and an almost exclusive focus on other less available sources seemed to me quite an unusual way of dealing with the swirl of information, opinions, facts, fictions, rumors, deceits, plots and counter-plots which run through Wall Street. Pursuing investing in his own way, for Bill "print" denoted something you hang on a wall with other art works, of which he owned plenty. Bill profited richly through his investing without reading anything except financial statements, stock reports and similar market-related materials. In Bill's book, books meant those with balance sheets, debits and credits, profit and loss and cash flow statements and other figures. The only thing I actually saw Bill read was lunch menus, and usually not even those because almost every day he ate at the Cafe Pierre in the ultra-upscale Pierre Hotel near his office where he ordered the same meal. Bill made his way through the Wall Street jungle not by his R.Q., his Reading Quotient, but with the help of his wide and well informed range of contacts, a network less visible but quite as functional as the tangle of lines, wires and pipes in the Park Vendome basement laundry room. He relied not on inside but on insight information. His large network greatly enlarged his net worth. Not print sources but personal Wall Street resources—contacts, connections, gossip, with much of the news and views collected by phone conversations—endowed Bill's investment decisions with highly informed judgments. He wasn't illiterate but a-literate. Print made no imprint on him. On Bill's desk rested in peace copies of the day's *New York Times* and *Wall Street Journal*, but near as I could tell the newspapers remained untouched, still in their hot-off-the-press pristine neatness. Although Bill left the unruffled papers in mint condition, he nonetheless managed to mint substantial profits in the market. By the time information appeared in the press, he'd already received and acted on the news, which for him were olds. Thanks to Bill, I came to realize that much of what appears in mass circulation publications represents lagging indicators.

By appearance and his aura of authoritative sophistication, Bill always presented an imposing presence. With his shock of white hair and aquiline or even leonine features he looked like he was indeed a kind of king of the Wall Street jungle, his throne in a palatial Upper East Side office where he sat at the nexus of information flows from many sources. Brooklyn boy by origin, Manhattan man by formation, colorful by character, shrewd by native ability, Bill was a widely known and highly respected Wall Street personality who, in turn, knew almost everyone in the trading trade. Unlike his physician father, Bill never pursued a formal education, which ended for him when he graduated from Erasmus Hall, a renowned Brooklyn public high school which produced such impressive figures as those in Bill's bank account as well as the even more impressive figures of a Miss Universe and none other than the contours of Mae West, more well rounded than suburbanites like me from the provincial Middle West flatlands. Other accomplished Erasmus Hall alumni included performers Beverly Sills, Barbara Stanwyck, Shirley Booth and Eli Wallach, world chess champ Bobby Fischer, Joseph R. Barbera, creator of the "Tom and Jerry" cartoon (who wisely put Tom ahead of Jerry), writers Mickey Spillane, Bernard Malamud and Arthur Frommer, medicine Nobel Prize winners Dr. Barbara McClintock and Dr. Eric Kandell, a governor of Georgia, a U.S. Attorney General, and the estimable New York City business executive Murray M. Stern. Adding further glory to the school are its founders, Alexander Hamilton, Aaron Burr and John Jay, who in 1786 established Erasmus Hall Academy.

Brooklyn was for me a land even more foreign than Manhattan. Apart from landing at

Red Hook pier in Brooklyn when arriving there on the *Queen Mary 2* from Southampton, only twice in my life have I been to the borough. Once I walked across the Brooklyn Bridge to enjoy the spectacular views, including the spine-tingling angle of vision looking down to the waters far below, a dizzying perspective which in a small way momentarily reminded me of how New York City's whirligig of activity was for me—a provincial far from his native habitat—also somewhat vertiginous. Years before when I was a boy my father took me to see the Brooklyn Dodgers play at Ebbets Field. I was curious about the team's fanciful name, so different from what the other major league clubs were called—for colors, such as Red and White and Brown, or creatures like Tigers, Cubs, Cardinals, Braves, Indians, Pirates, Yankees. I later learned that the team—once also known as the Bridegrooms, after seven players had gotten married in 1888—became the Brooklyn Trolley-Dodgers when electric streetcars replaced horse-drawn vehicles. Local residents soon realized that the new-fangled technology required them to take care to dodge the fast-moving cars, so pedestrians became trolley dodgers. As described in Section VII, along a grassy median in front of the big white house where I lived as a boy ran a streetcar, its route so fixed and predictable never did I have to dodge the trolley. After I grew up I realized that other means of travel, more flexible and random than fixed-rail trolley routes, represented more desirable ways of getting about.

Although Bill Jacobs entered that scholastic orbit at Brooklyn's Erasmus Hall which produced a galaxy of stars, he didn't shine later in life through academic accomplishments but by street smarts, mainly Wall Street smarts. The alchemy of real world hands-on experience transmuted him into star material. He started on the hard streets of Wall Street as a lowly "runner," an errand boy, for the long defunct H. Hentz and Company. By the time I met Bill in the 1960s he was CEO of a publicly-held investment company and also served as a private trustee managing money for wealthy families, among them descendants of the Sears Roebuck founders. An autodidact, Bill taught himself the ropes which he used to good effect to climb the Wall Street wall to great heights. Some aspirants start off as a front runner and never even approach the finish line. Overcoming his handicap of lacking a higher education, Bill ran a great race and proved to be a winner. Perhaps that lack of academic training brought him some advantages in the rough and tumble markets in which he operated.

As with everyone you meet by your own initiative, at the very beginning you're complete strangers. Little did I realize or even strategize how my business connection with Bill would over time develop into a gratifying and enduring relationship. Seeking a professional contact, I by chance ended up with a great friend. At the outset I introduced myself to Bill by letter, explaining that I'd come across in the course of my reading information about Abacus Fund, the stock exchange-listed investment company he ran, and wanted to learn more about the firm. Years later when Bill arranged for me to meet his friend Warren Buffett at Berkshire Hathaway's Omaha office, Warren told me that Bill had chosen the name "Abacus" so it would appear at the top of the listings for such investment firms and also to symbolize the promise "You can count on us." Those added up to two good reasons for the name, which Bill evoked by the small gold abacus cuff links he often wore. He displayed them to me as proudly as Guatemalan President Ydíoras Fuentes had years before shown me the silver cuff links received as a present from President Eisenhower.

My encounter with Buffett as facilitated by Bill and my few later interludes with Warren emphasized to me the value and importance of referrals. References from existing reputable contacts offer excellent network effects by giving you immediate credibility. But only with explicit permission did I ever use someone I knew as a door-opener. Cold calls, as I initiated to meet Bill, and warm calls might serve to get you access, but far better are hot calls based on a reference from a known and trusted figure. Such was the case for my meeting with Buffett. Back then, about 1980, a notice in the reception area of the Berkshire Hathaway office in Omaha stated: "On Wall Street there are a lot of bulls, a lot of bears and a helluva lot of bum steers." In Buffett I found a fellow New York-wary Midwesterner whose temperament

I immediately understood. By introducing me to Warren Bill gave me a good steer for which I was very grateful. He performed the same sort of introductions for me in other cases, so sharing his contacts with me. I'd advanced a long way since writing him out of the blue to ask Mr. Jacobs, as I then knew him, to send me some of Abacus's recent reports. In that opening letter I'd also requested a meeting with him on my forthcoming trip to New York City. The cost of the postage stamp on the envelope sent to Bill turned out to be a terrific investment.

You really never know how things will work out when you approach a stranger with a request. Very often the result of a cold call is to get a cold shoulder, but as it happened Bill and I took a liking to one another and we soon developed a warm friendship. Bill was quite responsive to my letter and said he looked forward to meeting with me. Even or perhaps especially, important figures appreciate attention, although they don't always reciprocate. I appreciated his prompt and cordial response to my inquiry. Viewed from his sophisticated natural habitat in the concrete canyons of New York I might well have seemed to Bill a provincial hayseed from somewhere out in the distant hinterlands attempting to brave the Wall Street jungle. If that was how the accomplished Big City operative viewed me he wasn't entirely wrong. Although not completely a babe in the Dantesque woods, at the time I was hardly a seasoned investor. I knew I needed help. I always tried to assess my limitations in a realistic way. That kept me busy, as I had so many. My interest in meeting Bill represented an effort to limit my limitations in the world of investing. My basic problem was that I occupied a remote position at the far end of the investing food chain. I was out of the loop, with no access to information beyond that which everyone else also learned in the press or from other widely followed sources. I needed a mentor, someone I could turn to not for specific advice but for how to think about investing. In Bill I found such a person, with the added advantage of networking possibilities and then, before long, the additional unexpected benefit of a great friendship.

Later, however, after I'd gained some experience and learned more about the financial industry, I realized that, although I'd never engaged in short sales of shares, I'd sold myself short. I eventually understood that I knew as much about how to invest—which was largely guesswork based mainly on widely available information—as did all but a very few highly astute market professionals, such as Bill. I made it my business to get to know some of those well-informed exceptional investors. As for all the rest of the vast mass of advisers, wealth managers, portfolio shufflers, stock jockeys, financial analysts, fund managers and similar characters, they offered opinions aplenty—buy this, sell that—aimed to activate trading and to justify charges, but I took no notice of such market pundits or commentators. Taking Care of Business in the financial realm led me to ignore all the noise and distractions which agitated Wall Street's ceaseless confusion of confusions.

As a pilgrim from the hinterlands to the Big City, I was quite impressed when I arrived at Bill's spacious upscale uptown Madison Avenue office, far more elegant than Buffett's modest provincial premises in Omaha, which occupied two floors connected by an internal spiral stairway. Of course, the Abacus stockholders were paying for the expensive space— Bill could count on them for that—but with him running the company its owners got value received. And if they didn't, Bill did. There in the plush surroundings at Abacus meeting the distinguished William K. Jacobs, Jr., I'd somehow managed to get my foot into the door of a sophisticated investing inner sanctum, one more exclusive than I'd known before. I wondered if I could stay there or if the door would soon slam shut on me. After one quick session Bill could lock me out. Bill's rather exotic pronounced Brooklyn accent accentuated my feeling of having entered a new world. I was hoping I could "count on Abacus" and its operator to help me navigate Wall Street, a short straight street but challenging to travel. With a market operator like Bill to guide me, a real pro would lead me along the Street. His impressive spiral stairway suggested that he could deal well with the ups and downs and convolutions of investing.

With his cordial and welcome reception, Bill duplicated the courteous treatment both

President Miguel Ydígoras Fuentes in Guatemala and Jorge Luis Borges in Buenos Aires had offered me. Those cold, or at least tepid, calls had somehow produced warm responses. This time, there in Bill's New York City office, my mission was far removed from political and literary encounters in Latin America. I was now taking care of business. With Bill, my business agenda gradually evolved to include the additional dimension of a personal relationship when we soon became friends. Generous with his time, advice, opinions, perceptions, information and contacts, Bill adopted a father-figure-like role as he mentored me. It was a lot more than I'd expected. Contrary to his name, Bill never billed me. Quite the opposite: he provided me with valuable mentoring I could not only count on but thanks to which I could also count the profitable results. He allowed me to use his name out in the investment community, an especially valuable credential. A reference from Bill or mere mention of his name operated like an Open Sesame, allowing me to enter many otherwise closed doors. Careful never to abuse the "Sesame privilege" or anything else in the relationship, I didn't impose on Bill with regular or overly demanding requests. But even without my asking, Bill often offered useful assistance.

As our relationship developed far beyond my original intentions, I realized that it would be necessary to bring Bill some reciprocal benefits so that the dynamics of our business connection and friendship didn't entirely flow from him to me. Wall Street was a two-way street. Participants traded not only financial instruments but also gossip, rumors, favors, information and other intangibles. This presented something of a challenge, as I could count on Bill to mentor me but what could I bring to the party? I addressed this problem in two ways. First, because I circulated, in part thanks to contacts Bill facilitated, in the so-called "value investing" community I was able to contribute to our professional dialogue some information, news, perceptions, observations and, what Bill most enjoyed, gossip, all of which he found of interest. In other words, I spoke his language, even if not with his pronounced Brooklyn accent.

The second and even more telling binding factor was a certain emotional connection somewhat like a paternal-filial feeling which slowly developed, so making our relationship more personal. Bill had no children. His sickly wife Edna, called Teddi, suffered from pulmonary emphysema and was seemingly too frail to risk or perhaps even to gestate motherhood. It seemed that by chance my connection with Bill lent itself in a small way to him viewing me as something of a surrogate son. This additional and quite unexpected dimension of our interaction exemplified the reality that close relationships ultimately depend on emotions and feelings, not facts and logic. You establish the best friendships based on affective factors rather than on purely practical or functional ones. Emotional rather than promotional considerations create the binding tie, as proved to be the case between Bill and me and my three other New York City friends described below. I was of course pleased to serve in a small way as something of a son substitute, especially because our father/son-like relationship didn't suffer from the well-meaning but often common censorious paternal intrusions such as discipline, criticism constructive or destructive, orders, judgmental pronouncements and similar sources of friction. From Bill I enjoyed only positive supportive behavior and, similarly, never did I burden him with any filial problems or bother him with worries typical of the kind even adult children often bring to their parents. Bill and I never really discussed this extra dimension of our relationship. He's long gone and, in truth, I don't know if he viewed our connection in the same way as I've described it here. I certainly don't want to exaggerate the friendship, so perhaps I should temper my description by downplaying the filial factor and restrict my comments by stating only that Bill and I were just very good friends. For sure Bill treated me well in many ways, and I always appreciated and valued his efforts to help me make my way in the world of investing.

Whenever I went to New York Bill and I met at his fancy office where I enjoyed seeing again that elegant spiral staircase. It seemed to symbolize that I was back in the big time, even if only for a very short time. Then, after my brief time in the big time in the Big Apple, back to the provincial Midwestern suburbs where I belonged. We always lunched at the same nearby restaurant, one where Bill had a regular table, and I also saw Bill and Teddi at their condo at the corner of Park Avenue and 79th Street. Bill and I bonded even more tightly when one day in the spring of 1987 I informed him that Edith and I had come to New York for a two month stay at the Park Vendome, a building he knew well as his mother used to live there. The Park Avenue-Park Vendome coincidence by chance served to establish another personal link between Bill and me.

In the early years Bill took me to lunch at the old Carlton House on Madison Avenue near his office. On one occasion I looked on in awe as Bill counted out a few hundred dollars in bills which he handed over to the head waiter as a tip. Bill's bills impressed me even more than his office spiral staircase. Back in the Midwest, a few dollars would suffice as an appropriate tip for a two-person lunch. I told Bill that his generous gratuity was the biggest I'd ever seen for a single meal. He chuckled and replied that the wad covered not just our lunch but represented a month's worth of tips. Even so, back home the amount he disbursed would have been enough for an entire year.

After the Carlton House restaurant closed Bill arranged for a standing reservation, canceled if not needed, at his regular corner Cafe Pierre table in the top-tier Pierre Hotel—Bill always lived in the grand style—where he and I invariably lunched. A secret door connected the hotel with the lobby of Bill's building on Madison just behind the Pierre. One of his proudest possessions and insider privileges was a key for that obscure door which enabled him to access the Pierre without going outside and around the block. Bill always knew his way around New York and many of the shortcuts there. The Cafe Pierre offered a special off-menu mishmash of a tossed salad chopped and mashed into insipid small pieces called the Jacobs Salad which Bill always ordered and which I occasionally also chose in order to humor him, even though I didn't particularly like the mushy soggy mixture. After Bill died his lawyer and I ate lunch seated at his traditional Cafe Pierre corner table and, much to the veteran waiter's surprise, I ordered a Jacobs Salad. The server was astonished that anyone still remembered the dish, which without ado soon appeared. Out of respect for and in memory of Bill, for once I savored the unappetizing dish. Bill and his beloved eponymous salad were not entirely forgotten there at his favorite lunch spot.

As time went by Bill included me in his social life, not only in New York but also when he was in residence over the winter at his vacation cottage on the grounds of La Costa resort in Carlsbad near La Jolla, California. Always a generous and convivial host, Bill (Teddi was by then deceased) invited guests both from the financial world and other friends active and accomplished in a wide range of professions. Bill's sardonic and sometimes slightly sarcastic but never malicious sense of humor produced frequent wisecracks which cracked people up. His acerbic humor surfaced one day to sting me when, arriving at a party in La Costa he hosted, I handed Bill a bottle of champagne, informing him, "It comes from a Midwestern winery but it's just as good as the French stuff." Holding the bottle up Bill examined it briefly, wrinkled his nose in slight disdain, and in his most pronounced nasal Brooklyn accent said, "I'll be the judge of that."

Always included in the La Costa parties were his next door neighbors there, Audrey Skirball Kenis and her second husband Charles Kenis, a Los Angeles importer of French wines. Audrey had been married to Bill's good friend, the late Jack Skirball, originally a Cleveland rabbi who on the west coast converted to a business career, becoming a wealthy real estate developer and movie producer. Jack's fortune eventually funded the Skirball Foundation in Los Angeles, the Skirball Museum there, the Skirball Institute of Biomolecular Medicine at New York University and the Skirball Auditorium there as well as many other charitable enterprises. As recounted below (Section VI, Part 4), my connection with Audrey and Charles survived Bill's death.

While in residence at La Costa, Bill died suddenly in his early eighties when a slip of the hand by a local doctor Bill knew well, a guest at the parties, supposedly (so I was told) punctured his patient's innards during a colonoscopy and *voilà*—or *voi*-not-*là*—the patient was no more. He still lay there on the table, but now he was not there: from Bill to landfill in one easy slip. The intricate life-supporting sub-skin tangle, that "wondrous jungle" with its "golden intestines" described by Nathanael West, suddenly stopped working, its delicate balance unbalanced by an accidental internal stab. Such is life, and its end. Plots and counter-plots threaten the humanimal from within as well as from without. The poor creature remains ever vulnerable to the unpredictable forces of chance, fate and luck.

In he fall of the year Bill died, his family, friends, associates and colleagues gathered in his memory at the Park Avenue condo, that familiar apartment where I'd spent many pleasant evenings with Bill and Teddi and which I'd now be visiting for the last time. I made a special trip to New York City to attend the memorial gathering, which drew together many strands of Bill's extensive network of connections, most of them from the East Coast. I represented one of the few Midwesterners at the remembrance for the quintessential New Yorker. Someone at the affair expressed surprise that I'd taken the trouble to travel from the Midwest to attend the gathering. Describing my long and close relationship with Bill, I explained that his abrupt and entirely unexpected disappearance had produced in me a lingering feeling of melancholy which required closure to bring about a sense of an ending. I imagine that my phantom memories of my late great mentor which haunted me after his death—and which, haunting me still, I record in my reminiscence of him—show that with a friend such as Bill you never really come to closure. I continue to remember with fondness the significant part he played in my life and hope that my presence in his life gave him an equal degree of satisfaction.

~ 5 ~

As with Bill Jacobs, the second New York City professional contact I made, in a completely different field, evolved from a purely practical career connection into a long and rewarding friendship. I met George Blagowidow, founder, owner and publisher of Hippocrene Books, in a way which differed greatly from how I reached Bill. In contrast to my approach to Bill by writing him a business-like letter requesting an appointment, my encounter with George depended almost entirely on chance and luck. In the early 1980s I wrote a book of travel essays for which I sought a publisher. Even for an already established author, by then having published two novels and some travel articles, writing an uncommissioned work purely on speculation represents a highly problematic enterprise. You've produced an orphan manuscript seeking to be adopted but with no parents wishing to acquire your offspring. Like all the doggies at the pound yapping for attention, your manuscript is one of many and only very few are chosen. Publishers deem most unsolicited submissions real dogs. Failure to find a home for my book would require me to "put it to sleep" like an unwanted pound dog. You create the work in a vacuum, laboring long and hard without the slightest assurance that a publisher would even read the thing, let alone publish your well-meant but most likely ill-fated effort. Finally finishing the project, the moment of truth arrived. How could I get the hard-won but unasked-for pages published? From concept to manuscript is a long journey, but the distance between the submitted material and a print version is even longer and, as I realized, often too remote to reach. The route to publication was in its way a road more challenging than the ones I took during my travels around the world.

I followed none of the conventional methods writers use to convert page to press and print. I didn't seek an agent, prepared no outline, submitted no sample chapters, sent out no queries ("Hi, I'm Tommy from the boondocks and I want to publish a book"), didn't toss the manuscript over the transom to land in the slush pile. None of those standard approaches did I pursue. I concentrated on writing the book and hoped for the best, even while fearing the worst—rejection. But I never deemed rejection of a manuscript as a rejection of me as a person, as my persona and being were not subsumed into the material. The two were different. A refusal to read or to publish my book pertained only to my efforts in a limited area. Other interests and opportunities existed for me, and my life would go on in spite of rejections by the publishing world. I adopted a Judah P. Benjamin-like stoical attitude.

Manuscript in hand, I ventured to New York to make some cold calls on publishers. You really don't know how cold the City can be if you've never tried to sell there a book manuscript for which you have no pre-existing contract or indication of interest. Shortly before visiting New York on my quest I happened to come across in the Sunday *New York Times* travel section a stray reference to a travel book published by Hippocrene, a rather odd name but not compared to the even more unusual "Blagowidow" fellow who I soon met at the firm's office. During the visit to the Big City I made my usual rounds, seeing friends (Bill Jacobs among them), business contacts (Bill Jacobs among them), museums, bookstores, *The New York Times* travel section to discuss further contributions from me for the paper, and various other meetings, activities and diversions. As I headed back to my hotel late one afternoon at the end of an already full day I paused to glance at my travel notes. To make trips as efficient and productive as possible, whether adventures in far foreign lands or closer-to-home specific agenda journeys, I always prepared a detailed *aide-memoire* (the only memoir-type writings I'll ever produce) of places to see and things to do. One brief entry referred to Hippocrene Books, a firm I knew almost nothing about. My unscheduled visit would be worse than a cold call: the climate would resemble Arctic temperature, probably freezing me out with sub-zero results before I could kindle any interest in my manuscript. I was simply a stray aspirant off the streets without any advance notice for my sudden appearance. I'd be like a nightmare—an unsummoned and unwanted intrusion which will hopefully soon vanish. Hard-bitten New York publishers are by nature skeptical of would-be authors who fancy that they know how to write. Nonetheless, I decided to nurture luck by giving that capricious and often decisive factor a chance to favor me. Although it was not like me to initiate a contact unprepared, I proceeded to engage in unlike me behavior, hopefully to put myself in luck's way. But it's one thing to try to attract Lady Luck and quite another to count on her appearance. To Glendower's boast (in *Henry IV, Part 1*), "I can call spirits from the vasty deep," Hotspur replies, "Why, so can I, or so can any man; But will they come when you do call for them?"

Like Bill Jacobs' office, Hippocrene was located on Madison Avenue, the lower end, but the two establishments were otherwise completely different. While Bill occupied spacious and fancy premises, complete with an elegant spiral staircase, on upper Madison Avenue behind the Pierre Hotel, Hippocrene, as I soon learned, was crammed into a small space with no staircase but many bookcases. This was the difference between Wall Street wealth and publishing penury, between the Pierre and Peter, between upper and lower, bucks and books. I was comfortable in both worlds and enjoyed their contrasting characteristics, which I found refreshing. If variety is the spice of life I favored strong condiments. Investing and writing establishments based in completely different parts of Madison, upscale uptown and lower-scale midtown, seasoned my life well.

On a whim that late afternoon in New York I retraced my steps and doubled back to venture an unannounced visit to Hippocrene. Admittedly, it's rather presumptuous to expect to gain access to a publisher just by showing up, but there I was. As I approached the Hippocrene office at about 4:30 the door suddenly swung open—I didn't have to lay a hand on it for an Open Sesame—and a dapper middle-aged gentleman on his way out stepped into the hallway. In accented English he asked me what I wanted, to which I replied that I'd come to see a Mr. Blagowidow, pronouncing the name by guess. He responded that I'd just met him. His accent as exotic as his name, this Mr. Blagowidow, however he pronounced it, for sure didn't speak a Brooklyn lingo like Bill Jacobs. Nothing local about this guy. He was obviously a transplant from somewhere else. I wondered what back-story lay behind the "Blagowidow." What kind of name was it, where did he come from, what sort of temporal and geographical hinterlands represented his past? Whoever this Mr. Blagowidow was and wherever he was from, he had excellent instincts, as he didn't brush me off. Retreating back inside, he invited his unannounced, unexpected and probably unwanted intrusive visitor into his corner office, and suddenly I found myself in the inner sanctum of a New York publisher, quite an advance from the hard sidewalks of Manhattan I'd been pounding just minutes earlier. Sometimes you don't even need to push on a door to open it: you just have to be there when it does.

George Blagowidow and I chatted there in his office for quite some time, a conversation which was to continue for the next three decades until he died, age 90, in 2015. A mutual interest in many topics sparked our first dialogue, which crackled with instant familiarity as if we'd known one another for some time. Our connection jelled immediately, and thus by chance began a long professional association and, as with Bill Jacobs, a great friendship. We never tired of discussing travel, languages, reading, books, travel writing, new ideas, old guidebooks (we both collected vintage Baedekers) and many other shared interests. George eventually published *Last At the Fair*, my book of travel essays, and then my seven other travel books, proof positive that he was an astute publisher with excellent literary taste and judgment.

My chance encounter with George shows the importance of being in the right place at the right time—among the most useless (one of many) comments in this book, as it's obviously impossible to know what the right place and time are. Random benefits and burdens can occur anywhere at any moment, and on occasion you find yourself at what might be a right place but at the wrong time, as luck is just then absent, or at the wrong time at a wrong and ever luckless place. It's never possible ahead of time to discern where or when chance might favor you with good fortune. But at times it's possible to put yourself into a situation which might prove beneficial. Such was the case when I decided to call on Hippocrene unannounced.

The very narrow margin of time which allowed me to meet George before he left his office provided a real life *Time and Again* moment. At the end of that novel, two characters who failed by a matter of seconds to meet altered the future by banishing from existence the child they eventually produced in the original version of the story before it was replayed as the "Again." With the re-imagined new scenario their son, a character in the book, would never exist. In a similar way, my chance encounter with George depended on crossing each other's path during an interlude of just a few seconds, after which he would have been gone. If you miss the narrow entry point to meet someone, each person continues on his own way, never knowing that the other exists. Only by chance did I, almost literally, bump into George. Had I been on the up elevator while he was already going down, or had I for any other reason arrived at his office a few seconds later, George would have been on the move and it's unlikely I would ever have met him. Our initial connection, which depended on such a precise temporal coincidence, seems in retrospect quite improbable. It was a near-miss encounter based on split-second timing. But somehow we met, a matter of a few moments which influenced what followed over the rest of our lives. Years of a professional relationship and a great friendship depended on split-second timing. Reflecting on this unlikely episode, I wondered how many other life-changing happenings—chance events, unexpected encounters, unplanned developments—eluded me by the minuscule margin of a few seconds. It's impossible to know what did not happen. In fact, almost everything which could happen didn't, the missing non-happenings by far out-numbering the very little—but more than enough for one lifetime—which actually occurred. Vastly more exists in theory as a non-event than in the unique one-off status of what took place to assume a format which forms part of reality. The play of realized infinite forms which comes into being represents only the tiny residue of all which might have taken place or materialized. Only that which has actually entered time do we experience, while all the rest, the vast everything else in the realm of potential and possibility, remains forever unknown. Only in *Time and Again*-type fiction can scenarios be re-imagined and time revised. In real life no refutation of time is possible: there is time but no again.

Colorful and charismatic like Bill Jacobs, George was an extrovert with a great ability to relate to people. We immediately began on a first name basis, but I made it my business to learn how to pronounce his last name correctly. He interacted with a wide variety of friends, acquaintances, colleagues, associates and even strangers, some of whom became friends. He took a personal interest in people. With George you were always a particular and special individual, not just a type. For him type was reserved for printing, not for people. Whenever George and I were together we always discussed not just the business at hand, publisher and author talk, but also the personal side of our lives. He enjoyed hearing about the activities, travels, experiences and relationships of his friends and associates, and he shared his own such stories with me. In addition to his gregarious manner, George displayed a wide-ranging interest in a number of varied topics and pursued an eclectic range of activities: chess, the cultivation of ferns, honey, foreign cuisine, travel, culture, history (especially military), languages (he spoke six or seven). Always a great traveler, he shunned tours and organized trips in favor of independent adventures, a more granular and rewarding type of travel which gave him direct contact with local people and cultures. As both a man of action and a man of thought, George brought to his life many experiences and also reflections on his ventures and adventures. One without the other lacks an essential dimension, resulting in only half a life, but so active and pensive was George he enjoyed a life and a half. Russian Orthodox by heritage, George was anything but orthodox in his colorful way of functioning as an earthling on the lonely planet. He lived in Technicolor.

George's choice of profession reflected his personality. Publishing was for him as much a way of life as a means to make a living. His interest in publishing no doubt stemmed in part from his view that the good life consists of experiences, relationships and intellectual pursuits rather than possessions. Through publishing George enjoyed access to interactions with both the outside world and the world of ideas. He was one of the fortunate few whose personal interests and professional activities coincided, a highly desirable symmetry which represents an important component of a charmed life. I always supposed that George's lack of materialism and his desire to favor the intangibles of life—people, relationships, travel, ideas, experiences—arose not only from his intrinsic personality and temperament but also from his early-day experiences in wartime Europe.

Born in Poland in 1923, as a young man George became a "figure of interest" to the Gestapo after the Nazis occupied the country in 1940. The Gestapo refused to allow George to remain in medical school at the University of Vienna. Had he been a doctor I would not have wanted to encounter him in his professional capacity. As it was I had a keen interest in his professional activity. In the spring of 1942 he returned to Czestochowa, his home town, there to learn "that the Gestapo was inquiring after me" (as George wrote in his memoir) as a suspect in the Polish Resistance. He eventually escaped to western Europe, then to the New World to make a new life in New York. In his younger years in war-torn Europe, George saw how material objects, houses, property, things, valuables and other tangibles were vulnerable to plunder and destruction. That experience no doubt shaped how he lived his life. As he'd so vividly seen, possessions remain subject to loss, so true wealth lies in experiences, relationships, the life of the mind.

George's war-intrusive boyhood experiences in Poland perhaps also contributed to his most striking and enabling characteristic—his optimism. Under any circumstances and whatever the occasion, George invariably retained a positive state of mind. Nothing seemed to shake him. He remained cheery, optimistic, forward looking, always warm, agreeable and pleasant. His early unsettled and unsettling years didn't embitter him or suppress his fundamental positive outlook. In spite of serious and even grave (some literal) conditions during his childhood, and regardless of being uprooted, dispossessed of his natural habitat and displaced from his native land, George somehow remained a great optimist. Even later in life, suffering the loss of his wife and a daughter, George retained a positive outlook.

Some people, George apparently one of them, are seemingly programmed, perhaps genetically, to view life optimistically. This personality trait, which George shared with Judah P. Benjamin, could in part explain his habitual equanimity. Apart from his innate character, George's early-day experiences in Nazi-occupied Poland may also have contributed to his later optimism and his forward looking attitude. Just as George's choice to pursue experiences, relationships and intellectual matters rather than worldly possessions may have arisen from his wartime-instilled awareness of the fragility and impermanence of material things, so his bellicose childhood surroundings perhaps served to give him a balanced and reasonable perspective. Having as a boy faced extreme and ultimate conditions—some life-threatening, others life-taking, and many destructive of his familiar childhood world—George gained the hard way a perspective on what real problems and threats are. His youthful wartime sufferings perhaps taught him to view with equanimity problems of the type which confronted him later in life.

George's life displayed a generosity of spirit and an open-minded sharing mentality. He viewed the world with a wide perspective, not tunnel vision, and he enjoyed pursuing new interests. By his positive outlook, how he dealt with people, and how he treated me, George served as a role model. His behavior taught me many lessons By the narrowest of temporal margins we happened to enter one another's existences, a fateful encounter brought about by chance and luck. If the measure of a person's earthly existence lies in his impact on those who knew him, it can be said that all who encountered George benefited from his presence in their lives. The publisher left his print and imprint on everyone who had the good fortune to meet him.

As a ladies' man, George had an eye for figures; as a numbers man, so did Bill Jacobs. That figurative contrast symbolized to me how those two accomplished New Yorkers influenced and enhanced my life in such different ways. One was bookish, curious and worldly; the other, book averse except for books of account, narrowly focused, sophisticated only within the limits of Wall Street and the Upper East Side. While Bill operated in the hurly-burly investment world, George functioned in the context of texts and ideas: balance sheets for one, a balanced life for the other. Only thirty short Madison Avenue blocks separated their two realms, close by geographically but far apart by their essential differences. Bill worked in the swish upper 60s with its fancy shops, salons, boutiques, *haute cuisine* and even "hauter" establishments. Down in the lower 30s George occupied a more work-a-day world where bridge-and-tunnel commuters braved Manhattan to earn regular paychecks. Up there in the more rarefied numbers they spoke French; prices at the Cafe Pierre, with its continental ambiance, would have been considerably lower had the place been called Cafe Peter. Down in Midtown George often took me out for Grub Street-type grub for ten bucks at a neighborhood Katz's Deli or McSorley's Saloon kind of eatery. Grub Street and Wall Street were different worlds, and Pierre types didn't mix with Katz or dogs (coiffed poodles up, mongrels down) or McSorleys and McDonalds in the lower parts of Manhattan.

Over the years I greatly enjoyed engaging with two such different Big City characters. However, never did I mix the two worlds. I felt it was better to keep my New York professional activities separate and confine each to its own place. My sense was that revealing my participation in a second and radically different field might tend to dilute my credibility with George or with Bill. I didn't want to diminish my standing, such as it was, with either of them by suggesting that I was devoted to more than just one pursuit. I thought that each of them might take me less seriously if I engaged in two challenging and time-consuming professional endeavors. If in fact they deemed me a dilettante that perception wasn't entirely incorrect. As an acutely curious character I wouldn't have been satisfied pursuing just one serious productive activity.

One characteristic of curiosity is that it makes you promiscuous, in a constructive way. Both Bill and George greatly enriched my existence in two different fields, and with each mentor I also enjoyed an enduring friendship. It was a serendipitous blending of two TCB elements—business and the business of life. Although those two blended, I always kept my investing and my writing life separate, and neither man knew the other existed. I was the only connection between the two of them.

Oddly, writing and investing are somewhat similar in that they both use familiar everyday grammar, so to speak—for writing, words and syntax; for investing, widely disseminated and easily quantifiable information. Would-be practitioners of those two disciplines often believe that familiar language and commonplace accessible facts are sufficient to enable effective performance in each of the two activities. But even though writing and investing aren't rocket science or brain surgery, it's a mistake to assume that simply knowing language or accessing market information will produce positive results. By real world experience, over time, in both fields I learned otherwise, often paying a price for the lessons I learned. Much to my sorrow and only occasionally to my delight—unfortunately, all too many sorrows for my mistakes and far fewer delights for my successes—I came to understand the substantial challenges presented by the two seemingly simple activities. But never did I discuss writing with Bill or investing with George.

~ 6 ~

Although my investing and my writing activities remained apart, I introduced my third New York City friend to George as the two of them had much in common. My great friend Murray M. (Mike) Stern and Bill Jacobs had only two things in common, and they never met. Both were esteemed Erasmus Hall graduates and each excelled in his professional activities. Both Mike and George were also successful and, in addition, they shared many interests. I knew they'd enjoy one another, and I would have felt remiss had I not introduced them. Although they officed within a few blocks of each other, the chances of publisher and garment industry executive meeting without me acting as a go-between were slim, so rather than leaving an encounter in the hands of fate I took the matter into my own hands. Although I'd met George on my own, Mike I knew through my family. They thus represented the created and the inherited categories of how you find friends.

Unlike for Bill and George, my relationship with Mike began and ended with friendship, as there existed no professional connection between us. My relationship with Mike predated the one with George by some 35 years. I didn't need that long to perceive that the two of them should meet, for as soon as I met George it was obvious that he and Mike were kindred spirits, and that commonality inspired me to introduce the two of them. Chess, military history, reading, travel and keen curiosity represented just a few of the interests Mike and George shared, in addition to which both were successful small businessmen in the Big City and with nearby offices, Mike's at 34th just off Fifth Avenue and George's at Madison and 32d. Maybe they'd even exchanged passing glances out on 32d or 34th or on the Avenues, or perhaps had lunched at the same neighborhood eatery at the same time but oblivious to one another, strangers when they eat. At times, even the right place at the right time doesn't help mature potential into the actual. A catalyst was necessary to bring the two of them together and that service I provided so as not to leave their encounter to chance. It was the least I could do for two exceptionally meaningful figures in my life.

Mike I met when I was a boy and he a young man, and so enough time then remained for us to enjoy an unusually long friendship. We knew each other for 63 years, my entire adult life and some two-thirds of his. Although of different generations, we were of the same mind. Some people you meet don't wear well. With Mike it was the opposite: the longer I knew him and the more I learned about him the greater my admiration and affection. He reciprocated my feelings and we developed an unusually close friendship. As with George, Mike and I shared various interests and our views on many matters coincided. Both Mike and George, and Bill Jacobs as well, proved to be top-notch teachers, by way of example and by their practical common sense, as well as by their perspectives on life. From each of the three, all a generation senior to me, I learned many useful lessons, ones unknown in classrooms. A year after Mike married in May 1945, he and his wife Elinor Green bought a spacious house on leafy Oxford Boulevard in Garden City, an upscale suburb on the Long Island Railroad Hempstead line. In its own way the Sterns' Oxford establishment brought me over the years memories as pleasant as those from my time in 1957 at another Oxford, the university in England. For something like 40 years the Garden City house served as a haven for me in the sprawling New York metropolitan area. I often stayed out there in suburban garden-like Garden City, its greenery along with Elinor Green Stern a welcome relief to the City's hard gray sidewalks. Moreover, the quiet leafy Long Island suburb garnished with trees reminded me of my own little green enclave back home in the provinces. The Stern house was familiar and comfortable, a treat and a welcome retreat from the crowds and ructions of the City.

Garden City was pleasant but THE City ever beckoned. That was where the action was. While in residence in the suburbs I temporarily became a commuter into Manhattan. Mike and I took the train in from the Nassau Boulevard stop to Penn Station, something of a madhouse, as I saw the frantic scene, where crowds of travelers rushed and jostled their way to and fro, just where to and where from I didn't know but they were all in a great hurry to get somewhere. No garden city, Penn Station was a crossroads swarming with humanimals, the terminal quite in contrast to my quiet dead-end street back home. On one occasion during the dead-of-winter—cold, dark, gloomy, forbidding January days when even Garden City wasn't very verdant—I asked Mike which train he planned to take to the City the next morning. "The 6:40," he replied, his jest eliciting from me a laugh as I awaited word of a more civilized departure time. But unfortunately Mike wasn't kidding, and so up and out it was for us the next morning, so ultra-early that not even the crack of dawn lined the darkness.

That up-and-at-'em self-starting pre-dawn departure typified Mike's personality. He was always an aggressive can-do businessman with great gusto and energy, not only for his company but also for everything else in life he tackled. At the same time, Mike was a thoughtful and pensive person whose activities and hobbies represented those of a well-rounded man of many parts. In addition to the interests he shared with George—such as chess, military history, travel, books, the life of a small businessman—Mike also followed current events and world affairs. He was curious about and interested in many things, as reflected in his main recreational activity—reading. Mike read widely and deeply and, with his retentive memory, he recalled much of his extensive reading. He often surprised me by remembering obscure details from something he'd read years before. His command of military history, domestic and international politics, intelligence (plots and counter-plots in a governmental context) and many other topics, along with his astute insights on a wide range of subjects, endowed his conversation with great substance. In the same way as for George, but quite in contrast to Bill Jacobs, Mike's interests transcended the purely pecuniary. He both earned a good living and made for himself a good life. A successful entrepreneur and executive, perhaps his commercial accomplishments in part resulted from his inquiring mind and openness to new experiences. Those traits exposed him to opportunities and perceptions a more limited mentality might have missed.

Other factors in Mike's success included his self-discipline, decorum and persistence, all of which conformed to his somewhat regimented military manner. Indeed, Mike once considered making a career of the Army. After enlisting early in World War II he entered Infantry training at Fort Benning, Georgia and in July 1942 was commissioned as a first lieutenant. He was then ordered to report to Quartermaster Corps headquarters in Washington, perhaps summoned to duty there by the unit's head of Research and Development, the renowned Harvard Business School professor Colonel (later General) Georges Doriot, whose legendary course Mike had taken while getting an MBA at the School.

Doriot exerted a near mesmeric effect on many of his students who, like Mike, were entranced by the famous professor. A larger-than-life figure greatly revered by his students and businessmen alike, Doriot was a professor of wisdom as much as of business. He brought to the classroom an unusual synthesis based on both book learning and practical experience. Doriot himself operated in both realms. He was an academic who also functioned as a highly successful entrepreneurial businessman. Mike often referred to the General as if his views represented the ultimate and conclusive authority on the matter at hand. Doriot's tenure at Harvard didn't tend to attenuate or to shelter the General from business activities or real-life bottom-line ventures. Apart from his teaching skills in the classroom, out in the business world Doriot in effect invented the venture capital industry when he was recruited after the war to run the American Research and Development Corporation (ARD). The professor could now practice what he professed. As Spencer E. Ante writes in *Creative Capital: Georges Doriot and the Birth of Venture Capital*, "it is Doriot's work running ARD that truly distinguishes him as a twentieth century giant." General Doriot's famous Harvard Business School course on "Manufacturing"—a misnomer, as it would have been better entitled "Living"—delighted and informed nearly seven thousand HBS students who took the course between 1927 and 1966. The course's eclectic content made a profound impression on many, Mike included, who learned from General Doriot more about running a life than running a factory.

General Doriot's lectures presented practical wisdom on both career and personal matters. His views covered not just the nuts and bolts of manufacturing but also the philosophy of business and the business of life, a complete TCB package. The title "General" defined his military-like behavior. The classroom commander sometimes locked the lecture hall's doors to exclude latecomers, and to avoid wasting time he took no questions. Mike and other HBS graduates I knew who had taken "Manufacturing" would often address problems by asking, "What would the General do?" For my own part, when it came to TCB matters I'd occasionally ask myself, "What would Elvis do?" Obviously, I wasn't an HBS graduate. As Bill Jacobs, who never went to college, showed me, effective TCB was possible without HBS and other degrees. Although Mike Stern's rather precise military manner arose primarily from his personality, perhaps General Doriot's similarly regimental bearing also influenced Mike's martial but not militant behavior. Something of the General rubbed off on all who encountered him. That was also true, I believe, for Bill Jacobs and for George Blagowidow. Their influence endured in everyone who knew them.

For those of us who never met General Doriot, much of his character, charisma and commanding presence remains indistinct. What we know about him derives only from how others describe him by their personal knowledge or by after-the-fact accounts based on research. Similarly, the remarkable personal magnetism and allure of someone like Judah B. Benjamin escapes precise understanding for people who never knew the man (and the same could be said about Snoopy). Both Benjamin and Doriot were compelling and influential leaders, but their essential magic disappears with the demise of the man. Later generations know such colorful characters only through derivative accounts which can never give a complete picture on how the dominant figure managed to exert such a hold on his contemporaries. But writings by and about a defunct wise man allow Tommy-come-lately latter-day observers to glean at least a notion of an exceptional character who, by force of personality, influenced others of his generation. I by chance came across such a document when I happened to find in a used book store a mimeographed copy of the outlines and notes General Doriot used to present his HBS "Manufacturing" course. The manual was published in 1993 by the French Library in Boston which, as from 1947, the General and his wife Edna avidly supported. The folio contains a treasure trove of the General's wisdom—but not wit: he was humorless—along with the maxims and principles and aphorisms which so intrigued and guided many of his students, including Mike. The General's lecture notes provided me with high-class HBS course material without the need to pay any tuition, a bargain Doriot would approve of. Many of Doriot's pronouncements, as gleaned from his lecture notes, match even Elvis's profound TCB doctrines:

- Do not be a good optimist with bad luck.
- Be wrong in the right way.
- Be aware of the "Cost of Time (as it applies to everything and every action)."
- "It is so easy to analyze and criticize...Analysis has no meaning if it does not give us a new direction and strength to pursue a new and useful task."
- "Without actions the world would still be an idea."
- "After a certain period experience may lead to lesser effectiveness—more fear—lack of courage. What is experience really worth?"
- "A good manager never has to make any decisions."
- "Ability to get Somebody's time may be an indication of effectiveness."
- When making a decision, "What else is there to be questioned beside fact?"
- Decisions "must be based on what one believes in."
- "In important affairs the intelligence (mind) is nothing without the heart... [but] the heart...is not enough without boldness of mind."
- And finally, to conclude this illuminating mini-collection of the General's practical wisdom and uncommon common sense, a great truth I've observed in many situations: "Boundary between success/mediocrity and oblivion is a very thin one."

As my frequent encounters with Mike Stern revealed, his thinking and behavior greatly benefited from General Doriot's enlightened approach to all manner of situations. In an analogous way, I greatly benefited from the experience and wisdom of Mike and my three other New York City friends and also from any number of other people I met and befriended over the years. Those contacts and connections served to expand my horizons, my knowledge and my opportunities and gave me access to a wide range of experiences which greatly enriched my life. Had I not connected with those people my life would have been considerably poorer. In many cases I met them only by chance, those lucky encounters somehow developing into a friendship. In the spirit of General Doriot, along with the professor I'd profess that "Boundary between meeting the people who enrich your life and never knowing they exist is a very thin one."

Mike's career greatly benefited from a real-life object lesson which reflected General Doriot's HBS lesson that a very thin boundary separates success and mediocrity. Mere chance brought to Mike's attention a life-changing idea which he creatively used after the war to establish and develop a highly successful business. Mike happened upon the concept while serving in (then) Colonel Doriot's Quartermaster Corps unit in Washington. After Doriot became head of the Research and Development Branch of the Military Planning Division in 1942, he oversaw the design and specification of new products and equipment for the military. The department published such best sellers, still awaiting treatment by Hollywood studios, as *Mosquito Proofing the Jungle Soldier* and *Doron: Fiberglas-Plastics Body Armor*.

Among the projects the Colonel undertook was one to develop cold weather gear, a program which led to a study entitled *Operation Frostbite*, clearly a title more promising for a movie than the other reports. For this project Colonel Doriot sent Mike to Canada to study severe winter conditions so he could submit some recommendations for uniforms suitable to prevent frostbite and other nippy threats from the extreme cold. While on the mission, Mike noted the papoose-like bags in which the native Canadian mothers held their infants, a carrier which later inspired him to invent bag sleepers for babies. After his military service Mike established a company to market sleeping sacks for infants under the Sterncraft brand name. Those innovative snug cozy and comfy nighty-night bags for babies brought Mike great success and respect in the children's wear industry. Although he didn't make a career in the Army, as he had considered, the Army made a career for him. Mike's business card included lines to the effect, "If you date someone that's your business; if you marry the person that's your business; if the two of you have a baby that's my business."

Twice a year Mike took a six-week-long selling trip, calling on department stores all across the country. With him he lugged large sample cases but always with room in his bags to carry books and an auto-chess set. When Mike came my way I often hosted him at home. His visits with me served to deepen and to solidify our friendship. We came to know each other well. Over the years Mike and I shared more confidences and personal information than I exchanged with anyone else. At the same time we often didn't converse at all, remaining silent as I read and as he either read or concentrated on his auto-chess device. He told me once that if I'd been a chatterbox he would have withdrawn to stay in peace and quiet at a hotel. I responded by saying that if he had been a motor-mouth I would not have invited him to stay with me. Long wordless interludes occasionally punctuated by conversation suited us both.

Mike's interest in reading led him to help others enjoy the same activity, and to that end he became active in supporting libraries. For some years he headed the Garden City Public Library board, and in Sarasota, Florida he upgraded and managed the library at the retirement facility where he lived with his second wife, Ruby, who he married after Elinor died. He and Ruby met by chance on the Long Island Railroad commuter train early, very early, one morning when Mike helped her complete a crossword puzzle. His prowess at words impressed her, a clear illustration of the romantic possibilities of high-class literacy. Mine is low-class so it doesn't help me. It may not have been love at first word, but mutual affection soon filled in the new relationship's blank spaces, both down and across, and before long the former strangers on the train tied the _ _ _ _ (clue: "a tightrope").

Although Mike's interest in books, reading and libraries represented some of his major interests he was by temperament not bookish, pedantic, intellectually pretentious, over-cultivated, abstract, professorial or a rarefied and refined ethereal type. Always solidly grounded, Mike was practical and businesslike. His compulsive reading derived from his broad interests, fascination with the world and its humanimals, and his intense curiosity. He lived long enough, to age nearly 98, and remained in good condition until near the end, April 2009, to see and respond to many changes. Not long before he died Mike told me how much he regretted that his life would soon end. Although I could imagine many reasons why he might express such a regret, I was curious to know more specifically just which considerations prompted him to lament his disappearance before long. He explained, "Because the future is going to be <u>so</u> interesting."

As some of that future has by now occurred, I've seen what Mike was unable to see. In my turn I, too, will be forced to abandon the future and leave it to those who survive and then, over time, to the newborns, hopefully swaddled in Sterncraft sleeping bags, not yet alive but destined to comprise the generations to come. The whirligig of time will spin on without the Mikes, the Bills, the Georges, the Toms, the you's and all the rest of us no longer alive to experience the play of infinite forms. They'll play on without us after we become formless.

Although gone, Mike remains very much in my thoughts. Still today I often reach for the scissors to cut out an article to send him as I did over our 63-year-long friendship. But now my hand stops short as no one remains to receive all those clippings. The packages would end up in the dead letter office. For all those years Mike remained a fixture in my life, a comfortable, companionable and compatible friend who, like the future he longed to see, was <u>so</u> interesting.

~ 7 ~

My fourth New York City friend operated in a world completely different than those of investor Bill Jacobs, publisher George Blagowidow, and businessman Mike Stern. Just as for Mike, it was through my family that I met actor Donald Buka, who I also knew for some 60 years. When I met both Donald and Mike in the late 1940s I was still a boy and they were both established professionals. The age difference, however, didn't matter and eventually, as an adult, I became good friends with both of them. At whatever age I happened to be at any particular time I always sought out both younger and older friends as I knew they'd bring me new perspectives from different generations. I'd like to think that, for my own part, I brought to the oldsters and the youngsters of my time useful attitudes and perspectives typical of my own generation.

As for my relationship with the four older New Yorkers who befriended me, I sometimes thought that perhaps each of them saw in me something of a son figure who perhaps brought to the fathers (and to Bill, childless) some qualities and satisfactions otherwise lacking from their own sons. For whatever reasons, I established with the seniors a special connection which I believe rewarded them with a few filial-type or other enjoyments.

Apart from the commonality for the four that I was perhaps in a limited and minor way a kind of surrogate son, they all also had in common magnetic personalities and colorful, vibrant and energetic personal characteristics. Donald's career in fact involved the role of being a character as he belonged to that fabled star-dusty profession show people call "the Business." The catch-as-catch-can trade of acting enabled him to make a living, but barely. Donald made it his business to introduce me to "the Business," giving me a back stage perspective on the profession. For a kid—and then later as a boy, young man, adult, older fellow—from the "fly-over" cultural desert of the provincial Midwest, Donald always represented to me glamour, excitement, the magic of the theater. He was show business incarnate, a performer at the core of the Big Apple's plays of finite forms and shows, each and every one with transient acts eventually to end with the final curtain. From America's heartland I gained access through Donald to the heart of the New York theater world where Broadway's bright lights illuminated for me behind-the-scenes views of those curtained corners. Maybe for every light on the Great White Way the shreds of a broken heart littered the audition halls, but those cardiac shards of thwarted ambition I never saw. For me the spotlight remained on Donald, a successful actor whose stage door privileges he shared with me so I could glimpse some of the hidden scenes where "the Business" played out. With Donald I was an insider, by observation although not by participation. I played no part in "the Business" except, thanks to Donald, as an occasional Peeping Tom.

Back in the early days Donald lived in modest but fanciful quarters near the theater district. *Playbill* covers papered the bathroom walls, the familiar faces of the actors gazing at you while you performed your functions. In a role reversal, the stars were the on-looking audience and I was the featured performer. Some 60 years on, a ghostly sort of symmetry haunted me when I read Donald's obituary in the online *Playbill*, the digital format unsuitable for wallpaper designs. I've occasionally wondered what became of those old *Playbill* covers Donald pasted on his bathroom walls, abandoned when he later moved on to a larger apartment on the Upper West Side. Perhaps the pages have browned and curled and peeled away, reduced to shreds, like all those Broadway broken hearts, and washed down the drain. Or maybe a subsequent tenant, a macho testosterone-fueled Wall Street type, papered over the *Playbill* covers with *Playboy* photos, later—after he made partner—replaced by more sober *Business Week* covers to evoke the business of finance or *Fortune* pages to refer to his New York-won fortune. "*Fortune*": that's a magazine title which certainly spoke volumes to someone like me who viewed fate, chance and luck as the dominant elements in one's good fortune or misfortune. *Playbill* show business figures and *Playboy* model figures formed part of the play of infinite forms, but those photogenic earthlings all remained hostage to Fortune and subject to the corrosion of Time.

Whatever the ultimate fate of Donald's *Playbill* covers, they seemed to symbolize one of the performer's most prominent traits—his theatricality. In a way Donald was always on stage even if offstage in the *Playbill* papered bathroom or engaged in other mundane everyday activities. Somewhat unpredictable, Donald often went off-script by acting in offbeat, zany, dramatic and occasionally temperamental ways, each mood imbued with considerable personal charm. Although, in truth, Donald never rose above the second rank in the acting profession, in part no doubt because of his rather volatile temperament and his temper, he seemed to know almost everyone in "the Business," from stagehands and gaffers to the biggest producers and directors and the most luminous leading lights. During some sort of gala performance we once attended at the Alvin Theater he introduced me during intermission out in front of the theater to a galaxy of Broadway stars—here, Leland Hayward; over there Jerry, as Donald called him (Jerome Robbins to me); then Dick (Richard to me) Widmark; a hug with Ros (Rosalind for me) Russell. Although Donald was on a first name basis with all of them and many others, to me they were only distant stars out in the performing firmament far from my off-off-Broadway Midwestern setting.

As I accompanied Donald on his rounds around the city it seemed that it was impossible for him to go anywhere without encountering people he knew or who knew him. Many were colleagues in "the Business," but on occasion someone recognized Donald from one of his stage, TV or movie appearances. The afterglow of the vicarious thrill I got, as a visiting Midwestern provincial, from being in the company of a New York show business celebrity glimmered a long time after I returned to the "stix." Such denigrating word was the designation "the Business" Bible *Variety* gave to the remote uncultured lands beyond the Hudson in the famous "Stix Nix Hick Pix" headline referring to movies for the hayseeds out in the wilderness. Contrary to its title, the trade publication seemingly didn't much value variety. Residents out in the sticks were viewed as hicks inferior to the creative metropolitans who peopled "the Business" with cultured city sophisticates. I played my part with a realistic Oscar-quality performance by playing myself, a bumpkin typical of that vast mass audience out in the boondocks. Whenever Donald brought me in contact with the fabled show business cast of characters he knew I'd act like a simple country rustic in the Big City—except that wasn't an act: I played the real me. Back home some of the magic revived when I saw Donald on TV, his phantom figure flickering across the screen something like a ghostly shadow play which served to evoke my vicarious show business connection through him. More of a stargazer than star-struck, thanks to Donald I saw an entirely new universe with a galaxy of show folk. Knowing Donald was a heaven-sent opportunity for me to reach new heights with my New York experiences.

Those highlights of my New York visits which included the bright lights of "the Business" brought me experiences completely different than the more buttoned-up and businesslike worlds of Bill Jacobs, George Blagowidow and Mike Stern. They were businessmen; Donald was a "the Business" man. Donald often took me to his TV and theater calls, to his performances, to rehearsals, and to other professional activities. I accompanied him to TV filmings, backstage events, auditions, callbacks, and once even to collect an unemployment check, truly an inside view typical of an actor's life, one ever fated to put him temporarily out of business in "the Business." His appearances were all too frequently punctuated by disappearances when he was out of work and on the dole. Through Donald I met a wide cast of characters, some active in front of the footlights while others worked behind the scenes. He shared with me anecdotes, gossip, inside stories and intimate tales and details of people who populated the small and close-knit show business community. Never, however, did Donald's stories originate from a boastful or braggadocio attitude. What struck me as exotic was for him merely work-a-day chit-chat. He was simply talking about his everyday routines and his colleagues and associates, much as my neighbors might talk about crabgrass, squirrels invading bird feeders, real estate tax assessments and other such major suburban subdivision concerns.

Donald's most stimulating practice, as far as I was concerned, was how he displayed to me a never-ending beauty pageant parade of attractive actresses, starlets, models, chorus girls and other well-rounded performers he dated. I always looked forward to meeting his next companion. For years there was always a next, and twice a next became an "ex"—a charmer he'd married, then divorced. Donald's third and last wife was Suzanne Sinaiko, an heiress who lived in a high style but was quite down to earth. Somewhat abashed at her good fortune in possessing a fortune, Suzanne once told me almost apologetically how her late husband Arlie, a physician, had become wealthy not from his practice but from shrewd investing. His initial stake originated with his brother's successful chemical business. Arlie later became an investor in one of the first hedge funds, run by Alfred Winslow Jones. A soon famous April 1966 article by Carol Loomis in *Fortune* magazine on Jones mentioned that a "large stake is held by A. Arlie Sinaiko, a physician turned professional sculptor, who, together with his family, has about $2 million invested with Jones; most of this amount represents portfolio appreciation." (The $2 million in 1966 would now be the equivalent of about $18-20 million.) Suzanne's money burned a hole in Donald's pocket. Using her bank account, he greatly enlivened her life. It was a fair exchange—her wealth and his vitality. If it was a marriage of convenience it proved to be very convenient for both partners and they each enjoyed the relationship until Suzanne died.

Women found Donald's charm and manner irresistible, quite in contrast to my own romantic persona which women found all too resistible. I wondered what Donald's secret was. Whatever his technique, it certainly worked. Maybe he was just acting, but it was a convincing performance. Donald occasionally discussed his love life with me in some detail. He once told me that of all his paramours Hedy Lamarr was the most erotic. He described her charms to me in graphic detail, as if relating how Hedy frolicked in the nude in her 1933 German film *Ecstasy*, the very title reflecting Donald's account of his trysts with the actress. Donald's description was a lesson in gross anatomy quite as revealing as medical school dissections. She titled her 1966 autobiography *Ecstasy and Me: My Life as a Woman*, a phrase which well summarizes Lamarr's ecstatic behind-the-scenes romantic career.

So fabled and colorful were Donald's in-front-of and behind-the-scenes inside show business career I often urged him to write a memoir but he never adopted my suggestion. He told me many of his best stories but refused to record them. I think he was too busy enjoying the living to engage in the writing. As far as I know he never kept any written accounts of his starring role as a long-time insider in "the Business." But "Leading a good life is better than keeping a good diary," said Siegfried Sassoon, and for sure Donald lived a good and even something of a charmed life. With his death in 2009 perished Donald's countless memories, never to be recounted. Now just dust, my great friend will in my memory always shine brightly with the luminous glow of star dust.

~ 8 ~

In all too many cases a life worth recording remains unrecorded. Donald Buka was one of the dozen or so of my especially talented and accomplished friends whose lives merited a memoir, a diary, a journal or some sort of permanent record. To each of them I suggested that they write such an account, but only a few responded and most of the life stories have been irretrievably lost. Among the New York City quartet in this Section, at my urging my two bookish friends did produce a written account of his life. George Blagowidow, not only a publisher but also an author and thus well equipped to write, published his reminiscence. Mike Stern, an avid reader familiar with written works, also responded to my suggestion and wrote a hand-written reminiscence of his life. Like Donald, Bill Jacobs never showed any interest in writing about his life. Bill wasn't even interested in reading about anyone else's life or, for that matter, anything except investment materials. He was numerate but not literate. Donald brushed off my prompt like a casting director dismissing a hopeless character at an audition call. As far as I know, my brief sketches of Bill and Donald represent the only writings extant which record a few impressions of the two of them.

Almost all the people I aspired to inspire to produce a memoir demurred with the excuse that they didn't know how to write. With the exception of George, the author of some five or six books, this was a valid response. But I tried to convince my reluctant friends that compelling content would trump style. Their lives included so much of interest that enough material existed to endow their memoir with a noteworthy and valuable narrative even if the writing lacked some fluency. Perhaps my own reminiscence is an example of the triumph of content over form, but that may just be wishful thinking on my part. In any case, I noted to my friends that an editor could improve the style but no one else could create the life and its material, and for good measure I added that something in the way of a memoir was better than nothing, which would be the case if no memories were preserved as dead men tell no tales and without a written version all would be lost. All of this fell on deaf ears, for only George and Mike listened and acted. No other potential memoirist found my arguments persuasive. My failure to convince my friends to preserve their lives in print reduced to smoke all their experiences, stories, events, happenings, vignettes, anecdotes, perceptions, thoughts, wisdom, Lamarr-type encounters, musings and amusings, feelings and everything else. Except for a passing interlude in the mortal memories of the survivors—themselves soon to vanish—all was lost. The play of infinite forms which each of my friends experienced vanished with them, all their earthly sensory impressions phantomed into ghosts.

In addition to shunning my suggestion to produce a memoir, each friend to whom I made such a proposal invariably countered by noting that I hadn't even written my own reminiscence, so why was I bothering them about the hassle such an enterprise entailed. My friends complained that I was attempting to task them with a laborious project I myself was unwilling to undertake. This stuck me as a valid point, and it prompted me to wonder why I was willing to let my store of memories from a long lifetime filled with various ventures and adventures expire with me. For a number of years I contemplated how the story might be told, and finally suddenly devised a format which suited my purposes. But I left the project for very late in the day and now, near the end, I hasten to record in this terminal work some selected episodes and thoughts from the play of infinite forms as I experienced them. But time may yet defeat my long delayed exercise in preservation.

Over the years I've often reflected on and always been grateful for all that my New York City quartet as well as many other friends elsewhere brought to my life. Each enriched my existence in a different way and all of them in worthy ways. Each of the four New Yorkers represented entirely different worlds while I—remote from them by age, by distance, by accomplishment, by life experience—was the connecting factor to form a kind of quintuple helix of our five disparate lives. They linked through me, the outsider. Those four—Wall Street Bill, publisher George, businessman Mike, performer Donald—are the only friends I discuss at length in my reminiscence, both out of appreciation for all they meant to me and as examples of how lucky I was to find such meaningful relationships. A few other friends I mention very briefly, but virtually everyone else I omit. Omission, however, doesn't imply indifference. My home-town friends and others around the world I greatly appreciated, as they represented highly valued and gratifying enhancements to my life, as I hope I did for theirs. Nor does my omission of women in this account of some New York relationships suggest indifference. It just happened that for me Manhattan was not a Womanhattan, as in the city I found the four mentioned noteworthy male friends and mentors rather than any womentors. I'll try to do better next time to bring some diversity (or dihersity) to my New York City circle. I've described the quartet of New York friends in detail for various reasons:

to preserve in print a few elements of their noteworthy lives; to pay tribute to their role in my life; to recognize that they were especially meaningful friends; to illustrate how each of them operated effectively in his own chosen field. Most of all I wrote about them because each of the four represents the kind of person I admired and in some ways tried to emulate. Moreover, they serve as examples of other accomplished friends of mine who lived meaningful and to some extent charmed lives unnoticed by the general public. Productive and successful, those talented and diligent contributors to society never sought or attained public acknowledgment or acclaim and were content to carry out their day-to-day useful activities of the kind which provide the sinews and substance which make the world function. Anonymous and unsung, those estimable friends lived and worked privately and effectively, leading lives more authentic than many self-seeking types who pursue fame. Each such friend was known in his field, but lacked wider recognition. Here I recognize how much they contributed to their mini-worlds and also to me so as to help me enjoy something of a charmed life.

Over the years I met a certain number of well-known people and enjoyed my brief encounter with such celebrities, a few of whom I describe in this reminiscence. But it was mainly curiosity, not not name-dropping potential or mere celebrity chasing, which motivated me to meet famous characters. I also hoped for an interesting conversation and a learning opportunity, but those potential benefits didn't always occur. I met some rather big names with small minds, which made me wonder how such well-known characters managed to gain some success, or at least fame which passes for success in our society. Miguel Ydígoras Fuentes was the only president I ever met, but he was small potatoes, or small bananas, as he ruled over a small country of little importance in the big wide world, unless you were Chiquita Banana or Jake the Flake. In my own land I met Richard Nixon in his pre-presidential phase, and another time I trod presidential territory on a White House tour, the too roomy and over-stated mansion of state so different from the down-home spacious but unpretentious white houses common in the leafy suburbs out in the hinterlands beyond the Beltway where real America lived. I knew a big white suburban house like that, one I lived in once upon a time. Always a provincial, I preferred White Castle to the White House, and my down-to-earth down-home friends to Washington's salons, political snake pits, cliques, factions, lobbies, the high powered clubs where the high achievers clustered and the wanna-be's infiltrated. The only famous mover and shaker I admired was Elvis Presley. A quiet life on a dead-end street in a Midwestern suburb, with occasional forays out into the big wide world, left me quite content.

William Faulkner said that to get a book written, "If a writer has to rob his mother, he will not hesitate; 'Ode on a Grecian Urn' is worth any number of old ladies." Similarly, I'd never exchange my treasured fame-less friends for the experience of meeting the various celebrities I encountered. My friends were worth any number of famous people. Although a certain number of my friends enjoyed at least some elements of a charmed life, few were charmed in a substantial way. To attain that rare status requires far too many unlikely factors to exist and persist, most of them depending on the capricious influences of luck and chance. The odds are long, time too short, and life too uncertain to enable any humanimal but the most fortunate few to live in its own way. Your fatal circle and then the structures and strictures imposed by the world and by your decisions, by chance happenings and also by the open-ended nature of the future all conspire to limit your freedom to chose your own life. To enjoy a charmed life many disparate factors, most beyond your control, need to go right, but for most people too much goes wrong. Much as you hope and plan for a charmed life only chance determines the outcome. A life which is "charmed" is one that closely matches your ideal of the good life, a status easy to define but hard to find. Somehow I managed to find that elusive and rare state of being.

As I've explained, years ago I determined that for me the good life would feature experiences and relationships. In this Section III I've described four of my especially meaningful relationships. In Sections IV, V and VI I'll discuss some adventures and activities around the world which brought me the kinds of experiences I sought. Relationships and experiences: there you have the essence of my experiment in living. Memories of the experiences I retain, but by now all too many of the relationships have come to an end. My four New York friends and many others elsewhere are gone. However, they all somehow linger, phantom-like, at the places we once frequented together. After people disappear, your memories of them in part become transmuted into the familiar settings where you enjoyed the company of your friends. You don't have to be a disembodied former person to haunt a scene: the place itself seems ghostly. In *The Ghosts of London* H.V. Morton described not defunct people but antiquated historic haunts and quaint bygone customs from the city's past: herb shops, royal waxworks, Hansome Cabs, snuff, curfew, lamplighters, all relics of yesteryear.

For me now as a passing visitor temporarily in New York, I from time to time pass by a place which one of my friends and I used to frequent. A kind of ghostly aura lingers at those by now, for me, abandoned settings—a phantom emanation which transforms into a ghost place the once convivial space my friends and I enjoyed together. The places still exist but are now forever empty of my friends. Only the phantoms survive, shadowy absent presences which prompt me to remember the long gone days when my friends and I spent pleasant times in each other's company. By actuarial realism, before long I, too, will become a ghost, my charmed life at an end. Memories of my friends and of the times we knew and the places we visited together will die with me.

Even for people you know extremely well, your expired encounters with them in the long-gone past somehow seem shadowy, ghostly. Time has distanced the defunct from life, and yet once upon a time the phantoms were all as real and as lively as am I, at least for a brief while longer. In his *Autobiography*, historian G. M. Trevelyan writes of "the poetry of history" which evokes the "fact that once, on this earth, on their familiar spot of ground, walked other men and women, as actual as we are today, thinking their own thoughts, swayed by their own passions, but now all gone, one generation vanishing after another, gone as utterly as we ourselves shall shortly be gone like ghosts at cock-crow." The generations appear, live briefly, then disappear. The life and times of each generation define for it the experiment of being alive in the world. Other eras, past and future, have brought and will bring completely different experiences to those living during such times. Members of each generation only know the fragment of reality which prevails during an individual's limited earthly tenure. The conditions, happenings, traumas, dramas, challenges, advances and declines, charms and curses, concerns and problems, glory and gore of other generations, past and future, remain distant from us. As de Tocqueville wrote: "[N]ew families are constantly springing up, others are constantly falling away, and all that remain change their condition; the woof of time is every instant broken, and the track of generations effaced. Those who went before are soon forgotten; of those who will come after, no one has any idea: the interest of man is confined to those in close propinquity to himself."

So it was for my generation and my vanished friends who once existed and now do not. So it is for me soon, and for you sooner or later. And so it will be for all, charmed or cursed, who come after.

Section IV
TWILIGHTS FAR FROM HOME

~ 1 ~

From my early years travel become part of my life. Never did I lose my interest in exploring and experiencing new places and lands far from my familiar native soil and routine home life. In my explorations of the lonely planet I didn't go as far as Robert Louis Stevenson, who noted in *Travels with a Donkey in the Cevennes*: "I travel not to go anywhere, but to go. I travel for travel's sake. The great affair is to move." My own trips differed in two ways. First, I preferred faster and more efficient locomotion, and one that didn't eat, than donkey, and second, my travels were meant not simply to keep me on the move but were always purposeful—not in a practical sense, to enable me to gain any advantages or riches, but to enrich my life in other ways. Simply moving about didn't move me to travel. I designed my travels and embarked on my curiosity-inspired adventures because I wanted to see specific countries and sights, observe characteristic customs and practices, and create stimulating and interesting experiences. I traveled to go places and do things, not just for the sake of being in motion. For me the "great affair" was not to move but to expand the range of my experiment in living.

Once I decided in my mid-thirties to diversify my life, adding to my provincial suburban base occasional trips to far lands, the world and the future—however long it might last for me—presented me with a vast playhouse of infinite forms. Ahead lay a chaos of possibilities which, as time went by, would eventual coalesce into reality and become my past, the outcome of my experiment in living. Where the encounter between my curiosity and the play of forms out in the world would lead me I didn't and couldn't know. Like all experiments, the results remained uncertain until, finally, they've taken shape as proof of concept or as a failure to yield the intended results. The future infinite forms which would come to represent my past were a puzzle yet to be solved, like a creative composition with countless possible versions. As Dieter Hildebrandt wrote in *Piano Forte: A Social History of the Piano* (translated by Harriet Goodman): "A keyboard after all is a huge, insane puzzle, a puzzle with a million solutions, allowing infinite permutations of melody and modulation and harmony, a puzzle that a composer will piece together a thousand times, only to be faced with it afresh the following day like some primeval chaos, waiting for the world to be created."

From the very moment of conception a creation, an agenda, an embryo develops through a maze of infinite possibilities and permutations. A humanimal's fatal circle begins with its fetal circle, after which mitosis inexorably establishes the creature's physical characteristics and then, once thrown into existence, the earthling becomes subject to a mitosis-like division of time which breaks the temporal bulk into seconds, minutes, hours, days whose content establishes the person's lived earthly experiment. Just as for a developing fetus, you can't go back to the beginning to start over. Time past, forever lost, is a hinterland no irrendentist efforts can reclaim. Unlike a composer at the keyboard you don't get a second chance, let alone a thousand. Once potential becomes reality the puzzle no longer remains "insane" but is completed, ordered into a set piece. But even when once set, the piece itself can in some ways seem insane. Existing does not necessarily entail meaning. The lonely planet may not be entirely meaningful or sane.

My two opposite but apposite ways of life offered a satisfactory solution to the puzzle of how to try to live a charmed life. The refreshing contrast between how I functioned at home and on the open road proved workable and pleasurable. One mode of being without the other would have been lopsided and lack equilibrium. Pure vagabond roamings severing me from home base would leave me rootless and unsettled. As a transient always on the move I'd go everywhere but belong nowhere. Remaining home-bound, always confined to a narrow provincial suburban life lacking the occasional novelty of experiences out in the big wide world, would limit my life to an overly cozy and comfortable repetitious existence without variety. I valued both ways of life. A change of scene and seen stimulated me and served to assuage some of my curiosity, but no matter how exciting the trip I was always glad to return home. Of all the many various infinite forms which happened to reach my consciousness, one of my very favorites was the sight of my little house after a long trip.

Largely uneventful, my home life merits no description in a reminiscence devoted to more worldly matters. Only a microscopic writer like James Joyce could produce a classic work based entirely on the myriad petty activities, events and thoughts experienced by one rather ordinary Dublin resident on a particular June day. My account doesn't attempt to duplicate that technique as I wish to remain a ghost writer rather than to become a Dubliner's double. It would be of interest neither to me nor to a reader to include mundane daily details based on a hermetic internalized personal account of insignificant day-to-day happenings at home. Dickens said trifles make the sum of life, but they only make some of life and for the most part usually lack interest except for the person trifled with. I prefer to focus on the outward looking outward bound part of my life, the activities far afield from home base. Those ventures and adventures in distant lands comprise the defining feature of my restructured life. The daily localisms of my regular suburban U.S. Midwest provincial existence in a Dublin (Ohio) kind of place offer little of interest, even to me.

At the beginning, travel was imposed on me. Like every child I went along for the ride wherever it might take me. I didn't ask to be taken anywhere. As a boy I really didn't know where anywhere was. My early-day trips somehow just happened. I realized that a lot of places existed beyond the big white suburban house where I grew up, and I knew that some of those places were a long way away. But I didn't know anything about such remote areas or exactly where they were. Other countries were out there somewhere, but all I knew was home—the big white house and nearby areas. The only foreign region I could see from the front yard that I might possibly visit was the moon. It seemed earthy, a *terra firma* kind of heavenly globe I could tour, but it was out of reach and in fact not even *terra*, just an empty orb floating in space like the earth had been before life originated there and would be—floating on and on through space and time—after life vanished from the lonely planet. The moon—truly lonely, even if not a planet—represented how the earth would look after its time, when everything on it had disappeared.

Once exposed to travel, even as a boy when I knew almost nothing about that diversion, I found the few different places I was taken to alluring. It seemed to me that visiting new regions served as an excellent way to satisfy curiosity. Travel provided a grab-bag of infinite forms. Although my curiosity then existed, at the time it didn't yet focus on specifics as I didn't know enough to be intrigued or attracted by particular topics or places. But even in those early years somehow a keen curiosity about the world possessed me. Where this innate trait came from I didn't know, nor did I much care. Like the big white house and the world beyond and the moon, curiosity was just there. I took it as a given that curiosity would be one of the major influences which would determine how I shaped my agenda and my life. Curiosity in fact proved to be the dominant volitional force which determined my experiment in living. In the end, however, it was such random forces as luck and chance which in fact most definitively influenced the outcome.

My first trip outside my homeland took me, taken without being asked, to Mexico in the mid-1940s for a stay of some months. When you're young everything is a command performance. To live a charmed life in later years, it's necessary to escape commands issued by others so you can follow only your own orders and order your agenda to suit your interests. The travels forced on me in the early years did not—as is often the case when a child gets older—provoke me to rebel against that imposed activity. I was a beneficiary rather than a victim of the parental marching orders which took me to other places. The Mexico trip gave me a seminal introduction to the world of foreign travel. The exotic country was strikingly different from the then very limited world I knew. Traveling south of the border far from the big white house to a country with *grandes casas blancas*—different by description from the one I knew but otherwise somewhat similar—greatly expanded the perimeters of my narrow existence. The boundaries of my fatal circle were delimited, but not the sky's-the-limit potential to roam the lonely planet. The moon high in the sky was off-limits, but maybe one day in the far future I could get there and see the earth as it really was, a tiny corner of the cosmos where, for a few fleeting tens of millennia, life in its multitudinous forms happened to exist.

Mexico's colorful scenes and novel practices fascinated me. Architecture, customs, clothes, food, faces and much else contrasted with how they all appeared back home. They even spoke a strange language down there, one I'd never before heard. I wondered what it all meant. All the novelties made me curious to learn more about them. Everything was thought-provoking. The new ways and byways and scenes made me realize that how things operated at home wasn't definitive but only facultative, a way but not necessarily the way matters might be arranged. You could accomplish the same ends by different and maybe even better means. Communication and responses down there involved words different than the ones I'd learned at home. In Mexico those familiar home-side words were useless. South of the Border "moon" became *la luna*—same dead orb floating in space but with a different name. In that foreign land south of home *tengo sed* replaced "I'm thirsty." You could be thirsty for hours, days even, without relief if you left unsaid *tengo sed*. *Sed* sounded just like "said" but meant something entirely different. This new language stuff was tricky. Although I picked up enough Spanish to communicate some basic thoughts, I was frustrated that a large hinterland of words and grammar unknown to me existed beyond my linguistic capabilities at the time. I was beginning to realize that many other hinterlands, far removed from my home base place, occupied remote areas unknown and unfamiliar to me. Later in life I would spend much time, energy, effort and some money to familiarize myself with such foreign places.

I imagine that for some young boys a country like Mexico would be too strange and foreign for comfort. Some places or experiences too alien or incomprehensible can be unsettling. But Mexico's strangeness didn't baffle or bother me. With its colorful, lively and unpredictable play of forms (maybe Paseo de la Reforma, the capital's main boulevard, referred to the ever mutating re-forming forms) so different from the ones back home, the country brought me a whole new set of sensory impressions. I wondered what the other lands down there in that part of the world were like. Just south of Mexico was Guatemala, an odd name. It sounded like some kind of tropical fruit, but among the papayas and mangoes on sale I never saw a guatemala. What did the name mean, how did the people down there live, were there big white houses or only small colorful ones, did the country have a president like we did or just a *jefe* who ruled the land with an iron hand? It was all a mystery to me. Unknowns of all kinds lurked out there in the big wide world. I had a lot to see and to learn. My early-day interlude in Mexico inspired in me an interest in seeing what the world had to offer. Back then I knew almost nothing of the lonely planet, except that it seemed alluring. My youthful Mexico experience proved infectious in a way effective to cause me to develop an incurable case of wanderlust, a malady medical science has never been able to cure. Only after more than half a century of travel was I able to eliminate most of its symptoms, but never the underlying condition—which would have required a curiosityectomy, a particularly serious operation for someone in whom curiosity is deeply embedded.

Over the years I occasionally pondered the origins of my compulsive interest in travel. My interest in how one of the driving—and flying, railroading, busing and boating—forces of my life originated was not at all a concern but only a mild case of curiosity. For all matters, what is and not why it is was always my primary focus. No one in my family set an example for me or even encouraged me to see the world. By and large my family members avoided the far corners of the globe. My relatives and also most of my friends all contentedly stayed closer to home. An occasional domestic trip and maybe a daring foray every so often to Europe broke their home-based routines but seldom or never did travel exotica tempt those in my circle to get off the beaten track. Nothing in my inherited setting, my circumstances, my surroundings or the interests of my family or friends, all of which defined my life as a boy and a young man, suggested that a wanderlusty (maybe better described as wonderlusty) life would be so compelling for me. My interest in travel was an aberration. Some people in my fatal circle traveled to Grand Tour countries, European lands which offered comfortable conditions and postcard familiar sights, but no one ventured beyond to more remote places.

I of course realized that many people never had any interest in or even the opportunity for foreign travel. Various factors restrained them: force of circumstances, financial limitations, work-a-day job obligations, family life which both contented and confined them, lack of curiosity, indifference to different places, unwillingness to give up the comforts of home and a familiar setting for the discomforts and uncertainties of travel, a reluctance to unsettle settled routines. Any and all of these considerations represented understandable reasons to remain a stay-at-home, but none applied to me. In any case, for whatever reasons most of my family, friends, acquaintances, neighbors, fellow suburbanites and others I knew remained in place. That way of life certainly offers more comfort and is less threatening than the problems faced by travelers who run off to strange new places. Things are different and more dangerous out there. Why go to a place where it rains *gatos y perros* when you can stay stay at home and get just as wet with drippy cats and dogs?

I well recognized the merits of staying home, a realization which made me wonder sometimes why I was so willing to abandon the suburban comforts of my conventional and satisfactory provincial existence. It would be far easier just to settle in and, like so many people I knew, enjoy the benefits of familiar surroundings rather than venturing trips to strange places where I was a complete stranger with a very incomplete understanding of the alien surroundings. Perhaps I was a slow learner, but it eventually occurred to me that travel is in fact a subset of curiosity, a notion which served to explain my interest in seeing the world. Travel represented an obvious response to my dominant trait. My compulsive curiosity both bedeviled and beatified me. At home my mind churned with unsettling thoughts of being in far places, while away my life was filled with memorable travel experiences even as I often longed for home. I was cursed with insatiable wanderlust, and blessed with insatiable wanderlust.

I would never claim that my travel life represented an attractive model for anyone else to follow. Adventure travel is an acquired taste, one which happened to whet my appetite rather early and acutely. But I can well understand why such an inconvenient activity of great fascination for me would for most other people be of little or no interest. In a way it's rather odd to pursue an enterprise which brings so many disadvantages. Travel, especially in underdeveloped areas, entails many problems. For one thing, it costs, not only money but time, energy and effort. It's inconvenient, uncertain, often hazardous, and it's really a lot of work. The very word derives from *travail*, French for "work," and indeed travel brings many travails, traumas, troubles and trials. With all those negatives and others it's a wonder why anyone subjects himself to such challenges and discomforts. What does travel get you, other than headaches, backaches, bellyaches, heartaches, walletaches? In tough areas you're always a moving target, a tempting prospect for plucking. Intrinsic in the enterprise of passing through strange and difficult places is a sometimes spooky vulnerability as well as a bewildering confusion of sensory impressions. You're everywhere a stranger and all the locals are for you foreigners who (most places) speak in sounds you never before heard and can't understand. Staying home offers many advantages. Why leave?

But I left. I left, which was right. In *A Cab at the Door*, V.S. Pritchett tell of the time his father asked him why he was so eager to go to France. Pritchett replied that he was attracted by the fact that France was different: over there, he noted, a street is called a *rue*. He went on, "I could not explain what an immense, strange world lay in that word <u>rue</u>." For a curious soul, a foreign word like *rue* exerts a kind of magical effect. You wonder what sort of unexpected delights you'll find in a place with *rues* rather than "streets." A home-body lacking interest in travel would rue the day he left his domestic comforts to see the *rues* of Paris, a city far too unruly and unsettling and foreign for words. Better to remain in your own neck of the woods rather than crossing *la mer*—the sea—to see a *bois* with similar trees in a far land. So Thoreau opined from his Walden Pond cabin in the woods (the *bois*) where he lived: "It is not worth while to go around the world to count the cats in Zanzibar." As H.L. Mencken would respond, with the handy rubber stamp phrase he entered on unsolicited letters he returned to sender, "You may be right." So I admit that Thoreau may well be right: why journey to Zanzibar, to Catalina, to Catania, to Catalonia, to the Catskills or anywhere else to count cats when you can just as easily tally them, and dogs as well, at home? But on the other hand Thoreau may be wrong. What for him was a worthless activity for me represented a compelling opportunity. In spite of the undeniable difficulties inherent in long journeys, to amuse myself with kitty tallies and other such frivolous and odd pursuits in far away places my curiosity urged me to travel. Seeing cats around the world seemed to me much more enjoyable than staying stuck at home taking feline inventory there. In fact, at home I spent no time counting cats or dogs. I devoted my reckoning to counting the days when my next trip would begin.

    As much as I enjoyed and appreciated my life at home, when there I habitually thought about, planned and prepared to take the next trip, while also pleasurably remembering past ones. Polar explorer Robert Peary's comments describes this sort of fixation: "Just think of the life of the travelers in an unknown country or an unknown sea. Then think of the nights and their thoughts of the morrow—what will it bring?" Peary goes on to meditate on "the constant expectation of some new strange sight...wondering what mysteries lie beyond the beautiful shores, what secrets the distant mountains are guarding and the distant valleys are hiding in their unknown depths." Not only realization—the actual experience—but also anticipation comprises a major factor in the delights of travel. In fact, travel is a three dimensional activity: the planning and anticipation; the excitement of the trip itself; and the memories. This potent combination of the yet to come, the now, and then the aftermath becomes addictive. As addictions go (or don't go: by definition they persist), travel and other compulsions such as curiosity, reading, thinking and ice cream rank among the more benign and beneficial kinds.

    Peary referred to the traveler's "nights and their thoughts of the morrow." In *Last Things* Lewis Eliot (not to be confused with Tom Eliot), the narrator of C.P. Snow's *Strangers and Brothers* novel series, recalls the "unfamiliar rooms ... The hotel bedrooms of a lifetime." I, too, occasionally remember some of the hotel rooms which housed me far from home. All around the world I stayed in countless away way stations of vast variety. Some were shabby, some dirty, many both. A few were elegant but most of them ordinary or nondescript and almost all of those temporary usually charmless cubicles impersonal, alien and somewhat alienating. You have a roof over your head but no roots beneath your feet. The rooms provided a facility for brief transient shelter, a pressing concern for a wayward traveler like me who often arrived in a city without prearranged accommodations. Whether at the very bottom or below the bottom—a real pit, the Sultan Mohammed Hotel in Ghazni, Afghanistan was subhuman—or the palatial establishments with every convenience and luxury, each and every hotels room where I stayed had one thing in common: it wasn't home. Everywhere I traveled outside my own country I was a stranger in a strange land. In a more cosmic sense, the world itself and everything in it struck me as strange, including my own homeland and home and even my very being. As I saw it, humanimals occupy a strange place beyond understanding. The sublunar world made sense, if at all, only in and of itself, not in any cosmic way. In their limited little world, earthlings liked to pretend that various earthly practices, seemingly endowed with significance, were meaningful. But it was all just play-acting, a shadow play performed by ghosts. On the moon things were much clearer and simpler: human civilization had not taken root there to confuse and complicate the lives of moonlings, if perchance any existed, with an insensate play of infinite forms.

Section IV: Twilights Far From Home

Hotel names sometimes self-flatter to deceive. Bombay's Ritz (for sure no Ritz-Carlton) where I stayed in January 1974 was as ritz-less as an ordinary Ritz Cracker. On the other hand the designation Bussaco Palace, the hotel in Portugal where I spent a night and a lot, represents truth in advertising. The majestic establishment, built in the late nineteenth century by the last king of Portugal and later converted into a hotel, offered truly regal facilities. For a hotel you don't know what's in a name until you're in the place. Just as I never counted cats in Zanzibar—would have, but I never got there—or anywhere else, I haven't counted the total number of days or weeks or months I spent on the road. I'd estimate that I devoted some six, seven or eight years in all to travel, spread over more than half a century, with perhaps about one-third, as much as I spent in sleep, in the Third World. In any case, many moons of my allotted earthly existence I spent in those "unfamiliar rooms" occupied in passing as I made my way around the lonely planet.

Even when I traveled alone I didn't feel lonely on the lonely planet. I was so busy getting around to see places and at day's end so fatigued I didn't have any time or energy left over to think about being alone. Sometimes a traveling companion accompanied me, but often I just took off on my own. But I must admit that on many a solo evening as I watched the day wane into dusk, the fleeting light edging into darkness and then night—twilights far from home—I felt slightly melancholy. "Twilights far from home": that evocative phrase captures the rather bittersweet feeling which at times came over me when I was alone at dusk in a distant alien setting. It's how Christopher Patten, the last British Governor-General of Hong Kong, described the twilight of his diplomatic career when the Crown Colony was transferred to China in 1997. Many "twilights far from home" shadowed my far-flung evenings as dusk darkened the day into night.

Capturing this feeling in Aphorism 638 at the end of Volume 1 of *Human, All-Too-Human*, Nietzsche describes "The wanderer," not someone who heads for a final destination but an observer, a roaming character who "takes pleasure in change and ephemeralness" and who at times "will have bad nights, when he is tired, and the gates of the town that was to offer him repose are shut." But then arrives "the blissful mornings of other places and days," as the Wanderer proceeds onward with his free-spirited roamings. Such describes the progression of my feelings as I passed a melancholy miserable Christmas Eve night in that gloomy class Z-minus New Delhi hotel (as recounted in Section II, Part 4) before awakening to a bright and blissful morning when I continued my trip to other places in better days. When traveling on my own far from home I was in an alonely but rarely lonely state. Somehow it seldom bothered me to travel alone. One reason was that as a singleton I could better concentrate on and absorb the exotic surroundings which enveloped me. In addition, I could set my own schedule, agenda and pace. The occasional tinge of melancholy when twilights far from home shaded the evening into night were a small price to pay for the enticements, excitements and pleasures of the open road.

As I discovered, there's "world enough" to see but never enough time to see it all. As Andrew Marvell described time's fleeting nature in "To His Coy Mistress," at my back I could always hear time's wingèd chariot hurrying near as I moved on and on toward a desert of vast eternity. Always, from the time I was boy, always I could hear the chariot drawing ever nearer, its rhythmic approach clip-clopping and clicking away the seconds, minutes, years. That sound accompanied me my entire life, starting back at Phantom Lake with the wood smoke in the air at Camp Ironwood in Maine. Those campfires have long since burned into smoke, into nothing. Robert Louis Stevenson said that the "great affair is to move," so surely he must have approved of the prime mover, Time. Time's a truly great affair for earthlings as it never stops moving. Since passing time relentlessly moves, better an elegant chariot to remind you of the message than a nag like Stevenson's donkey. Father Time deserves the best. I'd hate to think that the Old Boy who runs down our clocks was reduced to an ass. Lack of time prevented me from seeing more than just a relatively few of the play infinite forms on offer for a curious humanimal who roamed the lonely planet. Of those few which came to me I include far fewer in this reminiscence. Those few episodes represent only some selected examples of the whole. Perhaps it's possible for a reader to imagine some of the rest by reading between the lines to see in the unwritten blank spaces what lies behind the text. Because time and space limit what little I can include, most of what happened remains beyond the covers of this book. What I chose to recover from my past and put into my account will hopefully serve to reflect the entirety.

It's quite by design that my narrative is somewhat discursive, sketchy, eclectic and not chronological. A disjointed account, with its disjunctions and diversions, in fact represents just how travel experiences reach "the wanderer." Nothing comes, as at home, in a regular even flow. Away, a helter-skelter jumble of impressions swarm in at random. I've tried to suggest something of the confusing nature of travel, especially in underdeveloped areas, by presenting my account in a discursive way, bringing in collateral but not irrelevant matters in an attempt to reproduce how events actually happened. Rather than a complete and orderly *tour d'horizon*, this reminiscence is a "detour d'horizon." The detours transform the narrative from purely a narrow recitation of what happened to include various other matters, ancillary but to my mind related, on lessons learned and new perspectives as suggested to me by my experiences. As noted earlier, travel gave me a graduate course regarding one humanimal's experiment in living. Even now, however, as my roving seminars near their end, I haven't qualified to receive a diploma. There's too much to learn ever to graduate. At this late date, when almost all my travels and my thoughts about them are behind me, like time's wingèd chariot at my back drawing ever nearer, I can perhaps pass a semi-final exam on what it all meant. But I merit no diploma, as my report card reads "incomplete." Although I failed to fulfill all the requirements to graduate, from my travels I learned a wide variety of assorted lessons, some significant and other less so. Two of the most important I'll discuss in Part 2 below.

~ 2 ~

In reviewing, now near the end, what travel meant to me two major themes emerge. The first great lesson I learned, not long after embarking on my wanderings, informed me that a journey based primarily on curiosity should involve not only museums, places listed in guidebooks, cultural attractions, local traditions, natural habitats, native cuisine and other such substantive elements but also a lot more. For a complete travel experience, insights as well as sights and sites are necessary. Well-wrought travel should create a much broader and ultimately richer assortment of the available infinite forms than simply moving about to see stuff, stuffing yourself with culture and culinary specialties, checking off one-by-one your listed must-see tourist attractions. If travel were only that, then the enterprise would have held much less fascination for me. For me travel meant more than only travel. It provided real-life real-world experiences which showed how humanimals the world over arranged in so many different and creative ways their responses to the experiment of existing in a particular setting. What I found illustrated how earthlings made themselves less lonely on the lonely planet. The wide range of formats represented variations on a theme—how to exist, given the contours of three concentric fatal circles: your own; the conditions in the society, culture and country you inhabit; and the irresolute and ineluctable forces in mankind's natural habitat, its planet earth homeland. Everyone everywhere was subject to those three fateful elements, but people in different lands reacted differently to those factors, the first two situational and the third universal. Those variants interested me as much as the conventional tourist sights and activities. For me travel wasn't a vacation but an education. The only vacation was my effort to vacate long-held conventional views and replace them with new perceptions based on the novelties I found when out in the world. There was a lot of strange stuff, both from natural forces and the mind and hand of wo/man. All sorts of permutations existed out there, a great spectacle of fluctuating forms. I wanted to see, hear, taste, feel some of them. (I could do without some of the smells.)

By degree over many years my trips furnished me with a graduate course worth more than any number of academic degrees. It was educational to discover the commonalities and contradictions. How different countries, cultures and societies arranged their attempts to impose order and meaning upon the given conditions which prevailed in each place operated to create plays of many varied infinite forms. Circumstances greatly affected how locals dealt with the challenges of existence. Both situational and systemic forces shaped responses to the experiment in living. Highly developed countries functioned with intricate economic and societal systems underpinned by the efficient meshing of many complicated and disparate elements. Third World residents lived, mostly barely existed, in catch-as-catch-can basic elementary conditions. But whatever the standard of living, magnificent or miserable, palace or shack, cuisine or grub, high-livers or underdogs, humanimals everywhere attempted to arrange their societies to establish some structure and to cope with the intrusions and predations of a natural world indifferent to its inhabitants. A lot of effort went into making the lonely planet more hospitable, even if ultimately it was in fact an alien and forbidding place. As Carleton S. Coon wrote in *The Story of Man*, the story "is that man has been converting energy into social structure at an ever increasing rate. As he has drawn more and more energy from the earth's storehouse he has organized himself into institutions of increasing size and complexity." As I saw, size and complexity varied depending on the nature of each country and culture, but whatever sort of system happened to prevail, I tried during my travels to learn what little I could about how each society I encountered dealt with its own challenges. I tried to discern how people around the world—many of them forever confined to their own little native habitat—went about Taking Care of Business, given the local opportunities and constraints and, in a more existential dimension, the ineluctable reality of the cosmic fatal circle which confined earthlings to the lonely planet, with all its unavoidable conditions and its inevitable personal and systemic mortal conclusion. Each and every earthling existing only provisionally as a pre-ghost, how did those mortals arrange their societies and their own lives when everyone was but a phantom-in-waiting?

Among the factors which interested me were history, customs, beliefs, rituals, arts, architecture, the economic system—how human nature adjusted and operated in the particular and often parlous context which also included natural nature and its influence on the local institutions and practices. In many cases I glimpsed only erratically and randomly those often shadowy and indistinct societal components, for many remain hidden from the view of a passing traveler. Obvious sightings such as seeing tourist sights don't reveal the complete picture. Some of your impressions also depend on less evident sensory input enabled by an acute awareness of your surroundings, an awareness which often comes to you unexpectedly. In fact, the very nature of adventure travel entails random and unpredictable happenings and nuances. Out in the world things don't cohere too well. Intended arrangements seldom prove definitive, as the need to improvise often demands an on-the-spot spontaneous revision. Everything is in flux, including you. Plans take you only so far. To cover the full distance involves dealing with many unforeseen situations beyond the scope of your planning. Because of the essentially unstructured nature of independent travel, how you actually experience your adventures presents you with an unsettled inchoate crazy-quilt of impressions like a piano keyboard with a million possible combinations of notes or a baffling jigsaw puzzle with incomplete fragmented pieces. Only after your trip can you in a composed retrospective way try to arrange the confusions into something of a sensible—even if arbitrary—pattern. Part of the charm and challenge of adventure travel arises from the process of accumulating, assimilating and curating the chaos of sensory impressions which you happen to receive. Many of them defy your attempt to bring some order and holistic perspectives to what you've experienced while adventuring out in the big wide world. Much remains foreign, disorderly, resistant to categorization, incomprehensible and meaningless and even mysterious. In this way, among others, travel resembles life. The main difference is that you can realistically pretend that travel makes some sense.

In conformance with my interest in traveling with a perspective broader than simply the usual F-grade Frommer-Fodor-Fielding versions or packaged tours, during my trips I made a point of visiting and observing local places and practices beyond the conventional tourist sights and activities. Some of the obscure elements I found came from local suggestions, while others originated from reading travelers' accounts—recent and vintage—of their adventures. The unusual interested me as much as the well-known doings and seeings listed in guidebooks. I felt that as long as I was going to all the time, trouble, effort and expense to get somewhere I might as well try to extract from the experience as much as possible. In the spirit of Henry James, I wanted to be a traveler on whom little was wasted, even if my more grueling trips wasted me, and in any case probably more escaped me than I was able to absorb, process and retain. Existence is for the most part rather wasteful, especially at the end when all becomes offal.

One element of each place I found quite revealing was the nature of its complexity. As Coon observed, social structures evolve into ever more complicated formats. Just how these complications actually operated evidenced in part the way each society in reality functioned. For an outsider this sort of information remained much less apparent than the usual tourist attractions. But independent travel forces you to cope with local conditions, so bringing granular insights. The process of dealing directly with locals directs your attention to many subtleties otherwise missed. Those on organized tours or casual travelers often fail to discern nuances which can be quite revealing. During my brief passing visits I did the best I could to learn about those particularities and peculiarities. The underdeveloped areas presented complexities based on a bewildering chaos of languages, dialects, customs, practices, beliefs, tribes, clans, sects and sub-sects, religions, all greatly complicating efficient governance, the economy, and other elements needed to promote stability and well-being. Competition and conflict between groups and a struggle to access limited resources, grabbing for crumbs rather than sharing pieces of an expanding pie, led to disorderly and dysfunctional situations where complications arose from cultural fragmentation. As Milton Eisenhower observed following (in my footsteps) his summer 1958 visit to Guatemala, the country suffered from "under nationalism" due to "an absence of patriotism." Loyalty was given not to the country but to one's affinity group.

By way of contrast to the Third World, in highly developed countries the complications arose from the very factors which facilitated a high standard of living. Intricate supply channels, webs of communication and utility lines or waves, complex infrastructure, on-demand goods and services, regular and frequent transportation links, efficient distribution chains, just-in-time production lines, on-call medical facilities, always-on big data digital connections, easily accessible news, instant information and much else represent high-touch high tech complexity. Modern miracles of logistical efficiency bring further complexity to advanced societies. Supermarkets, pharmacies, hardware stores, auto parts warehouses and retailers, shopping malls and other such establishments are filled with thousands of curated SKUs, each "stock keeping unit" always available for customers to fill every need at any time. It's an amazing system of unimaginable complexity, one which took a lot of thought, planning and execution. Countries and economies so rich in accessible and functional goods and services are complex in ways quite different from fragmented societies complicated by unproductive rivalries, dysfunctional economies and oppressive regimes. They're mirror images: in the First World, things mesh; in the Third, frictions fray connections. In the Third World, complexity arises from fragmentation; in the First, from integration. On my far-flung travels I tried to discern some of the factors which comprised the ways which made a place complex. This information in turn brought me some enlightening insights about each area I visited.

My notion that travel represents more than just the trip itself greatly expanded the range of experiences I enjoyed while on the road. I veered into side streets to search out obscure corners of a city, trying to absorb as much as possible. If you stay on the straight and narrow, like a trolley track, the itinerary is confined to a fixed route. I detoured to visit places off the beaten track. I sought nuances, hints, eccentricities, subtleties in an attempt to grasp the essential elements of a place. This enriched my adventures. In an 1828 *Edinburgh Review* article English historian Thomas Babington Macaulay described how learning about foreign places from books—and of course visits are not only better, but necessary for a real learning experience—expands your mind by introducing you to a variety of new views: "Historical reading is analogous, in many respects, to foreign travel. The student, like the tourist, is transported into a new state of society. His mind is enlarged by contemplating wide diversities of laws, of morals and of manners."

This educational dimension of travel, as described by Macaulay, I tried to include in my experiments in experiencing the world. At its best, then, travel is learning, a kind of practical on-the-job training based on real-life out-in-the-world experiences. A good traveler draws from his adventures many lessons which transcend the actual journey itself. Like the education provided by reading, travel gets you out of yourself and introduces you to other realms—places, societies, cultures, people and peoples. This refreshes and enriches your life, nourishing it with new perspectives, ideas, concepts. But reading and roaming differ radically, as one may pursue only as an inactive armchair traveler, while the other takes you out into the real world. My basic mission was not to read or to write about the world but to experience it. Both ways, however, on-the-fly and on the chair, serve to unsettle your fixed ideas. The clash between your inherited givens—the common practices around you, your long-held beliefs, your native habitat and familiar surroundings, the conventions, the prevailing local routines—with all the variations and permutations you encounter out in the world serves to spark in your mind newly illuminated perspectives. These supplement, or in some cases replace, old certainties. Among other benefits, a trip should produce at least a few alterations in how a traveler views the world and life. From new lessons learned in old lands a globetrotter should in part reset his or her mindset.

In addition to the first major lesson I learned from my travels—to enhance the trip by transcending its basic nature as only a travel experience to include the additional educational dimension—a second major realization taught me about the transcendental benefits of world travel. Expanding visits to include educational elements commonly beyond the range of a typical travel experience represents a lateral extension of the enterprise. The second lesson entails another dimension, an in-depth awareness of the past. Macaulay's comparison of historical reading to travel describes this second aspect of adventures on the open road aimed at opening the traveler's mind. To get the most out of a trip, especially one to a foreign land, a traveler should supplement simply experiencing a place with an awareness of its place in time. To the on-the-ground sightings and impressions should be added the historical background. To understand a place requires that its current as-it-now-is status be enlarged by a deeper knowledge of how it was—its history. Only with some sense of the past can the present be grasped. As time unfolds, it congeals into history, a fixed back story which shapes the present and the future. By understanding how a place came to be you add to your visit a perspective beyond what you see. This enables you to transcend the immediate here and now and expand it into the past.

Awareness of the past is useful by adding depth which allows the visitor to get below the surface of a society, the better to understand it. Some of the missing pieces, the puzzle of a place, fall into order if you know something about its past. But even—or perhaps especially—the most detailed knowledge of history will leave you rather ignorant for it's difficult or impossible to disentangle the intricate tangle of past time which shaped a society. Like all other opinions, conclusions drawn from history are simply the historian's view based on his perspectives and his educated and at times over-educated guesses which may or not be valid. Differences and disputes over the meaning of the past are common. "Time will tell" is inaccurate, as often time has not told with clarity. Uncertainties over how what happened influenced later developments, the many mysteries and confusions in the shadowy past, and other imprecise events and influences veiled in the misty haze of dead time all blur accurate perceptions of what came before an epoch or an event and what happened afterwards. In his biography of Abraham Lincoln Lord Charnwood, commenting on how the president's advisers influenced his decisions, concluded that "If we tried to be exact in saying Lincoln, or Lincoln's cabinet, or the North did this or that, it would be necessary to thrash out many bushels of tittle-tattle." Similarly, to grasp even in small part the historical significance of past events upon present conditions would involve tittle-tattle tales and details of such complexity, never could you complete a definitive analysis. In any case, whatever you concluded would simply represent one person's opinion. What professors and professional historians profess as authoritative is only facultative—just another passing stray element in the vast play of infinite forms.

Although the past in many ways remains inscrutable and in all ways—other than your own past—lies beyond your immediate knowledge, with an actual journey you experience the world directly and vividly. But even immersion in that first-hand real-life experience remains difficult to comprehend. Much of what befalls you in the course of a trip really can't be easily understood, either as it occurs or with the benefit of hindsight. Everything—the past, the present, the future, humanimals and their myriad ways, the lonely planet, the desolate moon, the even more lonely cosmos with its churning galaxies, the shooting stars, comets and meteors, spinning matter and infinite space—consists of a bewildering confusion of tittle-tattle. No one knows what it all means. In February 2002 Secretary of Defense Donald Rumsfeld, asked if he knew of any substantial evidence that Saddam Hussein in Iraq was supplying weapons of mass destruction to terrorist groups, replied in what was to become a famous epistemological formulation: "There are known knowns; these are things we know. We also know that there are known unknowns; that is to say we know there are some things we don't know. But there are also unknown unknowns—the ones we don't know we don't know." Rumsfeld's observation applies directly to adventure travel, an enterprise which represents an amalgam of the three categories he mentions. In far lands you face the known, what you realize is not known, and the wild card of which you have no idea what you don't know. Trips in many ways represent a way of being which mirrors life in general.

Not knowing defines the very nature of travel. That stimulating uncertainty, which applies to human existence as well as to trips, represents one of the reasons travel attracted me. Life, however, didn't attract me. I didn't volunteer for the experiment. No curiosity motivated me to experience life. I had no choice in the matter. I was neither attracted to nor repelled by life. No one sought my opinion. Existence was inflicted upon me. I became a humanimal without being consulted. The strange phenomenon of life existed long before I did, and one morning I suddenly found myself part of the pre-existing format, which soon—a known known— will continue on without me. Existence is a sometime thing. Temporarily joining the play of infinite forms, I eventually decided to make the best of the situation and seized the opportunity given to me to explore some of those forms, an enterprise which came with various known knowns and unknowns as well as some unknown unknowns. But the uncertainties I faced in my adventures were in general only the relatively basic unknowns of how daily events would unfold rather than the extreme, challenging and perhaps life-threatening mysteries which confronted such early-day explorers as Mississippi River pioneers Marquette and Joliet. In his *Journal* Marquette described how in June 1673 their native guides departed, leaving the two explorers "in the hands of providence" as they entered the then little known waterway "that would thence forward take us through strange lands." So it is, in a much more attenuated way, for latter-day travelers who venture to visit even by now well known and well mapped exotic areas. Out in those "strange lands" surprises and uncertainties lurk, plots and counter-plots threaten. Venturing into those lands you become a plaything of providence, a toy of Lady or Lord Luck. Chance soon presents you with many of the knowns which comprise your trip. With adequate preparation you're aware before departure of the knowns and even some of the known unknowns. It's Rumsfeld's third category, what you don't even realize you don't know, which lends an especially sharp edge to the enterprise of travel. This is particularly true in Third World countries where unexpected random happenings occur with great regularity, so regular that in a way the random becomes routine. What you don't know is the nature of those surprises and how they will play out. Both the known and known unknowns can bring challenges, while the ones you're completely unaware of can be more shocking and present surprising opportunities or serious perils.

Trips operate in a mode quite different from how things work at home. The scenarios are almost opposite. In foreign lands remote both by geography and customs from your homeland you survive in an opportunistic way. It's an existence based on catch-as-catch-can-or-can't improvisations for transport, shelter, water, food, whatever and what-never. In the Third World you move according to the whims of schedules more fictional than timely. You eat odd things at off hours, at times a mid-morning breakfast-lunch, brunch-like grub grabbed as a hurried harried snack. On other occasions you gobble down "drunch," a late afternoon dinner-lunch. In underdeveloped countries I always carried some cans of tuna as emergency rations in case no clean or edible food was available. You stay here, there, anywhere, seldom knowing ahead of time—a known unknown—which roof will be over your head or the pillow under it. A clean water supply is ever in your thoughts, especially when it's not in your system. Water requires very careful attention. In the Third World regular access, as needed, to unpolluted hydration presents one of the most persistent and difficult challenges. Without drinkable water you soon run out of gas, even if some of the food gives it to you. When the supply gets low all else becomes subordinate to the need to find clean water. Your thoughts are diverted from external matters to the inner need for thirst-quenching fluids. It becomes an obsession. You can lose your luggage, skip a few meals, sleep on a train station bench and still keep going. Without water you're stopped dead in your tracks.

In underdeveloped lands the pathogens are very well developed. Nothing, but nothing, should enter your mouth without careful contemplation of the risks. In some places even the air is toxic. Ideally you'd stop breathing in cities laden with highly noxious pollutants. Unlike respiration, food and drink remain under your control. Every swallow and gulp comes freighted with potential adverse consequences. Even the most cautious practices for nutrition and hydration can result in stomach-churning upsets and outcomes. What goes in can come out in undesirable ways. Whoever coined the expression "happy-go-lucky" never traveled much. For a traveler the phrase presents a perverse word order which should be reversed to make it less adverse. Both satisfactory travel and life in general require that luck precedes and enables happiness: "lucky-go-happy."

To enlarge the travel experience "by contemplating wide diversities of laws, of morals and of manners," as Macaulay put it, presents a substantial challenge. It's difficult to make sense of the tangled mishmash of the past and of all the confused happenings which comprise the contents of the trip. The "wide diversities" of the three mentioned elements, both historical and as experienced day-by-day by the traveler, represent only a tiny fraction of all the components which form a society. Because of the vast play of infinite forms, to which both natural forces and humanimals have contributed, it would take the efforts of a lifetime, a lifetime of Methuselah duration, even to begin to absorb and make sense of only a few of the diversities of which earthlings are a part. To understand just one country or a single city, your own included, or your own mind and life would be a major undertaking far beyond the scope of any humanimal. One collection at the British Museum in one little corner of London would require years of study to analyze and understand. London itself,

England, Britain, Europe present much more extensive and complicated geographies. Places like India as it is and Egypt as it was would demand more time and study than even the most scholarly and diligent specialized researcher of great longevity could muster. Such deep-dive scholarly learning into the past was far beyond my capabilities or interest. To prepare for my trips I only skimmed the surface. Time ever pressing, I tried as best I could to educate myself before leaving home by reading selected guidebooks and some general works on topics relevant to the areas I intended to visit. From my research I produced voluminous pages of hand-written notes containing information and prompts which greatly helped me carry out efficient and relatively complete visits. All along the way I frequently consulted my notes to guide my moves. Of course, a lot of what I experienced couldn't be anticipated or reduced to helpful hints for the wayward traveler.

In conformance with my intention to see and to learn as much as possible, all sorts of obscure, eccentric and off-beat details found their way into the pages I'd prepared. In addition to the usual main tourist sights the entries included lesser-known corners, some of them more typical and suggestive than many of the obvious places which attracted most visitors. Even in the most travel-friendly and accessible places around the world—London, Paris, Rome, England, Disneyland, Graceland—I made a point of consulting the local tourist office when I arrived. Early in my travels I learned that a brief session at the visitor's center would more than repay the time spent there. Visiting the tourist office would allow me to reduce the number of unknowns and increase the knowns. In most such offices helpful materials abounded, many with information and details unavailable elsewhere. Back in my day no internet gave access to information which would serve to reduce the frictions of adventure travel. Investing a few minutes collecting the brochures and asking the resident attendants who manned and womanned the office about local matters yielded much actionable information saving time, effort and energy. I sought as complete a package of information as possible. My main goal was to reap as much as I could to compensate for all the many inconveniences, discomforts, hassles, bother and expense of getting to and then around far lands. It was a lot of work to see the world properly. I wanted an experiential return on my investment commensurate with the associated costs. As with any investment, at times my efforts yielded satisfactory results and in other cases the bottom line fell short of the intended return. In any case, in advance of departing on a trip I always tried to moderate my hopes and aspirations by tempering them with reasonable and realistic expectations. That kind of attitude helped to limit disappointments not only for my journeys but also in my life in general as well.

From my early days of travel my goal focused more on seeing many countries in some depth, to the degree possible for a passing wanderer, rather than on visiting most or all of them superficially based on a check-list mentality. I've known and known of a few world travelers possessed by the compulsion to visit many or all of the counties on earth. Often the visits would be so abbreviated in time and restricted in place it could hardly be claimed that the country collector actually saw the country. Such a flying visit represented a mechanical and ritualistic mode of carrying out the enterprise of travel, a practice which didn't appeal to me. Some places, not withstanding my curiosity, didn't much interest me. Had I managed to reach such remote and unappealing lands, the time spent in them wouldn't reward me enough to justify the effort to get there. Even keen curiosity has its limits, and time and energy are further limiting factors. Among the nearly 200 countries in the world enough of them did interest me, so I felt no compulsion to see each and every one. Over the years I restricted my visits to the places I thought would provide me with sufficient memorable experiences to last a lifetime—and by now they nearly have.

~ 3 ~

Without the Big Bang all would have stayed intact as one vast nothing. That certainly would have simplified matters: no explosion, no cosmos, no forms. From the big burst onward what was thrown into existence tended to shatter, fragmenting into pieces to create the playhouse of infinite forms. The process continued, from the cosmic scale on down to the whirling terrestrial pebble inhabited by earthlings. Their civilizations, as they rose and fell, divided the earth into countries, territories, states and statelets, provinces, kingdoms, empires, fiefdoms, principalities, electorates, grand (and not so grand) duchies, all manner and womanner of geographical units. Each enclosure was set apart and delimited by borders, frontiers, back tiers and other lines of demarcation imposed to control or to keep humanimals in or out. Sometimes the lines shifted, but always they existed. What was given to mankind as part of its fatal circle, the globe, became a crazy quilt of separate territories. Like the Big Bang, earthlings fragmented unto units the original pristine unity. It didn't have to be that way. The pre-cosmos could have remained inactivated by the blast or, following it, the world could have been left in the firmament as a single integral province, like the moon, undivided (so far) by boundaries and borders. But that's not how things worked out. By the time I was big banged into gestation and then emerged onto the earth, the variety of countries which checkered the globe's map had turned mankind's homeland into a greatly divided terrain. This made the planet interesting but also quite complicated and challenging to see. Over more than half a century, I tried my best to get around to most of the main territorial fragments which comprise the riven lonely planet's surface, but 50 years didn't give me enough time to see it all. Even with my wide-ranging travels I barely scratched the surface, beneath which I will soon repose forevermore. But at least I managed to get out and about to experience as much of the place—a strange, eccentric, fanciful creation with its known unknown of just why the place exists—as I could.

By common consensus, based on tallies by governmental and other official organizations and agencies, about 195 countries occupy the earth's surface. But an organization called The Travelers' Century Club lists 327 separate units, not all of them countries. I never joined the club, as I never needed or wanted to validate my travels with any sort of formal recognition or credentials. Established in 1954 essentially as an organization to give "bragging rights" to members claiming to have visited at least 100 of the names listed, the club based its inflated number of eligible places on a broad definition of polities: "Although some are not actually countries in their own right, they have been included because they are removed from the parent country." In any case, the bloated list offers a happy hunting ground for travelers who aspire to bag a century and for more ambitious types who aspire to visit every "country," a checklist sort of accomplishment very few people have managed to attain. One member visited all 3,146 U.S. counties (much easier these days when there are only a mere 3,114 counties), but even the Century Club doesn't consider those political units countries. Some of the most questionable entries on the list illustrate its quirky nature: Scotland and Wales; seven national areas in Antarctica; Pacific islands such as Easter and (part of the Line/Phoenix group) Christmas, Lord Howe (how did Howe get in there?), the Ogasawara group (never managed to get there), Juan Fernandez (the Robinson Crusoe Island), Nauru and Niue. One clear kind of overreach is the division of Turkey into Europe and Asia and Egypt in Africa and Egypt in Asia (Sinai Peninsula); in each country you can reach two qualifying listings for the price of one.

Not only can you become a centurian-like road warrior by visiting any number of point-worthy "removed" places, you can reach the century mark merely by setting foot in a country without actually advancing into it or seeing anything there, for "even the shortest visits would suffice—even if only a port-of-call or a plane fuel stop." That's really no way to run a travel life. Only one country in the world did I barely graze, and never would I claim to have visited the place. In the small building which straddles the border at Panmunjom I circled the table, walking in a few moments from South to North Korea and then, thankfully, back to the South. These were the fastest border crossings I ever accomplished. Was I "in" North Korea? Yes. Did I go to North Korea? No.

What, then, is a country? In 1933 the Montevideo Convention established the principles that four attributes were necessary for an area to claim the status of a "country": a permanent population, a defined territory, a government, the capacity to engage in relations with other states. Over the centuries varying numbers and types of geographical constructs have become known as countries, the mutations and permutations changing their dimensions according to how humanimals decided to chop up the earth's 197 million square miles. Civilizations and its discontents seem to favor dividing rather than merging the globe's political units. At the present time, but subject in the future to slight variations, the United Nations has 192 members, as does the Universal Postal Union, while the International Telecommunications Union has 193 (Line Island group not included), and the World Health Organization includes 194. The U.S. State Department shows "Total count of independent states" at 195, which excludes Taiwan, not officially recognized by the U.S. as an independent country. It's apparently a dependent country, but on what or whom? *The Economist* refers to 196 countries; Stanford's travel book store in London estimates "about 193"; Coca-Cola sells in 195 countries, while Nestlé sells in 189 (heads will roll in Vevey if Nestlers fail to enter the missing six soon). About 180 recognized currencies are used in the world. The (not quite) World Bank has 189 members. Consider all these figures and they add up to some 195-odd countries, some of them quite odd indeed, a mess of a map created by mankind on his allotment of some 200 million square miles, a speck of cosmic dust divided into smaller specks. For this lonely planet spinning in space the 2005 Lonely Planet's *The Travel Book: A Journey Through Every Country in the World* listed 230 countries, perhaps hoping to increase the total—the more the merrier—to make the world a little less lonely.

To visit enough countries—real ones, not just strays—to qualify for the Century Club you'd need all of the lands located in North and South America, Europe, all around the Mediterranean and most of Asia. That leaves for a second hundred the former Soviet Union, the Middle East, Africa, such remote corners as Bhutan, Nepal, Mongolia, many island countries, and other outliers. Although many places in the first hundred are difficult to reach, the second "century" countries are even more remote and much less accessible. Based on the Century Club list, over more than half a century I managed to visit some 165 places, but reached somewhat fewer of the 195 or so polities considered as countries under recognized international standards.

On its website the *International Travel News* lists 196 countries. On offer is a "Been There, Done That! All Sovereign Nations Award" for travelers who have visited all the countries. I managed to reach only 125 of the 196 total so am ineligible for the certificate, which I wouldn't request anyway as it costs seven dollars. But each of the countries I visited I saw in some depth and breadth. I covered most of the places listed for North America (all 3), Middle America (17 of 20), South America (11 of 12), Europe (42 of 45), and Asia/Middle East (34 of 47), but fell short in Africa (15 of 54) and Oceania (3 of 15, of which 11 are small islands). Of the 71 missed, I'll probably never see any of them as by now I'm eligible only for the "Will Never Go There or Do That" certificate.

Many of the far lands I visited I reached in the early to mid 1970s on my three five-month-long trips, one to South America (every country except Guyana), another to North Africa and the Middle East, the third to the Far East. That trio of long journeys greatly added to my tally of countries. Many others around the world I visited in a more piecemeal way on shorter trips, both before the 1970s and then afterwards over the next 40 years. Once I restructured my life, as of 1969, for starters I took the three long trips in the 1970s. Some dozen or so places I hoped to visit I never managed to reach. Because those lands are far too far to get to conveniently and would require a long hard trip limited to just the specific country (I've seen those nearby), it's unlikely I'll ever manage to set foot in any of them. By now such distant destinations are for me beyond reach, so my adventure travels are over. Although I still re-visit a few places to see friends for specific events, mainly funerals and weddings which some spoilsports might consider similar, I've by now come to accept that I won't be able to reach those dozen or so far places I remain interested in seeing. Until recently I was in denial, still pretending that I'd eventually travel to them, but the realities of old age and their effects eventually suppress even the keenest curiosity and reduce a person's willingness to give up the comforts of home to see new places. I traveled when the going was good and when I could. Although I'd enjoy seeing more, I'm satisfied that I saw enough.

Never did I engage in Century Club-like checklist travel. On the other hand, I greatly enjoyed converting fanciful place names into experiences. So many of the names were so strange, odd, unusual, exotic and irresistibly alluring I couldn't stay away from the places those names designated. All through the years maps intrigued me. Many of those oddly shaped colorful countries along with cities and geographical features labeled with black type fascinated and eventually attracted me. A city like Cox's Bazar in Bangladesh had such an evocative name I longed to go there but, alas, I never reached the country or the city. But was it a city?—maybe just an overgrown market town. I wondered who Cox was and what his bazar looked like. The name evoked vivid images of the British Empire, of the Raj, and of an early-day trading post in the mysterious East, a colorful and perhaps even bizarre bazar there on the Bay of Bengal coast. But then again perhaps the settlement in no way resembles how I envisioned it. I'd come to know the place, get the real feeling of the city, only by seeing Cox's establishment, whatever it was, for myself. No description can take the place of a hands-on, feet-on-the-ground presence in an exotic land.

Back in the 1970s, of course, no instant internet information existed available at the command of a tap or a click or a voice. These days I could look or talk online (but never have) to find out all about Cox and his exotic bazar. But that's just not the same as going there and being there, immersed in the place. The internet can give you information and opinions but not experiences. What's called virtual reality isn't at all real reality and no substitute for the real thing. Invented artifacts — digital, creative or otherwise—don't offer realistic surrogates for the real world. Only imagination based on experience can visualize life as it actually is. Even today when miraculous digital technology casts its spell on humanimals, the world-wide web offers images and visions far inferior to my real-life experiences around the world which wove the intricate web of my past. But earning those experiences required a lot more effort than would virtual visits.

Although I never managed to see the bazar town named for Cox, I did reach many other fabled places with legendary, exotic or evocative names. It would be a poor adventure traveler indeed who never saw for himself places like fabled Timbuktu or the many other marvels around the world which attracted me as if by a magnetic force. Their irresistible names teased, enticed, enchanted and drew me to them: Petra, Pagan, Palmyra, Penang, Persepolis; Easter and Devil's Islands, Angel Falls, Death Valley, the Dead Sea, and in Norway I went to Hell (Hell in Michigan, northwest of Ann Arbor, I missed) and to Paradis; Upper Volta, Middle Tennessee, Lower California; Tibet, Bhutan, the Khyber Pass; Sucre and Potosí in Bolivia, Khiva in Uzbekistan, Medina and Madain Salih in Saudi Arabia, Dire Dawa in Ethiopia, Ghadames in Libya. On and on goes my hard-earned list, names foreign and magnetic: Borobudor and Prambanan, Anuradhapura and Polonnaruwa, Chichen Itza, Chichicastengo, Nagh-I-Rustam, Kom Ombo, Altamira, Meteora, Iguaçu, Abidjan, Xi'an, Fatehpur Sikri, Bandariaga, Gonvie, Thingvellir, Split and Bled (places, not a happening). All over the map, those strange names: al-Kharga, Cairo, Djenet; Dawson City and Belize City; Tikal and Tiquisate; Manaus, Cuenca in Ecuador and in Spain; Yemen, Oman; Katmandu, Mandalay, Manchester, Manhattan; the ghostly remains at Pergamum, Machu Picchu, Angkor Wat, Hampi, Leptis Magna, and not to forget Phantom Lake at Camp Ironwood where the phantom campfires, soon ashed into nothing, still burn in my memory. All those places and so many others I visited I still remember, a mishmash mash-up like a mental gazetteer of the great wide world I came to know.

Even more familiar names of more accessible places piqued my curiosity, and over time drew me to them: Paris, Parma, Palermo, Pensacola (a Coca-Cola competitor?), Puebla, Penang, Paramaribo, Panama, Peoria—what on-the-ground real-world presence did each of those names represent? Each designated an entire mini-world where humanimals pursued their fleeting lives, pursued but never caught them as time, ever on the wing, pursued every mortal soul. Names of all kinds, foreign by evocation or by location, exerted a powerful and almost mystical influence on my imagination. They stimulated and activated my curiosity, agitating it into a driving—or flying—force, impelling me to see the places represented by the names. A kind of teaser, the name only denoted a particular location; the place itself was the essence of the matter. In *Tropic of Cancer* Henry Miller recalled how the name of Santos Dumont, the famous Brazilian pioneer aviator, mesmerized him: "it smelled of sugar and of Cuban plantations...Santos Dumont was a magnificent word which suggests a beautiful flowering mustache, a sombrero, spurs, something airy, delicate, humorous, quixotic," a coffee bean aroma. In December 1971 on my long trip through South America I inhaled for myself that seemingly sugary coffee-aromatic romantic name when I went to the Santos Dumont airport (used for domestic flights) in Rio de Janeiro (somehow, Sugar Loaf in Rio didn't seem as redolent of sugar as did "Santos Dumont") to fly to Brasilia. Later in my trip I visited the truly sugary-named Sucre in Bolivia, the country's remote nominal capital. But what's in a name? I didn't know if La Paz was peaceful, but the city was in fact in effect the center of government in Bolivia. Far from the big city, Sucre was an isolated provincial town devoid of the usual pomposities and pretentions found in purpose-built soulless capital cities such as Brasilia, Canberra, Washington or in a grim place like Albany where government is the dominant business. I unfortunately didn't manage to travel west from Canberra out to Wagga Wagga or to Walla Walla, two doubly colorful places I'll never see.

For me place names stirred all sorts of visions of far lands, colorful scenes, exotic surroundings and anticipation of all the novel, unpredictable and exciting delights travel can bring. In the names I read adventure, and I went the last mile to reach those places. Bernard Berenson reflects on the magnetic effect of place names in *A Matter of Passion*, a collection of his letters. Writing in September 1933 about his travels in Austria, the art scholar noted, "There may be nothing worth seeing at Graz, but names of towns tend to become 'motor ideas' demanding realization. I shall be an infinitesimal shade less dissatisfied with myself when Graz shall be more than a mere name." Such was the case for me, as by travel I aspired to bridge the vast gap between merely nominal knowledge of places to actually experiencing them. All too many "motor ideas" fueled my desires to see the names in person. As a major subset of curiosity, travel provided me with one way, a major one, to fulfill the needs of that compelling trait.

Oddly, a book on a localized and static activity quite the opposite of a globetrotter's wide-ranging roamings contains a description of the qualities needed for an ideal travel experience. In Izaak Walton's 1553 *The Compleat Angler, or The Contemplative Man's Recreation*, he observes that "he that hopes to be a good angler must not only bring an inquiring, searching, observing wit, but he must bring a large measure of hope and patience." This temperate attitude conforms to the characteristics desirable for a serious traveler. Walton's subtitle reference to "the contemplative man" especially resonated with me because my approach to travel attempted to combine the active and adventurous elements with a certain reflective and meditative response to the happenings and sightings. That synthesis allowed me both to experience the trip and to contemplate how its events might inform more general matters. I aimed to extract from travel not just the immediate sensory benefits but also additional perceptions and conclusions which might enlarge my experiment in living. This goal was in fact the basis for my decision to restructure my life. Walton's reference to "recreation" echoed my attempt to re-create my way of being by gaining exposure to many diverse inputs unavailable in my cozy little provincial suburban subdivision. My new subdivision entailed sub-dividing my life into two parts, adding adventure travel to the routines of my home-based subdivision so as to expand my horizons beyond the dead-end street where my existence unfolded day-by-day with few variations. That regularity both pleased me and motivated me to abandon it from time to time.

Overseas adventures, many of them random, contrasted greatly with the more predictable patterns which operated at home. Mercurial passing moments of an on-the-go trip characterized the activity. At the very outset of my independent foreign travels as a very young man, I realized that the best and, in fact, essential method to capture and organize a few fragmentary impressions, reactions, contemplative thoughts and tentative conclusions from such fleeting experiences would require keeping a journal. Starting with my first trip to Europe as a teenager in the summer of 1955, a lifetime ago, I began to write a daily account of my experiences and my reactions to them while on foreign trips. Although the journals are filled with detailed accounts of my activities and impressions for every day I was away in foreign lands, for this reminiscence I haven't consulted one line of those diaries, voluminous even if not luminous. One reason I avoided looking at them is because I didn't want to get lost in the past. More importantly, I shunned the journals as I opted to include in this account only those episodes, happenings and recollections which remain so vivid and memorable they truly represent the most striking and lasting impressions from the travels of a lifetime. I didn't keep the journals with the intention of using any of the material for my published travel writings. In any case, my books and articles came long after many of my trips. When in 1955 at the outset of my overseas travels I began the practice of producing a daily account of my foreign travel experiences, I never contemplated writing about my travels. For me the trips were merely gratuitous curiosity-driven adventures unconnected with any specific or practical purpose. Only nearly a quarter of a century later did I begin to produce travel articles and books for publication, and even then most of my journeys were motivated by reasons other than to write about the trips.

Potential books or articles didn't feature in my decision to use scarce trip time to make the effort to write detailed journal entries every evening summarizing the day's activities and my thoughts about what had happened. One benefit of this daily labor arose from knowing that recording my impressions required me to focus my attention on the ever-changing scenes and variety of events I happened to encounter. This made me more aware of the transient and shifting minute-by-minute sensory impressions which accumulated in a confused rush to form the day's contents. As they appear and then disappear before your very eyes and reach and then escape your other senses, you try to fix your gaze and ears and nose on the fleeting images, sounds and smells to capture a few memorable remnants of the plenitude of the play of infinite forms you happen to experience. The journal serves as a transformer to reduce the sensory overload to a more manageable format. A contemporaneous journal preserves fragments of the insubstantial, mercurial, will-o'-the-wisp, smoke-thin, phantom-like passing events which every moment disappear into time's residue, the past. With a journal, the few stray scraps you manage to salvage remain available later, long after you've recorded them. The entries jog your memory, helping you to revive, review, relive and re-enjoy some of the passing pleasures which momentarily delighted you during your travels. The on-the-go writings also prompt you to recall some of the miseries you suffered which at the time vexed you but have long since lost their bite. I believed that it was something of an abuse of privilege and a misuse of resources—physical, temporal, financial—to go to all the trouble and expense to travel, and especially to distant and exotic lands, without preserving some sort of record which captures at least a few shreds of the fleeting experiences. The same attitude induced me to produce this reminiscence. I felt that someone who enjoyed something of charmed life, including among its benefactions many twilights far from home, should share some of those travel delights and some of their lessons by preserving them in print.

The travel journals offer a complete account of the world as I saw it in the second half of the twentieth century and the first decade or so of the twenty-first. I was methodical about writing the daily entries every evening before the contents of the day slipped away and a new day brought yet another deluge of happenings and impressions. As for many other elements of my life, I was compulsive in the task, but producing the journal was a pleasurable compulsion. To give myself an incentive to complete the job, I'd delay dinner until the account was finished, and only then did I allow myself the grub. I consider the collection of travel notebooks as among my most valuable possessions, even if to anyone else they're to a large extent worthless, other than to use to start a fire, in which case my past would literally go up in smoke. As it is, the past only seems like smoke, an insubstantial vapor which has disappeared. Although hurried writings, caught on-the-fly from my trips, enable me to relive in my mind pleasurable passing past events, my journals are meaningless to other readers, unless someday in the far future someone will somehow be interested in a record of how the world appeared to one curious humanimal who saw the lonely planet during a brief span of time in a particular era.

My main advice to other serious world travelers was to keep a travel journal but, as for my accomplished friends who I urged to write a memoir, I imagine that few globetrotters followed my suggestion. It's a lot of trouble to keep a journal. Part of the art of shaping something of a charmed life is to decide just what's worth troubling yourself for. It's preferable to choose the kinds of problems you're willing to deal with than to suffer the problems which happen to come your way. Once in a blue moon I've wondered if the fresh-faced young Australian student I encountered in late 1973 in Katmandu followed my recommendation to keep a travel journal. We happened to meet one morning when renting bicycles in the Nepal capital. The two of us pedaled together out to Patan, not far from the city. Some 15 years older, I had by then completed many of the far-flung travels of the type she was just beginning. You see those young Aussies all over as they make their way around the world, some en route to London where many of them live in the Earl's Court neighborhood, known as Kangaroo Alley. The Aussies invariably carry in their backpacks a touch of home in the form of a jar of Vegemite. Although friendly, fun and outgoing, the Aussies' taste in bread spreads is quite questionable and no fun at all. Their beloved Vegemite is best left to them. The icky-gooey substance is definitely an acquired taste, one seldom acquired by non-Aussies. Understandably, only some 2 percent of the stuff is sold outside the home country, and most of that no doubt to expatriates.

As we wheeled out to Patan, my bike buddyette asked me for some travel tips. I told her that the most important advice I could offer, other than to leave the Vegemite at home, was to keep a journal. She was young, just beginning her travel adventures. Without a journal, she'd remember little of what happened in any detail. Our encounter and my advice—and all the rest: her travels, Kangaroo Alley, the people and places and the play of infinite forms she experienced—she'd probably soon forget. Life back home would envelope her, closing her off from the memories. They'd atrophy and eventually mutate into vague ghost-like phantoms from her long gone past. But I well remember our brief time together—one of the thousands upon thousands of the play of infinite forms which came to my attention over the years. I admired her form, physical and conversational, but doubted that she was yet informed enough to understand my advice. To someone of her age, time would seem infinite and memory certain. No way she'd forget any of the play of infinite forms which came to her. No way—she'd remember it all forever.

Her shapely legs pumped gracefully up and down and up, pleasing limbs and a pleasing rhythm, as we rolled along into the countryside. Her bilateral bicycle movements reminded me of birds twin-winging in the high blue windless skies through the Mississippi Flyway back home, far, far away. Time went by, our visit to Patan soon ended. We biked back to Katmandu, bid each other adieu and parted company to end our passing interlude together. I of course have no idea if the young Aussie ever followed my advice to keep a journal. By now, getting on to half a century later, the then young lady is now an old one and maybe even six feet under or more likely settled down Down Under, perhaps with her Vegemite-fed grandchildren these days off on their own world travels. If she failed to keep a journal, the old lady will be able to recall far fewer of her youthful global travels than can I of my long ago worldly adventures. When I occasionally read just a few lines, an entire experience revives in my mind and I can pleasurably recall some of the passing episodes which I so enjoyed over the long years during my many twilights far from home.

Much of what I failed to record soon slipped away, never again to be remembered. Like wood smoke melting into the air, the past and its events just vanished. Of course, even the events I memorialized in writing disappeared, but at least the journals—and now this book—preserve a few phantom traces of what happened once upon a time. In *Present Past Past Present: A Personal Memoir* (my reminiscence is in the nature of an impersonal memoir) Eugène Ionesco discusses (Helen R. Lane's translation) the significance of durable personal accounts: "Despite wars, however, despite revolutions, despite natural catastrophes, despite invasion, despite the centuries, despite the changes of societies and of languages and of ways of thinking, texts may still exist to bear witness to death but also to life. The memories of a Japanese lady at court in the eleventh century, the memoirs of a bourgeois of the sixteenth century have come down to us through innumerable disasters. It is necessary to leave traces."

It may not in fact be necessary to leave some traces, but it seems desirable. For people who don't enjoy the opportunity to adventure out in the world and to write travel journals, children and their descendants serve the purpose of leaving traces. Or you can spawn both writings and new humanimals. In my case I thought it might somehow at sometime or other be of some interest to a few other earthlings if I passed on some of what I saw and learned during the course of my travels, both around the world and through the dark Dantesque woods. Such is one of my purposes in devoting my little remaining time to produce this reminiscence. So many twilights far from home did I glimpse as a day of travel waned, shading into night, I decided to preserve in these pages a few memories from long gone time—life-vanquishing vanished, vaporized adventures—so that they didn't all disappear forever like a dying campfire's smoke.

~ 4 ~

As every world traveler knows, the most common question you're asked is a variant of which country is the "best" you've seen—the most interesting, the most enjoyable, the most memorable, the most unusual, the most exciting, your favorite. The answer of course depends to some extent on a traveler's own interests and tastes as well as on his particular experiences in a country. Great personal interactions and experiences in a lesser land can favorably color your impressions, while unfortunate events in a fabled or alluring locale can darken your views. If you get mugged or your pocket picked or dysentery in Shangri-La, the place won't seem too idyllic. No matter how the "best" question was phrased, I always responded with the same three names. Two of the countries no seasoned traveler would dispute, even if they didn't represent his or her own choices. My third choice, however, was more surprising and controversial and, I admit, disputable. That outlier I'll discuss later in this Section. As for the two more conventional choices, in my opinion both Egypt and India share top ranking, the former for its evocation of the distant past and the latter for its present-day fascinations. Each country offers in its own way an outstanding travel experience, the one typified by its place in time and the other by its rich contemporary ambiance. No travel agenda is complete without a comprehensive in-depth visit to Egypt and to India.

In the spirit of the millennia which Egypt incarnates with its monuments, a well-known age-old perspective on the country serves to convey the sense of great antiquity. In the first sentence of *Persian Wars* Herodotus (died 424 B.C.) states that he writes "in the hope of preserving from decay the remembrance of what men have done," as the famous Rawlinson translation puts it. The 1987 translation of the passage by David Grene (of the University of Chicago) offers a more colorful and evocative version: Herodotus writes "so that time may not draw the colors from what man has brought into being." Herodotus wants to grasp the past and give it color and life as best he can through words. In Ionesco's terms, the ancient historian hopes to leave traces—and he did, both of himself and of old Egypt. Without some sort of recorded account the past, whether humanity's or personal, becomes colorless and forgotten, a reality which prompted me to write my travel journals. Those materials—and this book—will of course inevitably eventually be forgotten, but at least I've slightly delayed the process. In the Prologue to *The Travels of Sir John Mandeville*, the famous 1499 hoax account of the author's invented journeys "through many diverse lands," he notes that because of "the fraility of mankind...things passed out of long time from a man's mind or from his sight, turn soon into forgetting." Herodotus' book of history helps mankind, frail as it is, to to keep in mind Egypt's long-time past.

In Book II on Egypt Herodotus writes: "I shall extend my remarks to a great length, because there is no country that possesses so many wonders." It's striking to bear in mind that by the time Herodotus was writing about Egypt in the 400s B.C. Egypt had already produced centuries of wonders. The royal line began c. 3000 B.C.; the first pyramid, the Step Pyramid, dates from around 2650 B.C.; King Tut was born about 1325 B.C., nearly a millennium before Herodotus. Herodotus is as close to us in time as the 2,500 years which separates him from the advent of Egypt's royal dynasties. To us one of the remote ancients, Herodotus wrote about Egypt when it was already an age-old land. Of few if any other countries can it be said that there exists a continuity of "many wonders" over such a vast range of centuries.

Although such a long history which produced so many significant traces of the past would seem to offer an untold number of opportunities for comment, in fact Egypt's very antiquity diminishes the prospect of adding new observations about such an old land. Latter-day visitors, be they in recent times or a millennium ago, which for a place like Egypt includes late-comers, have for centuries already remarked on the country's wonders. Little remains to be said, and I can add nothing. For so many years have travelers by now admired and described the great sights, only a few fresh perspectives remain and —for the life of me, a life soon to be over—I can't think of any original comments. But I can refer you to the opening lines of *Midaq Alley* (Trevor Le Gassick translation) by Cairo native Naguib Mahfouz, a Nobel Literature Prize winner (who did not attend Erasmus Hall High School in Brooklyn as did Bill Jacobs and Mike Stern). Mahfouz sets the scene for his story in the swarming teeming steaming streets of an old section of Cairo by emphasizing Egypt's history: "Many things combine to show that Midaq Alley is one of the gems of times gone by and that it once shown forth like a flashing star in the history of Cairo. Which Cairo do I mean? That of the Fatamids, the Mamlukes or the Sultans? Only God and the archaeologists know the answers to that, but in any case, the alley is certainly an ancient relic." The novelist's focus on an ancient relic pleased me, since I now fall in that category. Mahfouz infuses his opening comments with a temporal rather than a physical description by featuring past rather than place. Time is one of the story's leading, or following, characters. In just a few lines the author evokes his country's vast past, infinite compared to the passing scene in the alley, an "ancient relic" indifferent to the present-day hustle-bustle but meaningful to "the archaeologists" who live in the past. In Egypt even an ordinary back alley like Midaq exists in the context of its long history.

Deep time permeates everything in Egypt. The past is always present. Agatha Christie, married to archaeologist Max Mallowan, once noted that a husband who specializes in antiquity makes the best spouse, as the older you get the more interested in you he becomes.

By that standard Egypt is a country of extreme interest and, as such, one of the three most memorable lands I saw. After a certain point, long since surpassed in the case of Egypt, time becomes so vast you just can't comprehend the tremendous duration which separates the ancient past from relatively recent times. Although even yesteryears within memory seem somewhat indistinct (what happened last year on this day?), the yestermillennia of Egypt are so remote time shrouds them with a gauzy, hazy blur. So ancient is early Egypt its distant past seems a phantom civilization. Not only does everything which exists suffer the ravages of time but time itself grows old. Someday time as earthlings know it will come to an end. When the world finally ceases to exist Father Time will no longer afflict humanimals and all other living things with His degrading decay. Time's wingèd chariot will grow silent, its relentless whirring wheels finally spinning no more. Although all of Egypt's olden moments are gone, its monuments remain as tangible evidence of past times. The weight of the long centuries seems to inhere in the temples, pyramids and other solid remnants of the bygone dynasties. But for none of the long-mummified 320 Pharaohs could their pyramid schemes keep the monarchs in the game. Kings and Everyman alike were just passing transients in the play of infinite forms.

In contrast to the vast stretch of time which extends for centuries into the past, Egypt's built environment for the most part occupies a limited physical space, a narrow strip bordering the Nile. Although time in Egypt spreads back nearly forever like a vast temporal inundation, time's residue, the antiquities, remain largely confined to the vertical waterway band by which most of the population lives and where the monuments stand. This geographical arrangement in Egypt resembles that of Chile, delimited by the sea and the Andes. A less obvious but similar configuration characterizes Japan, and especially the main island, long and narrow Honshu, some 800 miles from top to bottom and ranging from about 30 to 140 miles wide. Around Tokyo the island expands to make room for Mt. Fujiyama, Honshu's centerpiece, and then narrows in the Kyoto, Kobe and Osaka section. The island's shape somewhat resembles that of Chile or the Nile River valley. In Japan as well as the other two elongated countries you travel vertically by length rather than by breadth. But the ups and downs of journeys in those places are the same as elsewhere.

The vertical alignment in all three countries arises from natural conditions, but—unlike for sea-surrounded Japan, confined by water—the limiting geographical features in the two continental lands aren't necessarily definitive or permanent. Their political boundaries are subject to the intervention of humanimals, as the hand and arms of man also play a part in how most borders are established. The geometry of geography depends on the way people deal with the layout nature established. Military and political forces might have shaped Egypt to extend itself in a much longer lateral direction west along and below the North African coast, while Chile could have expanded its territory by leaping the Andes to cross into the trans-mountain lowlands to the east. There was nothing inevitable about how each country's configuration took shape. But as history happened to develop, the sea and the Andes lock Chile into its familiar and logical but arbitrary vertical shape and restrict Egypt into an equally narrow convenient vertical strip centered on the Nile. Within those bands, traveling up and down rather than back and forth, a visitor to Chile and to Egypt can see almost all the main places of interest. In both countries, however, I followed my usual procedure—seeking remote areas beyond the well-trod vertical tourist route—by straying from each strip to see some outliers. In Egypt I deviated from the usual Nile-centric itinerary to travel west out to the desert oasis of Al-Kharga and also east to Suez City and the Canal, while in Chile I flew out to Easter Island and went to some areas near Santiago. It's always informative to get off the beaten track to visit the back, or the side, country.

Being accustomed on my native soil at home to the spacious expanses of the U.S. Midwest, the narrow formats imposed by geography in countries such as Egypt and Chile, as well as most islands—even large ones, like Japan—and low-ceiling New York City rooms, gave me something of a cramped and closed-in feeling. Unlike the less limiting and more roomy expanses near my home base where you can take off in any direction and soon reach open country, the rigid vertical in Egypt and Chile confines most of the local humanimals into a narrow space, hemming them in with crowded Midaq Alley-type conditions. Daily existence in a place like the Alley typifies how many people the world over live. Those cramped quarters in turn suggest the confinement suffered by the many earthlings forced to remain stuck in a kind of vertical existence, a narrow rut where their day-to-day ups and downs offer little or no opportunity to expand their horizons. Such circumstance-limited beings lacked freedom of action and mobility, privileges of the sort which by chance endowed my existence with attributes of a charmed life and which I never took for granted. What I'd been granted was gifted by happenstance and could at any moment be withdrawn by operation of the same fickle factor. Those unfortunates confined both in place and in life and who lacked choice existed as if restricted to a trolley line with its unalterable itinerary out and back and out again along the same narrow route. On a streetcar you reach the terminal stop on a set journey without being able to choose another path through the dark forest.

Back home I functioned for some years in a repetitious rut and route in some ways similar to how earthlings lacking choice lived. I rode the same rails day after day, on and on toward the end. But unlike most people, I enjoyed the luxury of choice, and finally I derailed my routine existence to find new lines of activity. I possessed not only the opportunity but also the will to make a change. By expanding my horizons while also retaining my home base routine existence I enjoyed both my limited and confined domestic space with its trolley-like regularities and the wider experiences and perspectives deriving from adventure travel. Unlike the trolley-dodgers in Brooklyn, I didn't merely dodge the streetcars but abandoned them for more flexible and adventurous ways to travel. This enriched my experiment in living, maybe even almost to the bursting point as my mind filled with a surfeit of sensory impressions from the infinite play of forms I encountered.

Veteran adventurers say a true traveler doesn't know where he's going, while a mere tourist doesn't know where he's been. For a tourist, Egypt and Chile offer ideal itineraries as for the most part you go north-south or vice-versa, making it easy to know both where you're going and where you've been. Wherever my wanderings happened to take me I always knew both where I'd been and in general where I was headed. My general direction was certain, but the devil in the details brought to the adventures many diabolic surprises and challenges. Uncertainties abounded, as I usually traveled in a free-flow improvised and *ad hoc* way with no fixed arrangements, structured plans, detailed itineraries or specific reservations, other than at times preoccupied by serious reservations about the wisdom of visiting some particularly problematic, dangerous or challenging areas. Although I knew ahead of time which countries I'd be visiting and in what order, the day-to-day details I left largely unplanned. Even with all of my preparation, after I left home much remained up in the air, not only my flights but also my daily flights of fancy which landed me here and there in a whirligig of action spinning me through place and time. I would even have ridden in time's wingèd chariot to get around, but for the fact that it constantly evoked too many intimations of mortality.

In a land configured vertically like Egypt your trip in general follows an obvious channel, a sort of visual assembly line of sights as the antiquities and other attractions lie mostly near the Nile strip. It would be misleading, however, to leave the impression that travel in Egypt includes only remnants of the past dating from ancient times. Just as India offers a colorful contemporary vibrancy along with a rich past, so Egypt presents a pulsating present in addition to a fabled history. In both countries the then and the now combine to bring unforgettable travel experiences. The buzz and bustle of the living generation in places like Midaq Alley offer the traveler in Egypt a dimension which transcends the past and contrasts with its moribund nature. Encountering present-day Egypt and Egyptians brings the story of the country up to date. A very long continuity seems to exist regarding how the people view foreigners. Herodotus tells of a Greek visitor who traveled up the Nile to Memphis (no, not to discuss TCB with Elvis) where "He met with the utmost hospitality [and]...friendly treatment." Some 2,500 years later, in the course of my two visits to Egypt, I found the same sort of friendly and welcoming attitude there. Most of the people I encountered were pleasant, gracious and accommodating. On one occasion when I paused to study a city map in Cairo, a well dressed and reputable looking gentleman asked if I needed directions, and he then proceeded to accompany me a few blocks to guide me to my destination. Another time, on the west side of the Nile in a poor section of Asyut, a city south of Cairo, an impoverished local fetching water from a well used sign language to invite us for tea. What little the poor man had he was willing to share with strangers, a touching touch.

Egypt suffers from many limitations and shortages unknown in my part of the world, inconveniences of the kind no doubt inconceivable to my friends back home. The rivers which provide the Midwest with plentiful water and irrigate the region's extensive fertile fields endow my land with benefits scarce in Egypt. Given the country's limitations, the problems of daily life and the challenges to eke out a bare existence, I wondered how the people managed to maintain their pleasant and friendly attitude toward foreigners. As I often found in many other impoverished countries, the locals tended to be rather sullen, irascible and even hostile to outsiders. I once asked an Egyptian why his compatriots seemed so welcoming and gracious. He replied that the Egyptians had been receiving visitors for some 5,000 years and that by now the residents were used to foreigners, liked the attention, and were proud that outsiders wanted to see the country. Maybe I did the Egyptians a favor by going there and giving them yet more attention.

Looking at the age-old monuments, I wondered what all the thousands of other visitors over so many years saw in the same stones I viewed, in passing, in my time What did those who came before Herodotus think about the ruins, already centuries old by the time he wrote about them? Each passing visitor, whatever the era, had his own personal perspective on the ancient structures. The fabled temples, the pointy pyramids, the Sphinx (would Thoreau travel to count that cat, if feline the beast be), the tombs, the statues, and all the rest of the venerable monuments left the on-lookers with both common and unique memories. Those of us today see the same old stones with a viewpoint colored by all the past, a now elapsed past which for long-ago visitors was still the unknown future. In a thousand years the then visitors will see the ancient structures, the same we see, with knowledge of times we can never know. The generations come and go, the stones remain. It's hard to get your mind around the leavings of a civilization as old as Egypt's. The Eiffel Tower, the Empire State Building, the London Eye, such lands as Graceland and Disneyland you can fit into some kind of cultural framework and into a human time scale, but it's difficult to comprehend the immense time span between now and the era back when Egypt first began to emerge from the mists of prehistory some 5,000 years ago. Yet even that dizzying earthly time interlude shrinks into nearly nothing compared with cosmic duration, a near infinite temporal span which dwarfs everything on the spinning mite which is mankind's homeland. The entire species of earth-bound humanimals (past, present, future), poor souls, exists in only a fleeting moment compared to the unimaginable duration of cosmic time. Even Egypt's five millennia are a mere instant when compared to the elapse of time between the Big Bang and the final entropy when everything comes to an end. Viewed in those terms, the Pyramids lose their weighty solidity and permanence. They seem insubstantial in light of the millions of years since the initial explosion hurled fragments (some of the material later to be formed into pyramids) through the void.

The night has a thousand eyes. From the night skies, with an infinite number of starry eye-lights looking down upon the lonely planet across thousands and thousands of light years, nothing seems significant. All that nature and man happened to create here is dwarfed into nothing by time and space. Those distant celestial gleams in the darkness gazing upon the Pyramid pebbles, the minuscule eyelet known as the London Eye, the slight terrestrial wrinkles known as the Himalayas, greater Cairo, the Great Gatsby, the Great Wall (supposedly visible from the moon) and all the other superficial earthly fixtures, figures and movement seem only a ghostly shadow play of infinite forms. Those remote glimmers from afar bring slight points of brightness in the eternal darkness to the benighted little planet where dwell earthlings lost in place, in space. The moon's hidden far side, which earthlings call the dark side, has somehow failed to suggest that the earth, too, has its dark side. Although the moon's dark part remains fixed in space, on earth darkness always shadows some part of the lonely planet as the ceaseless twists and turns of the daily cycle spin the globe. The night which every day shrouds some of the lonely planet serves as a reminder of the dark side of the world and all which inhabits it. The endless play of cosmic infinite forms takes place in light and in shadow on a stage of infinite time and unmeasurable space. The entire combined brainpower of all the humanimals who ever live wouldn't suffice to comprehend the unfathomable temporal and spacial dimensions which hold dominion over the fatal circle which circumscribes human existence.

In the incalculable vastness of the cosmic fatal circle, even a country as ancient as Egypt is a rounding error. On such a cosmic scale Herodotus was yesterday, and Egypt began only a few days ago. The pyramids which seem so solid and eternal, part of the scene for nearly 4,700 years, exist in only a brief provisional state. The most imposing constructions will, with all else, disappear as the whirligig of time spins everything into dust. Just another day at the office for Father Time. In his essay "Urne-Buriall" [sic] Thomas Browne writes: Time "antiquates Antiquities and hath an art to make dust of all things." Time "maketh Pyramids pillars of snow, and all that's past a moment." And so, he continues, "the iniquity of oblivion blindly scattereth her poppy, and deals with the memory of men without distinction to merit of perpetuity. Who can but pity the founders of the Pyramids?" Good question, Tom—like many Toms, Tom Browne knows how to cut to the chase and address fundamental matters. He's a true-story teller.

Things are getting serious here on the lonely planet: Father Time grinding everything into dust and Oblivion a Poppy not very paternal, as all is eventually forgotten. The stars and the moon, they don't care—they're just there, out in space, oblivious to earthlings and their lonely planet homeland. If even building the Pyramids represents a failed attempt to defeat the ravages of time, what chance is there for any other endeavors to capture a ghost of a shadow of the phantom phenomenon we know as earthly human existence? Not much. Tom Browne's observations present the reality, but his truths offer no clues to help other Toms, Dicks and Harriets for how to spend their meager time-allotment to live until the iniquity of oblivion shrouds their mortal being, deleting all memory of their quickly passing former existence. To this puzzling challenge, each human- and huwomanimal will react in its own way, always within the limits of his or her fatal circle and subject to the whims of luck, until the creature collapses into nothingness as if falling into a black hole. Meanwhile, before suffering that inevitable fate, there remained in my situation a lot I wanted to see and experience. Here I was, an involuntary earthling thrown into existence and stuck briefly on the lonely planet, with its unalterable alternating light and dark sides. What should I make of that opportunity—or was it an incubus? Looking on the bright side, I tried to make the best of it by deciding to pursue my travel pastime, as if time needs a mere mortal to help it pass.

~ 5 ~

Deeply set in time and with its long continuity, Egypt represents a world-class example of the "then," the quintessential land of the past. In India I found a country permeated with the "now." The two countries of course also share the characteristic which define the other, present-day Egypt colorful and picturesque and India with its own fabled past. But as I saw

Section IV: Twilights Far From Home

them, the dominant impression each brought me was the long "then" in Egypt and in India the scintillating "now" sparking unforgettable scenes. The very geography of each country seems to symbolize each land's essential characteristic—space emblematic of time. Egypt's linear alignment and unbroken time-line offer avatars of order compared to India's sprawl and the country's confusion of cultures, sects, castes, clans, classes, customs, practices, religions, beliefs, languages, dialects and the colorful chaos of sights, scenes, sounds in that land of cacophonous confusions everywhere. The never-a-dull moment play of forms which sharpen your sensory impressions in an effort to capture it all—really more than you can absorb—create an overwhelming sense of the "now." in India. Strange and exotic vignettes pass before you everywhere. In Egypt the set pieces, solid monuments and antiquities, seemingly time incarnate, dominate the scene; in India the country's history seemed a kind of backdrop to the present day fluid whirligig. In its baffling cultural and ethnic complexity, India appeared to me as bewildering as did Egypt's age-old history with all those pharaohs, dynasties, temples, statues, ruins and other residues of time, along with the invaders, occupiers, rulers, eras which transpired and expired over 500 centuries. Egypt's time strata present complicated layered leavings which in their way challenge the visitor's comprehension as much as do the complexities of India's 1.3 billion humanimals pursuing all manner and womanner of lives in that fabulous land, a vast buffet of a travel feast which you can never entirely consume. Indian presents a Technicolor panorama, a spectacle churning with a never-ending play of infinite forms on the stage of a crowded, cramped, congested, chaotic, curious country.

Seasoned travelers say that visitors to India either adore or hate the experience. My own reaction, based on three visits which took me all over the country, responded with both views at different times. The more I saw of India the better I liked it. On my first visit, some five weeks December 1973–January 1974 during my five-month-long Asian trip, I found the country so disorganized and difficult I struggled just to fulfill the basic requirements of travel. Electricity, flights, trains, buses, services all ran, or didn't run, so erratically I faced constant problems and disruptions in transportation, hygiene, opening times, sanitation, food reliability and other elements of independent travel. Schedules were fiction, and even the hotels weren't very accommodating. Coping with all the difficulties greatly depleted the time and energy I could devote to sightseeing. So enervating was my India interlude back then I was elated finally to see the last traces of that vast and at the time inhospitable land slip away into the distance, forever I hoped, as I sailed away. During my slog from Calcutta to New Delhi to Bombay to Madras via many places in between I longed for the day when India would cease being a hassle and become only a memory. With relief I gazed at the receding Indian coast as the ship leaving Ramerswam on the southeastern coast thankfully carried me away toward Jaffna in Ceylon (now Sri Lanka), a country, as I soon found, much less demanding and more manageable than India.

Completely by happenstance, my cabin mate in the two-bunk stateroom on the overnight trip was one of Ceylon's most important labor leaders. Before we disembarked the next morning he gave me his card and said I could contact him if I needed anything while in his country. His card was a welcome credential, even more useful than an American Express card I'd used to free myself from the gun-toting Córdoba character who'd stopped me on the street in that Argentine city. Based on my hundreds of twilights far from home in distant strange lands, I can affirm that having a connection in a foreign country gives you a certain sense of security. You never know when or for what reason you might need a local contact. Although I never had to call the labor leader, I felt more at ease traveling on his home turf where he could help me if needed. Meeting him was pure serendipity, a fitting description as "Serendip," an old Persian name, once designated the island of Ceylon. It was odd to experience there an episode defined by the very name of the place. Horace Walpole coined the word in a letter to a friend referring to the author's 1754 *The Three Princes of Serendip*. The odd word by chance came to life with the serendipity of my encounter en route to Ceylon with the labor leader.

This rare case of a place name coinciding with an experience at the location itself fortunately didn't apply when I crossed Death Valley, even if that conjunction of locale and happening would have provided a ghost writer with a truly terrific travel adventure to relate.

Some years after my first visit to India I decided to venture another visit, and in April 1999 I warily returned for a few weeks on my way to Bhutan. The second time around I was pleasantly surprised at how conditions for an independent traveler had improved, and I enjoyed my stay. The oppressive forbidding challenges which confronted me twenty-five years earlier and had so demoralized me had by now improved, so my second trip brought me only the usual grueling and inconvenient discomforts quite familiar to and expected by independent travelers in Third World countries. Such problems, known knowns, go with the territory. You don't venture to such underdeveloped areas expecting a soft life. In fact, difficulties comprise part of the experience. My enjoyment of that second trip made it easy for me to decide to join two friends (a couple) in early 2002 for nearly six weeks in South India, an adventure I found so stimulating and memorable I hoped to return yet again to India. But such travel is a young, or younger, man's game and I'm resigned to the reality that never again will I see that great country.

As my view of India gradually evolved and developed over many years from dislike to enjoyment, so too did the country itself substantially change and improve. But even with its more modern, efficient and workable latter-day advances the sub-continent in many ways remained the same, an eternal India retaining its fascinating congeries of colorful scenes. A rainbow of hues as bright as saris painted vivid city-scape sights and scenes, along with sounds, smells and a mix of other varied piebald sensory impressions, all exerting their fascination. Few other countries offer similar scenes. One which resembles India is Pakistan, which I visited on another trip, and another is Bangladesh which I never reached, so missing a chance to satisfy my curiosity about Cox's Bazar. By its size, variety and pluralism, however, India is unique. In that I never took an organized tour to India I experienced the country directly in an often delightfully disorganized and unfashionable fashion. Nearly every step of the way brought me to some sort of unexpected and unusual exotic setting or happening, ones which greatly enlivened my journey. My adventures around the world were in general entirely self-guided, not as part of a herd led around, but in some areas I took organized tours. For some countries or on-the-ground situations, group travel makes sense. But I always realized that a self-managed trip offers the only truly genuine form of travel, and especially in a country like India. Independent travel represents the highest and best form of the enterprise of seeing the world.

You really get to know a lot about a country when you're forced to cope with the logistics of travel there. Toughing it out on your own requires you to deal with the granular nitty-gritty problems common to all travel—meals, transportation, accommodations, safety, remaining healthy, seeing the sights and other needs and wants. Nothing happens unless you go to the trouble, at times a lot of trouble, of making the arrangements yourself. This exposes you to the hazards and benefits of contacts with the locals, often creating a certain abrasive friction which sparks many learning opportunities. Hands-on efforts lend great substance to the trip. The burden of providing for yourself is often aggravating, irritating, frustrating, exasperating, but at the same time such annoyances and vexations are educational and enriching. Coping with the problems helps to inform you about the realities of the country, adding a new and practical dimension to what you learn from books and word-of-mouth and at museums, monuments and other cultural and tourist attractions. Guidebook-listed or otherwise well-known places and events are insufficient to represent the essence of a country. The people and their way of life are more representative of the existential realities of the place. Not withstanding the many difficulties I faced, never would I trade all of what I experienced and learned from running my own travel adventures for the comforts and conveniences of an organized tour. Planned fixed itineraries lack spontaneity and opportunities for serendipity. Group travel shields you from informative interactions with local people and protects you from on-the-ground difficulties which, sometimes in an unpleasant way, round out your experience of a place.

Because tours are in some cases and places quite useful, I've often been willing to accept their disadvantages in order to gain the many benefits. Travel in some countries presents special challenges, uncertainties and difficulties, better faced as part of a group rather than individually. The known knowns, known unknowns and unknown unknowns can in some areas be too overwhelming for a solo traveler. Tours also operate faster than independent travel, so going with a prearranged set program enables you cover more distance in less time. Moreover, age limits your willingness, and for some people their ability, to address the many problems of free-lance travel. Under the belief that it's better to travel in a regimented way with strangers restricted to a trolley-line-like fixed itinerary than not to go at all, I've often opted for a tour, all the while realizing that with such a format I'm engaging only in a kind of quasi-travel. Travel in a country like India differs greatly for those who go independently as compared with tourists on a group tour. I occasionally use this distinction to discern to what degree someone is a true traveler. If you visit India on a tour, and especially with a pampering high-end operator like Abercrombie & Kent, Tauck, Geographic Expeditions, you have of course been to India but not really traveled there. My hands-on efforts require a lot more time and energy than for the tourists who benefit from hand-holding by a tour company. By smoothing your way and removing you from direct contact with the man in the street, the tour operator distances its travelers from the chaos and confusion which confront the independent wayfarer. I must admit that when travel-weary in challenging places like India, I've looked long and longingly at the fancy air-conditioned buses encapsulating tourists who enjoy all the arranged amenities for their twilights far from home. Those privileged characters get direct delivery to upscale hotels, pathogen-free meals, a steady supply of bottled water, bountiful breakfast buffets, transportation to the sights and then onward to the next city. But all those conveniences come with a price, in addition to the cost of the tour. The degree to which you're removed from the daily exigencies of a trip limits how much you derive from the adventure. Out there immersed in the streets, India presents to you a picture quite different than the perspective seen looking down through a picture window of a deluxe temperature-controlled coach. One's often a pain, the other's a protective pane.

Coping with India on my own brought me many treasured memories. One day in Calcutta in late 1973 I found myself among the thousands, millions maybe, of humanimals jostling, swarming and surging over the Howrah bridge across the Hoogly River during a *bahnd*, a general strike. The throngs typified in an extreme way one of India's most typical and oppressive characteristics: you're always surrounded by people. It's virtually impossible in India to be alone in a public area. Crossing the bridge, carried forward by the crowd's surge, I was, like the earth in the firmament, a speck lost among the thousands. It seemed a surreal experience: little noise (unusual for India), no disorder, no violence, only a vast mass of humanimals advancing relentlessly, moving on and on like an animated version of time's wingèd chariot, an enormous clot of people filling the narrow artery which spanned the river. You don't get that kind of experience if confined to a tour bus. It doesn't take cast or caste of thousands during a general strike to make you feel insignificant in India. Out there you're one in a million, one in a billion, not by way of distinction but by your commonality with all the other earthlings. There among the throngs, people everywhere, I often envisioned my distant cozy little house, my own private and tranquil enclave on a quiet tree-garnished provincial suburban street in the spacious verdant U.S. Midwest. I occasionally longed to be back home to enjoy the comforts of my privileged corner of the world where space, peace, privacy, greenery, familiarity, family, friends endowed my daily life with many amenities and benefits, all of which I temporarily gave up to roam the world. My Third World experiences made me appreciate all the more all those home-based advantages which contrasted so greatly with the conditions in countries like India. But in one way my life was no different than each and every one of the billion plus humanimals who inhabited India, or each and every one of the more than seven billion mortals who occupied the lonely planet. As the world turns, its dark side eventually suppresses for each person, whether charmed or hexed, the planet's brightness. My dead-end street at home represented the existential reality of humankind's fatal circle, for no matter what your station in life or where the journey takes you the ultimate destination was terminal. All routes end in a dead-end.

I first encountered India's crowds and all the rest of the country's kinetic energy when I arrived late 1973 in Calcutta, a formidable city which served as my introduction to the vast land. Based on the sights I saw there, it didn't take me long to appreciate all the more my privileged existence which afforded me something of a charmed life. My travelers' checks and check-list of sights and curiosities to glimpse as a passing visitor, my cash stash, credit cards, and air ticket back home all served to set me apart from the swarming masses. No matter how much I immersed myself in the steamy streets and the hyperactive life of the city, I was not really part of it. As everywhere, other than back home, I was only a passing observer, ever ready and able to move on. My presence in Calcutta was a half-way house, between a protected journey as part of a sanitized tour and more granular independent travel to a place I directly experienced but from which I could at any time escape. By day I saw *radiwala*, women and children scavengers, picking through smoldering mounds of garbage to salvage scraps of metal, glass, plastic, cloth. Barefoot rickshaw drivers trotted through the crowded streets pulling passengers along in the local *"unter"* version of an Uber-based on-demand transportation service. Beggars, urchins, the maimed, the diseased, the emaciated, idlers, dozers and other street people along with sore-ridden mangy mongrel dogs and wandering cows dropping turds presented a motley swarm of living beings. At night I stepped over or around the shrouded bodies of the street people, families sleeping on the sidewalks or huddled in alcoves or crammed into doorways. Kerosene lanterns which glowed and flickered in the darkness cast eerie jerky phantom-like forms over the homeless as if part of a shadow play. Such was my introduction to India. There in Calcutta, I wondered if the rest of the country was as colorful, as chaotic, as exhausting as the metropolis where I first experienced India. I soon found out as I continued onward: the answer was yes—it was all a spectacular and unforgettable play of forms.

From Calcutta I flew to Nepal for a week-long stay where in Katmandu I paid $2 a night for a hotel room (I was ineligible back then for a senior discount). Modest as the place was, the hostelry offered pure luxury compared to the conditions for Calcutta's street people. After Nepal I returned to India to continue my circuit there. Getting around the vast subcontinent presented a challenge. With Indian Airlines, arrangements were always up in the air, but not the planes. Over-booked and erratic departures caused inconveniences and delays. It was faster to travel by slow bus or train, even if they too were often unreliable. Because of the great distances I tried to use trains, but hard class was hardly pleasant and late departures and arrivals were all too common. Every car on every trip was jammed with passengers. In India some 23 million passengers travel by rail every day, and sometimes so crowded was the train I'd taken it seemed as if all those millions were riding along with me. You could never be sure just when a train would arrive, depart or, if underway, would continue. I was stranded once for ten hours when a train headed for Bombay came to a dead stop, no explanation given. On one occasion I was pleasantly surprised when a train I awaited actually arrived an hour ahead of schedule. When I complimented a conductor on the truly unprecedented ultra-prompt service he explained that it was the previous day's train running 23 hours late.

It would take the better part of a lifetime to see and to understand India, even to a small degree, and a series of books to relate all the adventures of an independent traveler there. So plentiful are the sensory impressions in the country, a single mind or a few pages hardly suffice to absorb and retain and then to record them all. Only a few of my adventures can I include in an attempt to give the flavor of the country. My trio of trips included one particularly enduring memory which in a way serves to epitomize my experiences in the sub-continent. One day in 2002 my two traveling companions and I went to Kanyakamari to visit the country's southernmost point at Cape Comorin, named for Kumari, the Hindu goddess who protects the shores there at the confluence of the Arabian Sea, the Bay of Bengal and the Indian Ocean. All things end, even vast India, and there at water-washed Cape Comorin India disappears into the sea.

Like my dead-end street at home, capes evoke an ending. They're a *finisterre* terminal point. At a cape you've reached land's end, beyond which earth's solidity gives way to a watery expanse alien to the *terra firma* habitable by earthlings. A cape is the spacial equivalent of time which eventually runs out for each humanimal. From a cape there's no escape: you're at the end, a terrestrial finality. During my travel capers over the years I collected many capes, more even than those in Bat- or Superman's wardrobe. Cape Horn I saw in passing from a ship headed to Antarctica. From a mile or so offshore there at the tip of South America you see a lonely rocky dot of land which marks the end of the Americas. To the north lie all of the hemisphere's people, places, cultures, civilizations; to the south, only the depths and breadth of the seas and the fearsome Drake Passage which separates all the familiar New World territories from the desolate Antarctica ice, a region inhospitable to humanimals. Nonetheless—irrepressible creatures that they are—earthlings have ventured down there to check things out. They seem unwilling to stop trying to understand what it's all about—the lonely planet and its play of infinite forms.

At the other end of the earth from Cape Horn in a far distant corner of the globe I saw North Cape, which occupies a high promontory in the upper reaches of Norway. There you stand only some 1,300 miles from the North Pole, the final north, the closest I'll ever get to the top of the world. Each and every cape is capable of stirring long thoughts about endings: Cape Girardeau, a rare inland cape overlooking the Mississippi River at a Missouri town of that name; Capes Cod, May and Hatteras; Cape Coast in Ghana; Cap Haitian; Cape Verde in the Atlantic; Cabo San Lucas at the southern tip of Baja, California; and many others, all denoting land's end which, in turn, evokes other finalities. Cape Finisterre represents a particularly evocative end point. In Roman times the peninsula on the west coast of Galicia in Spain was believed to be the end of the known world. It was a known known, so the ancients believed. But like many knowns, in the course of time it proved incorrect. Land's End, England's most westerly point where "Eng" loses its "land," designates where solid ground, terrain where man can gain a foothold, gives way to the sea. One minute you're standing on the hard granite headland in Cornwall, the next you reach the far point, the end, just where the earth mutates from solid ground to the inhospitable waters below. All lands if continued long enough end in depths and he who would keep that from you is no true-story teller.

Each such end point, no matter what the cape shape, represents a terminal earthly feature. Quebec's Cape Chat, population under 2,000 not including cats, takes its name either from the cat-shaped waterside promontory there or from a renegade feline turned into stone. If further interested, readers can discuss the name's origin in an online chat room, but it would be useless to consult Thoreau on the topic, as he disdained travel to count cats. The Cape of Good Hope, 30 miles below Cape Town, is Africa's best known ultimate point even if not the continent's southernmost tip. That extremity, Cape Agulhas, lies 100 miles east of Good Hope. Having seen both of those capes, I know that the splendid setting there at the Cape of Good Hope offers a much more dramatic, evocative and historic image than does the flat rather drab beach-like terrain at Cape Agulhas. At Good Hope a thin spit of sharp-peaked land razors its way out to the sea before disappearing, and the very name recalls the olden days when daring Portuguese navigators and explorers rounded the Cape hoping to survive the voyage and, if so, for a productive and profitable expedition to the mysterious east. But even good hope is hopeless at the finisterror dead-end point where life ends.

There back in India, where the country's land mass ends at Cape Comorin, Mahatma Gandhi's insubstantial mortal remains reached their end. After his assassination on January 30, 1948 in Delhi far to the north Gandhi's body was cremated, burning away in a few puffs of smoke as his earthly flesh and bones disappeared into thin air. On February 12 some of the ashed residue of what was once Gandhi was thrown into the waters at Cape Comorin where the three seas border India's *finisterre*. In his introduction to *An Autobiography, or The Story of my Experiments With Truth* (first published in 1927), Gandhi explained that "it is not my purpose to attempt a real autobiography. I simply want to tell the story of my numerous experiments with truth." Although completely lacking in Gandhi's authority or standing, and with a life vastly different from and inferior to his, like him my purpose in this reminiscence is not to present an autobiography or memoir. I seek simply to record a few stray remembrances of some places, people, events, experiences I happened to encounter in the course of my by now prolonged earthly existence, soon to end at Cape Mortality. Much—most in fact—have I omitted, for my account describes not a life but only some selected fragments of one, an attempt to sketch a few lines to outline a particular humanimal's experiments, not with truth but only in being an earthling for a brief time. In truth, unlike for Gandhi the butterfly of truth eluded me, and my earthly experiments didn't yield any existential truths. My professed wisdom or conclusions stated in my account are only opinions and subjective observations, not truths. Never did I seek and certainly never did I find any ultimate terrestrial certainties. I never managed to reach any cape-like definitive finalities which expressed absolute truth. I was just another humanimal, no different from the thousands swarming across the Hooghly River bridge in Calcutta, feeling my way through the shadowy Dantesque woods. Like all of the thousands of other mortals on the bridge and the world over, I was inexorably advancing toward the inevitable dead-end at the end of the road. Ghosthood was the ultimate neighborhood, a traveler's final stop. That was the only undeniable truth I ever managed to find during my experiment, a truth not fit for purpose as it stated an ineluctable fact which refuted not time, as Borges had written, but the value of existence itself.

There at Cape Comorin where Gandhi's remains disappeared into the swirling waters, the end is visible, tangible. At that remote point where Gandhi's ashes found their end even great India terminates. I made my way out to the very tip. There by the water I plucked the very last mainland stone before solid land gave way to the seas. Removing that sea-side stone diminished the sub-continent by the tiniest fragment, just as the disappearance of one mortal humanimal barely depletes the billions of earthly souls still alive. One little stone, one little life won't be missed. Back in 2002 when I stood at land's end maybe a billion or so Indians then existed, each and every one of them for a few fleeting moments now off to my north. For once, momentarily, I was alone in India, removed from the millions. No crowds swirled around me as I stood by myself there on the tip of the continent. I was at the end of the line, the extreme far reaches of a huge country brimming over with energy, activity, vitality, movement, color. But, like Gandhi, I was removed from it all, at least momentarily. During those few moments all the life to the north was for me ecliptic, the space like an intervening lunar passage which blocked me from the living India.

The final words to the nation of Jawaharlal Nehru, India's first president following its independence from Britain in 1947, after his death in 1964 capture something of India's essence and the flow of time and of its rivers through the country and on out to the seas. Part of Nehru's will was read on All India Radio by his sister, Madame Pandit. After directing that his remains be cremated and a handful of his ashes be thrown into the Ganga (the Ganges), Nehru went on, or at least his words did, even if he didn't:

> The Ganga, especially, is the river of India, beloved of her people, round which are intertwined her racial memories, her hopes and fears, her songs of triumph, her victories and her defeats. She has been a symbol of India's age-long culture and civilizations, ever-changing, ever-flowing, and yet ever the same Ganga. She reminds me of the snow-covered peaks and the deep valleys of the Himalayas, which I have loved so much, and of the rich and vast plains below, where my life and work have been cast.

> Smiling and dancing in the morning sunlight, and dark and gloomy and full of mystery as the evening shadows fall, a narrow, slow and graceful stream in winter, and a vast roaring thing during the monsoon, broad-bosomed almost as the sea, and with something of the sea's power to destroy, the Ganga has been to me a symbol and memory of the past of India, running into the present, and flowing on to the great oceans of the future.

Thus does the Ganga symbolize India. Like all rivers, the phantom past, the shadow show of the play of infinite forms, capes and much else, the Ganga—as it flows on to disappear into the sea—is emblematic of mysteries, memories and matters beyond the waterway's currently flowing observable currents.

~ 6 ~

All rivers flow through time as well as place. The waterways come from the stormy seas of the past and, like Nehru's Ganga, flow onward to the great ocean of the future. In many lands, probably most, there flows a river such as the Ganges which, as Nehru said in his ghostly postmortem testament, serves to symbolize a homeland in many of its varied dimensions. Such dominant rivers run rich with history, cultural significance, commerce, folklore, myth—the flowing waters of the waterways flooded with meaning beyond their purely geographical presence. The Thames, the Seine, the Rhine, the Danube, the Douro, the Tiber all represent in their European lands more than just a waterway. Like the Ganga, those fabled streams—all symbols of a country and its culture and history—also evoke the stream of time. Other rivers elsewhere educe a similar flood of historical debris in the form of jetsam and flotsam from the past left by time's ever recurrent currents —the Nile (at 6,695 miles the world's longest), the Amazon (6,516 miles), the Yangtze (6,380 miles). In the United States the fabled Missouri-Mississippi system (fourth largest, at 5,959 miles) waters a vast stretch of the midland as the rivers and their tributaries flow through the North, the West, the Midwest, the South, on their way to the sea, and also flowing on and on through time to the future.

Mentioning the United States and its dominant river system, so American in place and name, may seem a discordant intrusion in this Section IV which features such ancient and distant in time and place civilizations as Egypt and India. But in my view the U.S.A. deserves its place, along with India and Egypt, as one of the world's three most interesting countries. I realize that my homeland doesn't offer as many world-class attractions as Egypt or India, and I admit that a few other countries might well be chosen as one of the three top travel destinations. Italy, which I rank as Europe's most culturally rich country and with terrific *gelato* as well, could well be the number three. But notwithstanding the creamy ice cream, Europe's long and event-filled history, the many monuments and museums, the refined cultures and savory regional cuisines, the scenery and the natural and built environments, and many other attractions and distractions, in my view none of the Continent's countries offering those and other attractive characteristics qualify to the degree necessary to replace the U.S.A. in the top trio mini-list. Comparing America with so many other places around the world enables me to see my own country with a fresh eye. It's not only my own land but also one of many where I've traveled and explored in depth. What I found is that the U.S.A. offers a great variety of attractions, a wide assortment of different cultures, an unusual diversity of natural, ethnic, historical and societal components, and a broad pluralism, all of which lends the country a truly compelling and variegated fascination. Travel in the U.S.A., a callithump of a place, presents a great variety show, a gallimaufry mixture filled with a nearly endless range of surprises and oddities (among them, perhaps, your author), curiosities (ditto), eccentric people (ditto), places and practices, unexpected and unusual enclaves, out-of-the-way frontiers and back-waters, off-the-beaten track colorful corners, all a melting pot and a crazy-quilt mash-up of a country. Unlike the narrow configuration of places like Egypt, Chile and Japan, the States occupy a diffuse landmass which provides a hospitable setting for a play of infinite forms. While the U.S. lacks deep history, it excels by the country's range of territory, the extent of space and what the Continent offers compensating for the limitations of past time.

Just as Nehru deemed the Ganges in many ways represented India, so the fabled Missouri-Mississippi system serves as one of America's mainstream symbols. The Father and Mother of waters spawned much of the land's history and folklore. The river's vast catchment area includes a wide and in places even weird variety of sights, cultures, historical happenings, ethnic groups, settlements, landscapes and much else which evokes the country's great diversity. In the regions drained by the Mississippi system lie any number of memorable offbeat attractions to be discovered, by plan or by chance. Greatly deprived would be any world traveler failing to see the white squirrels in Olney, Illinois, or the world's largest catsup (so spelled) bottle, a water tower in Collinsville, Illinois as dominant and as eccentric there as is the Eiffel Tower in Paris. Roaming the river-watered regions will take you to small towns and obscure settlements based on a wide variety of beliefs, practices, religions, faiths, ethnic origins. Dotting the Midwestern map near the Missouri and the Mississippi are enclaves which evoke the wider world. A traveler will find, in Iowa, a Czech community and the Amana colonies; in Illinois and Missouri, Amish towns; Mormon Nauvoo and Swedish Bishop Hill in Illinois, and Stockholm in Wisconsin, Missouri's German Hermann and the 1840 Bethal German Colony, a utopian settlement similar by intention if not by belief to Indiana's New Harmony, founded by idealistic settlers who arrived by flatboat, a "boatload of knowledge," on the Wabash River.

Any number of saintly cities and rivers bless the mid-America river country, as if lending the region holy water sanctity. Minnesota boasts St. Joseph, St. Cloud, St. Croix, St. Paul (an improvement on Pig's Eye Landing, as the settlement was called when founded in 1829). In Iowa is St. Donatus, founded in the 1840s by settlers from Luxembourg, and in Missouri St. Joseph, St. Francisville, St. Patrick, St. Charles, St. Louis, Ste. Genevieve, while Louisiana offers another St. Francisville and another St. Joseph, along with St. Rose. So many saints consecrate the region a traveler needn't journey to the Vatican to find a beautiful experience. Around the rivers Native American names sing with a uniquely American voice, a poetry of the past: Winnibigeshish and Wabasha, Kickapoo and Keokuk and Kaskaskia, Minnehaha, Natchez, Ojibwe and Oquawka, Chippiannuck ("village of the dead," a cemetery in Rock Island, Illinois). State names such as Arkansas, Illinois, Iowa, Minnesota, and two states as well as the very rivers—the Missouri and the Mississippi—which vein their way through America's heartland also recall the early-day inhabitants of the region.

In those areas crisscrossed by the many waterways dwelled over the years famous and fabled characters who populate Mid-America's past. Along and near the Mississippi lie Mark Twain's Hannibal, Tom Eliot's St. Louis where as a boy he watched "a strong brown god" flowing by, Elvis Presley and his Graceland in Memphis, William Faulkner's Oxford, Mississippi. At the river's northernmost point in Bemidji, Minnesota, stand statutes of lumberjack Paul Bunyon and Babe the Blue Ox, while down-river in Illinois statues of Popeye in Chester and of Superman in Metropolis honor those two small town heartland cartoon characters. Riding along the River Road near the Mississippi you can dine at such Michelin no-star eateries as Sneaky Pete's Bar and Grill in Le Claire, Iowa, which boasts a salad bar in an antique bathtub, or Sikeston, Missouri's Lambert's Cafe, featuring "throwed rolls" tossed to diners at the family-style restaurant. Continuing onward to the rather backward hamlet of Onward, Mississippi, you'll see the area where Teddy Roosevelt's 1902 bear hunt inspired a Brooklyn, New York couple (having successfully dodged the trolleys there) to call their stuffed animal toy creation a Teddy Bear.

Visitors to and residents of the United States who restrict their knowledge of the country to the coasts and large cities miss the lesser-known areas which define much of what America is about. Many remote corners and sights tucked away on the byways and back roads offer quintessential American scenes and settings which give travelers an in-depth picture of the country unavailable to those who shun the more remote regions. To understand America it's necessary to see the inland areas beyond the coasts and the metropolitan clusters. For the real America you have to leave the interstates and roam the inner states where a traveler will find many typical and varied back-road sights and delights. In my roamings around the country I found the U.S.A. more diverse and multiform in its play of infinite forms than many other lands. It's perhaps something of a paradox that far older countries in a way seem much less varied. In Italy and France for example, smaller towns usually present the same format: the main square and a few smaller ones, and generic shops which look much the same from one town to the next—cookie cutter bakeries, featureless antiseptic pharmacies, cafes indistinguishable one from the other, the usual bar/coffee shop, similar street scenes, rather repetitious architecture and monuments. A few churches and a city hall represent Pope and state, the town plan repeats a duplicate of other towns. Often a kind of soporific lull dulls the ambiance in those provincial towns, each in many ways similar in both appearance and mood as if complying with a pattern decreed by the central authorities. To some extent Latin America presents the same regularities. In Mexico the only city of any size which on its main square lacks a cathedral, church, convent, monastery, chapel or other religious building is Cuernavaca. In other cities some sort of ecclesiastic structure stands on the Plaza. Of course, uniformity also blands many places in the U.S., but enough variety exists to provide a wide range of diverse attractions. Traveling over many years by car on the highways and byways and back roads of my homeland I found it a most interesting country, one which in its own way ranks among my travel favorites.

No matter how vast, every continent ends at *finisterre* and no matter how long, every river eventually finally flows on to its end and disappears. Even the mighty Missouri-Mississippi ends, the river drowned in the sea after its long and lively course through the middle of the country. On through the heartland streams the great waterway, past those oddly Native-named New World states and onward by the saintly settlements, past the New World's Cairo and Memphis, no match for the age-old originals, and on down to Venice, a name borrowed from the Old World to designate a true *finisterre*. There ends the highway, Louisiana and the U.S.A. The last mainland town, Venice perches by the wetlands at the southern edge of the continent where the States give way to the sea. Beyond, reached by boat, lies or floats Pilottown, the waterway's very last inhabited settlement, not very settled as it stands above the water on stilt-like supports. At Pilottown live pilots who take control of inward-bound ocean ships to guide them the 106 river miles up to New Orleans. Returning to sea, outward-bound vessels take on board at Pilottown pilots to navigate the ships out to a landmark or sea-mark buoy 22 miles offshore in the Gulf of Mexico where the regular captain resumes control for the voyage on the high seas. Out there where the last traces of the Mississippi disappear the river meets its end. For a distance the stream's brown currents streak the sea, and then those vague evidences of the great waterway lose themselves in the ocean's fathomless depths until finally all traces of the mighty river cease to exist.

Although the Mississippi begins as a mere trickle at a creek in the north woods of Minnesota, during its long journey to the end beyond Venice the river gradually gains volume and force. After its confluence with the Missouri just north of St. Louis the Mississippi indeed becomes "a strong brown god," as it floods on down to the distant sea. At that time-haunted confluence where the two rivers join, a liquid crossroads, flow currents of history from early-day America through the years and on up to the present day. Past that evocative point floated Marquette and Joliet on their explorations of the waterway, and from the confluence began the 1804 Lewis and Clark expedition to venture into the then wild west. Along the Missouri and the majestic Mississippi there in the heart of the American heartland once flickered in the darkness

Native American council fires, their warming glow now grown cold, the burnt logs ashed and all long gone with the winds and rains and the sands of time. Across the swirling waters at the confluence sounded the now silenced long lonely whistles of the vanished steamboats of yesteryear, and past this fabled corner traveled in bygone times pioneers, frontiersmen, backwoodsmen, adventurers, trail blazers, trappers and trekkers, river roustabouts, roughnecks and smooth talkers, grifters and drifters, confidence men and men of confidence, ne'er-do-wells and well-to-dos, faith healers and men of faith, settlers and the unsettled restless and all the rest, a long and by now long-gone stream of characters colorful and shady, a great ghostly procession of figures haunting the eternal waters, all of whom remind us that all rivers run through time as well as place.

My brief sketchy description of a few back road and backwater areas and attractions, as exemplified by ones located in the catchment region watered by the ribbons of rivers which flow through the center of America, will perhaps serve to suggest why I rank the U.S.A. among the world's most interesting countries. I realize that my choice, perhaps far too eccentric to be convincing, is something of an outlier. Let's face it: albino squirrels, a bathtub salad bar, a catsup bottle water tower and other such oddities really can't match the wonders of ancient Egypt or lively India's ever fascinating colorful motions and commotions, nor many of Europe's travel treasures. But taken all in all, American offers many unusual and enriching attractions to please even the most seasoned traveler. I could have pursued a satisfying and convenient visa- and hassle-free travel life, no passport needed, just roaming the States. For me my own native land in the New World matched the old worlds for curiosities and curiosity-quenching travel experiences.

A colorful and adventurous nineteenth century character, a raffish fellow named Constantine Rafinesque, synthesized in his person a combination of Europe and America. He personified both the Old World and the New, for his life incarnated in one individual the two related but different cultures. Born in Turkey to a French father and a German mother, Rafinesque emigrated from France to Philadelphia, then lived a decade in Sicily before returning to the U.S. in 1815 to stay in New York for three years. In 1818 he began his wanderings in the wilderness of the (then) West, finally becoming a professor of natural history at Transylvania University in Lexington, Kentucky. Rafinesque's varied experiences and peripatetic way of life endow his observations on the travel life with authority. In his *A Life of Travel and Researches in North America and South Europe* he observed: "A life of travels and exertion has its pleasures and its pains, its sudden delights and deep joys mixt with dangers, difficulties, and troubles." So it was for me, and so it is for every true traveler.

During my more than half a century of enjoyable roamings many pleasures did I find, but also many challenges. All along the way, and especially in Third World lands, problems confronted me. They brought not only occasional dangers and many difficulties and troubles but also invigorating—even if ofttimes vexing—experiences which sharpened my instincts and refreshed my intellect. Some problems I managed to solve in satisfactory ways, while others proved more intractable but thankfully not permanently damaging. Both the delights and the defeats contributed to the rich fabric of my travel life, and none of my experiences would I wish to delete. The discomforts, dangers, difficulties and troubles all added to my life experiment, even if at the time they happened such trials were unwelcome and disconcerting. Taking my travels as a whole, I enjoyed every minute and mile and in retrospect even every mishap. Never would I trade any of what happened to befall me to regain time or dollars spent or to eliminate an unfortunate interlude. Everything which occurred—the good and the bad— operated to bring me treasured experiences which I now recall with pleasure. By now long past, my adventures remain only as misty memories which lack monetary value and are weightless insubstantial phantoms of yesteryear invisible other than in my mind's eye. Those ghostly memories are worth nothing to anyone except to me, for whom they're precious and priceless.

~ 7 ~

By now, in old age, most of my travels, whether adventurous or more benign, have reached an end. I sometimes pretend as if they'll continue but I know they won't. Like the Mississippi at Venice, my life has flowed by and soon I'll be immersed in the sea of eternity. Looking back, up-river, I occasionally wonder at this late date what all my motion meant. Those journeys with so many twilights far from home I in part viewed as a post-graduate course in life and in how the world looks and functions. Now that my travels are very nearly complete, I've pondered what lessons my adventures taught me. Earlier in this Section I suggested that travel entails more than just sightseeing. The open road also opens your mind by bringing opportunities to observe a wide range of phenomena produced both by nature and by humankind, itself a product of nature. For me, not only tourist sights but also culture and society as well as history featured in my travels. All around the lonely planet play the infinite forms, a never-ending spectacle. Deserts, jungles, mountains, woods, rivers, plains and other terrains provide the settings where humanimals over the centuries established what became culture, history, civilization, society. It's all a great show. My curriculum out there lacked a major but offered many minors of major interest. I also noted earlier in this Section that travel involves not just the here and now of an immediate visit but also the here and then, for just as the past haunts every living being it haunts every place. All rivers flow through time as well as place, and all time flows through place as well as people.

The whole enterprise, the entire experiment of both travel and living, was quite a trip. A few of my friends in a position to live an independent life which might have—but didn't—include travel occasionally asked me why I spent so much time, effort, energy and money to engage in such an inconvenient and ephemeral activity as adventure travel. Inverting the inquiry, I turned it back on the questioner by retorting, "The real question is why you didn't do the same." Most of the people I knew who enjoyed the means to travel largely avoided it, preferring instead to cling to the comforts of home or to take only expensive group tours. The sort of security and relative certainty offered by staying at home and the predictability and comforts of organized tours didn't appeal to me. For me a high-class tour offered less interest than the class I attended while on the road in the school of hard knocks. The problems and discomforts of adventure travel were part of the experience, and for me preferable to remaining a stay-at-home.

My travels stimulated me, giving me a lot to think about. My experiences in far lands nourished my mind with much food for thought. Like the bathtub-filled salad bar by the Mississippi in Iowa, the world represented a great buffet where I could feed my appetite for novelty. To the historic dimension of the "now and back then"—the past as it influenced a place in its present incarnation—my thought-provoking journeys also inspired me to contemplate the "now and then what." That perspective recast my view from the present and its past to the present and the future. How would a place seem when seen from a far future perspective looking back in time as we now view Egypt's ancient past and antiquities? This sort of thought experiment led me to contrast the alternative scenarios I envisioned with the lively current kinetic scenes I observed, filled with humanimals and with an apparently permanent built environment, including age-old monuments and dominant latter-day structures. It all appeared to be so natural and normal—the people, the buildings, all the rest. What exists seems authoritative, as if it belongs. But forms which look like they belong don't last very long. Nothing could seem more permanent and dominant than Manhattan's towering World Trade Center. In a grisly sort of exchange, the destroyed structure has now been traded for a new tower on the same site. It's just a question of time before that edifice will in its turn disappear.

I tried to imagine how places would look decades or centuries later. Things change, sometimes so slowly they don't seem to evolve or degrade but in the end they disappear. Most places inhabited by earthlings will most likely become ghost towns. A cataclysm, probably man-made but possibly imposed by natural forces, will no doubt destroy life and dismantle most or all of the structures—physical, cultural, societal, economic, political. No more would humanimals surge through the streets or occupy the buildings. At some point in the far future, when the inhabitants are all deceased, their motion ceased, the entire city and system would be gone, everything vanished as smoke from a campfire. Like Carthage, like the Mississippi losing itself in the Gulf, like all defunct beings, no traces would remain. Whether land or liquid or life, everything ends. That's the nature of things. What by chance happens to assume a form is in essence a manifestation whose very nature is to lack existence, and eventually that normal and natural state will come about. Just as at Santiago, Chile, after I crossed the continent and turned North toward home, every step through life took me closer to my final home which is non-existence. As for everyone and everything the non-being of eternal oblivion (pre- and post-existence) represents True North and no true-story teller can keep that from you. Life is an exception, not the norm.

In "Deceiver Time" Spanish poet Góngora asks: "Carthage no longer stands assured, why you?" That's a reasonable question, even if it really doesn't apply to my way of thinking. Always rather tentative and uncertain regarding such existential questions, I really didn't take a very assured stand about such epistemological matters as the reason why defunct Carthage, now not so sure of itself, disappeared, or why I happened to appear and was for now (only a little while longer) still standing. After I by chance, out of nothing and nowhere, entered the realm of the living—why me?—I remained quite unassured that my being or anyone's made much sense or any difference to the universe. My involuntary existence didn't change the cosmos, nor will my soon to occur non-existence. So for me Góngora's question wasn't particularly relevant. And in any case I preferred not to deal in "why's" Why should I? "Why's" don't make you wise. "What's" are much more informative and useful. The world as it is, not why it or anything in it is, represents the important factor. With my musings under twilights far from home, I came to realize that nothing of what I saw all around the world needed to stand anywhere at all, not me, not Carthage, not the Pyramids, not the World Trade Center, not anything or anyone in the vast sub-continent known as India, not even its land mass, not the Mississippi River or the catsup bottle water tower or Elvis's Graceland or Snoopy's doghouse, nothing really had to exist, and whatever did exist could have been in an entirely different form. Any of the infinite forms could have existed in another format—a mustard bottle water tower or a Gracelessland, not a sub-continent but a submerged one like Atlantis, and no pyramid but instead a circle. Any fatal circle could have been different, or not been at all. Whatever happened to materialize and take its place in the world and in the entire cosmos arose at random simply as one version of what might have existed. All the vast number of versions which don't exist far outnumber the "singularity" (to borrow a term) which does. Technically, however, a singularity refers to the point where the usual laws of physics cease to apply. Surrounding such singularities are surfaces called event horizons which swallow forever whatever happens to cross the event area. A true singularity thus relates not to what is but to potentialities which instead of occurring or existing disappeared into non-events. All that never happens occupies a realm vastly more crowded than the play of infinite forms which in fact happened to enter existence.

Whatever by chance joins the lonely planet's play of infinite forms is simply the minute residue of nearly infinite potential. Random selection largely determines the content of that detritus. My comments on the contingency of everything obviously don't represent original perceptions. As long ago as the time Herodotus was writing about Egypt, his contemporaries realized how random the world is. The ancient sages from Greek, Roman and Biblical times contemplated, discussed and in some cases recorded their views on the great issues pertaining to the chance existence of humanimals on a tiny speck of earth spinning in the cosmos. All subsequent commentary represents only a long footnote to the original age-old old-sage observations. My own views presented in my account are quite unoriginal. What happened to me is original only to me. My brief experiment as a living being was unique for me, but brought me no general original insights. My reminiscence relates only to a single life and isn't intended to offer thoughts previously un-thought. I can add nothing to the wisdom of the ancients. They already contemplated almost everything (other than in the realm of science) regarding the lonely planet and the place of earthlings on it.

As long ago as 384 B.C. Eunonymous of Hydra, a shadowy (some say mythical) figure lost in the mists of the distant past, supposedly wrote a now long-lost manuscript covering all the points relating to what doesn't happen. He discussed the non-existence of everything which potentially might have come into being, the might-have-beens possible before in fact there actually materialized only the one version out of billions possible. Eunonymous's phantom manuscript, *On All That Which Never Existed and Never Will*, may itself not have existed. No copy of the text survived, so his original writings themselves serve as a kind of avatar of their theme. But the philosopher's thoughts did survive, as they were reconstructed and summarized from various early-day commentaries on the original, a kind of reverse engineering which revived the text by synthesizing what other ancient sages wrote about it. The surrogate narrative based on the commentaries on Eunonymous's original was published under his name in 1884 (by then he was a long-gone ghost writer) as a treatise by him in a very small edition (some scholars say 100 books or less) by Brill in Leiden, followed in the early 1930s by a somewhat revised version issued by a small obscure London publisher of esoteric works, a predecessor to and later acquired by I.B. Tauris, founded in London in 1983. That second revised version was also a very limited edition, each copy numbered, under the new title *Never Was, Never Will Be*. This was apparently thought to be a less formal and more marketable title than the earlier Brill volume with Eunonymous's original designation. Some scholars preferred the Brill edition, more complete and profound, while other readers favored the revised volume which was less elliptical and more accessible. It's impossible now to compare the two as, by a quirk of fate, no copies exist. Both latter-day editions have, like Eunonymous himself, become phantoms.

Not only are those two versions out of print, they're out of existence for no known copies are held in libraries, museums or archives anywhere. It's possible that a stray copy of the Brill or of the London volume remains hidden on an obscure dusty shelf in a private collection somewhere. It's said that a library in a castle deep in the Romania Carpathian Mountains holds a copy (unknown which edition), but in the Balkans people say a lot of things. A footnote (#248) in an article in the June 1904 issue of *The Journal of Speculative Studies* notes that a copy of the Brill book was once seen in a small private library in a baron's hunting lodge somewhere in Bavaria, but academic journals are filled with dreamy scholarly notions. The National Library of Albania in Tirana claims to have purchased a copy of the Brill volume when it was published, but for more than a century the book's been lost in the stacks or elsewhere (maybe stolen or perhaps sold by one of the library workers). A hand-written marginal note in somewhat faded purple ink in the August 1935 "Miscellany" catalog issued by London rare book dealer Bernard Quaritch Ltd. enigmatically mentions the I.B. Tauris predecessor publisher edition—not listed in the catalog—without further details. Other than those few hints, the modern-day versions of the original manuscript have, like Eunomyous himself and his work from more than two millennia ago, disappeared almost without a trace. The two editions remain only as ghostly phantoms of what once existed, or might have. Eunomyous may have been the original ghost writer.

Although neither the original manuscript nor the two derivative works based on the commentaries survive, the ancient wisdom of Eunonymous still remains, as some philosophy textbooks contain a paragraph or two, and in one case a long footnote, on his thoughts.

He discussed at length the idea that all that doesn't exist—the vast potential of possible outcomes, none of which emerges into reality— represents true reality because such potential possibilities far outnumber the one version which finally assumes a form and is. Each form we perceive, from the past or our on-going daily perceptions, represents only the single final phenomenon, a mere nothing measured against the millions of alternate possibilities which comprise the dense, even if invisible, true reality. What actually comprises the world as it's known to humanimals represents only one extremely insignificant random outcome among all those which might have emerged into being. Eunonymous holds that it's a mistake to endow what exists with too much meaning, for that which happens to come into being is arbitrary.

The contingent nature of existence greatly concerned Eunonymous. Because randomness and chance largely determine what actually occurs to create the reality that exists in time, it's not realistic to view the world as if it means anything, for no discernible plan or formula created it. Randomness doesn't represent the sole factor which governs what takes form. The philosopher recognized that what precedes any event or thing does establish some deterministic necessity which shapes outcomes, but he stated that it's impossible to know the degree to which the accumulated weight of the past determines what happens. If we could know how the past controls the future, the Greek sage pointed out, then accurate predictions would be far more common than such foresight actually is. So complicated is existence on the lonely planet, even hindsight provides few insights. In short, later thinkers have nothing at all to add to the topic, for some 2500 years ago Eunoymous illuminated the matter of inexistent potentials as he meditated his days away there on sun-drenched Hydra. There really wasn't much else to do on a remote Greek island back in those days—no TV or internet, no crossword puzzles or computer games, no cheap-edition paperback mysteries to read (but anyway enough existential mysteries to ponder), no Swedish beauties in bikinis to admire and desire. With all the main points covered centuries ago in *On That Which Never Existed and Never Will*, there remain only a few rag-tag remnants of after-comment relating to the Greek's original insights. Although almost all of humankind has lived following Eunonymous's long ago existence, no one since then can add much to what he said. Science is accretive and cumulative, but matters regarding the existential situation of earthlings on the lonely planet have long since been analyzed, even atomized into infinite detail.

It's easy to confuse Eunonymous with Euhemerus. In fact, every day all around the globe thousands of people get confused about those two ancient philosophers. They're similar but different. Like Eunonymous, Euhemerus was a Mediterranean islander, believed to have been born in Sicily, and they both lived more or less at the same time (Euhemerus about 330-260 B.C.). Rather like the ghostly works of Eunonymous, the *Sacred History* by Euhemerus came down to us only in summaries and fragments. Perhaps the first fantasy travel account, the book tells of the author's visit to Panchaia, site of a utopian society where Zeus supposedly once lived. As a community which could have existed but did not, such a society reflects the theme Eunonymous addressed: all that which might have but did not enter being. So similar are the two ancient sages, it's as easy to confuse them as it is to confuse Lady Gaga with Madonna. Which of them—the Lady and the Donna, not the philosopher pair—wore that famous bra with the pointy cone-shaped cups? Even more thousands around the world ponder that question daily.

It's just as well that almost everything which could happen didn't. By the law of averages, even more negatives than those which actually exist could enter reality to vex earthlings. We should be thankful for all the potential threats which by chance fate happened to omit. As it is, the play of infinite forms already includes enough challenges—among them, the wingèd chariot—that harass humanimals. In *That Scoundrel Scapin*, Molière says that it's useful to contemplate "all the dreadful things that might" occur and to thank your "lucky stars for whatever hasn't happened."

Someone with vivid powers of invention can visualize a few of the billions of phantom potentials which never ripened into existence. But such fantasizing produces visions which lack the essential existential corporeality of the one particular event, being or thing which by chance happened to come into existence. An invented formulation is a fancy, not a fact. But in a way even the actuality of facts is something of a fancy, as whatever does exist is in fact unlikely, unnecessary and a violation of nature. What does not exist represents the natural state; existence is the exception. Each humanimal which happens to be conceived is inconceivable. What exists is essentially an interruption in the natural order of things, a fissure in reality which is essentially indifferent to the creation or the continuance of any particular living thing or anything else. What transpires represents a violation of the natural state of affairs, which is nullity. Non-being is an analogue to how things were pre-Big Bang. Before the explosion nothing existed and everything was only potential. Maybe things were better that way. Both the cosmos itself and every earthling of any era lacked existence for millions of years before suddenly appearing, exploding into being before reverting to non-being. Whatever happens to exist is only an aberration, provisional, a momentary random example of infinite potential. Through cracks in reality emerge specific forms, but their very existence represents disorder as their absence, which lasts nearly forever, is the norm. The existence of any of the infinite forms lasts for only an instant in time. My life and its content were only momentary and arbitrary passing phenomena. What seems to me so settled, firm, fixed and familiar, me as a particular person so well known to myself, is in fact only kind of a brief living phantom, a spooky kind of ghostly character as insubstantial as the elapsed past or as a campfire's wood smoke melting into thin air. All which takes form will ultimately become deformed and eventually revert to its original formless state of non-being. The world as I saw it with its play of infinite forms, and the way I've described the lonely planet I knew, represents only one of a nearly infinite number of possible versions.

In a mere millennium or so, and maybe even sooner, the version of the world I experienced during my era will change into new and different formats as what now is gives way to what will be. Yet to be created consciousnesses, if any earth-dwellers remain to perceive the spectacle, will experience a world strange to those of us who lived centuries earlier, such as now. For the moment, my contemporaries and I share a particular era, fellow travelers who now exist on earth together during a brief passing interlude in time, place, space. For now we all occupy this little speck of cosmic dust, a particle of matter spinning in the void of a cosmos indifferent to our existence and fate. After my many twilights far from home I now see it all with an enlarged perspective: the humanimal's earthly setting and circumstances, the creature's fatal circle, how things happened by chance to turn out, this particular ghostly and lonely planet with all its confusions and illusions, its wonders and wonderings, its phantasmagorical evanescence. Now near the end, I see the world as if viewed from a great distance, much as described by Archibald MacLeish following the December 1968 Apollo 8 moon orbit when the astronauts saw their home planet from the vastness of space: "To see the earth as we now see it, small and blue and beautiful in that eternal silence where it floats, is to see ourselves as riders on the earth together, brothers on that bright loveliness in the eternal cold."

Section V
THE MISTY CHORDS OF MEMORY

~ 1 ~

Like rivers which flow through time and place, perspectives change over time and place. By my early thirties after some travels around the world, I'd lived long enough and seen enough to sort out my priorities but not so long as to make them unachievable. Time had not yet caught up with me, as by now it has. You can never turn back, but at a certain point in time it's too late even to change course going forward. I avoided such a deadline by contemplating alternatives while they remained possible rather than just as eventual retrospective regrets. I aimed to limit or eliminate "if only" remorse. Regret by commission bothered me less than by omission. For my first three decades my life had followed along the usual track like a trolley on a fixed route, each stop a way-station on to the terminal point. That line led through childhood, youth, maturity—such as I was able to achieve it: am still working on becoming seasoned—and continuing on with a conventional suburban life, and then on and on and on along the route into whatever future remained for me, finally ending at the last stop. I realized that to reduce or eliminate regret toward the end I needed to change course while time still remained for me to transfer to another route, one less traveled. I wanted the usual way-stations to give way to some less familiar far away places which would enliven my otherwise local-line life.

My agenda inspired me to envision and then to pursue a change of scene, both symbolically and literally. I hastened to restructure my life while time permitted a new direction, the ultimate goal being regret reduction. I hoped to shape the content of my life to accord with my interests and capabilities so as to live in my own way. I kept in mind my goal—that when I neared the end, as now, my remembrances wouldn't be suffused with remorse. It's of course a mistake to live in the future or the past, but nonetheless you can to good effect visit those outlying periods which flank the present. It can be useful to contemplate time-gone, rigid in its finality, and time-to-come, permeated with chancy potential. The liquidity of the on-going present as it flows along to absorb the mercurial future and convert it into the frozen past offers both threat and opportunity. Before the future slipped away I at least had a chance to fill it with some components of my choice.

Once you recognize and accept the constraints and opportunities of your fatal circle, along with the circumstances your life to date has created, you can then attempt to assess in a realistic way your potential to shape the future. Ideally this process will produce a workable balance between your present situation and its restrictions and how you anticipate evolving your experiment in the years to come. Such an equilibrium is by no means evenly balanced, for the present greatly outranks the future. You exist in the here and now, while what's to come offers only potential drenched in uncertainty. The flexible future, however, in a way outranks the past, which remains beyond change and lacking in potential. To accomplish this workable balance between the then present as of my thirties and the then far future, which—by elapse of time— is now, I visualized myself as an old man, an exercise no longer theoretical. Casting my thoughts forward, I pondered what sort of personal history I hoped to forge and then eventually to review (as I'm now doing) from my aged stage toward the end. This forward-facing view allowed me to imagine how I might attain the desired ends before the end. I am now at that point in time, and my reminiscence records how my experiment worked out for me.

By now the result of my enterprise has become clear to me. I can trace the entire process, progress and outcome and assess how closely reality conformed to what I envisioned a lifetime ago. This book represents that evaluation. Although a Doubting Thomas type not always sure of my decisions, back in my mid-thirties, nearly half a century ago, I knew I wanted to make a change and no dilemma, second thoughts, hesitation or doubt delayed or deterred me. Time was passing—the wingèd chariot ever on the move—and so I needed to move ahead myself with my new format. I suffered no "Hobson's Choice" dilemma based on having to choose (in picking a horse) only one possibility or nothing. For me choices abounded and were not mutually exclusive as both my existing way of life and the new one I added could co-exist. My situation was more like a "Morton's Fork"-type win-win case as described by John Morton, a medieval cleric. He reasoned that the king could effectively extract taxes from both high-livers and from more modest subjects. Those who lived well could afford to pay and the frugal possessed savings which could also be tapped for taxes. In any case, I always tried to avoid mutually exclusive decisions. If offered a chance to share a meal with Marilyn Monroe or with James Monroe, I'd dine with both of them—me seated as close as possible to Marilyn with Jim distanced on the other side of the table. For me no either/or binary choice in balancing my two ways of life or between the present and the future concerned me. Each had its place. I continued to live in the present while considering the future but without dreamy visions of what it might bring. I took my cue from yet a third Englishman, Dr. Samuel Johnson, who cautioned against postponing life by focusing too much on the future: "He who sees before him to his third dinner, has a long prospect."

An almost tangible tension spans the temporal distance between living in the moment and how to transcend that fleeting immediacy by imagining ways you might live in the future. Finding a workable balance requires a realistic assessment to evaluate how you could possibly shape the future as the present carries you forward. You know where you've been and where you are, but the far future—as time overtakes you day-by-day, forming the content of your life—remains indistinct and subject to many contingencies. "Present mirth hath present laughter/what's to come is still unsure," said Bill Shakespeare. No one knows if the days to come will bring laughs or tears or, to avoid the binary, both. The future is a foreign country, and uncertainty goes with the territory. I could to some extent mitigate the unsure elements by remaining in my more secure home-based existence which offered regularity, some predictability and relative security. But that default position, retaining what was already in place, would not in the long run bring me the sort of life I desired. I was willing to change my life even though that restructuring upset my stable situation and forced me to address a considerably more problematic and unsettled future.

I found it useful to approach this process with something of an optimistic attitude. Although I realized that what's to come is still unsure, I hoped that somehow the future would play out to bring me the results I sought. Such a view Judah P. Benjamin expressed when he referred to his trust in the future. Benjamin's positive temperament gave him the ability to believe that not only present but also future mirth would bring laughter. Horace's oft-quoted familiar admonition to "seize the day"—*carpe diem*—seldom includes the rest of his thought: *quan minimum credula postero*. But if you "put no trust in tomorrow," as he advises, it becomes a more difficult challenge to attempt to live life in your own way. And what of the night? Horace omitted *carpe noctem*. Is sleep part of life or only an inconvenient interruption of it? Sleep is the activity which perforce occupies the most time during an earthling's existence. But we don't seize the night—it seizes us, an odd arrangement. Only two-thirds of our time can we attempt to seize. The rest is lost to us, completely beyond our control. The relentless need for sleep suggests that the world is too much for a humanimal. The creature must nightly withdraw from life to recover from the trauma of daily existence, an unnatural state. Our little life is not only rounded by eternal sleep; to a large extent our existence consists of sleep, which greatly shortens the hours available to carry out our earthly experiment. Dead to the world in slumber, you can't experiment. Even a life which ripens into old age lacks sufficient time for an earthling to experience more than superficially the adventure of briefly existing on the lonely planet. A humanimal which reaches the age of 82 had at its disposal a total of about 30,000 days. Fully one-third was lost to sleep, so reducing the available days to 20,000, and if you deduct another 6,000 or so to account for childhood through age 16 and then about a year at the end for infirmities, the net result yields only some 14,000 mature and active full days during a long lifetime. At best, less than half of an adult earthling's total allotment of time remains for him to conduct his experiment. It doesn't take all that long for time's allotted days to run their course. Time's wingèd chariot, running onward at a fixed ceaseless pace, quickly reaches its terminal.

My enterprise to nudge the on-going present into a form which might enhance the chances of producing "the only true success"—to live life in my own way, as Christopher Morley put it—didn't violate Dr. Sam Johnson's warning against living in the future. My efforts were meant only to try to shape the future, not to substitute it for my present life. Visions of desired future outcomes didn't displace my focus on current activities. Their cumulative effect would represent my past, so I didn't neglect the granular day-to-day happenings which represented how my life's content, over time, took shape. For me back in my mid thirties, "over time" amounted to about half a century, based on my actuarial if not actual prospects. That temporal scenario, less some 17 years deducted for sleep— including old-age nap time and dozes provoked in part by reading (but not writing) boring books—served as my working hypothesis. I had no idea if my estimate would prove to be anywhere close to the mark. In any case, 50 years is a time span far beyond one's ability to predict or shape the future. But at least the span of half a century represents a human scale, unlike Stan Parker's forward-looking scenario, lasting 20 times longer, in Patrick White's novel *The Tree of Man*. Longing to live for a thousand years, Parker observed: "Why, you'd see things happen. Historical things. And you'd see the trees turning into coal. You'd be able to remember the fossils, how they looked when they were walking about."

My aspirations were considerably more limited and realistic than Parker's. I was curious about the future, but not as curious as Parker. For me a mere half century would suffice. I had no interest in surviving for an entire millennium, and in any case I didn't know if Medicare would cover me all that time. Such an Egyptian-type time-scale extended far beyond my imagination. The next five decades, if I lasted that long, would suffice to give me enough to experience as I made my way through the dark woods while continuing to contemplate how to shape for myself something of a charmed life. A thousand years would include far too many decisions and twilights far from home and over the millennium yield so many memories I could never sort them out to derive any sense from my ten century long recollections. The future indeed interested me, but not the remote periods beyond my time. Unlike the remote past, which was to some degree knowable, the far future seemed completely abstract. A more human time-scale, rather than a cosmic one or even a much shorter mere millennial perspective, represented the duration which was applicable for my own agenda. Both my allotted time and my ability to contemplate and carry out my experiment were limited.

Once I decided to try to rewrite what might eventually be entered in an account of my life, I faced the future which loomed ahead with the emptiness of a *tabula rasa* quite as blank as the pages or, more recently, screen faced by an author writing a book. The past was replete with content, a tangled jungle of intertwined happenings, but the future stretched ahead like an empty desert devoid of life. I wondered how on earth I could fill the empty days ahead which, like empty pages, seemed to mock my aspirations to create content with some meaning. I weighed the merits of occasionally giving up the existing home comforts and relative security of my life for the rigors and uncertainties of possible future pleasures. In "Empedocles on Etna" Matthew Arnold echoed Dr. Johnson's warning against present benefits for will-o'-the-wisp potential future ones:

> It is so small a thing, To have enjoyed the sun,
> To have lived light in the spring,
> To have loved, to have thought, to have done;
> To have advanced true friends, and beat down battling foes—

> That we must feign a bliss of doubtful future date,
> And while we dream on this, Lose all our present state,
> And relegate to worlds yet distant our repose.

The dichotomy between present existing realities and theoretical benefits of an invented future resembles the difference between actual experience and academic formal learning removed from the world as it is. A tension exists between intellectual preparation for life and living it. The degree to which an active life will benefit by book-learning and classes to help prepare for participating in the outside world is difficult to measure. Too many academic lessons can lessen your ability to respond to real-world conditions. What you learn in ivory towers and dreamy ivy-greened halls can only take you so far, and then experience becomes the best teacher. In *Farther Off From Heaven* William Humphrey recalls the contrast between being told "study, work, be good" and his father's example that it was preferable to follow the behavior of "those grasshoppers who lived for today, for themselves, who idled away their time." Humphrey summarized his father's mode of being: "The lesson of his life was: live, don't learn. For tomorrow you die—if you last that long."

I didn't need to read Humphrey's book to learn his father's lesson. From an early age I realized that tomorrow I would die, and by now for me tomorrow is nigh. But that early-day realization didn't lead me to a disjunctive either/or choice of either learning or living. I wanted to do both, and also wished to enjoy the present while keeping the future in mind. I aspired to extend my existence both in place, by combining home living with adventures away, and also in time by considering the present while remaining aware that what happened in the future would determine the extent of my experiment's success. I hoped to be both a grasshopper and a globetrotter, with each mode—mind and matter: contemplation preceding and following action—supplementing one another. Contrary to old man Humphrey, I didn't want to idle away my time, learn nothing and live only for today. I preferred both to study and to participate in the outside world in an attempt to synthesize the thinking and the doing. Each activity would enlarge the other, but only one without the other would diminish my life experience.

The transitional phase in my thirties by no means represented a mid life crisis. Rather, it was a carefully contemplated and willed change of course "midway through my journey," as Dante put it. In fact, I somewhat regretted that I never suffered a mid-life crisis. It seemed the thing to do back in my time. A serious existential crisis I could add to my collection, and later recollection, of my experiences. I was somewhat curious what that psychological trauma I'd heard about would be like but I never learned. I instead enjoyed a mid-life opportunity, a beneficial chance to change which allowed me to re-shape my life rather than having it re-shape me by a crisis.

A number of unlikely factors must somehow combine to create the rare opportunity which allows a humanimal to try to arrange its life in accordance with the creature's interests and wishes. A person's fatal circle, a preexisting factor which sets the terms of engagement, and then the way your life happens to evolve and, going forward, the influence of the random and capricious elements of chance, luck, fate, serendipity, happenstance, destiny all greatly affect the experiment. To facilitate the chances of living something of a charmed life, many elements need to fall into place properly. I recognized that what would happen as I embarked on my new way of life would largely lay beyond my control. Faced with the vagaries of luck and all the rest, I could only attempt to shape the results I desired. The actual outcomes remained problematic. I approached the challenge not only by casting my thoughts into the distant future by imagining the general scenarios which toward the end—by Jove, it's now! That was quick!— would eventually satisfy me. I also contemplated more specific kinds of personal history which would in old age, for me now, best serve to reduce regrets, remorse, unrealized aspirations, thwarted ambitions, unfulfilled hopes, disappointments, omissions and other such irredeemable failures to accomplish my goal. An ideal outcome is of course never entirely possible, for even the most charmed life filled with satisfying feats includes defeats and regrets. Moreover, as enchanting as a charmed life is, it remains suffused with a certain sadness and regret for the transient and vanished fleeting moments. You could even say that the more charmed your existence, the more you regret how evanescent it was. As Proust concludes in *Swann's Way*:

> The places that we have known belong now only to the little world of space in which we map them for our own convenience. None of them is ever more than a thin slice, held between the contiguous impressions that composed our life at that time; remembrance of a particular form is but regret for a particular moment; and houses, roads, avenues are as fugitive, alas, as the years.

Many particular forms did I experience and now remember. None of the forms do I regret, but at times I do lament how phantom-like and fleeting everything was.

Early on I quite clearly saw the future—not its details but its essence as a quickly depleting and irretrievable asset. For some reason, at a relatively young age I realized how fugitive the years would be even if back then a lifetime ago so many of them stretched ahead into the far future. I knew that no Borges-like refutation of time, once it passed, was possible. I couldn't tame time. Like the Mississippi, it would flow on relentlessly, a steady stream headed for the distant sea and oblivion. Gatsby believed in the "future that year by year recedes before us. It eluded us then, but that's no matter—to-morrow we will run faster, stretch out our arms farther....And one fine morning—So we beat on, boats against the current, borne back ceaselessly into the past." Borne back perhaps, but not born again for a second chance, for the past is but a congeries of disparate memories which bring regrets for vanished time. Like a river's current, current time flows on and on, onward such that neither boats nor chariots can change direction to move against the flow, reversing course to go back to the past.

Although I couldn't modulate time, I hoped to arrange it to suit my ends before I reached my end. Whatever fate decreed as my allocated quota, I aimed to reduce regrets by living a life which in retrospect would seem to me meaningful within the limits of my fatal circle and the ineluctable realities of human existence on the lonely planet. Having no choice as a resident there, I engaged with the general terms which, for unknown reasons—probably none, other than randomness—represented how the world and human life happened to be arranged. Time, mortality, fate, chance and all the rest were the common conditions ruling the benighted bright loveliness in the eternal cold. That's just how things were, and you had to deal with them in accordance with how mankind's fatal circle happened to have been formulated. All the existing forms and the formats set the terms of engagement. There was no way out: the rules were inexorable. You avoid them only by not existing, which is your normal state. Meanwhile, during your brief and fleeting incarnation as an earthling, you're confronted with conditions on the lonely planet as they are.

Only when you view your life in retrospect does time seem to fly. Looking ahead as a young man, the future seemed to me something of a monolithic temporal block which only slowly would be chipped away as Angel Falls would gradually erode the cliff-side. But I knew that time was more bedeviling than angelic. Now, as an old man, that Gibraltar of a temporal rock I faced as a youth seems to have crumbled into grains of sand in a desert of vast eternity. *Tempus fugit*, we're told, but that's not really how time travels its course. Time doesn't fly: it can't afford the fare and fees, and certainly not the delays or lost luggage. In the olden days before planes, time seemed a wingèd chariot drawing ever nearer. These days it's more like a greyhound, doggedly moving fast and not flying but dashing ahead as it leaps forward in the way greyhounds do, the years and leap years gradually accumulating second by second, minute by minute, hour by hour, day by day, week by week, month by month, a steady accretion, thin temporal slices which bit by bit swell passing moments into a life and then eternity.

Section V: The Misty Chords of Memory

It so happened that many years after I'd pondered how to articulate for myself the essential elements of my enterprise I came across a succinct statement which defines what I'd previously formulated only in general terms. A character in André Malraux's novel *Days of Hope* asks, "how can one make the best of one's life?" The other person replies, "By converting as wide a range of experience as possible into conscious thought." Therein lay the precise formula which had guided my agenda: to pursue an active, productive life, one rich with experiences and relationships which would give me a worldly education to endow conscious thought with practical real-life experiences. This format would enable me to distill for myself, even if for no one else, at least a modicum of sense or meaning. It would be a limited meaning but better than nothing, although "nothing" probably best reveals the truth of the matter. Almost everything earthlings activate represents an attempt to make something out of nothing. Descartes proclaimed, "I think, therefore I am." I reversed his thought into, "I am, therefore I think." Like Malraux's character, I didn't see the point of existing as an earthing without giving at least some thought to the experiment. I thought it would be a waste to let it all go by if I failed to contemplate the infinite forms which by chance happened to comprise my life. Action plus thought represented the two essential components. Formal learning augmented by real-world experiences followed by conscious thought would provide a tripartite synthesis allowing me to "make the best of" my life, as Malraux put it.

My reminiscence includes selections from my experiences and relationships and my reading and formal learning, all as informed or maybe deformed by conscious thought. Both the original sensory impressions and what resulted from the process of writing my account are purely arbitrary. The happenings which formed the content of my life and also my choice of what to include, as well as how I arranged the material, represent haphazard elements which by chance took shape to comprise what came to be. It could have all been different. Many random happenings were imposed upon me over the years, and to produce my reminiscence I imposed an artificial order to organize as best I could the confusion of events. The book's arrangement proposes no pareidiola, for I haven't presented a perception of a pattern where none exists. There is no pattern, not to any creative enterprise, to events, or to life itself. The only form available is that which a person chooses—among many possibilities—to format amorphous reality. Conscious thought may induce some structure, but the chosen arrangement represents only one arbitrary version. My hope is that by applying some conscious thought to the chaos of experiences life brought to me I can create a bridge which in a small way will connect my history and the text based on it with readers for whatever use or interest my account may provide, even if lacking in personal details which, as I've explained, are irrelevant to my purpose. I've eschewed what French historians call *égo-histoire*, which refers to how the historians employ professional techniques in their own memoirs. This kind of self-referential *moi*-ism I've avoided because I prefer to concentrate on the non-*moi*. In a way, however, *égo-histoire* does encompass something of my approach in this reminiscence, for the term refers to professional methods as applied to philosophical and political topics rather than strictly to personal, emotional and private matters. The contemplative elements of my account thus arise from my efforts to offer some general observations based on my own external *histoire*, derived from worldly adventures, rather than on the omitted self-referential *moi* elements.

Memories of my experiences and from my reading form the basis for this reminiscence. As time has passed, much of what happened to come to my attention over the years has faded into a misty dream-like haze, gradually losing substance like campfire logs burning into smoke disappearing into the air. Before the embers of my past become completely glow-less and cold, I've tried to preserve a few remnants in this life log. To title this Section V, I revised a word from Lincoln's First Inaugural Address: "The mystic chords of memory, stretching from every battlefield and patriot grave to every living heart and hearthstone all over this broad land, will yet swell the chorus of the union, when again touched, as surely as they will be, by the better angels of our nature." This must be one of the greatest single sentences (Lincoln's, not this one from me) in the English language, right up there with Elvis Presley's "Taking Care of Business," some of Snoopy's "Peanuts" wisdom, or the phrase uttered by a dinner partner after an expensive meal, "I'll pick up the check" unless, that is, he or she puts it down again. While many of my memories have grown misty, none are mystic in nature. My rather phantom past and my present musings about it aren't fantasized with mystical memories or meditations, for I'm not someone who transmutes the historical into the mystical or the phenomenon of earthly existence into other-worldly speculations. I was in no way moon-struck, my thoughts seemingly lunatic or deranged by airy extra-terrestrial speculations. However, in one brief passage (Section VII, Part 6) I do engage in something close to mystical musings. All else in my life and in this book is matter-of-fact, rational and curated by conscious thought, perhaps even too much so. My account in effect represents a self-motivated attempt to "play misty memories" to myself, an echo of the 1971 movie *Play Misty for Me* in which a lonely female listener frequently calls in to a disc jockey at station KRML to request "Misty." Like her, I hoped that my misty memories, swelling a chorus of fond recollections, would inspire "a thousand violins to begin to play," so stringing together some noteworthy chords out of the discordant past.

I've strummed those misty chords in an attempt to harmonize my memories and attune them to a certain rhythm which represents the play of infinite forms as I encountered them. The few samples included present not my life organized as it actually occurred but only an imposed authorial but not authoritative construct. Scribblers put words onto the page in an attempt to liberate and then forge some of the wordsmith's thoughts and memories, capturing them in written form. Writers imagine all sorts of ways to describe how the play of infinite forms might have or actually did play out. In 1884 the eccentric American "Baron" James A. Harden-Hickey, self-proclaimed ruler of the tiny desolate Caribbean island Trinidad (not the more familiar larger one off the South American coast), published his eleventh and last novel, *Fierpepin's Metamorphosis*. The book, which tells of a rather dysfunctional family, describes one ancestor who fitted his wife with a chastity belt, of little use as her lover was a locksmith. The title character Fierpepin attempts to silence a political rival, one Dr. Theodore Globule, by using a pump to extract a large dog's bark which he stuffs into Globule's mouth so the doctor can only bark when speaking at political meetings. Jonathan Swift describes an equally fanciful enterprise, recounting how Gulliver's travels led him to a character who operated "a project for extracting sunbeams out of cucumbers," the beams to be put into hermetically sealed phials and let out later to warm the air during cool summer days. Misty some of my memories may be, but I've tried to capture them more vividly, enduringly and realistically than simply as cucumber-sourced sunbeams, insubstantial and evanescent as campfire smoke.

~ 2 ~

My lifelong reading along with my various writings comprising in all some millions of words—the readings in thousands of books and the writings contained in two novels, eight travel books, some 200 or so travel articles and now this reminiscence—formed a good part of my existence. But in fact all those words represent only a minor element of my life. The wordy activities were only a side show. Although I wrote a lot of published paragraphs and pages and read thousands more, never was I very literary or scholarly. I read mainly for pleasure, not to access material for writing or for other practical purposes. Much of what I read related to attempts to satisfy my curiosity. My reading certainly didn't represent a way of life for me, and I tried to avoid emulating bookish types who viewed the experiment of living through printed type. I never confused the world of words with the world itself. From early days I realized that a life lived largely through writings is not a large but only a small life, as if one bound into the limited confines of a book. As for my own writing, it didn't reflect an inventive word-based construct of certainities. Little if any of what I wrote derived from a claim to authority or special expertise or insights. What I wrote was based primarily on life and on the world as I saw them, with no claim to include transcendent or definitive views. My experiences out in the world rather than a pretense to high culture or a high confidence in my opinions served to endow my writings with their subjective subjects and contents. Observations and conclusions based on my particular experiment in living represented only what I as a single humanimal perceived and believed, not what others should believe. I processed what happened to reach me with conscious but not conclusive thought. From the confusion of material randomly brought to me by life, I extracted and articulated into printed words memories and notions as evanescent but more weighty than cucumber sunbeams. In that way I captured misty memories, recorded in books and articles, in a format less fragile than the glass phials the character Gulliver encountered used to contain his solar gleanings.

In all honesty, it rather surprised me that against great odds my own insubstantial sunbeam-like gleanings happened to get published. Ever realistic, I'd supposed that my writing attempts would prove to be "Much Ado About Nothing," in which Dogberry says, "Truly, I would not hang a dog by my will, much more a man who hath any honesty in him." It wasn't to escape being hung by a Dogberry-type executioner that, as an exercise in candor, I honestly recognized that my writing life depended more on luck than on skill. Had my luck proved ill rather than good, I would have had no hang-up in hanging up my pen to pursue other activities. Such other pursuits would probably be easier than reviving and recording memories, perhaps better left as shadowy phantoms lost forever after sunbeams no longer warmed and illuminated my earthly existence. Untold and forgotten, those memories would never retain a place in the world. But somehow, luckily, against the odds, I happened to get published. This enabled me to fulfill one of my agenda items, which was in some way or other to keep active and to retain a place in the conventional world. I didn't want to become a vagabond, an exile, an expatriate, a floater, an insubstantial sunbeam of a figure, a transient ever on the move pursuing novel experiences never novelized or put into travel books and articles, or otherwise somehow put to some functional use beyond simply allowing me to live a charmed life. A footloose unproductive life would in fact lack charm.

In addition to retaining some sort of productive connection with the work-a-day world, I also wanted to cultivate and nurture my roots. My little house on a quiet dead-end street in a conservative Midwestern suburb and my old friends at home gave me a valued base which kept me tethered to regular life. To integrate those two elements, productive participation in society and normal suburban life, helped to bring some balance to my other life out on the road. Thanks to persistence, luck and chance I managed to accomplish my goal of remaining productive by being able to convert the vaporous mist of my misty memories into publishable hard print. I pursued this elusive and ultimately gratifying result by adopting the Wall Street expression , "Suck it and see." It's a Bumper Sticker-size Taking Care of Business approach, one which endorses giving things a try to see how they work out. As Jacob Bronowski observed, "The world can only be grasped by action, not by contemplation." If you never suck you can't find nectar, but sucking often yields nothing sweet. You take the bitter with the sweet, and if you don't try you're left with a flavorless taste in your mouth.

The writing life offers many advantages to an independent free-lance kind of person. I remember admiring my friend Carleton Beals' way of life, a roving correspondent who later became rooted in rural Connecticut companioned by a loving wife. Carolyn loved all animals, dogs, horses and especially Carleton. Although a writer's job description includes many desirable points, one—a kind of fatal bullet point—represents a mortal disappoint: never getting published. That step forges the essential link between the enterprise of putting words onto paper (pre-digital format) and then getting them transmuted into print. You could write all you wanted—words are free—but moving those pesky little black letters onto published white pages represents the big challenge. Although words served to encapsulate my memories reproduced in print, never did I conflate my writings with my actual experiences. Inky impressions cast only vague shadowy phantoms of the original happenings. Patrick White wrote disapprovingly in *Voss* that "words were not the servants of life, but life, rather, was the slave of words. So the black print of other people's books became a swarm of victorious ants that carried off a man's self-respect." I never permitted words either written or read to enslave or define me, so reducing my life to an anthill swarming with tiny insect-like marks. Seeking a wider perspective, I sought to transcend written works, whether produced or only read by me. For me words remained secondary to deeds. But when the time came to preserve a few memories, I indeed needed to use the ant-like letters to swarm thought onto the page.

Notwithstanding the pleasures brought to me by my word-rich life, I always realized that words could be a curse as well as a blessing. I perceived that in fact there were far too many words in the world, and especially in recent years when digital technology made it so easy to produce swarms of ant-like marks. A person's problems didn't begin, as Pascal claimed, with his inability to remain contentedly alone in his own room. Unrest and complexity originated not from the urge to travel but when humanimals became diverted from their immediate surroundings by the printed word. Although writings expanded the creature's horizons, books also diluted the lived experience. Movable type enabled humankind to learn something about the world without moving into it. At the same time, the great proliferation of words post-Gutenberg intruded on man's consciousness, instilling all kinds of unsettling notions, even if reading doesn't necessarily go so far as to unsettle the reader into becoming a traveler. By now far too many words have been cast into the world. To reduce the torrent, it would be useful if every earthling was given at birth a fixed life-quota of words for both written and verbal expression. Once exhausted, the depleted allocation would then end the individual's language communication, unless some sort of market enabling people to buy words developed. This sort of rationing would force earthlings, especially out-spoken ones, to think more carefully before expressing their thoughts verbally or in writing. I realize that after plowing through my reminiscence some readers may be persuaded to endorse my suggestion and apply it to me forthwith. But as originator of the concept, I'd ask to be exempt from any quota.

Admittedly, with my writings I've contributed to the plethora of words which infest the world. At the outset of my efforts, I really didn't expect that I'd contribute much to the vast number of written words, which far exceed carefully considered ones which are thoughtful and useful. In undertaking to become a published writer I was of course well aware that thousands of scribblers, neophytes like me, sat alone somewhere scrawling or (these days) scrolling their days away in an attempt to produce what each would-be writer, but no one else, believed would be readable "ants." Unencumbered by doubt, perhaps foolishly, I joined the crowd. I plunged into on-the-job training and embraced the challenge. I was both a student and a teacher, a self-motivated autodidact trying to grasp by hands-on action rather than by mere contemplation or classes the techniques and practices which might lead to publishable writings. I didn't pursue any of the conventional channels to help further my cause. I avoided the usual methods aspiring writers follow in an effort to become established. I focused mainly on how to write, hoping that decent material combined somehow with good luck might produce the desired result. The writing I could try to improve, but the luck was out of my control. I never took a writing course; never belonged to any writers' groups; never attended any writing conferences; never looked for or found a mentor, sponsor, first reader, adviser or anyone else who I imagined might show some interest in my efforts. I once sent some of my stuff to a reputable New York agent who represented among others three quite well known authors and, briefly, me. She never sold a word, and for all I knew never even read a word I'd submitted to her.

In the spirit of "suck it and see" I just kept experimenting, trying to write and rewrite and re-rewrite to learn the process. In order to master the job at hand I was a tough taskmaster as I viewed my work with an extremely critical eye. My "I" never got in the way of that eye. I removed my ego from the self-criticism I imposed. Viewing the work, as best I could, as would an outside reader, I tried to eliminate the infelicities and to improve the text to match the prose pros I admired produced. All these attempts proceeded by trial and error, with error predominating. I embarked on and pursued my writing enterprise without the slightest encouragement from anyone, but with skepticism and ridicule from a few people who disdained my project. I started without any courses, contacts or contracts, no agent, no publisher, no knowledge of the industry, no anything, noting but a stack of blank paper, some pens, and a grab-bag of travel experiences which, ant-like, swarmed through my mind. It remained to be seen if I could migrate the insects onto the page and then into print.

My general interest in travel and in each particular adventure long predated any notion that I'd ever write about my trips or, for that matter, about anything else. All through the years, both before and after I started writing, travel brought me many delights unconnected with possible books or articles. Just as I relegated reading to a subordinate activity—keeping it at arm's length so that books didn't come to represent my view of life—writing also remained ancillary to my main interests. Those included travel for its own sake—not to collect material to write about—and experiences and relationships, all the while retaining a comfortable settled home base. I was a traveler who eventually and occasionally wrote rather than a writer who traveled to get material. Other than for some purposeful trips to gather information, raw material to convert into writings was immaterial to my mission. Although I took various trips for the specific purpose of collecting actionable content, even with those journeys I focused not on the prospective word count but on what counted for me in other ways. Some travel writers seemingly run their trips in a contrived mode primarily to produce information and impressions to process into an article or a book. I had other things on my mind. That sort of purposeful travel didn't feature as my primary goal. I preferred to travel in a more free-form and random style with literal rather than literary considerations in mind. The implications of what I happened to experience provided food for thought, not nourishment to feed print works. So much did I value what a trip brought me, I didn't want to dilute the adventure with even a second-order factor such as viewing the journey as providing an opportunity to gather writing material. Words didn't feature in how I arranged my travel life. The primacy of the travel outranked the potential to write about it.

My new way of being, in fact, didn't originate for practical reasons so as to allow me to improve my place in the world. I started the wide-ranging travels only to improve my chances to live a charmed life. Utility rarely motivated either my curiosity or my desire to undertake a trip. I didn't want the "tale" to wag the dog. I deemed it a poor practice to seek experiences in a clinical premeditated kind of way. Such a constraint might operate to contort my perceptions, recollections and writings. Simply to collect material to be recycled into a written product contradicted my unstructured approach to travel. Better, I thought, to adopt Laurie Lee's blank slate attitude as he described it in *As I Walked Out One Midsummer Morning* to begin an excursion "to discover the world...a world for which I had no words, to start at the beginning, speechless and without plan, in a place that still had no memories for me." Similarly, I traveled with an itinerary in mind but with no ulterior plan. It was with a world- rather than a word-focused mindset that in the early 1970s I set my mind to take those three five-month-long trips, to South America, the Middle East and the Far East. Although written along the way, my travel journals conformed to how I viewed my travels as the diary entries represented no pre-publication sketches but served only to focus my attention on the immediate aspects of the journey.

Even after I started my writing I kept separate the two quite different ways of relating to the world. Travel brought a direct hands-on encounter, while writing and reading represented a derivative activity which reflected but was not itself the world. I never viewed the play of infinite forms as a spectacle produced for the delectation of an observer so he could write about the performance. The prism through which I viewed the experiment of being alive refracted experiences to respond to my curiosity rather than filtering them in literary terms.

Because I didn't live to write or, in truth, write to live, I differed from how many other authors seemingly viewed their travels or their entire lives. My attitude contrasted with the more literary outlook of a writer like Kenneth Rexroth who in *An Autobiographical Novel* described how he declined possible scholarships at Oberlin and at the Chicago Art Institute in the belief that "conventional education would only do me harm." He preferred "educating myself...instead of getting a good education and living a life which would have consisted largely of seminars in John Donne, about whom I would have written until I was old and gray. I had to go out and live, and so I had something to write about as the years went on." Although I completely approved of both Rexroth's endorsement of the need "to go out and live" and his views on the merits of self-education, I didn't accept his belief in living in order to have "something to write about." Nor did I go out of my way, as did Rexroth, to avoid a "conventional education." Such formal learning served to augment but not to replace my informal post-graduate course based on real-world experiences which, unlike for Rexroth, were not meant to accumulate material to write about. I never viewed the world as some sort of vast organism which existed for intellectual pathologists to dissect and analyze. Rembrandt's "Anatomy Lesson" as taught by Dr. Tulp illustrated how he cut up a cadaver, a useful service offered by a dead body. But a heavenly body like the lonely planet wasn't created to furnish earthlings with something to write about. I didn't aim to anatomize the world by surgically dissecting it. The place was for me an organic entity to be explored, not one reduced to words. I didn't live in order to write; I lived in order to live. As I saw the world, it wasn't a construct to engage with so as to dis-aggregate and reorganize the place into chapters. Raw experience arising from my travel I didn't filter through my consciousness in order to order paragraphs to fill blank pages. Any words which happened to reach the pages arrived there incidental to my main agenda, which was to experience the world.

In truth, a synthesis between various contrasting elements offers the most artful and effective combination conducive to informed views, as well as both a well-curated life and writings about it. Formal education along with the school of hard knocks, experiencing the world and thinking about it, reading and outdoor life out in the world all combine to facilitate the best perceptions and conclusions. Rexroth's dichotomy between conventional and experiential education suffers from the flaw of binary reasoning. More informative in some cases is to consider a trinary and maybe even a more "multinary" dimension, as up to a point the more ways you look at a situation the better your chances to perceive its implications, potentials and pitfalls. It's however inadvisable to anatomize a situation, as with dissecting a cadaver, into too many parts. A measured balance between analysis and closure should at a certain point end further contemplation of the matter. In *An Essay Concerning Human Understanding*, John Locke cautioned against overly analytical thinking: for interpretation "there is no end; comments beget comments, and explications make new matter for explications." Too often words—those of the thinker or notions proposed by "wrangling philosophers"—are manipulated "to support some strange opinions." In some situations, the more curated a decision the less desirable the outcome.

In any case, real-world experiences often don't lend themselves very well to a reductive exercise which expresses them in terms of contrived descriptions. To view travel or life in general as a medium for expression in literary terms restricts your vision and limits your experiences. Early in her affair with Virginia Woolf, Vita Sackville-West in 1924 accused the author of using the relationship for literary material: "you like people better through the brain than through the heart." Woolf recognized her propensity to live for the page, noting in her *Writer's Diary* (August 12, 1928) how she was cursed for viewing the world and life through words: "The look of things has a great power over me. Even now, I have to watch the rooks beating against the wind, which is high, and still I say to myself instinctively, 'what's the phrase for that?' " Rooks, books; words, birds—poor Woolf, a wolf-like predator feeding on words. It's indeed a curse to go through life constantly verbalizing into words what you experience or quantifying into figures what you think counts. It's just too reductive. For me it would be somewhat— but not always—disconcerting to view birds on the wing as flying phrases to be netted and turned into inert ant-like words caught on a page. Life doesn't consist of sentences, except for the the final conclusive sentence with no reprieve or sequels: The End.

Writers of course have to write about something. Well do I know the feeling of emptiness which confronts an author staring at an empty piece of paper or wordless screen which somehow must be filled. The blankness of the page induced into Woolf's perceptions constant contemplation about how the experiment in living might be reduced to words. Although an author needs some awareness of how her sensory impressions, thoughts and feelings might inform his writings, it's unsettling to look at the white emptiness and imagine that you can fill the paper with a page or marks comprising remarks worth reading. Years ago when a producer for NBC radio called comedian Fred Allen after his broadcast to complain about the script the performer responded, "Where were you when the pages were blank?" A novice would-be writer faces all too many blanknesses: pages or a screen, his fantasized books or articles empty of content, his mind, his bank account, his mailbox empty of responses to query letters or submissions. Yet another emptiness prevailed for me, as for a long time I lacked material to write about. Any account of my sedate, settled and repetitious routines at home would yield a soporific text useful only to put even insomniac readers, and myself as well, to sleep. My travels up to the time I started writing in the late 1970s were not, as I've noted, meant to furnish me with information or impressions to write about. Only curiosity and not publication potential motivated my journeys. So there I sat, staring at blank pages which like the future remained empty, offering potential but no hint as to the eventual contents.

As time went by I managed to fill some pages, drawing from a few misty chords of memory some material I converted into a published novel, then some travel articles, and a second novel. My persona, at least as seen by others, changed from that of a footloose and fancy-free peripatetic character purposelessly roaming the globe to the more esteemed status of published writer. But that designation somehow seemed odd to me and, in truth, still does. I saw myself then and now as a traveler, a curious fellow seeking experiences, an adventurer in a minor way who also happened to write. I was somehow always uncomfortable when others defined me as belonging to any particular fixed category, even one as admired as that of "author." My self-image was as a free-spirited free-form kind of character. About the only defining description I accepted as accurate and informative was "curious," a word whose double meaning, inquisitive and eccentric, expressed the main characteristics of my personality. My published works, however, did allow me to maintain an identifiable role in the workings of the world, one of my agenda's goals. Although I occasionally accepted as descriptive the designation "writer" I always demurred when someone called me an "author." The word seemed too dignified and pretentious to describe my modest efforts. "Author" denoted someone authoritative, and I was anything but. "Authors" I viewed as word-centric types like Virginia Woolf or T.S. Eliot or Jorge Luis Borges, thinkers who pursued a creative existence based on a literary attitude related more to style than to content. I was less a Woolf than a canine like Snoopy, whose practical street-smart views were undogmatic and whose words were rooted in real-world experience. Wordsmiths like Woolf invented texts and labored to refine them, while my approach entailed producing as best I could clear, plain, simple imaginative—not invented—prose based on lived experiences. A big difference distinguishes worldly imagined writings from wordy invented works.

Many "authors" seem satisfied to derive their material from invention rather than imagination. To invent texts emphasizes more the technique of writing rather than the content. Academic and professorial types are especially prone to featuring such literary elements as language, diction, style, voice and other textual artifacts, often at the expense of clarity and content. Such wordy authors (I exclude academics dealing with empirical science-based matters) often lack granular real-world material, so for them form trumps content. This hollowness arises because those authors—their prose and cons rather too academic—have limited contact with the world beyond books, seminars, classrooms, faculty lounges, libraries, writers' circles or the writing studio. Those who invent rather than imagine have little to work with other than their words. Such types operate in the restricted realm of type, bindings, tomes rather than "out doors." Bookish characters cling closely together in cloistered cliques, clubs, classes, comfort zones which offer protection against the discomforts of the outside world which exists in all its vivid confusion beyond and independent of words and of the ivory tower. A reader need only compare the thinly sourced works of all too many professors or writers' workshop graduates with what activists like Conrad, Hemingway and Melville—schooled out in the real world—produced. Although those accomplished authors—and other sages like Snoopy and E.A. Presley—never attended college, what they say offers more realistic and profound thoughts than many notions professed by (soft-subject) professorial types.

In *A Personal Record* Joseph Conrad makes clear the distinction between the two approaches to writing. Invention arises out of thin air, while imagination derives from experience and reality. "Inspiration comes from the earth, which has a past, a history, a future, not from the cold and immutable heaven. A writer of imaginative prose," Conrad continues, offers a "deeper sense of things" stemming from his attitude toward the world, not toward an abstract heaven. As J.B. Trend observes in *Civilization of Spain*, Cervantes favored "short sentences drawing from long experience." Writers who invent rather than imagine often produce long sentences based on short experience. Although all non-verifiable (non-scientific) writings are artificial and arbitrary, invented works are synthetic in their very conception and in that way less connected with real-world experience than imagined texts. Based on imagination rather than invention, Conrad's fiction is far superior to most novels made up by academics or house-bound writers. To the best of my knowledge, Conrad never attended the Iowa Writers' Workshop or any other such writing classes—only the school of hard knocks. But I could be wrong: perhaps, out of embarrassment, he enrolled under a pseudonym like Józef Korzeniowski or some other disguise. If so, he succeeded in spite of what the shopkeepers tried to sell him.

It's of course much easier and more convenient to remain in a cozy Park Slope condo or a Brooklyn Heights brownstone, at the Iowa Writers Workshop (Iowa City very pleasant), at a secluded writers' colony or other such word-shop, in a book-filled ivory tower or an academic library with nothing more threatening than bookworms. In those sheltered spaces you can sit around and invent stuff to write about, but those comfort zones never appealed to me. It was the world, not words, that I wanted to experience. Experiences out there all around the lonely planet enable you to imagine writings related to reality rather than just making things up. Writers content to invent understandably remain sedentary as in that way they can enjoy many home comforts abandoned by more adventurous creative types who bestir themselves to see the world and from it produce more substantive works based on imagination rather than invented notions drawn from thin air. I valued my home-based conveniences but they didn't serve to nourish my thoughts. I needed to get "out doors," as Andrew Grove advocated.

Henry James, who aspired to be someone on whom nothing was wasted, observed in *The Art of Fiction* that for an artist experience provides an "immense sensibility, a kind of huge spiderweb of the finest silken threads suspended in the chamber of consciousness, and catching every airborne particle in its tissue." No "artist" am I but just a well-traveled provincial suburbanite who has recorded in these pages a few worldly adventures. I'm indebted to Mr. James for his confirmation of the importance of experience for a creative type, even if he thinks a cobweb occupies my mind. Someone once described me as an "egghead," but "cobweb head" is going too far.

Inventive writers lacking experience confect works of artifice which lack depth. Graduates of writing programs or "workshops" often tend to employ rather mechanical workshop-like tools and techniques to bang words together as if writing is fabricated like a carpentry job produced from a blueprint. "Authors" in a classroom or an ivory tower, removed from the Conradian "deeper sense of things," commonly write wrong rather than write right. Verily, the only campus which interested me for my writing was the hippocampus where the brain forms memories. With rare exceptions, fiction written by academics or academically trained students lacks inspiration "which comes from the earth." No doubt some worthy novels by academics who profess to be writers exist, but I know of only a few, one of which I mention later in this Part. To invent a book's content reduces it to an unworldly make-believe story. Imagination endows the tale with reality. In his soon famous January 1961 "Farewell Address," President Eisenhower warned not only against the "military-industrial complex" but also cautioned that "the nation's scholars" might present the "danger that public policy could itself become the captive of a scientific-technological elite." A similar warning might apply to non-science professor authors and academic types. Aspiring writers should look not to scholars but to Conrad for inspiration on how to write by using imagination rather than invention.

Writing unsupported by real-world experiences lacks authority. In one of his letters (August 24, 1819) Keats noted: "I am convinced more and more day by day that fine writing is next to fine doing, the top thing in the world": first the action, then the poetic reaction. He earlier (in an October 8, 1817 letter) distinguished between invention and imagination: "a long Poem is a test of Invention which I take to be the Polar Star of Poetry, as Fancy is the Sails, and Imagination the Rudder." Invention originates in the firmament, out of thin air, while imagination is grounded (or immersed) in the earthly world. A writer needs the rudder to guide his course. More than ethereal invention, imagination combined with practical logic enables substantive written works. In *The Intellectual Crisis* (1919) Paul Valéry says that the European psyche of the time includes "ardent and disinterested curiosity, a happy blend of imagination and logical precision, a skepticism that is not pessimistic." This formulation serves to summarize the ideal mindset of both a traveler and a writer. My enterprise involved an attempt to encounter the world as it is, not a retreat into the virtual invented world found in a graduate seminar on writing or faculty lounge chit-chat. I sought hands-on eyes-on experiences to meet the lonely planet on its own terms, not term papers or literary terms produced by closeted students and inventive authors. Books I liked, looks I cherished. Nothing matched being out there in the world.

All too often books divert literary types transfixed by printed type from the surrounding real-life environment. In *Look Homeward, Angel* Thomas Wolfe describes a character who read Euripides at a meal while "all around him a world of white and black was eating fried food." The poor guy didn't want to put the book down to pick up a drumstick and enjoy the picnic. While more enlightening, experience can be less comfortable and sometimes even fatal as compared with theoretical academic inventions. In his *Brief Lives* chapter on Francis Bacon—whose modernizing ideas pioneered inductive reasoning, fact-based research and experience—John Aubrey describes the savant's not too brief life (1561-1632) and its end, which occurred while he "was trying an experiment." Bacon wondered "why flesh might not be preserved in snow, as in Salt." He bought a hen and stuffed its eviscerated body with snow which "so chilled him that he immediately fell so extremely ill that he could not return to his lodgings" and "in 2 or 3 dayes [sic]" died. By chance a later fatal chilly experience included Bacon's essays when 27-year-old Henry Elkins Widener, an ardent rare book collector, was last seen as the *Titanic* sunk while grasping a book believed to be a leather-bound 1598 volume of Bacon's works. Young Widener's immersion in books represented dry reading compared to his immersion in the icy North Atlantic waters where he disappeared. Experience can deaden as well as enliven a humanimal's existence.

From what I saw, writers of either kind—inventive or even the imaginative ones—as well as most other so-called creative types like poets, artists, composers, performers, film director "*auteurs*" and similar are much less original and creative than the entrepreneurs and small business people I encountered over the years. As opposed to those in the arts dealing with airy nothings, the work-a-day owners and operators of commercial enterprises coping with tangible and earthly practical matters exhibited great creativity. Focused on the bottom line rather than the story line, the real economy risk-takers faced real-world challenges, many of them resolved with ingenious responses which in their way were more creative than thought workers who deal with malleable abstract concepts. Based on my experience in both worlds, the business people were considerably more creative creatures than were characters engaged in aesthetic endeavors.

Both wordy and worldly, Goethe was one of the great exemplars of the belief that "What you live is better than what you write," as a friend in his youth told him. Goethe's famous aphorism spoken by Mephistopheles in Part One of *Faust* that all theory is gray conveys his concept that experience offers more vivid and valid impressions than mere theory. By now all gray myself, I hope my hair color doesn't suggest that I'm just a theoretical rather than an actual being. Although a ghost writer, I still remain alive—at least for a little while longer. My repeated emphasis in this reminiscence on the role of luck and chance represents not a gray theory but describes with specificity how things worked out for me. My account is not meant to suggest an abstract notion about how other lives happen to develop, a process of which I have little knowledge. I've tried to present a specific case—the only one I know in intimate detail—of how luck influenced one particular life and contributed to making it somewhat charmed. I recognize that other humananimals may attribute any charmed elements of their lives to factors different than the ones which happened to benefit me. In my case it was luck—more than ability, talent, effort, judgment, knowledge, information, intelligence and other such subjective or indeterminate elements of influence—which shaped my destiny.

Goethe's life presents a primer on how to confront the challenge of relating to existence as a humanimal. He describes how actual conditions, the world as it is, force the creature to respond to the given earthly fatal circle. As a *Weltkind*, a "child of the world" as he occasionally described himself, Goethe attempted to live so as to reconcile in a practical way being connected with the world while thoughtfully trying to understand it. He notes that his "Italian Journey" was meant to transform knowledge from books into learning by experience. Returning from Sicily back to Naples after his spring 1787 visit to the island, Goethe experienced a near-death episode when near Capri the boat was about to crash against some rocks, managing to avoid them only at the last moment. A positive pessimist, Goethe saw life as a slow shipwreck, although it wasn't very slow for Henry Elkins Widener. Meanwhile, before the boat sinks you should accept and enjoy your existence for what it is. There's nothing like a near shipwreck almost sending you to the deep to provoke some deep thoughts. Charles de Gaulle said old age was a shipwreck. Maybe some of my late in life meditations in this book arise from the reality that my voyage has almost reached its final port. The journey for me was shipshape, as my boat wasn't wrecked, but now I must reckon with the wrecker known as Time, whose wingèd chariot draws near.

Although Goethe recognizes and values the need for a conventional comfortable homestead, a stable personal relationship, interesting and productive work, sufficient financial means, he objects to a way of life which views those advantages only for what they are rather than amenities which help facilitate a wider conception of and participation in life. He particularly disdained disconnected creative and cerebral types lacking experience who focused on the world's ills and miseries, an "abuse of art." You might as well make the best of the world as it is, Goethe believed, rather than lamenting how it is. In his own artistic works Goethe didn't abuse his creative powers. He based his writings on worldly experiences, warning in his "Conversations with Eckermann" "against great inventions of your own." Better to base writing on your imagination playing over what you've experienced rather than on notions invented out of thin air. A poet "deserves not the name while he only speaks out his few subjective feelings." He must "express the world." Suffering from lack of experience, a thinker with "a subjective nature has soon talked out his little internal material." He must instead turn his "attention to the real world, and try to express it."

For my part, my writing aimed for clarity and communication, not for stylistic effects. To this day I really have no idea what "voice" means, a term which would concern only an "author" or a professor of writing. For me, doing the the writing ranked far above professing theories about it. In truth I must confess that I never really aspired to be a writer, let alone an "author" and for sure not an "artist." I simply tried to record a few misty chords from my experiences so they wouldn't one day all just vanish. Although I wasn't a particularly writerly type, works preserved in type motivated me to write. My curiosity brought me some adventures and some thoughts about them which I managed to transmute into published works. Piles of discarded drafts evidence my attempts to produce sentences which would merit the inky "ant" marks on the printed page. I didn't aspire to produce works of Shakespeareian quality or even ones up to the standards of "Peanuts," although I did hope occasionally to match the quality of Snoopy's words and those in *Mad Magazine* and *The Onion*. I was content to create content sufficient to interest, inform, entertain or otherwise benefit readers, ambitions enough for my purposes. No pretentions to any literary merit inspired me. Service to the reader was paramount. No less an authority than Robert Gottlieb, former top dog at Knopf and himself an author, opined in *Avid Reader* that writing "is a service job." A writer works for the reader. A scribbler is essentially no different than a dentist who fills holes or a plumber who opens holes, for a wordsman aims to fill or to open minds. Each of these professions is at the service of the customer, whose teeth, pipes or interests must be catered to. If forced to choose, society would no doubt prefer to retain dentists and plumbers rather than the less essential "service job" writers perform. You can live okay without writings but not with chronic pains or clogged drains. Moreover, a dentist's drill or a plumber's bill might well inflict less pain than a book like mine, especially if read to the end. I realized that whoever happened to read my words was investing in them a very valuable commodity—scarce time. Any reader gives precious minutes or hours of his finite life to connect with writings. My aim was to give the reader a return on his investment. Anyone willing to devote time and attention to my offerings deserved my best efforts, which were based not on what I invented but on what I learned from my real-world experiences. As things turned out, a few quite astute publishers and perceptive editors deemed my efforts worthy of print and so, to my surprise, I finally became a "writer."

It was largely luck which enabled me to gain such an honorable and enviable designation. "Writer" endows you with a status thousands seek but few find. My "suck it and see" efforts happened to yield results. By chance I got published, an unlikely achievement given all the aspirants. "Random House" indeed describes the process, for randomness plays a major factor in which submissions get selected to receive a contract rather than a rejection letter. But even having against the odds attained the coveted status of published author, I never adopted the affectations, practices or persona of a writer's life as some authors do. I didn't seek out or hang out with literary types, didn't join any professional organizations, wrote in pen but never applied to PEN, attended no writing programs or seminars, never participated in writers' conferences. What did they confer about anyway? Writing is a solitary, not a communal activity. I avoided the distractions at gatherings where people talked about writing rather than staying home alone doing it. I operated not as a Thomas Wolfe or a Tom Wolfe or a Virginia Woolf type but as a lone wolf, and in any case I was out of their league and completely unqualified to be part of any literary scene. I didn't become involved in any bookish circles or participate in an in-crowd but was always an outsider. Sidelined from the writing crowd, I deemed that the initials MLA designated My Loner Attitude rather than Modern Language Association, the academic trade group at whose meetings professors and would-be's network to trade favors, find jobs and scratch backs. Firmly placed in a chair next to my writing table, my back was unscratchable, and anyway I occupied no position which would enable me to exchange career advancing courtesies with academic or little magazine editors or others who could favor me with favors in exchange for mine. I was on my own.

The circles I traveled in didn't revolve around writing. Both my temperament and my life—whether at home or for adventurous travels—were remote from the narrow hot-house confines of the big-time Big Apple publishing community. Only once did I eat at Elaine's, that Upper East Side New York City feeding (lousy overpriced food) and watering hole where members of the publishing establishment (the authors should have been home writing) went to see the scene and to be seen. For me the Middle East and the Far East were more interesting than the East Side. I preferred McDonald's to McSorley's, a lower East Side saloon where editors, publishers and other book-land figures gathered. I didn't take mind-altering drugs with the hot-shot cool characters at arty parties and at the Chelsea Hotel, although at home I did occasionally pop a baby aspirin. Avoiding all the clans and cliques was no deprivation, as that stand-offish attitude conformed to my personality. I was never a joiner, not the Book of the Month Club, Diners Club credit card, AAA, AARP, Costco, Sam's, country club, city club, any club. I was a card carrying club-less member of nothing. In truth, I'm completely unclubbable, although, as every undertaker knows, quite spadeable. I didn't want to be a member but only a rememberer, to hear in my mind's ear a few misty chords of memory, remnants played back to me from my adventures far beyond the hermetic word world of writing. I paid my dues, not to social clubs or literary coteries but with many rough-travel twilights far from home and also suffering plenty of rejections by undiscerning editors and obtuse publishers. The perceptive George Blagowidow and his brilliant staff at Hippocrene were of course notable exceptions to how other less capable firms viewed my submissions.

It was the outside world which interested me, not the narrow circles of bookish types. Most of the writers I happened to encounter seemed to me rather uninteresting and too focused on words, especially their own. Gazing inward, or in-word, they saw things from a solipsistic perspective, somewhat oblivious to the big wide world beyond sentences, paragraphs, chapters confined to the narrow bounds of bound books. I never even aspired to become a member of the great transnational community known as "the Republic of Letters," a large amorphous coterie of like-minded bibliologists consisting of critics, reviewers, academics, pundits, professors of literature, writer colony elites, commentators, interpreters, translators, agents, little magazines, big shots and big mouths, cliques of insiders, professional net-workers, the publishing guild, teachers at graduate writing programs, graduates of how-to-write classes, literary has-beens or never-was'es or never-would-be's, the hangers-on and fallers-off, each of them in his or her own way somehow participating in and living from (or trying to) the creation and dissemination of the printed word. Those ancillary characters—who'd be out of luck if it weren't for the creative types who somehow managed to produce the published material—were both facilitating and feeding off what those of us who confronted the blank pages or screen finally, with great effort, filled. Silent and removed from the world like ghosts, we labor alone in our lairs to capture the elusive words which serve to form the content of what strangers distant in place and time might eventually read. The overpopulated and often somewhat pretentious lettery Republic was one whose passport I didn't carry and a place I didn't want to visit. My preferred destinations avoided the Republic of Letters and its insiders in favor of the outside world. In essence I was not a writer but simply a curious fellow who in part satisfied his curiosity by travel. Whatever I somehow happened to publish did not in my view credential me for citizenship in the Republic of Letters, which in any case didn't represent my true motherland but, as Arthur Koestler put it in another context, only a step-motherland. The derivative nature of the word world, removed from life even if attempting to reflect it, interested me far less than did the real world with its granular textures and frictions, also known as experience. For me the literal and not the literary offered more meaning.

A need for self-expression or a desire to create literary artifacts didn't motivate me to write. My writing instead originated from a more mundane purpose: to portray how one keenly curious character happened to experience his particular experiment of being alive on the lonely planet during a certain era in a specific context. I felt that if, perchance, I existed and

Section V: The Misty Chords of Memory

happened to be lucky enough to live in a way which gave me some opportunities to experience the world and life more broadly than most people I should say something about what I was fortunate enough to find with my experiment in living. This late-in-life reminiscence attempts to preserve a few impressions and observations from a humanimal who somehow happened to benefit from what turned out to be something of a charmed life. My advantages gave me vantage points which widened my perspective. As I stated in the beginning, my account is not about me, for I was simply the intermediary between one tiny finite consciousness, a singular but unimportant brain-box of passing sensory impressions, and the chaos of stimulation which sparked the synapses and the neural receptors. Like a transformer, I changed the voltage from what powered my experiences, vivid in the immediacy of their once current real-life illuminations, into shadowy memories which, in turn, form colorless inky black lines on the printed page. Colorless by hue but hopefully not by content, which is meant to offer a chromatic picture of what befell me. As noted in Section IV, Part 4, Herodotus wrote his histories "so that time may not draw the colors from what man has brought into being." In a much more modest and individual way, I've tried to depict my adventures in a way which paints a vivid picture of what chance happened to bring into being to comprise my earthly existence. My hope is that after my disappearance time will not too quickly draw the colors from my life experiment, although I do realize that eventually the hues are fated to fade away into a blank nothing. That reality has not deterred me from producing my reminiscence. In other words, I refuse to let time's nearing wingèd chariot govern how I choose to spend my final days, weeks, months or years—which will it be?—as an earthling.

Although I happened to be the receptor which collected and accumulated the impressions which formed the content of my life, I was only an accessory to the process, a mechanism rather than a central figure. For that reason, among others, I don't feature in this book, other than as a facilitator. The centerpiece of the reminiscence as its main character is the world and its phenomena, the play of infinite forms created both by nature and by humanimals. Because the account is not at all about me, the reminiscence by design omits almost all personal information, details which for my purposes are irrelevant and, in fact, uninteresting compared with the big wide world I experienced. The "I" in the text denotes only an observer, the eye which viewed the play of infinite forms, not an "I" which merits inclusion in the story, which in essence remains "I"-less.

In the back of my mind, rather far back most of the time, I'd for some years contemplated writing some sort of account, a summary of how I saw the infinite forms, in order to preserve a small part of what would otherwise be forever lost. One spring morning a few years ago when I awoke, in my mind I suddenly and surprisingly found the format and much of the content which, by some mysterious process, had finally occurred to me. The same day I without delay began to write, and by the end of the long hot summer finished the first draft. To produce a substantial narrative over the course of a single spring and summer entailed regular long hours of writing. But the main explanation for the relatively fast pace is that the reminiscence was informed by more than half a century of experiences, experiments, happenings, travels, reading, thoughts and other components which comprised a hinterland of material in the storehouse of my memory. The challenge was to select the most representative samples and then to structure them to evoke the entirety. I deemed it a great privilege both to enjoy the opportunity to live something of a charmed life and to retain in my final years the time, ability and temperament to produce an account of that life. From the perspective of the balmy days of May when I conceived of and began the book, the summer seemed to stretch ahead into the remote future. But over time the long hot summer reached its end, autumn arrived, the days became cool. The days do grow short when you reach September and the autumn leaves do turn to flame before they fall to earth, there to decay and disappear into nothing, like campfires which flame the logs into smoke that then vanishes.

I realized that beginning a long book with my short remaining life-span put me into a race against time. In the background time's wingèd chariot rumbled ever closer as I raced to complete the book. Although no known ailments, specific terminal conditions—not to be confused with the general terminal condition of being alive—or other health problems afflicted me, my contest with time arose both because of my advanced age and from the reality that at any age a humanimal, whether the creature recognizes it or not, is always engaged in a race with time. As I realized from a young age, my entire life in effect entailed a race against time, a race to see, learn, experience, discover as much of the infinite play of forms on offer as possible. Free time, among my most valuable endowments, represents an essential component of a charmed life. Free time is priceless. As a young man I learned to appreciate the time value of money, but as time went by I treasured even more the time value of time. It was the currency I needed and which I by chance possessed to expend in my own way to enrich my life. An acute sensitivity to the scarcity of time, its fleeting nature and its crucial importance in facilitating a charmed life greatly influenced me. Not only young people but also many seasoned older ones all too often fail to recognize that even long hot summers end, that autumn comes on much sooner than you expect, and that winter is not far behind.

The moment you're born the race with time begins. In *All the King's Men*, a magnificent exception to the usual inability of an academic to write readable and enduring fiction, the narrator, Jack Burden, watches his heartthrob Anne Stanton playing tennis with another beau. Burden thinks it was a "terrific and fundamental injustice" that "I wasn't out there on the court with her." He feels like telling the two of them, "You think you'll be here playing tennis forever, don't you? Well, you won't." The girl replies, "Why no, not forever." The fellow would say, "Hell, no, we're going swimming this afternoon. Then tonight we're—." Burden would say, "You didn't get it. Sure, I know you're going swimming, and you're going out somewhere tonight and you'll stop the car on the way home. But you think you'll be here this way forever," and the guy would reply, "Hell, no, I'm going back to college next week," and the girl would say, "I'm going off to school." Burden realizes that "They wouldn't get it worth a damn. There was no use in giving them the benefit of my wisdom." But from a young age I understood just what Jack Burden meant, and now in my old age I hurried on with my book just as I did with my life as from my thirties because I never supposed that I'd be here forever. I understood what Burden said—I got it.

I really can't explain how it was that even as a young man I realized that I wouldn't be playing tennis or remain part of the play of infinite forms forever. Long before I was both benefited and burdened by Jack Burden's wisdom, I profited from my own realization of time's finite duration and I knew that game, set and match would soon end. From this perception arose in my mid-thirties the corollary that I should without further delay revise my life with a new agenda and proceed to activate it. Now, years later, my impending end even more urgently impels me to record some elements of my soon-to-be deactivated life. Before long will be tallied the final score of the match, the net result not of a tennis game but the match between my consciousness and the play of infinite forms which happened to come to my attention to comprise my earthly experiment. I recognize, however, that my hard-earned and hard-learned lessons will little help other humanimals, for each undergoes its own apprenticeship to learn life lessons and to find out by experience what best works for that individual. My account presents only what I found useful as I groped my way through the dark woods toward the end of the journey to arrive at the desert of vast eternity.

A few basic lessons I learned about writing which also apply to life in general may be of some use to guide other earthlings through the woods. Those rules include: sketch out some ideas to reduce the need to rewrite; revise; don't revise to excess, as too much revision can degrade your original vision; omit as much as possible, as effective writing is reductive as well as additive; appreciate a good editor and value her suggestions; be persistent for both your writing and your efforts to get it published; view rejections as opportunities; think clearly and write accordingly based on imagination, not invention. Once formulated for my writing, those principles helped me with living. As such, the double-duty rules guided both my imaginative and my applied real-world experiences, as follows: I planned ahead, reviewed and revised as necessary, simplified my life by eliminating many ancillary elements, appreciated and listened to my mentors and trusted friends, was persistent and considered rejections as bringing new and better possibilities, and strove for clear thinking based on real-world considerations rather than invented fantasies.

Of the listed factors, finding reliable and sensible "editors" represented the most challenging task. To befriend someone who can usefully "edit" you requires a rare mix of unusual qualities. Your interlocutor must offer understanding, trust and confidentiality and be a confidant with intelligence, judgment, wisdom, experience, knowledge of your situation and history and an instinctive comprehension of what you're about, all operating to synthesize realistic perceptions and effective advice. As for persistence, this obviously plays a large role in all aspects of life, and I thank you dear reader for having persisted enough to read this far. Keep up the good work. Finally, I compliment you on your clear thinking in having acquired this volume, even if by borrowing but—preferably—by buying. Either way, my thanks for your passing visit to my world as presented in my reminiscence.

Much of what I learned about writing originated from my reading. I compulsively and incessantly read anything and everything, texts in all sorts of contexts. Books, magazines, newspapers, journals, periodicals, advertising, product labels, billboards, bills, court opinions, bubblegum wrappers, menus, comic strips and comic books, fortune (never misfortune: the Chinese must live charmed lives) cookie slips, ticket stubs, bumper stickers, warranty cards—you name it I read it, a blur of ant-like black-on-white or more colorful letters of words in sentences which sentenced me to a lifetime of reading. I wish I could claim that I remember most of the millions of words which passed through my mind, but in truth almost all have been forgotten. At the time of reading the words, however, they brought me many pleasurable hours which over time amounted to book-years of enjoyment.

But as much as I enjoyed the reading which so pleasured my years, never did I confuse the printed word with the real world beyond the book. No matter how realistic or how much based on imagination rather than invention, the artificial versions of life in the bound volumes greatly differed from how things were out in the vast trans-book realm which transcended the limits of words. Books represented the world writ small. Outside lay a raucous wonderland far different from the invented world of Alice's adventures. Her land existed only on paper. Unlike reality's confusions and disorder, volumes were organized range-bound constructs neatly contained in squared-off pages filled with aligned type. Out in the world churned an agitated confusion with masses of people spilling every which way in rough-edged noisy motion and commotion, a chaotic play of infinite forms—also known as real life—irreducible to tidy angular paragraphs or coherent chapters. Pleasurable as reading is, that sedentary activity offers only a vicarious experience, not the real thing. Words present an ancillary side show to the main performance—the great spectacle of ever changing fleeting infinite forms. Burying your nose in a book limits your nostrils to inhaling only a few wordy whiffs, mere surrogates of the pungent real world aromas a roamer inhales. Reading can inform, entertain, amuse, educate and bring many other benefits to readers, but in the end books serve only to enhance life, not substitute for it. Bookish types such as academics, scholars, compulsive authors, escapist readers and similar word addicts spend their lives buried in words until one day those bibliofanatics end their days buried not in tomes but in tombs after living in a paper world.

Although I pursued both reading and writing, I believe both those activities are somewhat overrated. It's of course gratifying for a writer to manage to tame unruly raw material, converting a discordant and diffuse messy mix into an organized settled text. But even the most accomplished text presents only a second-order reflected version of reality rather than reproducing it in a truly non-derivative way way. Writing and reading each represent only a surrogate and vague approximation of life, not the real thing. Too much reading or writing will distance you from living. The degree to which you devote your time and concentration to books represents the proportion of your life spent in an artificial world. Some people no doubt seek an escape from life by finding a refuge in the virtual world of words. Pity the character described by Richard Condon in *Winter Kills* who "must have learned about life from a correspondence course." There's of course no correspondence between the mail order lessons and the graduate curriculum real life teaches you. In *All the Strange Hours* Loren Eisley, referring to doing research in the old University of Pennsylvania Furness Library, reflects on "The scholar who descends into the catacombs of the past," so removing himself from life. "He has lost the realization that the flesh-and-blood inhabitants of his chosen period went about their affairs as living and limited human beings." The bloodless pages of a book can't capture what Spanish philosopher Miguel de Unamuno called men of *carne y hueso*, humanimals "who are born, suffer, and, although they do not wish to die, die." In reality the world as it is and earthlings, those strange haunted creatures who inhabit the place, can only be found and, to a certain degree, understood when experienced beyond books, not in texts but in the contexts where and how those benighted creatures actually live day-by-day as I saw them both at home in ordinary life and far afield under twilights far from home.

While books often answer questions they also raise others. Only scientific works based on experiments and early seventeenth century Francis Bacon-like inductive reasoning, not latter day artist Francis Bacon's blob-like artistic constructs, can provide accurate answers. Bacon's methods yield verifiable results, while Bacon's technique represents only one person's opinion about how to look at humanimals. Apart from scientific works, other writings—fiction, essays, memoirs, soft-subject topics such as in this book—contain mainly speculations and opinions. Even if based on facts, as is my account, any seemingly wise or useful observations never offer absolute and definitive insights. No matter how well informed or thoughtful, an author's personal views (in non-scientific fields) represent only one person's opinions. This very book now in your hands—I hope it's not in your face—or on a screen or an audio player contains many factual items from which I offer some observations and propose some conclusions, but they're only my interpretations. Readers may find my perceptions imperceptible or unacceptable. Because written works may raise as many questions as they answer, printed materials aren't particularly effective to assuage curiosity. Reading creates an infinity of mirrors which often reflect or refract only a disjointed confusion of artificial images. No matter how much you may learn from books, they usually imply that a lot more remains to be discovered between the covers of other volumes. In *Curiosity* Alberto Manguel discusses how curiosity can run amok, burying bookish types in an avalanche of words. He describes the International Institute of Bibliography, founded in Brussels in 1895 by Paul Otlef. This mad scheme, a Google plus Wikipedia-type enterprise conceived long before those digital resources, aspired to establish a universal encyclopedia. The staff amassed 2000 index card entries a day as workers attempted to record all knowledge. Meanwhile real life, and its constant explosion of happenings, lurched on outside the workshop.

The contrast between print and practical exerted a great influence on me. I became a believer in the merits of an active life, one informed and enhanced by reading but not dominated by it. Although I enjoyed the life of the mind I knew that it couldn't substitute for life in the outside world. Reading served in part to expand my perceptions and perspectives, but most of them originated from experience. Fritz Zorn's description in his novel *Mars* (translated by Robert and Rita Himber) didn't introduce me to a new idea but confirmed in vivid terms what I'd already realized:

> The true intellectual may well be a man who considers every question in all its different aspects, and therefore never comes to a decision and never acts ... [A] man who does nothing but think and who is too clever to stoop to anything so gross as action will be a failure in life. The opinions of a man who never does

anything but examine issues "thoroughly" and who never takes a stand on them are ultimately useless, and they collapse like a house of cards.

It was heart-warming and mind-pleasing to come across a passage which so precisely articulated what in general terms I already knew. To be sure, books had their place, but at the same time you had to keep them in their place and not let wordy works take over your life.

Not only as a reader but also as a writer I realized the limitations of words. I well recognized that the few selected samples from my travels I described in print, in this reminiscence and elsewhere, really presented only skeleton versions of the real episodes. The writings represented ghostly pale copies of the actual original happenings, which themselves had become ghostly after they occurred. My accounts were phantoms based on a shadow play of infinite forms. You can never quite capture on the printed page what you in fact experienced during your travels. You had to be there. Journeys bring far too many nuances, shades of implications, ambiguities, subtleties, refinements, gradations, hints, obscure details, shifting tones of twilights far from home to reduce to print. What you experience in travel remains elusive, invisible to anyone else. Even someone who accompanies you may see things in a different way. It's an illusion to believe that a book, one you write or read, presents reality as experienced. I never supposed that my accounts, or those in other books, contained descriptions which reproduced actual existential elements of travel. Words can only suggest them.

It's of course more convenient to access life through books. Reading about travel avoids the many inconveniences which confront people who encounter the world directly. Authorial control serves to arrange written works, while the world is deranged and disorganized. In its orderly alignment a book's text deceives, as ultimately life escapes attempts to order it. A world traveler knows that because what happens on a trip is uncertain and contingent, no neat chapters can really relate the story as it actually happened.

Because adventure travel entails such great uncertainty, one of the most important attitudes needed for that enterprise is a sense of optimism. In *Corridors of Power* C.P. Snow refers to "a man of action's optimism. The optimism which makes a gulf between men of action and purely reflective men." This echoes Judah P. Benjamin's view that taking action requires you to trust the future, a belief anyone embarking on a travel adventure must hold. A world traveler has to have faith that things will somehow work out as you make your way among strangers in strange lands far from your home base. You need such optimism especially in exotic Third World areas where you encounter all sorts of mysterious unknown foreign customs, beliefs and practices, ones completely alien to your own familiar way of life. Without a positive attitude it's more difficult to imagine that you'd enjoy traveling in such strange and challenging places. At the same time, it's advisable to temper optimism with a realistic sense of caution. A conservative and careful wariness helps to lower the likelihood of unpleasant situations and surprises.

You can certainly trust the past, forever set in time, immune to change and no longer hostage to fortune. But the future can be dangerous. A Benjamin-like trust of the future comes with a mind-set that hopes for the best while recognizing that the worst might happen. As for the present, it's happening right now, right before your very eyes as you read these words— long after my own "right now" as I'm writing them—and also before your ears, mouth, nose and with a passing ticktock touch on your mortal skin as time's wings waft upon your being a subtle, gentle caress which ages you second by second, the ever so stealthy gossamer moments vanishing like smoke from a dying campfire, the fleeting right now—every instant becoming a then—coming at you and coming at you wherever and whoever you are, whether you trust the future or not. The future doesn't care what you think or do—it's just there, waiting to enfold you, to age you, to end you.

For life more generally, apart from travel, trust represents only half the formula, for a trusting attitude must be tempered with some mistrust. That entails not pessimism but a certain skepticism which involves awareness and wariness, a kind of cheerful caution. This realistic attitude will help you recognize that random happenings can both break your way and brake your way, enhancing or slowing, delaying or stopping your travel adventure or life experiment. Luck and chance never announce in advance their plots and counter-plots. In travel as in life you take your chances. You leave the comforts and regularities of home to venture abroad at your peril and discomfort. In can be cold and alienating out there. Few places are more lonely than a train station or a big airport late at night. Those big spaces, normally so busy, languish empty and forlorn. Deprived of the dynamic daily bustle and hustle, the night-shrouded halls and concourses are ghost towns. From time to time I found myself stranded in a foreign station or airport far from my familiar and comfortable little corner of the world, a forlorn fellow alone in post-twilight nights far from home—home, where I longed to be.

In *Let Us Now Praise Famous Men* James Agee described the protective warmth afforded by the "shell and carapace" of home, a haven where we "cling in her skirts by night," a far more secure and welcoming comfort zone than soulless alien airports, empty train terminals and uncongenial hotels under twilights far from home. "All over the whole round earth and in the settlements, the towns, and the great iron stones of cities, people are drawn inward within their little shells of rooms, and are to be seen in their wondrous and pitiful actions through the surface of their lighted windows by thousands, by millions, little golden aquariums."

It was my own cozy little aquarium, my protective shells of rooms, I from time to time forsook to venture out into the big wide world, there to make my rounds around the round earth, mankind's terrestrial fatal circle, to see what I could of its play of infinite forms.

~ 3 ~

The nineteenth century symbolist French poet Jules Laforgue said that the worst punishment would be "to die without having seen the planet." Life must have greatly punished him, for Laforgue died at age 27, leaving him with only a brief opportunity to see the lonely planet. His sentiment expresses a thwarted ambition rather than, as for me, a well fulfilled aspiration. It's unlikely that many people would endorse Laforgue's opinion, as most earthlings seem content to stay within their limited comfort zone rather than venturing out to discomfort zones. It's far easier to remain at home in the protective shells of familiar and familial rooms, settled safely and comfortably surrounded by known knowns, than to shuck the sheltering shells and emerge unprotected into remote areas of the globe with all its unknowns. True punishment for homebodies would be to engage in wide-ranging travels to strange lands and twilights and mornings, noons and nights far from home. But the perceived security of the coddling comforts and familiar routines at home are something of an illusion, for even in its natural habitat a humanimal remains ever subject to the whims of chance and luck. Home or away, there's no escape from those governing forces. Plots and counter-plots are intrinsic to being alive.

Threats—from nature, from humanimals—loom everywhere. This is not a dark or a disturbed attitude. I believe it's realistic to realize how sudden setbacks can at any time damage or destroy plans, hopes, expectations, seemingly settled situations, entire lives. Wherever an earthling may be, capricious influences and random happenings can intrude, shattering the shell of home-based security or interrupting travel plans. But away is more risky. Once you abandon your routine life and venture out in the world, a multiplier effect of hazards makes you even more a hostage to fortune than at home. Misfortune lurks at every turn. At home the known knowns and known unknowns are more likely to be predictable or at least more easily dealt with. Adventure travel entails a plot and counter-plot-filled mystery story, as you don't know how it will end or if you'll end before the trip does. Until the trip is over, leaving you with misty chords of memory, the adventure remains shrouded in a haze of "mistery."

Like life itself, venturing into the unknown represents an experiment. Opportunities disguised as threats, hazards which initially seem to be delights, put an edge on the experience, an exhilarating activity which makes you feel more alive if it doesn't kill you. The passing events and impressions which seem so vivid quickly fade after the fact. Time moves on, the trip continues on, life goes on. Nothing stops, not until you eventually reach the temporal equivalent of *finisterre* when time runs out. Once you've transitioned from wary to weary by completing the trip and returning home the entire experience takes on a tincture of unreality as it begins to recede into the mists of the past. Back home from a visit to Vienna, Thomas Mann wrote in his *Diary* (December 15, 1919): "Strange to sit at one's accustomed place in the clear-minded state of morning and recollect the unreality of travel." This may be a Thomas thing, but it's more likely a common reaction: one Mann is every man.

My compulsive curiosity and desire to diversify my life motivated my willingness to forsake the comforts of home to see the world. All too many places out there, strange and enticing names on maps, stimulated my imagination, a mental facility I preferred to nurture and feature rather than to stay at home and rely on invention to conceptualize the world. As the travel-sophisticated Conrad observed (Joseph Conrad, not Conrad Hilton), the best source to understand reality derives from imaginative contemplation based on worldly experiences rather than on speculations invented by abstract thought. That creative Thomas, Edison, illuminated the world with his enlightened inventions, but I favored imaginative inspirations rather than ones created in a vacuum. Adopting Conrad's attitude, I ventured forth to satisfy my curiosity, to broaden my experiment on the lonely planet, and to stimulate my imagination, with the hoped-for additional benefit of super-charging my already somewhat charmed life. I took to heart Bronowski's view that the world could be understood, if at all, only by action, not by contemplation. You in fact need both hands-on experiences and mind-on thought to synthesize some sort of comprehension about your experiment. My reminiscence represents an attempt to achieve such a synthesis.

Studying the map was for me an unsettling exercise as I noted all too many alluringly named places. They beckoned and also seemed to mock me. Here we are, they said—where the hell are you? A kind of obsessive-compulsive force drove me to see as much as I could but this was no "disorder." My condition was not "OCD" but "OCO," Obsessive Compulsive Order, a trait which enabled me to effectively address Presley's Taking Care of Business mantra, for me the business of how to live. My business-like OCO procedures enabled me to put into orderly comprehensible formats carefully planned journeys to guide me through the confusion of countries and regions which erratically checker the map. Methodically dividing the world into itineraries, one by one I converted the beckoning place names into on-the-ground experiences. It's said that curiosity killed the cat, but in my dogged pursuits of curiosity-quenching experiences I was more canine than feline. But I was also like a Kilkenny Cat—nothing left but their tails: I tailored my wanderings to respond to my wonderings, the trips disappearing into misty memories but with the tales, like the tails, remaining.

Mystery not only pervades travel adventures but also inspires them. Until you see the world, it remains a rather mysterious place. After you actually circuit the lonely planet, some of its mystery dissipates. Paradoxically, however, the more you know about the world, both by experience and by contemplating it, the stranger and less comprehensible the place becomes. It's an odd sort of homeland humanimals occupy. For lack of time, interest, need, funds, curiosity or other reasons, many people disregard the mysteries, preferring instead to deal with more mundane matters. It's simpler to refrain from thinking about the world's strangeness and to ignore its fundamental oddity, and anyway daily duties distract and divert from such thoughts. But somehow curiosity jarred me into action and compelled me to question and to interrogate life and the world.

Existence is for every earthling, a benighted creature who arrives from the mystery of dark pre-life into the momentary brightness of being before passing on to mystery of forever darkness. As for every other earthling, I had to accept the given existential terms—like time, the conditions of earthly existence were irrefutable—but I didn't have to leave them unexamined. I felt like the Native American inhabitants of the mid-Mississippi River Valley as they viewed the mysterious waterway which appeared from somewhere, then disappeared to somewhere, as described by John Finley in *The French in the Heart of America*: "To those midway of the Valley it came out of the mystery of the land of Frosts and passed silently on or, in places, complainingly on, to the mystery of the Land of the Sun." A native American resident myself in the Mid-Valley, I too wondered about the Mississippi's mysteries: where did it come from, where did it go, what was the great river like beyond my limited perspective on it? Driven by curiosity and various cars, over time I managed to drive the entire length of the waterway to see the river on its course from where the flow began as a trickle in the Frostlands in the north to the Sunlands down south where the Mississippi, at the end a torrent, disappeared into the sea. The mighty waterway which seemed so robust and alive where I saw the lively flow at my mid-river native habitat eventually reached its end, like all earthly forms. My riverine travels showed me the Mississippi in all its varied forms. Endless shifting and ever-changing moods of the churning river stirred the waterway as it flowed through time and place, turbulent and angry in the great spring floods or leaden and sullen under gray winter skies, mellow and misty on a crisp autumn eve or serene and soft in the dusk of a long summer's day. Just as I journeyed all along the Mississippi to solve the river's mysteries, so did I travel further afield to unearth some of the terrestrial secrets around the globe. Its remote lands pulled me from home and drew me into worldly adventures all over the map. Frostlands, Sunlands, highlands, lowlands, ancient and Newfoundlands, many lands beckoned me to visit and I did.

Even after I become a seasoned traveler I remained a rather anxious one. Cautious and wary, I somewhat nervously made my way through strange lands. Unlike a fixed-line trolley with its set route or the capricious Mississippi, no rail or channel directed my path through time and place. It's something of a paradox that the two categories of travelers who most suffer from anxiety while on the road represent completely opposite types. It's only natural that neophytes who lack much experience overseas are often intimidated by foreign travels, even to the most benign places. One such debutante once asked me if it was safe to drink the water in England. Water okay, I replied, but I'd hesitate to recommend the warm soapy liquid which passes for beer over there.

Oddly, extremely experienced travelers are also subject to high anxiety when away as they've faced any number of unexpected difficult situations and are thus acutely aware of potential hazards. Based on various vexations from earlier trips, seasoned travelers can envision all too readily all too many challenging happenings which might present threats and problems. The more you've traveled the more you can conceive of worrisome situations of the sort which previously concerned you and which may well recur. An endless jumble of problematic "what ifs" preoccupy you. Hoping for the best you prepare for the worst, assessing contingency plans and alternative options to react to untoward developments which interfere with progress toward your goals. I was sensitive to small variations and exceptions which differed from ordinary procedures and routines. An errant tone of voice, a stray frown or other facial expression, a too curious official, a conversation in a foreign tongue as the conversationalists stared at me with hard looks could raise concerns in my mind. Such aberrations sometimes seemed ominous, a precursor which stirred concern over possible problems. All sorts of oppressive scenarios spun through my mind as I weighed how one or another potential happening might impinge on well-being, health, safety, security, plans or other factors which shape a trip's pleasures or perils. In far strange lands you never know what unexpected and sometimes weird events might from one moment to the next disrupt your journey. A stray pathogen, an errant car, a local functionary, plots and counter-plots at hand or afoot and much else can damage or destroy all your plans. Stay-at-homes may find it strange that experienced globetrotters suffer from such anxieties. Armchair travelers escape all such concerns and problems, which for the sedentary are limited to how to upholster their chairs when it needs more support. An independent traveler out in the world has to be self-reliant and self-supporting.

One challenge a world traveler faces is the need to make immediate improvised on-the-spot decisions based on very little information. An emergency room physician or an in-the-air airplane pilot must function with something of the same haste and lack of relevant data, but at least some tests and a quick review of systems serve to guide the doctor's and the pilot's responses. Out in the Third World, many situations suddenly develop with no systems to review or to back you up. Matters which demand a nearly instant response suddenly confront you. Back at home in the easy chair at your familiar native habitat you usually enjoy the luxury of time, the benefit of information and a support system to formulate many decisions. In early August 2017 on his last day as General Electric CEO, Jeffrey Immelt told the company's employees in his farewell message that "every job or decision looks easy until you are the one on the line." Displayed on the door between the cabinet room and the room of the prime minister's private secretary at #10 Downing Street is the slogan Prime Minister Harold Macmillan devised to guide him when conducting the business of state there: "Quiet, calm deliberation disentangles every knot." At home, whether at #10 or less exalted premises, knots can be untangled in a deliberative, contemplative way, while out on the road you often confront an entangled "Gordian Not" kind of situation which cannot be easily unraveled by even the sharpest of cutting-edge thinking. Quiet, calm deliberation is for homebodies, not adventure travelers.

So fluid and varied are many of the circumstances which arise as you travel in difficult countries, there's no chance to contemplate options in a leisurely way. Sudden happenings jolt your equilibrium, forcing you to rely on resources other than calm and rational thought processes. More salient under such uncertain confusions are less cerebral and rational and more tenuous will-o'-the-wisp considerations, such as intuition, instinct, gut feeling, emotions, fleeting perceptions, passing impressions, fact fragments, hunches, guesses. Those amorphous elements stem from your accumulated experience, both in travel and to a greater degree from your experiment in living in general. To facilitate any sort of decision you can draw on your memory bank, which contains all kinds of stray deposits, a few of which somehow surface to furnish some guidance. As I traveled I found that many on-the-fly split second judgments pertain to an ability to read people quickly and accurately, a skill more difficult than reading books. You somehow have to integrate very fragmentary information into a format allowing you make a decision based on little evidence. Many decisions, in travel and otherwise, depend on making judgment calls based on incomplete information. Not book learning but only experience can teach you the technique, which in any case is subject to many misjudgments. The more unfamiliar and exotic the lands and the cultures where challenging situations arise, the more uncertain and problematic is the context and your decision. It's that arbitrary and random quality which defines the very nature of adventure travel. The uncertainty—both thrilling and concerning—which characterizes the enterprise represents its attraction.

It would be greatly misleading for me to exaggerate my travel travails. In truth, with a few exceptions the challenges and problems I faced during my trips were benign compared to the truly oppressive and often dangerous conditions confronted by real adventure travelers. Those navigators, explorers, discoverers, pioneers and other daredevils who ventured into unknown or dangerous regions ran extreme and unavoidable risks. Against their bravado and hardships my much softer adventures were almost easy-chair easy. Well, maybe dental chair as I did undergo some discomforts connected with my version of the adventure travel drill. But my mishaps were nothing compared to those suffered by, for example, the famous South American explorer Percy Fawcett, who exemplifies not merely an adventure traveler but a driven explorer in regions completely inhospitable to humanimals. Apart from a few visits to North America, he restricted his trips to the wild areas of South America where in Bolivia, Brazil and Peru Fawcett carried out boundary-line surveys deep in the jungle and searched for the lost city of

Z. Compared to him, my travels ran the gamut only from A to B and never anywhere near Z. Fawcett was no casual Thomas Cook tourist traveling with a group or like a self-described adventure traveler like Thomas Weil, a cautious fellow who always carried in his money belt a return air ticket home. The going for Fawcett was tough, but he was even tougher. He never gave up. The explorer didn't reach for a return ticket to escape the forbidding toxic jungle. He seemed to enjoy privation and seemingly insurmountable difficulties. For Fawcett, the tougher the better. Extremely strenuous and dangerous expeditions into the jungles of South America energized Fawcett, presenting him with an irresistible challenge. His punishing way of travel was for me completely resistible. Of his 24 years of married life, Fawcett spent a total of ten on his jungle adventures. The jungle was his mistress, and in the end his master. While at home in England over Christmas 1907 for a break from his fate-tempting treks Fawcett noted that it seemed "as though South America had never been, yet deep down inside me a tiny voice was calling. At first scarcely audible, it persisted until I could no longer ignore it. It was the voice of the wild places, and I knew that it was now part of me forever."

Three years later, after a month or two back at home, the travel bug again bit Fawcett, as "thoughts of the wild places—with all their pests and diseases, their misery and discomforts—disturbed the ambient peace and called me back." The bite of the travel bug which provoked the explorer to return to the wild places was nothing compared to the biting stings, things and other threats he faced in the jungle. Page after page of *Lost Trails, Lost Cities* contain fearsome descriptions of bite-based and other hazards which lurked in the back country: cannibals, blood-thirsty tribes, fierce natives, poisoned arrows, "wild beasts, venomous snakes, savages and insects," tarantulas, vultures, beriberi, espundia, malaria, electric eels, tetanus, worms, stinging flies, fire ants, anacondas, fever, intestinal parasites, cobras, piranhas, stingrays, influenza, bees, wasps, boa constrictors, crocodiles, spiders, mosquitoes, fleas, hurricanes, leprosy, bushmasters, wild bulls, blood-sucking cockroaches, conjunctivitis, jaguars, scarletina, whirlpools, waterfalls. And I thought some of my trips were hazardous!

Out there in the jungle it was man versus nature. Mother Nature, which created mankind in all its glory and vulnerability, perversely seemed to want to destroy Her creation. To the humanimals She somehow brought into being She also brought all sorts of miseries and maladies, sorrows and sadness, disease, decay, deterioration, demise, disappearance. If Mother Nature had realized what sort of world she'd engendered, she would have suffered from post-partum depression. What was all this vale of tears—tearing the creatures away from their earthly existence—about? What the good Mother hath joined together to incarnate the humanimal form She also conspired to render asunder. Surely Nature with all Her creative ingenuity could have designed a better format for earthlings. Why go to all the trouble to create the creature only to destroy it? For some perverse reason, Nature distorted and degraded mankind's existence on the lonely planet by creating far too many plots and counter-plots which harass the poor humanimals with a plague of pathology. Mother Nature victimizes her own children. The predators, the poisons, the parasites, the bitings, the diseases and all the other natural hazards constantly put people in harm's way. Being an earthling was dangerous and terrifying. Much safer to dwell in the comforting realms of prenatal or postmortem nothingness. Trapped on the lonely planet, a life could meet its end as randomly as it began with the gamete. Your very origin was a shot in the dark, as is your journey through the dark woods, as was Fawcett's trek through the jungle. Fawcett eventually fell victim to the hazards of his venturesome life when in 1925 he disappeared into the jungle, never again to be seen. To this day his fate remains a mystery.

As related in *River of Doubt* by Candice Millard, none other than Teddy Roosevelt embarked on an expedition in the South American jungle similar to Fawcett's brutal hackings through the toxic terrain. Not only for a Doubting Thomas like me does "River of Doubt" describe the uncertainties of high-order adventure travel. Roosevelt's very existence was in doubt during his dangerous and, in truth, foolhardy float down the Doubt River. A former

U.S. President shouldn't be floating down unexplored rivers in the South American jungle. He should be back home, lingering in the spotlight, reminding the voters how great he is, and cashing in on his presidency. Why go to all the trouble of becoming a President only to end with death in a remote jungle? It's a poor exit strategy. Departing for South America by ship in early October 1913, Roosevelt wrote to his daughter Ethel that if the expedition had "good luck" it would be successful, but because so much "chance" was involved the results could not be predicted. Roosevelt faced many of the same kind of threats which so vexed Fawcett. The jungle is an equal-opportunity killer. The President accepted that he might not emerge from the adventure alive. "I have already lived and enjoyed as much of life as any nine other men," he wrote in a letter. "I know I have had my full share, and if it is necessary for me to leave my bones in South America, I am quite ready to do so." For my far less challenging travels, the only bones I'd leave behind were from chicken or steak dinners.

All sorts of strange and dangerous creatures lurked in the jungle Roosevelt dared to enter. Although pirhanha are renowned for their cutting-edge biting technology, more fearsome were the tiny nearly transparent candiru, a sharp-spined catfish which is the only predator other than vampire bats to survive solely on blood. Where does Mother Nature come up with these great ideas? On February 28, 1914, Roosevelt's party set sail on the River of Doubt. There was no doubt they'd face serious difficulties along the way. Termites ate a leg of TR's underwear; rations ran short; canoes were lost to the river; an expedition member drowned; cannibals threatened the men, with the portly Roosevelt a prime catch, although during the trip he lost a quarter of his weight, so progressively becoming a less desirable feast; ever more restricted meal rations imposed a near-starvation diet; insects incessantly pestered the intruders. Apart from those minor inconveniences and many other such trivial problems and challenges, the expedition was largely uneventful. Roosevelt suffered from fever and injured his left leg, which became infected, and he contracted malaria. He finally gave up, telling his colleagues he wanted them to go on without him: "You can get out. I will stop here." But they all kept going, even Roosevelt, still feverish and his leg wound infected. He suffered intense pain, and a potentially fatal bacterial infection weakened him as he approached death. Stating that he had the shortest life span in the group, Roosevelt again suggested that he be left behind. But on they all continued, and finally on April 26 they reached a relief party which had awaited the expedition and Roosevelt there received care and gradually recovered his health. As an advocate of living the "strenuous life," he certainly followed his own advice.

It could all have been different for Percy Fawcett. Like Roosevelt, he could have emerged from the jungle, acclaimed and honored. He might have found the (still) lost city of Z. Even the jungle itself might have been different, maybe a desert or verdant fields there in South America, and also perhaps Mother Nature Herself might have behaved differently. Her originality could have produced a more benign climate for earthlings without all the natural nuisances which plague mankind. Human beings themselves brought enough misery to the lonely planet; they didn't need any help from Mother Nature to torment, to pain, to harass fellow residents of the planet. She really didn't have to create all those bothersome insects and animals and pathogens which victimize humanimals. There was nothing inevitable about any of it. It all just happened to turn out that way rather than otherwise. The world could have been an earthly paradise, but chance and fate had other plans for the place. Maybe next time things will be better.

Once they exist, both natural systems and what mankind has formulated, tangible or intangible, appear to be normal, but each component is only a random arbitrary excrescence which mutated into existence from all other possible forms which might have developed. Whether from the hand of man or from Mommy Nature, seemingly fixed and familiar earthly formats are in fact not at all normal or natural. Such existent accidental manifestations represent only brief interim phenomena, a fleeting play of infinite forms in an unstable universe churning with change, every moment mutating what exists in the spinning cosmos, a monstrous random chaos with lifeless matter careening through the void, pinwheels of gases twisting in vast contortions, a whirligig of dizzying eternal motion.

On the tiny pebble which hosts earthlings, mortal humanimals cling to the familiar lonely planet and its ambient solidities. Surely the Pyramids will never crumble into the desert sands, the Rockies tumble, the Ganges or the Mississippi cease to flow, the World Trade Center in New York City crash into rubble. Those existing earthly features, everlasting—until they disappear—in fact exist by chance, not by necessity. They each represent just one of many possible arrangements. Most of the mules on the autumn 1844 Santa Fe Trail expedition died of starvation after an unseasonable September snowstorm. Later trail trekkers over the years arranged and rearranged the bones in odd patterns which baffled subsequent travelers, for whom the various random and meaningless alignments at the "Bonehenge" were as mysterious as the stone pillars at England's Stonehenge. Each passer-by on the trail saw only the particular version of the bone designs which existed at the time. Earlier and later travelers found different patterns. So it is for travelers anywhere at any time. You perceive only what's there at the time you happen to visit a place. At other times in other eras, in long ago yesteryears and still to come far future years, other travelers have found and will find different perceptions and experiences. As I roamed the globe I always realized that what I saw and whatever happened to me was specific to the time and the place as they and as I then existed when I visited, and it all could have been completely otherwise.

Like travel, life itself essentially involves an arbitrary experience and a one-off irreproducible experiment consisting of a shadow play which Indian poet Rabindranath Tagore described as "a playhouse of infinite forms." The Bengali sage known as Gurudev ("saintly teacher") blessedly furnished me with the defining image of how I viewed all of existence, on earth and in the heavens. Each form, no matter how solid, permanent, familiar, natural and inevitable it might seem, really represents only a passing entity, an arbitrary and provisional phantom-like apparition. It's all a ghost dance. Whatever exists is nothing but an experiment created by a conjunction of random components which for a time cohere into a particular temporary reality. It's nature's way of engaging in a playful "suck it and see" exercise. Humanimals are part of the toys. The play of infinite forms keeps reforming and performing, a never ending spectacle. All which seems so real and set in time and place is really just a fleeting provisional form, pre-phantom ghosts in waiting. Take your pick, it's all evanescent— the Eiffel Tower and all of Paris, the Rockefellers and Rockefeller Center and the Rock of Gibraltar, Mount Everest and Mount Vernon, the South American jungle, the Sahara Desert, Plymouth Rock and rock-and-roll, you, me, Bill Shakespeare, John Vermeer, Mike Montaigne, Elvis, Snoopy everybody and everything. Paris's Notre Dame might at any time burn, *votre dame* could disappear, any and every *Mann, mujer* and *bambino* decease and desist from existence. Whatever happens to materialize into a form occupies only a way-station on the way to non-existence, as transient as the Santa Fe travelers and the shifting mule bones on the wayside. Like a campfire's smoke, existence evaporates into thin air.

Travel for me accentuated the reality that everything is contingent and provisional. But it's not necessary to travel to reach that conclusion. If stay-at-homes think about life as it is, they can just as easily perceive how tentative and tenuous are people, places, all that exists, all that ever did and ever will. On a less macro and more personal basis, it's truly difficult for an individual to dehumanize himself by realizing that for almost all eternity he or she doesn't exist. It's hard to see yourself as not being there in plain sight, as you're so used to. Your non-being in fact by far predominates, as it's infinitely longer and much more natural than your existence, which is a fluke. What happens to be a person's corporeality is only a momentary random reality, and a highly improbable one. Just as everything which happens to come into being was once mere potential, so any individual's incarnation depended on an arbitrary mishmash of far-fetched factors which somehow combined to produce a certain temporary form. Against all odds, a specific humanimal appears. But all the billions of such creatures—those born and died and the more than seven billion who now populate the pebble and all still in the womb of time and yet to exist—are nothing compared to the trillions who could have existed but never did or will. What is not and never will be represents the natural order of the universe rather than what random natural forces happened to have in fact produced.

Although at times I contemplated serious matters such as those I've just discussed, I didn't take them or myself seriously. Such grandiose topics didn't much divert me from my TCB business at hand, my experiment in living, and I was not at all perturbed by the realization that it was only by chance that a certain someone who happens to be me (could have been someone else) is just now writing these words (could be other words at another time in another place in a different language) which the improbable you, a chance earthly apparition, will later (later for me, your now as you read) somehow somewhere some time or other be reading. A bewildering matrix of contingencies had to coalesce into actuality for this particular writer-reader combination to exist. Both humanimals had to appear, the one to undergo all the many experiences which inspired the content of this reminiscence, the other somehow to come across and then to read the account, all a very unlikely juxtaposition of random factors. Survivorship bias induces us to assume that the world's arrangement, as reflected in what happened to come into being, represents a definitive picture of reality. But what exists is simply that which by chance comprises the cosmos and its components at any one time. The play of infinite forms plays on and on, with ceaseless new variations.

Frivolous and fanciful thought experiments about cosmic matters might in a way be passingly entertaining as a mind exercise, but such speculations are essentially irrelevant to the way things are for earthlings, who have to cope with what exists and not what might have. It's easy to imagine how the cosmos and the lonely planet might have been different. Science fiction imagines the future; alternative history re-imagines the past. If Shakespeare had lived in our day he might've written about Lady Gaga instead of Lady Macbeth. All those archaic (to us) kings, queens, princes, knaves, varlets, maidens, the insubstantial spirits and ghosts and the other airy nothings to whom Shakespeare gave a local habitation and a name wouldn't feature as his current-day characters. Married to someone other than Anne Hathaway, the old boy would now be writing about Berkshire Hathaway and about, alas, poor New Yorick gilded Goldman Sachs types (rather than Gildenstern) trading away their lives for East Hampton rather than Hampton Court palaces. Shakespeare as our contemporary would not be the one we know but a new model modern version. His timing might well have been different, alive now rather than then. When Father Time mates with Mother Nature anything can happen. But whenever Shakespeare happened to exist and whatever he wrote, the time will come when no humanimals at all survive to remember that he and his works ever existed.

Eventually no longer will Will be remembered, nor will anyone or anything else. Memories distant or misty, twilights at home or away, charmed or miserable lives—all will vanish, all mankind and womankind and their works all gone. Vanished will be every last brick and nail, tooth and toenail, museum, donor memorial plaque, paragraph, painting, symphony, library, Broadway (on or off or off-off) theater and repertoire, every church and temple and mosque and stupa, every investment portfolio and gold bar and Goldman Sachs and pretentious East Hampton palaces, Rolls-Royces and Rolls and Royce and rock-and-roll, Elvis, Snoopy and his "Peanuts" chums, all hospitals and mortuaries and cemeteries, mountains and rivers, jungles and deserts, pyramids and all structures and their builders and occupants, dog and baby pictures and dogs and babies as well, all their parents and ancestors and descendants—all, all will one day cease to be or be remembered, all lost in eternity as if they had never been. Meanwhile, I'm taking a break to have lunch.

The ineluctable reality of transcendent transience and impermanence which permeates everything I somehow recognized from an early age, a realization which served to inform and shape my new agenda. The existential puzzle posed by my perception boiled down to a few simple questions. Given the finite nature of a humanimal's life, its random nature, the transience of everything which exists—then what? In the face of those unalterable realities, the fatal circle of human existence, what sort of life makes sense? Those modalities comprised the inescapable contours of the experiment of being alive. Then there was each person's fatal circle and his circumstances and situation, all of which would influence the answers to the questions. In my own case, an acute curiosity about what was out there in the world shaped my response to the terms of engagement imposed by being alive. Somehow the world –and, very briefly, me until soon ghosted into nothing—existed, to no particular end but there it and I were, a twosome—an odd couple. I thought it would be interesting and worthwhile to experience as much as I could before, perchance, it all disappeared or, inevitably, before I did. I figured that during the brief interlude between the something that was me and the post-me nothing I might as well get around to observe some of the diverse play of infinite forms, of which I was one. I was in the play, a player, and also played with by chance which toyed with every earthling. I played the part assigned to me which was, as for every humanimals, to make the best of a fragile fleeting existence.

By mere chance I happened to exist as a momentary worldly form, so why not suck it and see by pursuing the experiment of being in the widest range possible, given my circumstances. Before long the time would come, as by now it has, when the experiment would near its end, the pre-ghost ghost-writer soon to be phantomed into a ghost without any further earthly existence. In *The Three-Arched Bridge* (translated by John Hodgson) Albanian novelist Ismail Kadare tells of a certain character who collects local folktales, legends and ballads dealing mainly with "the division of mankind into two great tribes of the living and the dead. The maps and flags of the world bear witness to dozens of states, kingdoms, languages, and peoples, but in fact there are only two peoples who live in two kingdoms: this world and the next. In contrast to the petty kingdoms and statelets of our world, these great kingdoms have never touched each other." I took to heart, and mind, the reality that those two realms represented vastly different states, one in which people existed and the other populated by stateless expatriate ex-people. For the moment I was a citizen of the living, but from early days I knew that I'd soon cross the border into the land of the dead. For a play title *All's Well That Ends Well* will do very well, but along with the lively play of infinite forms out in real life mortality plays a leading role, one which prevents things from ending well.

As a Thomas ever loyal to my name-clan, I endorsed Thomas Jefferson's view that "The earth belongs always to the living generation...The dead have no rights. They are nothing; and nothing cannot own something. This corporeal globe, and everything upon it, belongs to its present corporeal inhabitants during their generation." For the time being I was a dues-paying and tax-paying member of the living generation, a subject (one without an object) in the kingdom of this world, and while still on the lonely planet as part of that lively realm I decided to see the corporeal globe. What were things like over there in Spain where people face death in the afternoon and mornings and nights as well, and what about life on the Mississippi down in Mississippi with its sound and fury and the Elvis sound in Memphis, and how did people live in the heart of darkness (no electricity there), and what is the way we live now here, there, everywhere?

The inexorable rules of the game decreed that it would all come to an end, for me and in a cosmic way. Meanwhile, before I succumbed I'd leave home to roam about to see what it was all about. I didn't view Tagore's play of infinite forms as a spectacle meant for the delectation of humans. Those forms simply existed as random transient sources of sensory impressions, and so I made it my business to experience them while and where I could. Carl Linnaeus, the father of plant classification, compared "the earthly globe" to a museum or a "splendid theater" created by "the Maker of all Things" who included Man in the scheme to play the role of spectator. People were necessary in order to observe in the Creator's work "the evident marks of divine wisdom." From what I saw around the earthly globe, the scenario indeed presents splendid theater, a play of infinite forms and forms of plays—comedies, tragedies, dramas, farces, an all-time Tony award-winning spectacle played with cast of millions, each character performing his or her part in the show. Contrary to Linnaeus, however, I didn't believe that divine wisdom featured in the performance, nor did I think that the spectacle was created for humanimals to witness. Reviews of the presentation were useless. The show was just there, like it or not. Too many tragedies shadowed the scene for the play to seem divine. A chiaroscuro mix of all too many sorrows, sadnesses and folly along with some joys, pleasures and beauty brought plots and counter-plots no divinity would script. A force more devilish than a divinity must have co-authored the play of forms, a shadow play more pathological than theological. It appeared to me that chance had randomly created a defective terrestrial format, as if the earthly organism suffered from systemic pathology like miscoded DNA. The earthly globe, as I saw it, seemed something of a random tumor-like Big Bang mutation, not the best of all possible worlds, certainly not the worst, but a benighted planet which, for better or worse, served as mankind's only home.

For me divinity didn't figure in the cast of characters. As a non-believer, the only religious holiday which struck a spark in my belief system was Lent. Giving up certain pleasures for Lent suggested to me the reality that nothing is given but only lent, and that in the end everything is given up. Every last element of existence as well as existence itself is all only lent. The entire experiment—all its events, experiences, memories, the illuminations ignited by synapses to connect up what the neurons hold—is just a flash in the pan. All is lent with no ultimate pay-back or redemption of the loan. The lending operation yields nothing for the lender or the borrower. It's a write-off. All the French subjunctives I labored so long and hard to learn when studying at the Sorbonne have no ultimate staying power. It was all just borrowed, soon to disappear. Like my very existence, the subjunctive was only tentative, conditional. The only given in life was that it would end; everything else was lent.

Although a non-believer, I enjoyed attending all sorts of religious services, which for me offered cultural rather than devotional experiences. I was an ecumenical non-believer, not non-denominational but none-denominational: none of the above, as for me all the heaven-sent doctrines which illuminated other earthlings were mere shadows. I'm glad that so many others found solace in religion and glad that I didn't. I saw every sort of religious ritual under the moon. There was no way I could reconcile so many widely varying beliefs, doctrines and ceremonies other than to conclude that, except for the true believers, none was authoritative. So inconvenient were some of the services which my curiosity impelled me to attend, I needed to have faith that the benefits would compensate for the disadvantages. At some of the Mount Athos monasteries in Greece I arose at an ungodly hour to attend the very early morning, or very late night, service. At the Orthodox mass in the Sofia, Bulgaria cathedral the rather unorthodox lack of pews and chairs forced me to stand during the service. In some countries you have to pay to visit a house of worship, a fee which seemed to me to desanctify the church and convert it into a commercial establishment, degrading it from a sacred religious place into sacrilegious one. In other countries you can enter the temple or mosque but are pestered to pay to get out. Religion came with a price.

My lack of belief in divine matters typified my tentative and doubting mindset toward things in general. At the time I recast my life when in my early thirties I wasn't completely sure that this was the right decision. Although I had no second thoughts, my first ones weren't entirely persuasive. At the beginning, doubt rather than conviction predominated in my outlook. I felt somewhat certain that some sort of change was necessary for me to attain a charmed life, but less convinced that adventure travel should feature in my new agenda. Maybe that would be too radical a change—too few comforts, too much culture shock, not enough quietude. Especially under twilights far from home when I occasionally fell into a melancholy mood, I sometimes regretted that I'd chosen a path which at times severed me from the security and considerable comforts of home. In exchange for my cozy corner in the suburbs I faced the uncertainties, challenges and inevitable problems of adventure travel. Although many potential delights could make the twilights less gloomy, going forward the trip's benefits were uncertain while its difficulties were inevitable. Unavoidable hassles plus theoretical pleasures didn't add up to a favorable cost-benefit equation. I could only hope for some sort of return on my investments in time, energy, effort and money. But since compensation for leaving home was unsure, adventure travel was a speculative enterprise. The only certainties in exotic far lands are vexations, hazards, hardships and problematic situations. As Percy Fawcett discovered, plots and counter-plots, risks and threats lurk out in the big wide world. My travails were nothing compared to his, but even so I sometimes wondered if I should have stayed at home, watching the Discovery Channel and enjoying the familiar routines and setting there in the suburbs of my provincial hometown where I was contentedly settled. The Middle West presented fewer challenges than the Middle East, America's heartland fewer hassles than Swaziland or Thailand. I could have more conveniently gotten the idea of foreign countries at Disney World and Six Flags (did they flag counterfeit amusement park countries or real ones?) than by traveling out into the wider world.

Claudian, a late fourth century poet, described in "The Old Man Near Verona Who Never Left His Farm" how happy is he who remains on his native land "where all his days from youth to age are cast." There on familiar ground where played the babe stands the old man who never sought "tumultuous misery" or "unknown waters." Savoring his inexperience, the old man clocks the calendar by his crops and the changing seasons. There's a lot to be said for this regular way of life, a long-standing and settled repetitious existence, earthy rather than worldly. The "old man near Verona" was like El Viejo who I'd encountered and conversed with on Easter Island. That oldster had lived a completely insular life, one which contented him. I pondered why I should spend my limited earthly existence seeing the earth, a greatly inconvenient hassle compared to the ease of remaining in my own little corner of the globe. For Shakespeare, England and The Globe were enough. Bill certainly didn't lack an understanding of the world, even though he never saw much of it. If he was a stay-at-home, why not me? Twilights far from home and strange beds in soulless hotels lack the cozy familiarity and pillowed comforts of your own cushioned nocturnal setting. Giving up the soft life at home for the rigors of the road might not really represent a good trade-off. And yet I went. In his poem "A Lot of People Bathing In a Stream" Wallace Stevens evokes the post-swim pleasure of returning home: "How good it was at home again at night" where the familiar bed beckons and life resumes in "the rooms which do not ever seem to change." But for me, that was precisely the problem: the familiar bed and the changeless rooms became too confining.

In contrast, by its very nature travel entails change. That's in fact precisely what you seek when you leave home. Like every trade, what you bargain for involves give and take. You can only hope that what you give up will be rewarded by the chances you take. The crepescular melancholy I at times felt during dusks far from home made me wish that I could just then be transported back to my remote familiar home-based comforts. With all the amenities close at hand, in my little corner of the world I never had to lift a finger for the basics. When away, everything required constant effort. Food, shelter, transportation, laundry, navigation, hygiene and all the rest, all so accessible on my home turf, presented unrelenting challenges. Everything readily available at home became a matter of will elsewhere. It was ever an effort to produce while away the necessities which at home already existed. That marvel of civilization, the ingenious system which brings you an on-tap supply of drinkable water, doesn't exist in many countries. In those lands clean water comes not from a faucet but from a container you carry with you. You quench thirst in the Third World with the contents of the heavy and bulky bottle burden you bear. It would strike my stay-at-home suburban cohorts as strange how in under-developed countries the well-watered facilities taken for granted in the U.S. aren't on tap. An earthling reflecting on the civilization of Mars in Ray Bradbury's *The Martian Chronicles* opines that "anything that's strange is no good to the average American. If it doesn't have Chicago plumbing, it's nonsense." Why Chicago, Ray? But he may be right, perhaps Chicago toilets are special—they don't call it the Windy City for nothing. (When I met Bradbury in March 1991 in Los Angeles where he signed for me a copy of his delightful novel *Dandelion Wine*, the author said nothing at all about American plumbing, Chicago or otherwise.) It was a bittersweet trade-off between home ease and the dis-ease (and diseases) of adventure travel—somewhat bitter, as you exile yourself and often find yourself alone in distant strange lands, but sweet as while away you enjoy experiences and adventures offered by the lonely planet's play of infinite forms, unavailable at home, which give the place its strange fascination.

Often I took off on my own. Traveling alone brings various advantages. Although sometimes a fellow or fellowette traveler accompanied me, I never hesitated to go alone. I learned that it's a mistake to delay a trip, or for that matter anything else in life, until you find someone to join you. As an independent wayfarer you can choose where to go, how long to stay, which activities and sights to devote your always limited time to. In that I was interested in getting around to see and experience as much as possible, free-lance travel suited me as I could set my own itinerary and schedule. Traveling alone also improved the likelihood of meeting other kindred spirits along the way. Sometimes people you happen to encounter travel with you for a time. Moving as one of a pair is quite different than a solo journey. With a companion by your side you become enclosed in your own self-sufficient micro mobile unit with little inclination to reach out to people you happen to encounter. Even though a two-person social circle is small, it acts as a barrier which encircles you as a couple and tends to keep others out. A singleton, however, remains more open to meeting someone, at least for a passing encounter and perhaps even a kindred spirit who might join with you for a time to form a doubleton.

Finding a fellow traveler along the way who will join you is a rarity. Most everyone you meet on the road is just another face in the crowd, a passing stranger you glimpse momentarily or briefly chat with before moving on. You soon forget the transient interlude as you continue on your own way, never again to see the other person. Although not all change is travel, all travel is change. As it happened, however, a few of my chance meetings during the churning ups and downs and ins and outs of travel over the years ripened into lasting and meaningful friendships. Those infrequent on-the-go and then on-going connections represent one of travel's most gratifying benefits. A traveler you happen to encounter in a remote off-beat far corner of the world automatically represents the kind of person with whom you might have something in common. Being at the same place—especially one distant and unusual—at the same time operates as a kind of filter, a curated selection process which makes it more likely that the two of you share some of the same interests. The commonality of place and time evidences the potential for a continuing connection rather than simply crossing one another's paths before parting ways.

Solo travel brings the further advantage of sharpening your perceptions. Alone, you focus entirely on the business at hand—to see, absorb, enjoy, understand. With someone else along your experiences are diluted by being shared as your attention is divided between an outward-looking perspective and focusing in part on your partner. Although in its way comforting, sharing is also distracting and subtracting. Concentrating directly on your immediate experiences enables you to develop more vivid images of the kaleidoscopic whirligig of adventure travel, creating mental and sentimental views which enable you better to recall later some of the infinite play of forms you've seen. Memories may be less misty if they originate undiluted rather than by creating them with a companion by your side. Although I appreciated the advantages of traveling alone, I didn't necessarily prefer to adventure on my own but often that's just how things happened to work out. It's not easy to find someone with compatible interests and personality as well as the time, the will and the means plus the initiative and energy to join you on visits to far exotic lands. I occasionally tried to tempt someone who I thought might be suitable, but usually one or more of the necessary elements failed to fall into place. It's rare that curiosity, temperament, free time, interest, an adventurous spirit, willingness to rough it, financial resources and other required components would cohere in another person as, by chance, they did with me. Adventure travel is an acquired, not a required, taste. Few people I knew were willing to extend their tastes beyond the plain vanilla—Dairy Queen rather than *gelato*—certainties and comforts at home. I never counted on finding someone to travel with. I took off anyway, and every so often along the way luck brought me someone willing to join me for part of my trip before we then parted, each of us again going off alone.

When no one who cares companions you, you bear all the cares of travel alone. Out in the world on your own no one shares the burdens, nor the memories. Alone, the twilights far from home seem longer and darker, a charmed life seems less charming. As it happened, one memorable day as I neared the end of my five-month-long South American trip in March 1972 I chanced to meet a particularly pleasing traveling companion. Suddenly there she was, a statuesque stewardess, a TWA beauty off on her own far from home in the same place at the same time as I happened to be. It was a promising encounter which kept its promises. It's rare to come across an unattached beauty, but somehow our paths crossed and our two separate solos merged to form a duo. What happened after our surprise encounter I relate in Part 4 below. If you've read this far you certainly deserve a little spice in my reminiscence, which I'll now enliven with my account of a dream made incarnate in the person of a gorgeous blonde straight from central casting who came to me even though I didn't even have a casting couch. It was just luck.

~ 4 ~

Tchaikovsky explained that his Fourth Symphony pertained to the theme of Fate: "the fatal power which prevents one from obtaining the goal of happiness." He deemed the theme to represent his belief that "all life is an unbroken alternation of hard reality with swiftly passing dreams and visions of happiness." That's one way of looking at things. But as one who manged to live something of a charmed life, it would be ungrateful for me to adopt Tchaikovsky's morose viewpoint. Although fate and luck operated as uncontrollable wild cards and in spite of the fact that hard reality was implacable, happiness of sorts was attainable if fate and chance cooperated. Perchance, they somehow offered me the opportunity but not the certainty to live life in my own way. Contrary to the composer's belief, happiness isn't always relegated to the realm of airy dreams and visions. Tchaikovsky obviously was never fated to meet a beautiful TWA stewardess traveling alone far from home. And anyway, back in my day the Aeroflot Russian stewardesses were anything but beautiful. I imagined they moonlighted in the MFWL—the Moscow Female Wrestlers League.

As the first of my three early 1970s five-month-long travel adventures (later followed by trips to the Middle East and the Far East), my circuit all through South America was an adventure in the nature of an experiment. I chose that continent for openers because South America was America, the name consolingly familiar even if not the terrain, and also because its main languages and partly Europeanized culture were more familiar to me than practices which prevailed in more distant exotic lands. I hoped those factors would to some degree help mitigate culture shock and perhaps reduce some of the rigors connected with such an experimental long journey to foreign regions. Unlike in many other areas, in South America I'd be able to communicate and to an extent understand the Hispanic-based way of life.

Severed from home comforts by my disruptive new agenda, it remained to be seen if my great initial travel adventure into the hurly-burly of an extended trip through the distended continent would prove satisfactory. If not, I could always abandon the journey and the entire enterprise, use my return air ticket to retreat to home base, and then revise my new agenda with less unsettling activities. But I stayed the course, and by the time I'd made the rounds—down the east coast, then west across Argentina and the Andes to Chile, and on north toward home through Bolivia, Peru, Colombia and Ecuador, visiting every country in South America except Guyana—I knew that I'd made the right decision to feature foreign travel in my new way of life. In spite of the hassles and a considerable expenditure of time, energy and effort, I greatly enjoyed the experience, which met my hopes and needs. By the time I'd reached my final stop, Ecuador, I'd already started to look forward not only to returning home but also to the next two anticipated extended trips on my agenda, the Middle and the Far East.

In March 1972 when I arrived in Quito I appreciated that welcome enclave of civilization after traveling in primitive Amazon jungle areas around Iquitos and Leticia. After a brief stay in Quito I temporarily abandoned the capital's comforts to return to the jungle, there entering the back country in the upper Rio Napo region where Jívaro head-hunters lurked. Just as was Teddy Roosevelt for the River of Doubt cannibals, around the River Napo I was fair game. For all I knew, the name derived from the Spanish word "kidnapo." That was just speculation, but I did know that some of the Amazon tribes coated blowgun darts with toxin from *phyllobates terribilas* frogs whose venom is indeed "terribilas" as the substance is fatal when touched. I suppose that each country and culture gets the kind of head-shrinkers they deserve. Where I came from, such heady professionals as psychoanalysts, psychologists, New Age gurus, faith-healers and other touchy-feely types treated the mental and sentimental problems suffered by civilization and its discontents. Out in the Amazon jungle the tribes cut to the chase, severing heads and shrinking them to a grotesque doll-like thing and, *voilà*, no more head problems. Before venturing into Jívaro country I'd seen at the museum in Quito specimens of the shrunken heads, ugly sinister looking little balls with cameo human features. My own head would offer a choice addition to the display—a fitting final fate for an errant *gringo* who tempted fate by entering the land where head-hunters lurked. I suppose it would be something of an honor to end up as a museum piece, a status even Percy Fawcett didn't accomplish. It's said that if you can keep your head while everyone else around you loses theirs, you're ahead of the game. By that standard I managed to stay ahead, returning to Quito with my head intact.

Lingering a few more days in the capital, one afternoon I went out to visit the Equator monument where a metal strip embedded in the ground made the imaginary line visible. I enjoyed stepping back and forth, easily crossing the Equator. For the first time in months I'd returned, even if only momentarily, to the Northern Hemisphere. I took this brief re-entry as an indication that my long trip was nearing its end. For a few moments I straddled the line in a symbolic effort to represent how my two separate and quite different ways of being—my home and away lives—seemed to position me on both sides of an otherwise rather narrow existence as straight as if running on a trolley track. Just as at Cape Comorin at the southern tip of India years later, where all the sub-continent's billion or so souls were to the north, at the Equator I savored the curiosity that both the entire Northern and Southern Hemispheres lay beyond where I was standing as I straddled the line. I've always enjoyed visiting extremities, such as end-points, capes, land's end, backtiers as you reach frontiers, terminals, boundaries, dead ends. Those finalities seemed to bring some sort of order and closure to otherwise open-ended and vast perspectives. Endings were conclusive—final places where the dazzlingly inclusive plenitude of the infinite forms became marginalized. The Equator's marker, which indicated where the hemispheres ended and also where they began, showed the earth's midpoint, a fixed reference in the huge continent through which I'd just traveled. By now travel weary, I favored looking in the direction of north, North toward home.

North of the Equator it was almost spring, while to the south autumn was coming on. I could change seasons simply by stepping across the line. One balmy early autumn day in March I went to see a large handicraft-folklore store in Quito called Akios. The spacious emporium brimmed over with a wide selection of tasteful treasurers, one of which, not part of the regular assortment, I was by chance about to encounter. The well-known store, more like a museum than a shop, took its name as the reverse (slightly modified) of "Sojka," the local Czech family which owned the establishment. As a non-consumer, on the infrequent occasions when I went to a store and a clerk asked, "How may I help you?" I'd almost always answer, "Just looking." Little did the attendant realize that my response applied not only to the store but to much of my experiment in living. "Just looking" described my enterprise of traveling about, keeping my ears to the ground and my eyes on the passing scene as I pursued a "just looking" mode to observe the lonely planet's infinite forms. So fleeting was my earthly tenure, I was largely limited to just looking, as nothing would provide more lasting in-depth experiences. A few passing glances, and then you were gone—that characterizes the life experiment.

While browsing among the colorful handicrafts at Akios, just looking at them, I saw nearby a tall golden-tanned beauty also examining the wares. Her filigree-like shimmering blonde hair outshone the store's merchandise. My browsing immediately became arousing as I gazed at the statuesque beauty who by chance had suddenly appeared before me. Among the play of infinite forms I experienced over the years, with Carol her pleasing finite form firmly stands out in my mind. Her curvaceous figure still figures among my fondest memories. Was it possible that she was on her own? Unlikely, as surely a boyfriend or companion lurked in the wings. It's different now, but back in those days I rarely encountered a good-looking female traveling alone in a far land. Almost always a guy was around nearby. Lucius Burch, in his day a renowned Memphis lawyer, observed in *Lucius: Writings of Lucius Burch* that "whenever there's an attractive woman around there are apt to be offsetting disadvantages—usually in the form of a perfectly adequate man." And even if the object of my attention (and distention) were alone, perhaps she preferred it that way. Seldom do guys stand a ghost of a chance with a gal of her allure. As it happened, however, it turned out that for a time I became Carol's "perfectly adequate man," for she was indeed traveling alone and receptive to my suggestion that we join forces to continue our journey together. A lone *gringo* and a *sola gringa* far from home, we paired up there at Akios and spent a delightful few weeks in Ecuador and then in the Caribbean, an unexpected and memorable climax to my long trip through South America.

Pure chance had drawn us together. Somehow Carol and I happened to be at Akios at the very same time. A few minutes either way and, destined never to meet, we would have missed each other. Apart from her attractive appearance, I was attracted to Carol by her obvious sense of adventure, for she'd taken off alone to visit a foreign land. Her presence at Akios also indicated an interest in the local culture. These factors evidenced that Carol and I had at least a few things in common. With its double agency, operating on two travelers, Fate had somehow also joined us together by its mutual intervention in both of our lives: that commonality occurred in a random but appropriate manner, as Carol and I turned out to be quite compatible, and for sure she was extremely patible. The timing was oddly precise, not only how we overlapped at Akios but also how I met Carol just when my long trip was nearing its end. Until then, all along the way ahead of me loomed a demanding and grueling itinerary which occupied all my attention and energy. In Ecuador, my last stop in South America, my driving push to see the continent was finally over and I now had time for the softer side of life. For that unexpected human touch there at the Equator, I managed to equilibrate my life by renouncing compulsive travel in favor of rather compulsive attention to Carol. She improved the geometry of my trip: to my up-and-at-'em on-the-move on-my-toes vertical and kinetic motion around the continent, Carol brought a more laid-back horizontal dimension. Meeting Carol illustrated one of the benefits of traveling alone. Had I already been with a companion in Quito, never would I have approached her—looked, yes; dreamed, yes; but not approached. Carol was the first, but not the last, woman I encountered by chance while away who expanded my horizon to the horizontal. Over the long miles and years I occasionally happened to pair up for a time with a lone female traveler before we each went our separate ways. Engaging with them but without an engagement, I hope I gave those sole travelers as much pleasure as I derived from their companionship. A few I kept up with for a while, but most just came and went.

After a few days together in Quito, Carol and I took the old *Autoferro* down to Guayaquil on the coast. Discontinued after torrential rains and landslides damaged the rail line in 1997-8, the narrow gauge Toonerville Trolley-like train descended on its fixed route from Quito, more than 9300 feet high (was it Carol or the altitude which took my breath away?), to the steamy equatorial jungle in the lowlands. Along the way, the *Autoferro* laboriously negotiated the diabolic zigzag tracks cut into the steep rock face just below Alusí, a point on the line known as the Devil's Nose. Months before, toward the beginning of my South American journey, I'd visited Devil's Island, at Oruro in Bolivia I'd seen "Devil's Dance" masks, and now at trip's end the Devil's Nose. Thankfully, at Angel Falls I'd managed in passing to escape the devil and in Quito had found an angelic companion. My stroke of good fortune in meeting Carol suggested that, contrary to Tchaikovsky, "dreams and visions of happiness" could at times soften "hard reality," as my nights with her did for me during our ardent horizontal connection.

From Guayaquil Carol and I flew to Jamaica, there to enjoy a restful stay in Montego Bay. My long and strenuous South American journey now complete, with Carol at my side I finally took a break from intense sightseeing for a relaxed vacation. We enjoyed a romantic Caribbean island honeymoon without the bother and expense of a wedding ceremony. It seemed to me more desirable to have a honeymoon without a wedding than to have a marriage without a honeymoon. My impression was that Carol enjoyed being a honey even without the moon included. Our informal relationship reminded me of Mark Twain's observation that old-time multi-deck white riverboats resembled a wedding cake without the complications. I endorsed the Chinese tradition of the half-phantom marriage whereby a still living "ghost bride" would participate in a wedding ceremony with a deceased son. With Carol, I was content to remain a ghost groom. As noted in Section II, in Chile earlier on my trip my post-graduate real-world education included some practical lessons in applied economics. But it was Carol who taught me to appreciate that a curve-shaped belle offered a more enticing figure than one on a bell-shaped curve. Strangers when we met, Carol and I quickly came to enjoy our togetherness and the interests we shared. Mooning over Carol, if not honeymooning with her, in Montego Bay made me appreciate the soft ending to my hard trip. But my travels weren't quite over yet. I proposed to Carol—no, not to formalize the honeymoon with a wedding ceremony and a ship-shaped cake—that we fly from Jamaica to nearby Haiti for a visit to that exotic island. Carol agreed, and together we flew on to Port-au-Prince.

In the Haitian capital we visited the markets, art galleries, shops, colorful corners filled with Technicolor scenes and sights, including the garishly painted "tap-tap" truck-buses which rattled their way through the crowded streets. Pétionville in the upper town was more upscale. There in that calm green clean oasis removed from the commotion, poverty and slums down below lived the city's upper classes. I saw the same everywhere in the Third World—the few on top and the many down at the bottom. Back in the whirl of the city the legendary gingerbready Hotel Oloffson offered a welcome retreat from the crowds churning through the nearby streets. Th famous establishment figures as a character in Graham Greene's Haitian novel, *The Comedians*. Greene, a colorful character himself both by name and career, describes the country in his book: "The land was stormy with color," he wrote. "The deep blue shadows sat permanently on the mountain slopes, the sea was peacock green. Green was everywhere." So was Greene—all over the map. In his *Journey Without Maps*, his account of traveling alone through the Liberian wilderness, and in many other books Greene set his stories in exotic lands around the globe. He in a way seems to represent the beau-ideal of a traveler-writer, not a travel writer as such of the kind who, with or without maps, journeys to gather material for his books but a curious man who travels to experience the world. Greene's trips formed the real-world basis for him to imagine, not merely to invent, his fictions. Greenland he apparently missed, but Greeneland—as is called the the imagined realm of "see"-scapes depicted in his fictional worlds derived from what he saw—represent a land he brought to life from life experiences, not armchair pipe dreams. For Greene the heart of the matter was to travel, then imagine into being realistic evocations of the places he visited, the people he met and the events he witnessed out in the world.

From Port-au-Prince Carol and I crossed the island to Cap Haitian on the north coast. We traveled well together. Tchaikovsky might have even revised his pessimistic thoughts about Fate had he seen how well the two of us related. Maybe our mutual enjoyment would have even inspired him to write his own *Ode to Joy* (which I'd revise to *Ode to Carol*). The trip across the island brought us even closer as we traveled local style in a vehicle jammed with passengers and with a low ceiling forcing us to contort our posture by tilting our head and hunching down during the entire trip. Along the way we passed through landscapes "stormy with color," views with hues offering a post-storm rainbow of vivid brightness. Arriving in Cap Haitian after our colorful but uncomfortable trip, we were finally able to unfold and unwind outside the cramped confines of the bus. We both agreed that we should explore other modes of getting back to Port-au-Prince (we flew, barely, on a rather rickety vintage plane). Some travel experiences, like busing between the two Haitian cities, are quite enough in the singular, but not our honeymoon which we resumed and duplicated there in the provincial sea-side town, much more relaxing than the hectic capital.

One day we visited in his studio Philomé Obin, dean of the primitive-style school of Haitian painters. Fluffy cotton-white hair tufted the distinguished dean's head. Giving Obin some attention by our visit didn't seem to flatter him. By then the painter was world famous, even though not very worldly, isolated as he was there in remote Cap Haitian. To my question as to who the major influence on his style had been, Obin replied, "*Personne. Moi, je suis le maître.*"—"No one, I'm the master." In that I was in the presence of "the master" of Haitian painting, I decided to buy one of his masterful works. The memento would eventually serve as a unique souvenir of my long travel adventure, now nearing its end, in the southern latitudes.

The laws of economics operated there in Obin's studio, as they do elsewhere. In a way it seemed that he tried to use those laws to support his apparent attitude that he really didn't want to sell me a picture. But maybe what I saw as his indifference was really just a bargaining technique. The marginal utility of the artist cranking out yet another painting was low. Producing yet more work represented diminishing returns for him. As if to discourage my purchase, Obin informed me that he was about a year behind in fulfilling commissions. That didn't deter me. Back then I had time and could wait. By now, however, a year represents a huge quantity and maybe all of my remaining time. I asked him what he charged, to which he countered by asking me what size painting I wanted. He apparently charged by the square inch. Or maybe it was by both the size and the assumed prosperity of the buyer, in which case Americans bargained at a disadvantage. Well, I was rich. Compared to the locals out in the street at Cap Haitian I was worth a fortune, and enjoyed great good fortune. When I selected the dimensions he quoted a price, probably more expensive on a space basis than prime Beverly Hills or Fifth Avenue retail facilities. The asking price was a fortune there in Haiti, but by a kind of arbitrage alchemy the painting would acquire much greater monetary value when transferred from Cap Haitian to the U.S.A. Third World to First, a magical transformation for the work of art and a transition I was looking forward to for myself after a hard slog through South America.

I didn't bother discussing the price with Obin. I knew he wouldn't bargain. He had the upper hand, as it was the one which would produce the painting. I wondered if there was a quantity discount. Maybe I should have ordered a mural or triptych. A trip triptych would for sure give me a souvenir of the journey. To pay Obin I peeled off and signed a stack of American Express travelers checks. Just as my American Express card had effectively functioned in Córdoba to free me from the gunman who'd detained me there, so the company's paper served to gain me eventual ownership of a museum piece painting. Possession of sufficient means means accomplishment of desired ends. After I paid, Obin asked me what sort of image

I wanted. I replied, "A typical local scene with lot of color," a completely ridiculous answer as it exactly described the painter's traditional style and subject matter. Obin declined to give me a receipt—a *maître* doesn't bother with such mundane paper work. I wasn't concerned at my lack of documentation for the money I'd handed over, and about a year later my trust was rewarded when Obin sent me a painting depicting five colorfully clad local women in a procession, each balancing on her head a milk pail and with the title *Marchandes de Lait* painted in red at the bottom of the scene. Enclosed in the carton containing the painting was a small bill for shipping charges which I promptly paid. Obin was a *maître* not only as an artist but also in the more practical and worldly realm of business. So much have I valued that travel souvenir—with its related memories of Carol—I don't begrudge him one gourde of his charges.

Carol and I devoted a day to the classic Cap Haitian excursion, a trip on horseback past the ruins of Sans Souci Palace in Milot, completed by King Christophe in 1813 and destroyed by an earthquake in 1842. We continued on up to the 1,000-foot-high Citadel fortress perched above the sea-level landscape below. A year later on the other side of the world the Krak des Chevaliers in the Syrian desert would remind me, by the castle's solidity and elevation, of the Citadel, of my year-earlier South American adventure, of Carol. But for the moment my five-month-long Middle East adventure remained in the future and yet to take form. Only later, with a long look back, could I connect the two experiences. As we rode up to the Citadel along the rugged stone-strewn path, rag-clad barefoot boys carrying pails of soft drink cans buried in ice chunks ran along with us. At the top the kids swarmed around us to peddle the drinks which, at $1 in gourdes, offered a much appreciated bargain. Buying two, I didn't even ask for a volume discount, although these days I might request a senior discount. Lugging cans and ice in pails up the steep path to the Citadel in sweltering temperature seemed a hard way to earn a living. Back home, boys that age would live a more pampered existence. The kids, all wearing shoes, would be out at the country club, drinking the soft drinks rather than selling them. In my suburb even the dogs lived better than the unfortunate street children in Haiti. The drink boys there at the Citadel made me reflect on and appreciate all the more the advantages my fatal circle had given or at least lent to me. There atop the mini-mountain I perched above it all, riding on my high horse which had carried me to the peak, companioned by a beautiful young woman. At Obin's studio I'd spent more on the painting than what all the kids together would earn their entire lives. Meanwhile, the boys labored to earn a few gourdes carrying heavy pails up the steep slope. Fate might have reversed our roles. I could have been the unlucky one, a rag-clad urchin selling soda at the Citadel day after day rather than a passing traveler on a brief visit, a wanderer free to continue on to adventures around the world. There's no explaining how luck and fate allocate their favors or why they happen to charm or to harm a life.

The barefoot boys on the Citadel trail knew as well as Starbucks the value of a choice location. By the time visitors reached the top in the enervating Haitian heat they were captive customers for hydration. I could never explain to people back home the water problem faced by travelers in most of the world. At home, the turn of a tap brings you clean drinkable water. Away, access to water often presents a chronic nagging concern. Without water your trip ceases to be a trip and becomes an obsessive quest for hydration. Water is the mother's milk of travel. Lacking it, you can't continue. You have to constantly monitor your supply so as not to be caught short. Because of its bulk and weight, it's necessary to replace water frequently as it's too heavy to carry around very much. When my supply ran low I carefully limited my intake to a few gulps, rationing my consumption no matter how thirsty I was. You can never be sure when more water will be available. As I didn't want to risk completely running out when I at times arrived at a low-star or no-star hotel with an inadequate supply, I passed a restless night tormented by thirst. In such places I never tapped the tap water. Like a drained bank account, sometimes it's just not possible to maintain your liquid assets at a comfortable level. As all world travelers know only too well, one of the great challenges of the enterprise is keeping clean water with you at all times.

Not simply such variances as the water problem or the ill-fated Citadel soda boys but also much else differentiated how things were back at home from the "hard reality," as Tchaikovsky put it, of daily life out in the Third World. What "dreams and visions of happiness" could the unfortunates in the underdeveloped lands possibly hope for? My journeys in those areas greatly reinforced my appreciation for my good fortune and for the comforts and advantages available to me at home. Even my modest life and standard of living there brought me benefits far beyond those available in most other countries, even in developed ones. Having so often been immersed while away in the hurly-burly cacophony of jammed noisy and polluted cities swarming with humanimals and clogged with traffic, the peace, privacy, greenery and quiet at my little corner of the world satisfied and gratified me. Based on my temperament and personality as influenced by what I saw out in the world, and especially the primitive and impoverished regions, I was content to live in a simple way, grateful for and satisfied with basic comforts. People in my neck of the woods seemed to take it all for granted, but I knew that the amenities available at home didn't exist in many places around the world. My worldview and passing views of the world greatly affected how I chose to live. My mode of being was frugal. I lived in a simple way, avoided waste, remained indifferent to style and fashion, emphasized self-possession rather than possessions, was disinterested in consumerism, shunned materialism, ignored fancy and luxurious services or things, favored an unadorned way of life, valued relationships and experiences rather than stuff, gave priority to intellectual pursuits rather than pursuing retail bargains. I lacked all concern with image, standing or externals of the kind my fellow suburbanites often employed to evaluate cohorts. My habits and practices were meant to facilitate living small rather than in a grand style. For me, the simpler the better. Well-traveled and worldly but a provincial at heart, I contentedly maintained a basic comfortable suburban bourgeois existence without frills or even without what some of my co-equals deemed necessities.

Upon returning home after completing a long hard journey in difficult areas, I always kissed the floor—gratifying, but not the same sensation as kissing Carol—the moment I arrived in gratitude for getting back to my haven, and for managing to do so safely. At the front end, when I departed, a safe return was not an assured outcome. It was almost a miracle, and a very welcoming one, to find at home such ordinary conveniences as clean water on tap, hygienic food, an uninterrupted flow of electricity, the trees and bushes which garnished my property, changes of clothes beyond the few garments which fit into my suitcase, a car of my own to take me when and where I wanted, even twilights close to home, and all the other normal amenities lacking out on the road. It all contrasted with how people lived in so many other places I saw. There was no way I could explain to my friends and acquaintances at home the vast gap between how they (and I) lived and how so many unfortunate humanimals existed in degraded conditions typical of underdeveloped lands. No matter how robust my powers of description, I just couldn't portray how things really were all around the lonely planet.

Ideally, intrinsic to the enterprise of travel is that you internalize much of what you experience. You can't very well share a trip with someone who wasn't with you, but you can make another use of your journey by retaining and pondering your impressions. Although they remain invisible to others, well do I remember the Citadel drink boys and so many other such vignettes of unfortunates, such as the women and children in Calcutta pawing through the steaming stinking refuse heaps to salvage cans, bottles, a few scraps of cloth. That hellish vision has stayed in my mind ever since, and never again did I fail to recycle cast-off items of a kind which, half-way around the globe, enabled indigents in India to survive. I adopted that practice long before the environmental movement at home promoted green behavior. I turned green before the color became fashionable. What I'd seen—not just read or heard about—far away on the other side of the earth, where luckless locals were forced to scrounge a basic living, in reality sub-basic, from garbage heaps, plucking from the festering mounds what at home was casually discarded, compelled me to recycle. I couldn't recycle my life—no refutation of time was available—but by recycling trash I could in a small way help the waste to find a new life, a resurrection of sorts. Based on the sub-human scenes I'd seen, I just couldn't toss my trash away. Conditions out in the Third World, so different from my own fatal circle, made an indelible impression on me and have haunted my memories and greatly influenced my life.

After our interlude in Haiti, Carol and I returned to Montego Bay for a brief stay, then we parted company and we each went our own way—she, based in New York City, back to TWA and me home to recover from my marathon trip and to start to prepare for the next one, a five-month-long adventure to the Middle East. Back in the States, Carol and I kept in contact for a time. She came to visit me, and I saw her a few times in New York. Over the spring and summer as I planned my trip to the Middle East I asked Carol if she wanted to join me for part or all of the adventure. Although she expressed some interest, she kept delaying a decision. I followed my usual practice by sticking to my plans and schedule, and—time and Tom both on the move—the following November I took off on my own without Carol. I knew that waiting for someone else would most likely result in an altered, shortened or even canceled trip. It was risky to defer departure in the hope that patience would yield companionship. Time was on the wing but, with a delay, I wouldn't be. As time went by, Carol and I went "bye," not with a clear-cut separation but by gradually drifting apart, a conclusion such as frequently occurs with chance acquaintances you happen to encounter far from home. Somehow, most of them lack staying power. A few chance encounters, however, managed to persist and they brought me very gratifying long-term friendships which greatly enriched my life, which would have been much less charmed without such friends.

By now lifetimes—Carol's and mine—have gone by. She could be a grandmother, even a "great" one. Maybe, like me, she occasionally thinks back on our long ago chance meeting in Quito, our Jamaica "honeymoon" and our Haiti trip, ancient misty memories from the distant past. Carol no doubt concluded, and rightly so, that at the time I was too foot-loose and fancy-free to be a keeper. Or maybe she sensed, also rightly, that I wasn't a high-flyer motivated to rise to great heights in life so that I could keep her in the high style befitting someone used to high altitude living. Carol probably realized that for her, TWA at the time represented a better choice than TW. I certainty would have liked to keep in touch, especially like the touching moments we enjoyed during our romantic Caribbean "honeymoon," but we never saw each other again.

~ 5 ~

Months before my lucky encounter with Carol in Quito, I'd begun my long South American trip in Venezuela. At the time I didn't know she existed, nor she me, and the chances of our ever meeting one another were virtually nil. Many contingencies and an entire continent distanced me and Caracas from Carol. But nearly five months later, chance somehow happened to bring the two of us together in Ecuador. Back at the beginning in Venezuela, a long hard and uncertain road ahead separated me from the surprisingly soft landing at journey's end with the high-flying TWA stewardess who happened to land in my life. Caracas was not only the start of my trip but also represented the inauguration of my new agenda to see the world. Although I had to start somewhere, scruffy Caracas didn't seem to be the most auspicious place to initiate my new way of life. Sprawling over an elongated area, the oppressive metropolis offered little of interest. But at least it was a start, and I supposed that things could only get better. The scenes, settings and cities would improve, so I hoped, as I continued down the east coast and then on the west side to pivot back to the north toward home. Indeed, in Ecuador when I met Carol the scenery greatly improved. At the outset there in Venezuela I also had in mind my later anticipated five-month-long trips to the Middle and Far East, but first I'd need to experience and then evaluate my initial experiment traveling alone in South America for weeks on end in unfamiliar lands a long way away from my native habitat. I wondered how I'd react to all those twilights far from home. Staying put for twilights at home would have been much easier, but my life there was too settled and predictable. My desire for change and an interest in satisfying my curiosity about the world motivated me to adopt the positive attitude of Richard Burton, one of the greatest travelers ever:

> The gladdest moment in human life methinks is the departure on a distant journey. Shaking off with one mighty effort the fetters of habit, the leaden weight of routine, the cloak of many cares, the slavery of home, man feels once more happy. The blood flows with the first circulation of childhood. Excitement lends unwonted freedom to the muscle, and the sudden sense of freedom adds a cubit to the mental stature ... A journey, in fact, appeals to the imagination, to the memory, to hope—the sister Graces of our mortal being.

I embarked on my first adventure with those three appealing factors in mind. Not content with simply reading about places, cultures, the world at large, and then inventing abstract visions about the play of infinite forms on the lonely planet, I sought real-world experience to stimulate my imagination. I also aimed to create from my fleeting travels a store of memories which in times to come, such as now, would furnish me with many pleasurable recollections from long past adventures. Hope also endowed my enterprise with a Judah P. Benjamin kind of optimism that things would go well. But, as always, luck and chance would play decisive roles in the particular play of forms which by chance would comprise my trip. Notwithstanding my hopes, looking ahead as I started off in Venezuela I felt the problematic future pressing in on me with an almost tangible force. It was intimidating to face all the uncertainties which lurked on the long road ahead, fully five months of relentless travel fraught with indeterminate challenges. With more than a little anxiety I wondered what the days and weeks and plots and counter-plots would bring me. Filled with an infinity of possibilities, each day held peril and potential. But which would it be?

The past was a jungle of confused overgrowth, history a tangled proliferation of how mankind and nature had over millennia co-existed on the lonely planet, perhaps to the disadvantage of earthlings. The world was a hot-house kind of place, noisy with perfervid humanimals clashing and also raucous random natural forces, all operating to create the dense thicket of the past. Although time past was a intricate maze as impenetrable as the jungle, the future, empty like a desert, lay blankly ahead open to what might happen to fill the vacant space. Miles and miles of sandy nothing awaited some sort of content, and farther on at the end lies a desert of vast eternity where eventually the endless permanent future of all which once existed would be spent. That was the True North toward home to which every earthling was headed. Only the ever-evolving present could a person experience. The future was an open question, the past a closed answer. The present and the future, where they interface in the now, vanished like a smoldering campfire's smoke burning away into nothing. What happens at the moment represents the granular essence of a humanimal's life. Then the moment is gone. In Caracas I knew only that I was suspended between the cluttered unalterable past and the empty future, its story still to be written.

Toward the beginning of the trip I experimented how best to get my "see-legs." By trial and error I taught myself some of the ropes and gradually developed various useful routines and practices to smooth my way. Over time I felt more comfortable and secure by in part managing the trip with repeatable procedures that helped me reduce some of the vexations and discomforts. To some degree I learned what to expect relating to basic routine matters, how to cope with the known knowns, and how to anticipate some of the known unknowns. For the unknown unknowns no preparation was possible. By regularizing my expectations and responses to standardize a few of my procedures, I mitigated some potential difficulties. It was on-the-job training of the kind travel books or any others could not teach. Printed pages were too flimsy to deal with the press of challenges which commonly confront adventure travelers.

One procedure I adopted early was that in every country I visited I made a point of leaving the capital or dominant city, often one and the same, to see something of the provincial areas. I felt that I couldn't really grasp much about a country's essence without visiting some of its outlying regions. In Venezuela my destination outside Caracas was the fabled Angel Falls, lost in the remote wilderness and accessible, without days and miles trekking through the jungle, only by small plane from Caracas. I arranged a flying visit with a service that offered a long day-long excursion to the Falls, isolated in the Gran Sabana of the Guyana Highlands. The fragile little plane reminded me of my death-defying flight in a similar craft years earlier when I flew just ahead of a violent tropical storm from the Tiquisate banana plantation back to Guatemala City. I hoped I wasn't tempting fate by yet a second flight over the jungle. Remote Angel Falls would be a devil of a place to crash, an adventure not on my agenda but it might be the destiny fate had in store for me.

My far-flung travels over the years included not only twilights far from home but also many late nights and early starts, with pre-dawn departures and midnight arrivals. How you moved about in many lands, seas and airs depended on inconvenient departures and erratic transportation services. Circumstances, not clocks and operative schedules, governed your mobility. The Angel Falls excursion entailed a particularly devilish departure. I rose about 3 A.M. to proceed by taxi—an expensive means of transportation I seldom used—through the dark streets of Caracas and on down the steep route to La Guaira by the coast below the city to catch the ultra-early flight. I hoped my efforts and the expense would be rewarded with views of the Falls. Clouds, fog or mist often veiled the tumbling waters from view. The trip involved an expensive, inconvenient and uncertain venture which might result only in a long flight for a brief visit to see mist, which wouldn't leave me with the kind of misty memories I sought. But never again would I have the opportunity to see Angel, so I decided to risk disappointment. Along with visits to places like Los Angeles or the San Angel quarter in Mexico City, Angel Falls was the closest to the heavens I'd get. Devil's Island, which I saw not long after Angel Falls, most likely symbolizes my eventual postmortem habitat.

Two other passengers and the pilot loaded but hopefully didn't overload the four-seat plane to capacity. The excursion didn't operate daily but only when at least two or three people wanted to see the Falls. How the weather would be on any day the flight departed depended on chance. We took off in darkness and as first light slowly brightened the landscape, below appeared the jungle which extended to the far horizon. Having found Haiti notably green, Graham Greene would no doubt have been equally impressed by the seemingly endless verdant surface, a smooth and unbroken green expanse with no evidence of a human presence. It was pure nature down there. From above the landscape looked soft and benign, but beneath the serene green tangle hidden from view all sorts of struggles and turmoil bloodied the jungle, staining the greenery with blood and gore from the ceaseless biteings and fightings as life and death vied for primacy in a never-ending battle in the chaotic battlefield below. Percy Fawcett would have loved it down there, a hellish paradise for him.

As we drew near the Falls, the pilot offered a welcome informed opinion: the day would be clear and the Angel would bless us with with visibility. The angelic designation derives not from the heavens but from the more earthly dare-devil bush pilot Jimmy Angel who in 1933 spotted the cascade while searching for the "River of Gold" deposits. Presently the silvery vertical river, no precious metal but only water, appeared in the distance, and before long we crisscrossed back and forth and back in front of the Falls to gain striking close-up views of the famous landmark, or watermark. Over the lip of a highland stretching behind fell the narrow flow to the greenery more than 3200 feet below. As for the Native Americans mid-course along the Mississippi, the source of the watery ribbon was something of a mystery. From somewhere unknown beyond the towering escarpment where the Falls fell flowed a hidden waterway out to the edge where the stream plunged into the dense vegetation below, ending there as the angelic waters vanished into the jungle. Even Angels reach their end.

The remote area where the river originated and flowed occupied the top of the Auyantepu *tepuy*, one of the area's age-old rock formations dating as far back as 2.5 million years. In *The Lost World* Arthur Conan Doyle described the "strange land" there, a raw and remote country, nature at its wildest compared with the refined and civilized precincts in central London where Doyle's Sherlock Holmes lived at 221B Baker Street. In the book, Professor George E. Challenger relates how, during his Amazon region expedition in South America, "we came at last to a tract of country which has never been described." There he saw "a long and enormously high line of cliffs exactly like an immense cataract" which he explored by climbing part way up the rock face. From the heights "I had a better idea of the plateau upon the top of the crags. It appeared to be very large; neither to east nor to west could I see any end to the vista of green-capped cliffs. Below, it is a swampy, jungle region, full of snakes, insects, and fever."

Doyle invented his description, but it was as good as imagined as he based it on actual experience, even if not his. The novelist had never visited the *tepuy* ("table mountain") area where its jungles teemed with death-life. The living things which infested the jungle with dangers brought death to humanimal intruders. Although the author invented the fictional unworldly setting of his story, if Percy Fawcett's claim is true Doyle created that lost world by finding it in the explorer's account of the area, which he describes in *Lost Trails, Lost Cities*. Fawcett discusses his experiences in such regions, noting the "flat-topped and mysterious" high areas, "their flanks scarred by deep *quebradas* [clefts]. Time and the foot of man had not touched those summits. They stood like a lost world, forested to their tops, and the imagination could picture the lost vestiges there of an age long vanished. Isolated from the battle with changing conditions, monsters from the dawn of man's existence might still roam those heights unchallenged, imprisoned and protected by unscalable cliffs." Doyle's imagination based on Fawcett's real-life description indeed pictured the rugged scenery the explorer had seen. Noting this, Fawcett continues, "So thought Conan Doyle when later in London I spoke of these hills and showed photographs of them." Interested in writing a novel set in South America, Doyle asked Fawcett for further information, some of which appeared in the author's 1912 *The Lost World*. The two of them, adventurer and Londoner, played their own game of lost and found.

From what I saw on my travels, much of the world is indeed lost, lost in the sense that it's unfit for human habitation. Waters, jungles, deserts, forests, ice, tundra, glaciers, wilderness, peaks and other inhospitable lands, sands and bands cover much of the lonely planet, even more lonely in those desolate places with few if any humanimals. Vast areas remain empty of human traces. Only a relatively small part of the world, a few tiny corners of the universe, mere nothings in the cosmos, were reserved for humanimals. They occupied minuscule islets in space and only a few moments in eternity. Oceans alone—apart from wetlands like Phantom Lake in Maine, Walden Pond and rivers—cover a large part of the earth, about 70 percent, and earthlings are ill-equipped to survive in that element. Only a narrow oxygen-rich band between *terra firma* and the rest of the cosmos offers some breathing space to permit human life, and even that favored terrestrial area was greatly limited. Jungles harbored snakes, insects, spiders, toxic plants, pathogens, monsters and other predators, while lifeless deserts lacked life-sustaining molecules in the form of food and water. The deserts and jungles and many other regions were not fit for purpose if, unlikely as it may be, nature's purpose was to people the earth. Mankind clustered within the confines of his little built environment, a construct meant to keep nature at bay. Those artificial enclaves represented aberrations, exceptions to the dominant conditions around most of the world where raw nature, ever red in tooth and claw, predominated. In fact, in the large scheme of things the entire phenomenon of human existence and the earthly play of infinite forms were an aberration, as the normal state of being for everything is non-being. That which happens to exist is an exception to the nothing which represents the norm.

What humanimals had created over the centuries was a glorious doomed experiment. Civilization was a passing presence, eventually to be forever absent. With its functioning societies, culture, governments, rules and order, defended borders, its rails, pipes, lines and wires, the machines and robots and equipment, science medical and otherwise, and all the rest of mankind's inventive accomplishments and institutions, civilization didn't represent the natural order of things. Nature did. A state of nature prevailed all around the cosmos and in most places on earth, and even where humanimals lived they often existed, barely, in impoverished primitive and uncivilized conditions. A pebble of a place spinning through space, the lonely planet was indeed a lost world.

After our fly-bys near the Falls ended the little plane landed at Canaima, perhaps the world's tiniest airport at the time. By the crude landing strip stood a small one-room box-like container which served as the terminal. It seemed truly terminal: out there in the jungle we were seemingly at the ends of the earth, a *finisterre* lost world remote from cities and occupied civilized places, those found worlds founded by humanimals. Those urban areas where earthlings clustered to escape the perils of the jungle were far away from where we'd touched down. A few locals loitered by the terminal. Those hardy souls lived and somehow survived there in the remote back country, a lost world of Easter Island-like isolation. One of the jungle dwellers observed how lucky we were to enjoy such good weather which permitted full views of the Falls, often veiled by the frequent mist and clouds.

According to the travel materials I'd read when planning my trip, a local character named Jungle Rudy operated a service to guide visitors on a jungle trek lasting a few days to reach the base of the Falls. There below the cataract, an intrepid trekker would gain a sensational perspective looking up at the towering tumbling torrent. The son of a Dutch banker, Rudolf Truffino Van Der Lugt had arrived (I'd learned from my reading) in the area in 1956 and established in the jungle a camp for visitors to Angel Falls. When I asked a grizzled weather-beaten fellow there at the hut if he knew Jungle Rudy, the man replied that he was the very same. Sensing a customer, he eagerly inquired if I sought to engage him for the jungle trek. I would have enjoyed the adventure, but I didn't have time, nor did I fancy remaining grounded out there in the remote jungleland, stranded in that hard-to-reach and hard-to-leave lost world as the little plane, my only lifeline back to civilization, departed and then gradually disappeared as it winged its way back to the Caracas metropolis. Time was always on the wing but not the plane, which might not return—and with an empty seat—for days. It was my one-off umbilical cord back to the big city. By remaining with Jungle Rudy, I'd be forced to temporarily cut the cord and be tied down out there in the jungle. Instead, Jungle Rudy would unfortunately have to wait for Jungle Jim or a Tarzan type to find a customer for the long trek to the base of the Falls. I was only a passing on-the-fly Peeping Tom, satisfied with airborne rather than on-the-ground views of the Falls. I'd seen Angel from on-high and didn't need to add a bottom-to-top perspective to my experience. It was quite enough to manage to reach the remote famous natural feature at all. My curiosity had been satisfied, and all of South America still remained for me to see.

The designation "Angel Falls" represents one of those fabled famous legendary place names which exert on an adventure traveler an almost magnetic force. Such adventurers would hardly be worth their salt if they never reached renowned sights like Angel Falls, Devil's Island, Easter Island, Coney Island and any number of other such world-class attractions. Among the most alluring of all is Timbuktu, virtually a synonym for a remote mysterious destination coveted by every true adventurer. In mid-October 1987, toward the end of our month-long West Africa trip, Edith and I did reach Timbuktu, a hard-earned visit to the famous desert settlement in Mali. We began our journey in Senegal, then made our way down the coast to Sierra Leone, Togo, Benin and—although our trip included the Ivory Coast—the adventure was no ivy-covered ivory tower bookish enterprise but a real-life encounter with the big wide world.

As usual during travel, especially in Third World areas, we faced in West Africa various vexations and vicissitudes. Annoying inconveniences came with the territory. On the Air Afrique flight from Abidjan to Ouagadougou, capital of Upper Volta (later renamed, but probably otherwise unchanged, Burkina Faso) my suitcase failed to arrive when we did. No matter how long any of my trips lasted, I always traveled with only two bags—a suitable suitcase, small but large enough to hold a few carefully curated items, and an even smaller shoulder bag. At the Abidjan airport the attendant forced me to check the suitcase. I should have given him some "dash," so he'd let me take the bag on board. Undashed, he insisted I check the bag, and when we landed in Ouagadougou my luggage was missing. Edith received hers, but my precious though valueless possessions languished somewhere in a lost world like Conan Doyle's hinterland.

With my meager possessions lost, I found myself stranded there in Upper Volta lacking almost all my necessities: no clothes, except what I was wearing; no razor; no personal hygiene items; no alarm clock; none of the little doodads I'd so painstakingly selected and included for my long trips to help me in small ways cope with the rigors of the road. I had nothing at all except a few stray items in my small day-bag. I liked to travel light, but this was almost weightless. Amateur travelers and other types who prefer to rely on service rather than self-service can be identified by the large amount of luggage they lug with them. In *North America* Anthony Trollope described his winter 1862 arrival by train in Rolla, a scruffy railroad town in Missouri where no porter was available to carry the Englishman's bags: "Burdened with a dozen shirts and a suit of dress clothes, and three pairs of boots, and four or five thick volumes, and a set of maps, and a box of cigars, and a washing tub, I confessed to myself that I was a fool." My kit was much smaller. Long ago I decided not to travel with a washing tub. Over the next few days Air Afrique promised that the bag would soon show up. Meanwhile, I existed in conformance with local standards—almost no possessions. To my pleasant surprise, a few days later the airline informed me that the suitcase had been located in Dakar, Senegal, to which the luggage had by mistake been carried when the plane flew on from Ouagadougou. Our trip had started in Dakar and my bag, perhaps longing to return to the beginning, had ended up there. The suitcase, Air Afrique assured me, would arrive the next day on the return flight from Dakar. Suspense unnerved me as I watched the conveyor belt at the airport deliver suitcases off the arriving Dakar flight. One by one the jumble of bags tumbled onto the belt until finally my familiar case appeared. Back at the hotel I celebrated by changing my clothes, shaving and brushing my teeth—using bottled water.

After our stay in the capital we flew on to Bobo-Dioulasso, off to the southwest toward Mali. On this flight I kept my suitcase with me. As in almost every country I visited, I wanted to get out of the capital to see a provincial city or area. Bobo served to extend our trip beyond the main city, and the town also satisfied my curiosity about a place with such an usual name. How could I depart my earthly existence without ever seeing what a place named Bobo-Dioulasso actually looked like? Jonesboro and Smithfield, Plainview and Normal I could live without, but not Bobo. Now that I've seen the place I'd say that Bobo is okay to visit but I wouldn't want to live there. I'd prefer a more Normal-American Jones or Smith type town.

At the end of our brief stay in Bobo Edith and I were resting in our hotel room—so strenuous was our West Africa trip, rest featured in our daily allocation of time—waiting to proceed later to the airport for our flight on to Bamako in neighboring Mali where Timbuktu, a name even more evocative than Bobo-Dioulasso, beckoned. Loud frantic banging on our door suddenly signaled some sort of urgency, or perhaps an intruder trying to force his way in to rob us. He wouldn't find much in the way of worldly goods. I cautiously opened the door to encounter a highly agitated fellow who insisted that we needed to leave for the airport immediately. It was a replay, years later, of the episode at the Tiquisate banana plantation when the pilot made an emergency departure to beat the tropical storm. I told the fellow that we'd arrive at the airport far too early for the scheduled flight to Bamako, but he continued to repeat his urgent entreaties. By instinct I decided on-the-fly to comply with the man's pressing insistence for us to prepare to fly. Helter-skelter, Edith and I threw our things together, checked out and jumped into a waiting car which sped out to the airport where we immediately boarded the plane which took off well ahead of schedule. No one aboard explained why we were making a forced departure. Unlike Guatemala, this time it wasn't the weather which threatened as the skies were clear. A storm of a different sort was brewing, but we were up in the air as to what was going on. During the flight all we knew for sure was that before long we'd be out of Upper Volta and down in Mali.

On landing in Bamako we proceeded to our hotel. Two local acquaintances, an American couple then living in the city, soon showed up. They expressed surprise and relief that we'd managed to reach Bamako as planned. He explained that a *coup d'état* was underway that day— October 15, 1987—in Upper Volta, that president Thomas Sankara had been assassinated by the rebels, that a new regime had taken power, that the borders were closed for a month, and that all flights out of the country grounded. It was quite a mouthful. Oblivious to all the plots and counter-plots, we'd escaped at the last minute, so avoiding an unsought month-long stay in Bobo. Next time I'm in Bobo-Dioulasso I'll be sure to keep up with all the breaking and braking news which might stop our trip dead in its tracks.

In Bamako my first order of business was to book plane reservations for Timbuktu. Forewarned that space on the few small Air Mali planes out to Timbuktu was limited, without delay I secured our seats. As mentioned earlier, two completely different traveler types suffer from anxious, nervous worrywart concerns. Both inexperienced neophytes and seasoned globe-trotters get up-tight about fearsome problems which might arise. Belonging to the latter category, I always worried about everything. From experience I knew that anything might go wrong at any time—losing your suitcase, getting caught in a revolution in Africa, suffering a sudden accident or sickness, being detained by a gunman, targeted with rocks or more hands-on violence, and a wide variety of other such annoyances. Anything could happen and a lot did, but ahead of time I never knew what challenges lurked ahead. As I traveled I constantly imagined all sorts of unfavorable scenarios. Most never materialized, but all too many did. I never let my guard down, hoping that my guard wouldn't let me down.

Cautioned by my experience-based travel anxieties, I knew it was necessary to confirm and reconfirm our Timbuktu reservations. I kept returning to the shabby hole-in-the-wall Air Mali office in Bamako to be sure that Edith and I were still on the list for the flight. They didn't nickname the carrier "Air Maybe" for nothing. The flight manifest was written on a blackboard at the office. With the swipe of an eraser your name could be deleted, reduced to chalk dust. After our third or fourth visit to Air Mali Edith, by now rather exasperated, said she didn't understand why I was so concerned about our reservations. She noted that so far our West Africa trip had gone smoothly, with no problems. Admittedly, the loss of my suitcase, the slight matter of a *coup d'état* in Upper Volta, along with a few other matters beyond my control, were somewhat bothersome. But why worry over routine flight reservations? I now asked Edith if she happened to make a connection between my constant diligence based on my preoccupations about travel arrangements and the fact that we'd enjoyed a relatively problem-free trip. Grasping the point, she no longer complained about my fretting and compulsive monitoring of arrangements. In fact, on our very next visit to Air Mali I found that our names had indeed been erased from the blackboard manifest. Even without "dash" I managed to persuade the clerk to bump two other passengers off the list and restore us to the flight. Edith complimented me on my persistence. A few days later Air Maybe reverted to Air Mali as we winged our way across the desert out to the fabled city of Timbuktu.

For all true travelers few thrills can compare with finally reaching Timbuktu, a mecca coveted by every globetrotter. Arriving by air it's exciting to catch a glimpse of the airport sign:

"Tombouctou." The name (in French) marks your arrival by evoking many of the settlement's characteristics. You think of Timbuktu's mystery, remoteness, antiquity, of the French Foreign Legion, camel caravans crossing the Sahara, early-day explorations, desert tribes, hidden treasures, secret corners, exotic rituals. Of course the name itself is not the thing. It represents only an arbitrary designation, not the underlying reality. The name is like a badge which doesn't define a person, who remains the same no matter how the label designates him or her or what he or she is called. Were the Bard named Elvis Shakespeare, people would still want to read and see his works; if the singer was named Elvis Shakespeare, he'd still be a crowd-pleaser; and if Snoopy was called Fido, his canine wisdom would still appeal to millions of admirers. (If Elvis had been known as Fido Presley, he might not have been so successful and—who knows?—the same for T.S. had he been called Elvis Eliot.) Should Timbuktu bear the name Levittown, I'd still want to go there. But it so happened that "Timbuktu" designates the place, and under that name has spread its fame as a must-see remote desert outpost.

By any name, an almost tangible mystique permeates old Timbuktu, a destination destined to unsettled curious curiosity-seekers until finally they manage to reach the remote desert city. After reading about Timbuktu for years and inventing images of the place, the town now came to life for me. On-the-ground, or on-the-sand, experiences there in the fabled Sahara settlement stimulated my imagination more than books had endowed my easy-chair abstract inventions about the place. "Go-to-know" is a mantra every real traveler will endorse.

A passing visitor of course experiences a place in a very different way than do the local residents. Although novel and colorful, the reality I perceived didn't really represent the true nature of the desert settlement. What a traveler views as an exciting and exhilarating travel treasure is for the local residents only base metal. There's nothing exotic about the place for those who live there. They live an impoverished day-to-day existence, unrelieved by variety, un-charmed by freedom of action and choice, unendowed by wealth, unenriched by travel. They're stuck out there in the sandy wasteland, confined to a landscape with no escape. For the locals the empty desert doesn't represent a future open to possibility but the empty years which stretch ahead, on and on to vast eternity Outsiders experience a place in a way alien to the people there who occupy their native habitat, mundane to the residents even while fascinating to visitors. When Miranda marveled in *The Tempest* at the "brave new world/That has such people in't!" Prospero counters, " 'Tis new to thee." So it was for our visit to Timbuktu—new to us but for the locals old and weary.

Novelty arises not only from place but also from time. Discoveries in travel represent the geographic equivalent of the series of novelties you experience as time goes by during your lifetime. For the young almost every category of happening or situation is one time and only once new and then never again so. Over time a new world grows old. On a second visit, Miranda would not be so wonder-struck. The contrast between how new arrivals and how residents perceive their surroundings is striking. Age-old settled customs, practices and routines common to the locals in a place like Timbuktu present visitors with fresh views. Perspectives and privileges shape how reality is seen. As brief passers-by with an escape route out, Edith and I saw our own solipsistic Timbuktu. Thanks to the plane ticket, better described on Air Mali as a lottery ticket, we could leave the desert settlement at will, or at least as Air Maybe decided. That departure to end our visit a few days later concerned me from the moment we'd arrived. We'd barely escaped a forced extended stay in Bobo-Dioulasso; an imposed layover out in the Sahara might prove to be Timbuktoo much.

At the time of our visit only the French-run Azalai Hotel offered western standards of accommodation. In many places we enjoyed staying at a low-level hotel characteristic of the area so we could get something of the flavor of how the locals lived. In some cases, such as at the Sultan Mohammed Hotel in Ghazni, the flavor proved rather too acrid. In Timbuktu we opted for the European establishment but at least its name evoked localisms, as the designation referred to the age-old Azalai camel caravans which carried cargoes of salt,

Section V: The Misty Chords of Memory   245

slaves, gold, manufactured goods across the Sahara. Back in the old days as many as 10,000 camels formed the convoy. These days the caravans still operate but with fewer ships of the desert. As I saw in many desert areas, horse power in the form of Toyota pick-up trucks has replaced old-fashioned camel power.

    Just outside the Azalai perimeter Tuaregs with camels in tow offered to take us on a mini-caravan excursion out to their tribal encampment in the desert. A Tuareg camp at an unknown location in the Sahara reached by camel indeed represented for us a brave new world. Edith and I discussed if we should be passing brave and undertake the adventure, or if we should play it safe by passing up the opportunity. Edith hesitated. We gazed out at the empty wasteland. It seemed unthreatening. But were the Tuaregs? I couldn't check their Dun and Bradstreet rating. Dusk began to shadow the desert, so we'd be traveling out there, and hopefully back, in darkness. It was a "suck it and see" situation. The only way to know how the excursion would work out was to try it. The worst that could happen was to go out and end up with a truly memorable obituary—"American Couple Disappears at Sahara Tuareg Camp Near Timbuktu," the headline would read. When contemplating a particularly dangerous and strenuous trip (our autumn 1977 adventure to Pakistan, Afghanistan, Iran) Edith said to me, "Anything you can do I can do." So off we went into the desert dusk on our camels, each mounted by a Tuareg escort who sat behind. In the setting sun the camels cast shadowed forms which dappled the empty desert sands with passing dark-light-dark patterns. Evening soon shaded into darkness. The comforting lights at the Azalai disappeared. Night falls fast in the desert. Daylight quickly abandoned us as yet another twilight far from home shrouded me there in Timbuktu, a long, long way from the distant corner of the world where I belonged. We were on alien turf now.

    The camels loped their way across the sands. The Tuareg who sat behind each of us handled the rope reins. With the tribesman I spoke my Sorbonne French, while he answered in his less *bonne* but understandable version of a language foreign to us both. Like jungles with their proliferation of forms, deserts offer artful if empty natural features—shapes, contours, curves of pleasing but arbitrary designs and proportions, a plasticity arranged at the whim of the winds and other elements. Such natural areas were inhospitable for humanimals. I wondered how the Tuaregs managed to survive in the Sahara. Maybe they lived by plundering gullible tourists who rode with them out into the desert. Jungle Rudy lived okay in the wilds of Venezuela, so maybe Desert Mohammed sitting behind me existed equally well in the Sahara. We'd soon see his native habitat, far from mine, where he belonged and where I'd be a complete stranger.

    After the twilight far from home shaded into night Edith and I could no longer see each other. We called back and forth to make sure we remained close enough to communicate. I hoped our desert escapade through the night wasn't worrying her. It was my job to do the worrying, and I was hard at work. There we were, bouncing along on our camels in the dark Sahara, crossing the empty sands in the hands of Tuareg tribesmen taking us to an outpost who-knows-where for who-knows-what. We really weren't sure what we were getting into. At least the Tuaregs weren't headhunters like the Jívaros in Ecuador whose territory I'd invaded, or cannibals or a tribe which practiced human sacrifice. But the Tuaregs might well have been bandits. Our money belts contained a treasure trove of wealth, enough to buy a whole fleet of camels and some Toyotas as well to run large Azalai caravans. To the Tuaregs our trip money represented a fortune. Maybe they were taking us for a ride. Our brave new world would prove to be a lost world and the adventure a misadventure and a misfortune if the tribesmen took our money belts. We'd even lose our air tickets back home, grounding us—or sanding us—in the Sahara.

    Second thoughts churned through my mind, but it was too late to have made the safe choice to stay in the protected precincts of the Azalai Hotel. We were sucking the Tuareg mini-camel caravan and soon we'd see. The few lights of Timbuktu now well behind us, only the far stars and a fragment of moon broke the darkness. We plodded on through the night, the camels' rhythmic motion taking us farther from town. When our errant ruminants every so often drifted apart I feared that the Tuaregs might be leading Edith away from me. My overactive worry imagination visualized that she might be abducted, kidnapped and inducted into a Tuareg tribe and then forced to marry the chief, nevermore to be seen back at the Azalai or anywhere else outside the Sahara. As the wife of a chief, she'd certainly be upgrading her status. People back home would be impressed. But her newly elevated outstanding standing would in fact be a terrible fate and, besides, if Edith was forced to stay in the desert we'd lose the value of her return air ticket home.

After a time appeared a camp fire flickering in the distant darkness. The Tuaregs were preparing a hot reception for us. As we neared the kinetic flames I could see some of the robe-shrouded figures sitting around the fire, its flickers casting ghostly shadows over the tribesmen. We dismounted, finally ending our bumpy ride on the humpy camels. We joined the party, sitting there on the sand near the fire which warmed the cool night air. The desert can get mortuary cold at night. The flames' heat and light seemed feeble against the all-encompassing chill and night. The fire writhed as its smoke rose into the air and disappeared into the darkness.

Presently one of our hosts removed from the fire a pot of boiling water which he poured into glasses containing tea leaves. As the confection steamed and steeped Edith and I looked at each other with the same unspoken thought: the water was boiled okay, its pathogens purged, but a visible scum of filth covered each glass. Smears, secretions, residues, blobs, spots, blotches and other ominous deposits of unknown origin coated the glass with an ugly and threatening layer. The noxious streaks and marks reminded me of the similar stains of unknown origin which disfigured the sheets at the Sultan Mohammed Hotel in Ghazni. We'd slept there in our clothes to avoid contamination; now, with our Tuareg hosts, I had no gloves to avoid touching the contaminated glass. Next time I'm in the Sahara I'll be sure to bring gloves. I examined the container more closely: with the filthy residue covering most of the glass, it seemed like a poisoned chalice. The glasses had probably never been washed—water was scarce in the desert. We both perceived the problem in silence. The residues could leave us with all sorts of ailments or, worse, permanently in a desert of vast eternity. The Tuaregs had probably become immune to the microbes which lurked on the glasses, but as strangers we were "mune" to the germs. Holding the glasses, we might be about to take our lives into our hands and into our stomachs if we drank from those toxic containers. This was one case when "suck it and see" didn't apply.

The Tuaregs gazed at us. What would it mean to them if we refused their hospitality? Is this a hanging or a stabbing offense in their culture? Or just ill-bred behavior, poor form so that our hosts would never invite us back. We'd be on the Tuareg's black list—no more camel rides or tea parties. The daggers slanted into the tribesmens' belts gleamed and glistened as the fire's flames danced. A long sword sat across the lap of one Tuareg. Was the weapon displayed to add a touch of local color or a touch of the blade? The glasses of tea awaited our decision. We'd have to sip it and see or refuse and await the consequences. No TripAdvisor was available to guide our decision.

The tea rituals I saw around the world represented a variety of traditional cultural ceremonies. In tribal areas around Africa and the Middle East it's said that the first glass or cup of tea indicates that you're no longer a stranger; the second denotes that you're now a friend; the third, that you've become an honorary member of the tribe. If we shunned the first glass would we be considered an enemy? *Chanoyu*, the Japanese tea ritual, was more civilized and for sure more hygienic. In Japan the cups, utensils and everything else were clean. But no matter where and how the formats played out they all shared one thing in common: never again would they occur in exactly the same way. Everything happened just once. We could participate many times with the same Tuaregs in the same place for the same sort of tea party, but the experience would always be different. Nothing stayed the same (except maybe the filthy glasses). As Ii Naosuke wrote in *Chanoyu Ichie Shū*, his nineteenth century book on the

Japanese tea ritual (as quoted in *Tea in Japan*, edited by Paul Varley and Kumakura Isao), the gathering involves "'one time, one meeting' (*ichigo, ichie*). Even though the hosts and guests may see each other often socially, one day's gathering can never be repeated exactly. Viewed this way, the meeting is indeed a once-in-a-lifetime occasion." Whatever happens, no matter how apparently repetitious, is a unique passing event. For a participant to discern the nuances and shadings of meaning for that singular happening represents one of the highest forms of perception and requires great concentration. But first you have to realize that the ceaseless play of infinite forms never repeats.

For sure, out there at the Tuareg camp near Timbuktu that night it was an *ichigo, ichie* one time, one meeting episode. In all of eternity there would never again be another time like this. Shrouded in darkness, the desert that night as we crossed the wasteland out to the Tuareg encampment didn't seem like a desert of vast eternity, as that poet Andrew Marvell wrote, and behind us time's chariot wasn't winging ever nearer but only a few camels. But ever on the move, time winged on. With measured gait the camels had moved across the sands as regularly and relentlessly as a ticking metronome or an hour-glass through which the grains of sand slid away. Tick-tick, tock-tock, drop-drop slipped away the seconds as we advanced toward the tea party, time passing on and on as our passing brave excursion crossed the desert sands. The clock was also running on our decision whether or not to risk letting the toxic tea glasses engage with our mouths. Edith and I now faced the moment of decision.

We continued to delay as I weighed in my mind the merits of drinking or abstaining. Should we accomplish the transition from stranger to friend to honorary tribeswoman and man by sipping three portions of tea from the disgusting glasses? If not, would our uncouth behavior be grounds for summary execution on the spot, sworded to death in the Sahara night, or just immediate deportation, turned loose in the desert to find our camel-less way back to Timbuktu? In that remote corner of a far land we really didn't know what or who we were dealing with. Odd shadows danced over the tribesman's sword. Ii Naosuke had suffered a "one time, one meeting" death by sword when samurai assassinated him in 1860. Our little tea party in the desert might cut our own lives short. Finally I whispered to Edith, "Let's give it a try"—a drink-it-and-see attitude. Out of the foul glasses we sipped the requisite three separate portions of the potion, earning our brevet-type promotion to tribal status. Next time we returned to that Tuareg encampment we'd surely be received with great warmth around the camp fire, again served tea in the same still never-washed glasses. But this first time would still be a one time, one meeting experience, and if toxic would also be a last time experience. As our digestive systems coped with the sudden shock of tea infused with a potpourri of pathogens, the fire burned down as the embers gradually reduced to ashes and the smoke vaporized into the night.

Inducted into the clan thanks to our daring unhygienic sips, our brief interlude with our fellow tribesmen drew to a close and we remounted the camels to return to the Azalai. As we started off toward Timbuktu the encampment behind us faded into the darkness. Above sparkled stars with a clarity unknown in light-drenched cities. Dots of light from the far stars above the empty desert, devoid of content like the future, reflected Nate Silver's thought in *The Signal and the Noise*: "Think of the future as speckles of probability." But clear as the stars were out there in the Sahara, those distinct speckles threw no light on future happenings, for it was probable that what lay ahead consisted only of possibles rather than probables. Too much randomness and chance affected outcomes to endow the future with probabilities. We rode silently through the sands. After a time the cameleer spoke. Complaining that he suffered from a severe toothache, the man asked me if I could suggest a remedy to alleviate the pain. Maybe drinking from the polluted glasses had caused his infection. When I proposed that he visit the local dentist the tribesman replied that he couldn't afford the fee. I then suggested that he try taking some aspirin, explaining that although it might reduce the ache the pill wouldn't cure the problem. The Tuareg asked me what aspirin was. His astonishing question unveiled an entire world to me. Suddenly I realized how truly strange and primitive life was in the far reaches of the world, those remote deprived areas, and how far they were not only by distance but also by their conditions from my own advantaged way of life. Timbuktu was even more remote than I'd imagined.

The tribesman and I, both humanimals living on the lonely planet in the same era, were separated by a vast distance in how we experienced existence. Here was a fellow, a fellow human maybe about my age, who had never heard of aspirin. He'd probably never heard of Elvis, Snoopy, T.S. Eliot or, most likely, even of Shakespeare. I worried that it might be dangerous introducing aspirin to someone who'd never taken the compound. Who knows how his system would tolerate the foreign substance. Maybe he was allergic to acetate of salicylic acid and would suffer an adverse reaction to it. In *Tristes Tropiques* Claude Levi-Strauss tells of a Protestant missionary in the South American jungle who gave an Indian suffering from a high fever two aspirin tablets. The patient later developed congestion of the lungs and died. This led his fellow tribesmen, expert poisoners, to believe that the deceased had been murdered by the missionary's pills, provoking the tribe to retaliate by killing six members of the mission. I hoped there would be no blow-back from my remedy for the camel driver's pain.

Out in the Sahara I was in a lost world, a desert as empty of modern ways as the lost worlds in the Amazon jungle. Even a substance as simple, basic and common as aspirin hadn't reached those outlying areas. Life in those places would be unimaginable to First World inhabitants like those where I lived. My friends and neighbors at home had at least heard of—even if never read—Snoopy and Shakespeare, and residents in my suburb enjoyed easy access to aspirin, Dairy Queen, Oreos, Big Macs, M&M's, up-scale shopping malls, a cornucopia of consumer goods, hedge funds, stock brokers, wealth advisers, psychologists and economists and other self-assured social scientists, along with all the other pleasures, services and conveniences which make life in my neck of the woods worth living. The poor Tuareg would never know those modern-day wonders.

When we reached the hotel I retrieved from my health kit some salicylic acid tablets, so introducing the camel driver to the new world of aspirin. We now parted, the Tuareg to return to his tribal life back in the desert and Edith and I to continue onward to the delights and delectations of our world travels and our home-based civilized comforts. Ever after, when I occasionally took an aspirin (baby strength) I recalled that tooth-aching Tuareg tribesman whose fatal circle so contrasted with mine. Our existences on the lonely plant were completely different. We occupied two different worlds. But in the end our destiny as mortal humanimals was identical.

Our time in Timbuktu came to an end. As I'd anticipated, Air Mali reverted to Air Maybe as a question arose about our reservations for the flight back to Bamako. The return ticket in my money belt, my security belt, became a dodo: flightless. I consoled myself that even if we were stranded for a time in Timbuktu we could at least claim that we were honorary tea-Tuaregs with all the advantages that status might bring us. But I managed, without "dash," to get our booking restored, leaving two other passengers grounded.

Out at the airport one last contact with a local fellow tribesman served as my farewell to Timbuktu. A Tuareg offered to sell me his sword. He removed the blade from the worn leather scabbard to display the weapon to me. It looked something like the sword I'd seen glistening in the light of the campfire out in the desert a few nights before. Then, the blade seemed threatening; now, by light of day, it was only a local handicraft item. But I didn't want such a memento, even if it was a genuine antique made not for the tourist trade but an authentic piece, the real thing. The well-used weapon had no doubt dealt death to animals and probably to humans as well. But the sword didn't interest me. Its owner held the blade closer to me, turning it from side to side so I could examine the artistry of the piece. Clearly the poor fellow wasn't trying to hawk a mass-produced souvenir. The weapon was his own personal possession, probably one of the very few he owned. I explained that I wouldn't be able to carry such an item onto the planes we'd take during the rest of our trip. But he had mouths to feed—camels, wives, kids, relatives—and other expenses. The tribesman needed to convert base metal into spendable coin. The creatures in his life must have been pretty hungry, as the Tuareg was willing to sell what might have been a family heirloom handed down through the generations for almost nothing. The asking price wouldn't amount to a hill of beans or a sack of tea leaves. He now handed me the sword, which I examined. Elaborate curlicue and arabesque designs engraved in the metal embellished the blade. I wondered how much blood over the years had colored those intricate lines as lacerated soft skin reddened the hard-edged piece. The shiny long blade—a Gillette on steroids—reflected the line in Borges story "The Dagger": "It wants to kill, it wants to spill sudden blood." What a contrast between the well-read, word-drenched, bookish Borges and the poor aspirin-ignorant, illiterate, knowledge-parched Tuareg—extreme humanimal variances in the vast play of infinite forms. I was somewhere in between, although I fell short of Borges' erudition, at least I knew about aspirin and was something of a Snoopy and Elvis scholar.

Encouraged by my close inspection of the sword, the Tuareg again lowered his price. Although I protested that I didn't want to buy from him such a meaningful personal possession, one he should hand down to his eldest son, he reduced the price even more. I explained that my resistance didn't represent a bargaining technique but that practicality rather than price was the main factor. I really couldn't fly around West Africa and then on across the Atlantic armed with a Tuareg sword. In any case it wasn't functional: I already had an electric razor and didn't need the long blade to shave, nor did I know anyone whose throat I wanted to cut, and my yard service at home used its own mowers to slay the grass.

The tribesman wouldn't take no for an answer. He was a can-do *oui* kind of guy. I probably could've bargained the price even lower by telling the swordman-salesman that I was an honorary Tuareg tribe member eligible for a family discount. Acquiring such an inappropriate trophy, however, would be like buying a 10-room Timbuktu mansion surrounded by a 20-dune yard for almost nothing, a bargain purchase with no follow-on utility. Such a misbegotten acquisition would entail sunk cost, lost time value of money, continuing operating expenses, negative cash flow and all that sort of real-economy stuff a buyer-beware agenda should consider. A cut-price sale for a cut-throat weapon of utility to a Tuareg would transform the sword from its highest and best use to an inferior function as an inert memento displayed in my house. Back home in my Midwestern suburban subdivision, people cut toenails and meat and grass, maybe now and then cut corners, but rarely if ever throats. In any case, I didn't want to exploit my extreme bargaining advantage. The seller desperately needed cash and I didn't want his vintage Tuareg sword. It was a completely uneven dialogue.

Our discussion brought back memories of my experience in Santiago, Chile some 15 years earlier, when I engaged in a similar reverse auction. The antique store seller continued to lower the price of the vintage book I was reluctant to buy, not because of the cost but because of inconvenience. When he finally offered to give the book to me, still I demurred. Once again in Timbuktu the law of supply and demand prevailed, as a highly motivated seller faced an indifferent buyer. Whether in the desert or the jungle, cities, the high seas, on the moon, wherever goods and services were exchanged for money or by barter, economic principles applied. Such laws aren't repealed in primitive lands or sands. Fundamental common-sense rules, whether recognized and applied or not, affect peoples' behavior everywhere. When humanimals parley, the same laws rule unless a sword's coercion is used to determine the outcome. As in Santiago, I represented buying power faced with a seller desperate for a buyer, a great imbalance in bargaining positions, and even more so as I really didn't want the bargain on offer. Like the book, the sword was bulky and heavy, and in any case no matter what the object I was never very acquisitive, only inquisitive. I could easily live without an antique Tuareg sword on display in my house.

But there at home looms the tribesman's heirloom. I can see it right now from where

I'm typing these words, for I ended up buying the thing. Much to the surprise of the insistent seller I paid his original asking price, a family surcharge rather than discount. In spite of my thwarted buying resistance, I suffer no buyer's remorse and hope that seller's remorse hasn't left my sword supplier with regrets. The sword serves to sharpen the misty memories of my long ago and far away visit to the fabled city of Timbuktu. My "one time, one meeting" time in Timbuktu, along with many other singular memories of the olden days when I roamed distant lands, remind me as my days draw to a close of all the twilights far from home I enjoyed over the long years, now for me grown short as twilights close to home and all other lights, home or away, gradually dim into darkness.

~ 6 ~

For any widely traveled and worldly person, memories of visits to such legendary destinations as Timbuktu and Angel Falls occupy a special place among collected recollections. Time was, back in my time, when remote areas still remained relatively little visited, well off the beaten track even for many adventurous travelers. Islands difficult of access like Devil's and Easter lacked hotels and other usual tourist facilities. By now those two once primitive places have been upgraded, or degraded, with civilized amenities which make things easier for visitors while at the same time depreciating their experiences. Both islands offer hotels and other services which bring passing observers comfort and convenience at the expense of diminishing the previously more authentic and original setting and ambiance. It's become harder these days to find places which retain their unspoiled pre-development flavor. A.C. Jowett, an American engineer in charge of a project to install a hydroelectric plant for Kabul, capital of Afghanistan, wrote regarding his May 31, 1911 passage toward the city through the Khyber Pass that "when one goes to Afghanistan he is generally spoken of as having 'gone off the map'." Exiting the map describes an increasingly rare sort of trip, one to a destination in a mysterious uncharted area, a "strange land" as Conan Doyle's Professor Challenger described "the lost world" he found in the South American jungle.

Exotic places "off the map" promise novel travel adventures more stimulating and memorable than trips to less remote and more frequently visited areas "on the map." Even sixty-six years after Jowett's journey to Kabul, when Edith and I reached that city after passing through the Khyber Pass traveling from Peshawar in Pakistan to Afghanistan, we felt that in a way we were going "off the map." By then Afghanistan was well mapped, but for us the country represented an unknown and possibly dangerous region. Once you venture off the map you enter a *finisterre* kind of place, an ends-of-the-earth territory where you become subject to especially unusual, uncertain and perhaps even grave travel hazards. Familiar practices and procedures are left behind, back at places on the map you've temporarily abandoned. You hope you'll see those familiar places again, but that prospect is a known unknown.

Risk and uncertainty represent key elements of adventure travel. Although it's exhilarating to take something of a calculated chance, the calculation offers no proof of concept. Your assessment represents only a guess which doesn't tell you if you've pressed your luck too far. That you learn only in the doing. Even the most seasoned and well-traveled adventurer feels both excited and anxious when on the verge of a trip to a country like Afghanistan. Your keen anticipation is tempered by an underlying tension which adds to the adventure a certain spark, but hopefully not a shock. Almost everywhere we visited on our autumn 1977 trip took us to places "off the map"—Afghanistan, Pakistan, the Khyber Pass, Iran. Such areas were tough places, tough both to travel through and for their rather raw rough-edged sense of menace. You didn't find too many stray Westerners roaming around on their own in those challenging Professor Challenger-type lands, but that lack of fellow travelers was one of the attractions. More settled and accessible corners of the globe where tourists were more common and travelers less so offered fewer adventures. In *A Traveler in Southern Italy* (1969) H.V. Morton describes his visit to Capua Vetere (Old Capua) north of Naples where he found the deserted ruins of a huge ancient amphitheater. We "wandered all over the grassy giant [ruins] as if we were solitary figures in some print by Piranesi, an unusual experience these days when ladies from Bradford and Kansas City are so often to be found seated upon the most remote altars." On our 1977 trip we didn't come across many ladies from Bradford or Kansas City or, for that matter, from other Western lands. There were the usual coddled and cosseted foreign government diplomats and agency bureaucrats along with the Non-Government Organization functionaries, most of whom enjoyed living standards far higher and jobs far less punishing than the locals. The NGO and public-payroll personnel were privileged characters dealing with disadvantaged locals. Although Edith and I were not the only developed world travelers in Afghanistan, the Khyber country, the North-West Frontier region of Pakistan and Iran, we were rather rare fauna in those less developed lands.

In Peshawar, commercial center of the wild tribal areas in northwest Pakistan, we stayed at Dean's Hotel, one of those classic English-ish colonial era establishments (opened in 1913, demolished in 2001) found East of Suez (also slightly west, at Shepheard's in Cairo) which evoke the days of Empire, on which the sun has by now long since set. Twilights far from home now shadow the imperial memories of the world-wide lands Britannia once ruled. The former colony we were visiting, Pakistan, was now more unruly than ruled. Through the colorful streets of Peshawar swirled a spin of humanimals—tribesmen, clans, races, sects, traders, buyers, sellers, lookers, idlers, ne'er-do-wells, pick-pockets and an assorted mixture of other characters ladies from Bradford and Kansas City would find exotic, as did we. Smuggled goods and bads churned through the busy frontier entrepôt, a city seething and stewing with a certain suppressed but tangible underlying tension. You felt as if things might at any moment explode, ignited by friction between the various fierce-looking clans and factions which co-existed, even if uneasily, in the teeming city.

Based in Peshawar, we took a few excursions to nearby areas. These were not your ordinary Gray Line tours. One day we hired a car and a driver to go to Darra-Adam-Khel, a village 27 miles south of Peshawar known for its gun fabrication workshops. Perhaps it was foolhardy for two lone Westerners to venture a visit to such a weapon-filled town there in the wild North-West Frontier area, but my curiosity made me a hardy fool. Small machine shops lined the main street, about half a mile long. In the dark cramped cubicles locals fabricated copies of Enfield rifles, Berettas, AK 47s, revolvers duplicating Colt, Browning and Westley hand-guns, and various other firearms, an arsenal of death devices. All were for sale to the tribesmen— and probably anyone else—in the rugged and unruly back country frontier area. I could have started my weapons collection, initiated and completed a decade later with my purchase of the Timbuktu sword, there in Darra-Adam-Khel. The clansmen who coveted the guns lurked in forbidden hidden enclaves, remote valleys, fortified redoubts, obscure mountain areas and who knows where else, a wild region infested by smugglers, bandits, militias, warlords, gangs and other trouble-makers trying to scratch or snatch a living from the inhospitable country. The Afridi, one of the main tribes, dominated the gun emporium where 40 or so workshops produced the counterfeit weapons. The British administration supposedly established the trade in 1896 by giving locals the right to make copies of Enfield .303 rifles in exchange for access to the Khyber Pass. It takes an estimated 10 days to produce a prototype of a rifle, then two days to fabricate a duplicate. Handguns, more complicated, required longer to copy. Death devices came in various formats. Products included not only conventional weapons but also guns disguised as pens and as walking sticks. In the unsettled Khyber country exploded a lethal play of forms, not infinite but deadly ones causing finite finalities.

We felt uncomfortable wandering up and down the street peering into the gun workshops. Staring and glaring, the artisans eyed us warily as we passed by. Although foreigners were rare—we saw no others there—guns were plentiful. We hoped none would be turned on us, triggering our departure one way or another. The shops were alive with noise: whirring and grinding as the machines bit into metal, banging and clanging as the pieces were shaped, a sinister symphony of deadly sounds as the weapons took shape. Although we didn't encounter any other foreigners in Darra-Adam-Khel, it would have been comforting to see just then a few ladies from Kansas City, a welcome Midwestern presence from my homeland there in the wilds of the North-West Frontier. But Edith and I were alone, strays far from home in unfriendly territory, a gun-filled village with hostile tribesmen scrutinizing us. Unlike our tea party at the Tuareg desert encampment near Timbuktu, no one stepped forward to offer us the traditional welcoming three cups of tea. In the wilds of the North-West frontier, three rounds of lead seemed more likely. I ventured to take some photos, each shot provoking the workmen to turn away or to cover their faces with their hands. Nervous traveler that I am, I began to sense danger. Perhaps shooting pictures would cause a shooting match. We decided we should leave immediately to get out of harm's way and retreat back to the more secure precincts of Dean's Hotel in Peshawar. For once a Dean, even if not an academic one, offered an attractive option for my post-graduate course out in the real world where trigger warnings involved rifles rather than trifles.

After our unsettling excursion to Darra-Adam-Khel to the south of Peshawar, Edith and I ventured another day trip out in the countryside to visit Swat Valley to the north, a remote far corner in the already remote North-West Frontier. A scenic and verdant area known as "the Switzerland of Pakistan," Swat somehow hasn't merited a return compliment from the Swiss, as they haven't designated any area of their homeland as "the Pakistan of Switzerland." In 1897 General Binden Blood, his name evoking the bloody conflicts which raged in the region, led the Malakand Field Force to rescue a fortified British camp attacked by Pathan tribesmen in the Malakand Pass forty miles northeast of Peshawar. Tribal uprisings began to spread until British troops suppressed the unrest. In his 1898 *The Story of the Malakand Field Force, An Episode of Frontier War*, based on his time at the front as a war correspondent, Winston L. Spencer Churchill, as he then styled himself, described the Valley: "[A]fter crossing the Malakand Pass the first turning to the right leads to the Swat Valley. The traveler is now within the mountains. In every direction the view is restricted or terminated by walls of rock. The valley itself is broad, level and fertile. The river flows swiftly through the middle. On either side of it, is a broad strip of rice fields ... It is a beautiful scene."

To his idyllic description of the picture-perfect serene scene Churchill adds a cautionary note: "The reputation which its present inhabitants enjoy is evil. Their treacherous character has distinguished them even among peoples notoriously faithless and cruel. Among many Pathans it is a common saying: 'Swat is heaven, but the Swatis are hell-fiends.' " Was I once again destined to encounter in heavenly Swat a fiendish hell-raising place as I had, at least nominally, at Devil's Island, the Devil's Nose in Ecuador, Hell in Norway? The pervasive sense of menace which seemed to permeate the North-West Frontier continued to unsettle us. Pathogenic Pathans who lurked in the area could prove fatal.

Churchill's view of the Swat Valley Pathans was shared by other adventurers in the region. English visitors to the badlands often commented on the murderous habits of the Pathans. As recounted in *Sir Aurel Stein: Archaeological Explorer* by Jeannette Mirsky, after the explorer's dog Dash III was killed by a car, his death "at least painless," Stein observed how he too "would like to disappear in that way; but if the choice rested with him would prefer a sudden end by avalanche, Pathan knife or the like." Stein named his dogs "Dash" perhaps because he was a dashing ultra-energetic early twentieth century adventurer, traveler, explorer, discoverer, surveyor, archaeologist, geographer and historian in Central Asia and nearby areas. During his expeditions he encountered many wild, ferocious hostile clans, tribes, bandits and militias, but of them all he singled out Pathans as the most blood-thirsty cut-throat group. In March 1926 Stein noted in his *Personal Narrative* how locals in Swat "walked about fully armed ... Even the humblest Pathan ... aims here at the possession of a rifle, revolver or pistol."

Weapons abounded and danger lurked in the Northwest-Frontier country. The Pathans typically carried "a fearsome looking bandolier full of cartridges" and a rifle, and the tribesmen weren't reluctant to use their weapons. *Plain Tales From the Raj*, edited by Charles Allen, includes an account by an English nurse who in 1929 arrived in Peshawar, "the city of a thousand and one sins." She recalled the bloody family feuds: "You would get a case coming in with all the intestines sticking out. They used to get the skin of a chicken and wrap the intestines in this skin to keep them fresh." She described the anatomy of violence: "We used to have jealous husbands cutting off their wives' noses, breasts amputated, even pregnant women with their abdomens ripped open." There at the Frontier— "never merciful"—the Pathan "had a very, very, cruel streak in him, and if you left a wounded man behind he was not only killed but frequently mutilated in the most obscene manner."

Both Stein and Churchill represent the man of action and of thought *par excellence*. They were great adventurers who contemplated and wrote about their in-the-field experiences. A synthesis of a hyperactive doer and an original thinker is a rare combination. Each man based his opinions and observations on the real world, not on invented notions about it. Paul Johnson in *Churchill* noted the contrast between a mentality which invents and one which imagines, the difference between the theoretical and the actual. Applying this to Churchill, Johnson states: "He had a historian's mind, eager to grapple with facts, actualities, to answer the who, how, where, when questions rather than a philosopher's, mesmerized by abstractions with their whys and wherefores." When Churchill and Stein described the Pathans the two men knew what they were talking about, as did the English nurse. Experience lends authority and informs imagination; invention borrows chimerical vagaries.

It's always unsettling to venture into areas known to be dangerous, such as the Khyber country and Afghanistan. At the time of our visit I wasn't sure if the Pathans had settled down or if they were still trigger happy, which might make us unhappy. Awareness of possible hazards and risks, known unknowns, puts an edge, perhaps one rather too sharp for comfort, on the experience of venturing into "off the map" dangerous areas. A gnawing subliminal anxiety darkens your mood—a vagrant ominous feeling other than fear, as no immediate threat confronts you. Rather, an amorphous and abstract sense of danger tinctures your outlook as you scan the ever-changing situations for problems. A cautionary free-floating sort of wariness based on the "evil" reputation of the region possesses you. Hoping such fears didn't already bother Edith, I kept my worries to myself. Reaching Swat Valley nearly 80 years after Churchill, I expected that by then the local Pathans were less violent in earlier times and presented no mortal threats, at least for foreigners even if not for rival area tribes, which frequently clashed. In contrast to the sinister ambiance we'd encountered at Darra-Adam-Khel a few days earlier, in the Swat Valley we attracted little attention. We traveled largely unnoticed through the fields and on up to the Swat capital at Saidu Sharif, there to see the white marble palace where lived and ruled the valley's chief. Unsure of his title—perhaps *wali*—I referred to him as the Sultan of Swat as the ruler, like Babe Ruth, was a big hitter who starred with the Valley Pathans, the home team tribe.

In a pleasant green setting by the Swat River we ate a picnic lunch before heading back down the valley to return to Peshawar. Up there beyond Saidu Sharif the road became something of a cul-de-sac. My imagination roamed into the back country even if we didn't. I envisioned that in the remote highlands to the north spread a web of obscure Pathan paths, trails, hidden byways, redoubts and secret routes snaking through the wild hinterlands where smugglers, bandits, tribesmen, traders and traitors, desperados, travelers eluding border control points and other irregulars lurked. The tough mountainous country there was no Switzerland.

Back in Peshawar we arranged to travel on through the Khyber Pass toward Afghanistan in a shared inter-city taxi up to the border. Like Timbuktu, Angel Falls, Petra, Angkor Wat and other such fabled names in the world of travel, all on the map even if in their isolated remoteness "off the map," the very designation "Khyber Pass" denoted an exotic mix of mystery, history, folklore, legends, all adding up to the romance of adventure travel. Our anticipation built as we approached the famous mountain pass. In the same way that I was thrilled to catch a first glimpse of the "Tombouctou" sign when we arrived at the desert city's airport, so it was exciting to see the "Khyber Pass" sign as we entered the renowned route between western Pakistan and eastern Afghanistan.

History lingers with an almost tangible presence there in the Spin Ghar mountains, the whirligig of time having spun long skeins of plots and counter-plots, fortune and misfortune over the centuries. Through one of the world's oldest known passes, part of the ancient Silk Road, advanced Darius I, Alexander the Great, Genghis Khan and now, less militant, Edith and Tom, two independent travelers among the many—maybe even including a few ladies from Kansas City—who fleetingly passed through the Pass linking Central Asia with the Indian sub-continent. Any number of warriors, adventurers, conquerors braved the Khyber, a strategic check-point whose control enabled the local tribes to exact tribute, tolls, tariffs and other payments from the wayfarers who sought to pass over the mountains. Bandits, highwaymen, warlords, brigands and other "evil" (as Churchill called them) characters once infested the area. Back in Peshawar we'd been assured that safe passage now prevailed but that occasionally a rogue bandit or itchy trigger finger still might at any time threaten outsiders who transit the Pass.

As we slowly wound our way up the contorted route I kept a watch for any road blocks or armed tribesmen aiming to disrupt our passage through the Pass. Advancing up the sinuous road, we climbed back and forth toward the high point at 3510 feet. Along the way stood in the near distance bunkers and watchtowers, lookout redoubts, vintage martial outposts housing tribesmen who guard and control the Pass. It was unnerving to realize that we were being scrutinized. We wanted to avoid attention. We wondered what the sentries saw in us: perhaps prospects ripe for plundering. With anxiety but without incident we reached Landi Kotal, the border point in Pakistan three miles before the Afghan frontier where we soon arrived at Torkham and then crossed to proceed on to Kabul.

Border crossings in the Third World and in out-of-the-way countries often present problems. Frontier officials can be capricious and unpredictable. I've occasionally been arbitrarily turned away, even when my papers were in proper order. For an American passport holder many borders around the world these days represent only imaginary demarcations which can be crossed relatively easily. When I traveled in Europe as a young man in the mid- to late 1950s, entering each country entailed official border control. Now you just pass across the borders without ado and without the need to exchange unusable money for a new currency, losing a few percent in the forced transaction. Even with open borders, transitioning from one country to another separated by narrow distance but wide differences often abruptly brings you to a completely new set of cultures, customs, conditions, languages and other striking variances. Frontiers represent both barriers and gateways. All around the globe, partitioned into some 195 countries, each possessive of its territory and proud of its individuality, immigration authorities stand ready to scrutinize your passport and, if required, the visa. At land crossings horizontal poles across the highway blocking your advance show and symbolize the barriers travelers face. Such obstructions and the officials who control them represent both endings as you leave one country and the beginnings of your journey onward to a new land, often with a different lay of the land. A mere mile or less can separate two completely different ways of life. Such is how the earthlings have carved up the lonely planet into enclaves.

Difficult frontiers where you're turned away deceive, for they don't front an entry point but instead a dead-end "finis-territory" which blocks you from advancing. At some borders I reached around the world, tempting but forbidden areas across the way—"off the map" or on—beckoned but were at the time inaccessible. At the Lok Ma Chou border post in Macau in late 1967 I saw China in the distance, but only in 1981 did I manage to enter the country. East of Kotka in Finland I looked across into Russia, which I'd previously visited. In late 1972 at the Tripoli airport where I was denied entrance to Libya I looked longingly at the immigration booths which remained off limits, at least until late October 2010 when I finally entered the country. When the Iron Curtain still hung I gazed across the Adriatic from the coast of Italy toward Albania, hidden behind the barrier until the Curtain was lifted, so allowing me to enter, which I did in October 2005. On separate trips in different years I saw from one side and then from the other the border post on the causeway connecting Bahrain with Saudi Arabia. Each of those two countries I reached by air, not via the causeway route. Borders opened and closed in unpredictable ways as politicians, strongmen, armies, combat, circumstances, whim, changes revolutionary or evolutionary happened to lower or raise barriers to permit or prohibit entry. At times a play of nearly infinite official forms, visa applications, questionnaires, red tape, paper-work and other formalities delayed or blocked access. Often, the less significant the country the more detailed and particular were the conditions imposed to gain entry. The visas and entry and exit stamps in my passports mark a long trail of arrivals and departures recalling many twilights far from home.

Transitioning across borders, whatever their difficulties, always fascinated and energized me. The process signified moving on from a place now familiar to me to a different territory which represented a refreshing change. Sometimes I passed through quickly with only routine formalities, while elsewhere long delays and red tape entangled me in slow and convoluted official and often officious behavior. Never was I asked for, and nowhere did I offer to pay, a handout to get me through. Highlighting the transition between countries, usually on each side of a border officials wore different uniforms and in most places spoke a different language. In a few areas, such as Lome where Togo and Ghana meet, a border runs right through a city. Passports or visas aren't needed to cross all borders. Traveling from Anaheim into Disneyland, a journey from Southern California to a fantasy land, requires no visa other than a Visa. And you can enter a Wonderland simply by reading about Alice's adventures. But those are invented rather than imagined. The frontiers of some geographical units conform to natural features, such as a river, while elsewhere the hand of man has imposed arbitrary lines to set the borders. American states west of the Mississippi bear angular shapes which occupy on the map rectangles like postage stamps neatly arranged in an album. But the world is really not so regular or orderly. The lonely planet is a wonderland rather more woeful than the fanciful land Alice occupied.

It occurred to me that perhaps my compulsion to see the world, to make some sense out of it based on experience, arose from a desire to bring some order out of the chaos of infinite forms. Maybe expanding and diversifying my impressions of the lonely planet would somehow offer me enough components to piece together a sort of picture which would cohere to offer some meaning. Perhaps my reminiscence represents the same sort of effort to arrange a selection of random happenings which shaped my life experience into something of a pattern so as to lend the experiment of living at least a kind of bare-bones skeletal structure. As a boy I enjoyed playing with stamps from foreign lands, places then unknown to me but many of which I eventually over the long years came to know. I arranged the little pieces of tinted paper by size, shape, color, country, design, denomination, content or some other categories. It was easy to rearrange the world as represented by sovereign stamps, but as I later found out when out in the real world frontiers were more intractable than bits of paper and often realigned by force of circumstance, usually force of arms, rather than by boyish arms and hands.

Perhaps it wouldn't be too far-fetched to say that in some ways earthlings made a mess on their little home planet in a tiny corner of the cosmos. Over the centuries humankind, in all its unkind belligerence, unleashed the dogs of war, fierce and fearsome attack beasts, animals which forcibly snarled the earth's surface with a tangle of separate domains. Cynocephaly characters, dog-headed humans of folklore and legend, populated ancient mythological tales, but perhaps human-headed dogs would be more true to life. Either way, the semi-canine creatures were mimetic examples of humanimals which doggedly fought in turf wars to guard and secure their own kennels. In any case, a dog-headed dog, like Snoopy, represented a much more attractive and less threatening fellow. Those fought-over territories blanket the lonely planet like a crazy-quilt, a chaos of kingdoms, fiefdoms, countries, principalities, dominions, domains, states, statelets, enclaves and other battlegrounds which dis-unify the world into disparate lands. Divisions multiply; additions subtract, as some lose what others gain. Land grabs score the earth's surface with lines in the sand, the soil, the sea, all cut into pieces and brought under control by one group or another. Then the disputed areas change hands, with new forces exerting dominion. Threading my way around the patchwork planet required a lifetime of border crossings. Dealing with borders by reading travel books from the now defunct Borders bookstore chain would have been easier. Although a book lets you cross a bound volume, a boundary, to enter a new world, one with a text and a context different from the ones you know, crossing borders presents more difficulties than reading about frontiers in Borders books. Works on paper such as travel books or postage stamps from around the world entail fewer problems but are less informative and authoritative than globetrotting or globecreeping across frontiers.

Although prepared for delays, hang-ups and demanding border officials at the Pakistan-Afghanistan border, Edith and I crossed without difficulties. In Afghanistan we visited Kabul, the capital, Bamian (the original Buddha statues were still standing), Mazar-i-Sharif in the northeast and other areas before proceeding southwest on the long ring road which loops through the desert via Ghazni (and its fabled Sultan Mohammed Hotel) and Kandahar and on to Herat before continuing on across the border to Iran by land. Along the southern desert route in Afghanistan we passed through desolate scrub land, a mostly flat and featureless terrain quite different from the more scenic Sahara where formations artfully shaped by winds rearranged the shifting sands into fanciful forms. Compared to the Sahara, the Afghan desert was a drab wasteland. Like jungles, deserts are forbidding regions, areas humanimals inhabit or even only enter at their peril. Inhospitable oppressive natural conditions operate to make life in the desert and the jungle difficult for man and beast. Other more habitable regions suffer oppressive forces not from nature but violence and destruction imposed by humanimals. Nature is red in tooth and claw, humanimals in sword and guns. Plots and counter-plots natural or man-made are hard-to-escape occupational hazards of the business of being alive. Unkind mankind somehow seems destined to degrade, demolish, destroy places, people, things. The lonely planet's national sport is conflict. In many places I've visited, such as the North-West Frontier of Pakistan, Afghanistan, Middle Eastern lands, countries in Africa, others, the strong man rules. Whirl, not tradition, is king.

Traveling through the featureless Afghan countryside made me appreciate all the more nature's artistry as it confected the Sahara's sand-scapes. Over the years I've seen the Sahara from various vantage points and never have the landscapes failed to fascinate. In the summer of 1969 I first entered the Sahara at Zagora in southern Morocco where my first camel ride gave me a truly uplifting experience as the creature rose on its front and then hind legs with a bumpy, jerky motion. I visited other areas of the Sahara in Senegal, Tunisia, Egypt, Algeria and for the last time in Libya, which after years of waiting I managed to enter in late October 2010. A window briefly opened for non-business travelers to visit, and through the opening I jumped before it soon shut with a bang—the bang of guns as extreme and bellicose disorder succeeded Ghadafi's rule. After visiting the great antiquities at Cyrene, Leptis Magnus and Sabratha our tour group of 11 or 12 Americans drove some 375 miles southwest from Tripoli across the desert, at times leaving the road to take an off-the-beaten track overland or over-sand route as a shortcut to reach Ghadames. Believed to be the oldest habitable *medina* settlement in the Sahara, the Berber oasis served for centuries as a cross-roads trading town. Still perfectly preserved, the walled enclave became a busy entrepôt and a way-station and watering hole for camel caravans crossing the desert sands. Inside the walls huddle white houses lining a network of narrow streets, passageways, alleys, squares. Between 1975 and 1983 the Libyan government removed the town's 10,000 or so residents and resettled them outside the walls in a new city built in the architectural style of the original, now preserved in pristine condition as a national monument. Little-visited Ghadames remains an outstanding "lady in waiting" travel gem, a truly unspoiled picture-perfect original Sahara outpost far removed in appearance and ambiance from the modern era.

Late one afternoon we took an excursion from Ghadames out into the desert to climb Ras-al-Ghoul, "Mountain of Ghosts," about 10 miles northwest of town. Atop the rock hill crumbled away ruins of a fortress of unknown origins, perhaps a Roman frontier redoubt or possibly remnants from an even more ancient pre-Islamic era of desert castles built across North Africa. When Islam arrived in the Ghadames area about 667 the local tribesmen who refused to convert retreated to Ghost Mountain. Nearby the Muslims established a camp and then a cemetery for their fallen warriors named Jebel ash-Shohada, "Mountain of the Dead." The ghoulish name Ras al-Ghoul arose when the invaders noticed strange lights at the fortress atop the Mountain of Ghosts. The dis-incarnate characters haunting the hill seem still to spook the desert redoubt.

The spectral Mountain of Ghosts designation also in a way described nearby Ghadames, now a Ghostdames as the empty depopulated ghost town remained only a skeletal shadow of its once lively character. So too, over time, will be the eventual ghost towns of London, Paris, Rome, Vienna, Bobo-Dioulasso, Calcutta, Cairo, Kabul, Kansas City where the traveling ladies come from, and every hamlet, village, town, city, subdivision and other peopled place, for the world is a phantomland wonderland. Hard as it may be to conceive, one day in the far future the lonely planet will become truly lonely after every last human habitation becomes empty and deserted like a desert ghost town. From atop Ghost Mountain I gained an extensive view across the sands. Although, unlike post-life time, the Sahara is not a desert of vast eternity, the sands there seemed limitless as I stood on Ghost Mountain that late autumn afternoon looking out at the landscape. At my back I could passingly hear the twilight far from home drawing near with a soft hushed rustle as dusk shrouded day, or maybe it was only my imagination or perhaps just a passing desert breeze which momentarily stirred the lonely sands before passing on. But I felt something was gaining on me, crowding me forward toward the end. Camels and Toyotas crisscross the desert, and always and everywhere also relentlessly advanced the wingèd chariot. In the far distance four miles away I could just make out the area where the boundaries of Libya, Tunisia to the northeast and Algeria off to the west met, joining together with imaginary man-made lines drawn in the sand. Travelers in that remote spot face a frontier post, a construct marking the artificial lines carving the desert into different lands and sands. Tribesmen avoided the border formalities and exactions by crossing elsewhere at deserted points in the trackless desert, nomads in a no-man's-land where no official controls interfered with free movement. In those uninhabited and unmarked areas nature, not man, ruled.

Such trackless sands with no evidence of a human presence are typical in the Sahara. Unlike the jungle, a riot of life, only deathly silence and empty vistas embellished with wind-sculpted dunes characterized the desert. Like the future, the desert was empty. You never knew what might materialize to fill the space, and in any case whatever appeared in the wasteland would only be temporary like the passing winds. We now gained another perspective on the empty scene as we gave up the Ghost by leaving Ghost Mountain to drive to nearby dunes which we climbed as dusk began to darken the desert sky. It's slippery-tricky to climb a dune, as with every step up the unstable sands you slide backwards. To reach the top you travel twice as far as a straight-line climb. After sliding our way up we arrived at the sharp-edge-shaped soft summit, and there we perched atop the dune to watch the desert sunset, a memorable twilight far from home. Shadows slowly crept over the landscape, casting strange moving shapes onto the empty sands. A slight chill cooled the early evening air as daylight, gradually and then quickly, drained from the sky. In the desert off in the distance, no signs of life enlivened the panorama. Our little group sat up there atop the dune alone looking down on an empty world, a remote corner of the lonely planet. No one spoke. Night soon began to shroud the scene, curtaining the infinite play of forms. A ghostland as silent, as lonely and as deserted as old Ghadames surrounded us. At last light we started to descend, slip-sliding down, back to earth, or sand, soon returning to the new Ghadames, a tiny illuminated enclave of humanimals set in a remote corner of the empty sands stretching away into the enveloping darkness.

The distant view of the Algerian frontier I espied from atop Ras al-Ghoul evoked my visit to that country exactly three years before. In early November 2007 I toured Algeria with a group, traveling first along the coast and then a long flight inland to Djenet deep in the Sahara not all that far southwest of Ghadames in Libya. When we'd traveled along the north coast from Algiers to the east before our visit to Djenet, armed guards in cars in front of and behind our bus escorted us. In that unsettled region tourists represented tempting targets to terrorists. Down in the Sahara where conditions were less threatening we traveled freely around the desert without security personnel guarding us. Dehydration, sunstroke and sand storms represented the main hazards. If it's not one thing, such as humanimals as terrorists, it's another. Natural threats, plots and counter-plots, unknown hazards lurk everywhere at all times. Guns, the sun, deadly thirst, stranded in the sand lost among the dunes—one fate's as good as another to do you in.

Near Djenet some of the Sahara's most impressive and beautiful sand-scapes make the desert a gallery of enchanting forms. Large sharp-edged peaks razored across the blue sky atop great sandy hills shaped and reshaped by the shifting desert winds. Those splendid stately structures stood as grainy temporary forms soon to be changed into new alignments, randomly designed but graceful natural shapes. Curves, lines, crevices and hollows formed an outdoor museum of sculptural beauty, the exhibits changing every so often at the whims of the wind. Arbitrary and temporary, each dune was a plaything of nature—just as were humanimals.

Our last evening in Djenet we returned to the desert to dine among the dunes beneath the star-sprinkled sky. A campfire blazed into the darkness a tiny patch of light as our meal cooked on the flames, the smoke curling into the air before disappearing. Out in the dark desert the night sky glowed with scattered glimmers of pinpoint starlight. I briefly left the campfire's comforting light and warmth and the aroma of the cooking meal to advance into the desert alone to gaze at the more distant lights in the cold cosmos above. Somewhere up there in the vastness of the heavens glowed the Dog Star, at least nominally a sign of earthly animals. The night had a thousand eyes, but I had only two and not world enough or time to count all the glimmers overhead in the firmament. Anyway, I'd have plenty of time for the heavens in my post-earth life—the next ghost stage, if there was one, after my present phantom-like earthly incarnation—provided I ended up in celestial comfort rather than in warmer climes below. For now, though, I was terrestrial, not celestial, an occupant of the halfway house between never before and never again. For a brief while I remained earthbound, a local lad tethered to the human homeland, to the here and now, a finite fellow momentarily on the lonely planet in a time-desert of vast eternity. Post-life didn't concern me. Only my earthly existence was sure, and that surety was brief and fleeting. I was just another passing form. Pie in the sky, a moon made of green cheese, the Milky Way (for me a chocolate bar)—none of those distant infinite cosmic forms involved me. Only the infinite play of forms down on earth implicated me—physical terrestrial realities rather than metaphysical or stellar matters.

Back at the campfire the aroma and the sound of the dinner meat sizzling over the flames represented the real and immediate world, the only one which mattered to me. The animal flesh I was about to eat and then, as a humanimal, transmute into internalized flesh and blood provided a nourishing organic connection with the outside world. All the rest—stars, planets, the galaxies, meteors, comets, the black holes and white dwarfs, the constellations, the whys and wherefores of it all—were far beyond me.

~ 7 ~

English poet Andrew Marvell compared deserts to "vast eternity." He was the guy who wrote that at his back he could always hear time's wingèd chariot hurrying near, while before him lie deserts of vast eternity. I shouldn't have looked at so many deserts, the Sahara and others, as in them I saw what Marvell perceived. His poetic power of suggestion haunted me. The desert which stretched out beyond Ghost Mountain near Ghadames was spooky as were all deserts, for they suggested vast eternity. Whatever thoughts they evoked, the deserts, the jungles, the inhospitable regions contrasted with the almost edenic conditions where I lived. Back in the comfortable and familiar setting where I came from, eternity seemed remote. On my native turf—verdant, not sandy—rains caress the fields, bins burst with crops, rivers water the earth, tame woods nurture wildlife, trees and plants and flowers garnish the landscape. I lived in a nature-favored garden of delights. Playground sandboxes and jungle gyms were the only desert and jungle-like spots in my natural habitat. We used cars, not camels, to get around. These civilized amenities contrasted with the way many dysfunctional places in the world I saw functioned. In the Third World masses of humanimals lived like dogs or, in fact, less well than did doggies at home who were well-fed, well-treated and pampered. Every pooch pet where I lived enjoyed a pleasant home, regular meals, free medical care and a fully funded 401 (k) (9) retirement plan. A dog's life in my suburb topped the lives of humanimals in the underdeveloped lands around the world I visited. In those deprived areas people existed in abject conditions, and dogs wallowed even lower in the pecking order. Many of those benighted places were desert lands, areas which, if lacking oil, don't lend themselves to lush cushy standards of living.

Not only all around the Sahara but also in other regions where I saw deserts, they were not as vast as eternity but suggestive of it. They dominated countries such as Yemen, Saudi Arabia, the Emirates, Syria, Jordan and elsewhere. In Jordan I visited the Desert Palaces, three isolated fortress-like outposts in the desert beyond Amman, the capital. I also entered Wadi Rum, Lawrence of Arabia country, with rock cliffs and other formations reminiscent of Zion National Park in the U.S. West, a long way from the land of Zion in the Middle East. I made this comparative observation to my driver, who replied, "Every American who comes here says Wadi Rum looks like Zion Park." I was glad I didn't disappoint him. Beyond Wadi Rum to the south I continued on to Petra, another of those fabled places unmissable by anyone claiming to be a world traveler. My time at "rose-red Petra, half again as old as time" made me curious about its sister city, Madain Salih, hidden away in a remote corner of Saudi Arabia, at the time (early 1973) closed to tourists but a country I hoped to see. For years casual visitors weren't permitted to enter the kingdom but finally, as for Libya, a window of opportunity opened and through it I accessed the country in February 2000 as part of an American tour group admitted to the long forbidden land.

In the interior we visited Riyadh, the capital, and Medina, not to be confused with Mecca, off-limits to non-Muslims, and then Jedda on the Red Sea coast, and later in the southwest reached mountainous Abha, a region quite unlike the desert which predominated elsewhere in the country. Over on the east coast in the main petroleum-producing region we dined at the world headquarters of Aramco, the Saudi oil company. During the meal I learned that extracting money from tourists' pockets couldn't compare with pumping oil from the Arabian sands. No need to bother chasing a few dollars or riyals from foreigners when right under your feet below the ground riches abound. Next to me at the dinner table sat an Aramco executive-vice-president, a Saudi who'd studied petroleum geology at a Texas university. Curious about the economics of the Aramco operation, I took the opportunity of close proximity to one of the company's top executives to ask how much it cost to "lift" (pump out) oil, then selling for about $30 a barrel. He at first hesitated to reply but I persisted, and finally the official revealed the secret number: "about $1.50 per barrel." I then asked if that included transport to the terminal, processing and shipping the crude to markets around the world. He informed me that those additional functions would add another half dollar or so, increasing the all-in landed costs to about $2, "a pretty good business, don't you think?" he asked me. I agreed.

During the trip I finally realized my long-held ambition to see Nabatean Petra's sister city Madain Salih. For more than a quarter of a century, ever since my visit to Petra 27 years earlier, I'd supposed that I'd never manage to get to that desert site to see Saudi's ancient Nabatean settlement, as remote to me in its earthly way as the Dog Star which at least I could see even if never visit. But early one warm Arabian day our group started off on the drive out to Madain Salih, a day-long excursion from Medina. The famous Arabian area traveler Charles M. Doughty, the first European to visit Madain Salih in modern times, had also become curious about the Saudi settlement when he visited Petra in the 1870s. In his 1888 *Travels in Arabia Deserta* Doughty wrote: "I visited Petra; and at Ma'an Settlement, which is a few miles beyond, heard of other Petra-like sculptured cliff monuments bearing many inscriptions at Madain Salih...[and] I resolved to accept the hazard of visiting them." It took Doughty a year to find a way to get there, 26 years fewer than I required. He finally managed to undertake "the adventure of journeying thither in the great Damascus Caravan. Arrived at the place after three weeks' tedious riding." Once there, Doughty found "Madain Salih to be an old ruinous sand plane [sic] with sand rock cliffs" which reminded him of "the strange half-pinnacles of the Petra monuments," while the "sculptured architecture with cornical columns" is "as all those before seen at Petra."

Our group spent a day out at Madain Salih, which lies on the old caravan route between Petra and Mecca. Like Libya's Ghadames in the Sahara, that other venerable caravan crossroads, Madain Salih is a desert gem little frequented by visitors. Located in countries difficult of access, both desert settlements lie in remote hard to reach areas. Travelers who manage to get to places like Ghadames and Madain Salih find especially striking and memorable unspoiled well preserved relics of the distant past. (In a more modest way the same description could apply to me, half again as old as some of my friends but not yet completely decayed—getting there though—by time.) Madain Salih's more than 100 tombs, as at Petra carved out of large rock formations, were mostly built between 100 B.C. and 76 A.D. After the Romans took control of Petra in 106 A.D. the trading routes shifted and both Nabatean settlements began to decline. As a sepulchral city to begin with, Madain Salih was in any case already a ghost town even before the camel caravans stopped passing through, after which the settlement became moribund, a doubly ghostly place.

Our party posed for a group photo in front of Qasr Farid, the largest tomb. Carved out of a massive free-standing rock formation set apart from the other monuments, the monolith's shape and isolated position recalled to me the similar but much larger fortress I'd seen at Sigiriya in Ceylon more than a quarter of a century before, the shape of the set-apart hill on which perched the Italian hill town Civita di Bagnoregio I often visited, the Citadel at Cap Hatien, as well as, never visited but seen in photos, Ayres Rock in Australia. Much later (in February 2019) the massive monolithic hillock Spanish San Felipe Fort in Cartagena, Colombia, reminded me of all those other travel high points. The longer you've lived and the more you've traveled, the more dense becomes the matrix of associations and memories. Supplementing your own reminiscences with worldly tales of highly experienced travelers, such as those on the Saudi Arabia tour, expands your horizons. It's delightful to exchange travel stories with such seasoned globetrotters, comparing notes and trading tips. All widely traveled adventurers have known many twilights far from home and well understand the rigors and pleasures of far journeys to exotic destinations. Those travel-addicted kindred spirits, veterans of the open road, are curious characters interested in seeing the world. Joining with them on tours brings good fellowship when on the road together, and sometimes even good friendships which last beyond trip's end.

A third little-known ancient Nabatean settlement visited even less than Madain Salih lies in the Saudi desert 140 miles northwest of Tabuk, which is about 200 miles northwest of Madain

Salih. Madain Shoaib takes its name from a prophet mentioned in the Koran who became the father-in-law of Moses. After Moses fled from Egypt he went to Madyan, as the settlement is called in the Koran, where he happened to see two girls fetching water from a well. Pleased that Moses had helped them, the girls told their father, who then invited the visitor to live with the family. Staying a total of 10 years, Moses became the quintessential "man who came to dinner" (as well as breakfast and lunch, and snacks and coffee breaks). After Shoaib offered one of his daughters in marriage, she and Moses tied the knot. Whether or not the marriage turned out to be a promised land for Moses only he could say. From photos—not the wedding photos but those depicting the site—it appears that the tomb facades carved into the rock faces in Madain Shoaib are smaller, less elaborate and more weather-worn than the ones I saw at Petra and Madain Salih.

Although the camel caravans of yore stopped passing through Madain Salih centuries ago, the latter-day desert dwellers there earned a living by mining the pockets of travelers. Like Churchill's warning about the rapacious Swat Valley Pathans, Doughty cautioned that the Madain Salih area was said to be "the most thievish station upon the road to Medina," adding that friendly people "bade me [to] have a care." No thievish threats vexed our visit to Madain Salih, but elsewhere in the far reaches of the Saudi Arabian desert we faced another sort of threat, not robbers but assassins. When we traveled in the remote region around Al Jauf in the northwestern part of the country, armed escorts accompanied our group everywhere. Curious as to the purpose of the gunmens' company, I learned that the guards were meant both to watch over and to watch us, protective and detective. They wanted to keep us under surveillance to track where we went and what we did. At the same time, the sentries also served to defend us from terrorist squads which might slip across the border from Jordan to attack a rare and choice target, Americans in the virtually unvisited Al Jauf area. We were said to be the first foreigners to visit the town in decades. We attracted many stares and one sanction by a *mutawa* enforcer on patrol for the Commission for the Promotion of Virtue and the Suppression of Vice. One of the virtue promoters caught out a tall blonde in the vice of letting her offensive long pig-tail braid dangle provocatively outside her *abaya* gown. The suppressor flicked the braid with his whisk-whip, forcing the offender to tuck the tail back inside her gown. I was sorry to see her neatly braided appendage disappear from view, but in Saudi Arabia it was altogether too lewd to remain on display.

Our tour leader for the trip was a Saudi living in New Orleans married to an American. Perhaps he was attracted to his wife by a sensuous and ever visible pig-tail, a porky sort of body part anathema to Muslims. It's just not kosher for Muslims to be exposed to pig tails. Originally from Al Jauf, he invited us for dinner at the spacious ancestral walled compound occupied by his locally prominent family. As a memento of our visit they gave each of us a small carpet woven by local artisans. This souvenir especially delighted me, not because it was free, although I appreciated that, but because the rug fit in with others I'd collected on my travels over the years. I favored those small handicraft items, both because they were practical and also small enough to fit into my modest-size carry-on suitcase. Carpets also conformed to my belief that typical local wares used by residents represented the best sorts of souvenirs. Such items reflected the culture and traditions of the areas where the pieces were made and where I bought them. Like over-sized postage stamps, the carpets I acquired on my trips were colorful rectangles which evoked foreign lands and, as for the stamps, I arranged the rugs in orderly patterns, at times changing them. The mosaic of memories which carpets my living room is always at hand under foot to remind me of my trips. Over many years I acquired small rugs in such places as the Covered Bazaar in Istanbul, Algiers, Tripoli in Libya (not the one I'd visited years earlier in Lebanon), Brasov in Romania, Cuernavaca in Mexico and in Tirana, Albania's capital.

Like other long-closed countries I longed to visit—Libya, Saudi Arabia, China, Russia, lands in Eastern Europe—Albania for years remained off-limits to tourists. The Iron Curtain there was at its most impenetrable. Unlike the highly evanescent Camp Ironwood campfire logs, which ashed into nothing, the Iron Curtain seemed rust-proof, impervious to breach and invulnerable to decay, an enduring barrier behind which Albania remained untouched by the outside world. I wondered what such a hidden protected place was like. Finally there came a time when the curtain lifted, and in October 2005 I traveled to Albania to visit the land long-shrouded from view. I must admit that I was somewhat abashed to tell people that I planned to visit Albania. Everyone knew it was a weird place, one which in that way perhaps conformed to my strange travel obsession—including seeing such oddities as Albania—which some of my friends viewed as rather weird. But Albania's weirdness was for me precisely its attraction. Eccentric countries appeal to eccentric travelers. New York, Paris, London, Rome, the National Parks, beaches, cruises, Graceland, Disneyland—those all made sense as a destination, but a land like Albania? You gotta be kidding! But I wasn't, and so off I went to see the strange land which for a generation had remained a *terra incognita* curtained off from the world.

Albania was indeed somewhat strange, but as an Irish proverb puts it, "What is strange is wonderful." In that sense Albania was a wonderful wonderland, in some ways even odder than Alice's. The quirky country was delightfully odd, a *Brigadoon* of a place displaced in time, asleep for years, an anachronism. It was a land which exemplified Borges' refutation of time, for Albania's long isolation had enabled it to refute the modern era. It was the only country I'd ever visited where I found no one, other than our local tour leader, who spoke English or, for that matter, any of the four other languages I know. I greatly enjoyed my hard-won ability to converse in other tongues, as it enabled me to indulge my curiosity in five languages. In Albania, however, I was out of luck. Edith once told me that I offered a dial-a-language service, but in Albania no matter how I calibrated the dial I couldn't communicate with anyone. Even at the supposedly literate National Library I couldn't find anyone who spoke English or French, Spanish, German or Italian, nor were there any leaflets or brochures in those language, and not even a posted notice, "No English spoken" or "Speaking English forbidden" or "Yankee go home." Because Albanian wasn't offered as an elective in high school or college, I never had an opportunity to study the language back in my student days. Maybe these days "Albanian as a second language" courses are available for travelers to that land. As it was, I remained unprepared to communicate and was stranded in a linguistically lost world, an other-worldly Borges-like limbo at a library filled with thousands of books, a jumbled jungle of learning which was for me a desert of knowledge with works written in an arcane code I couldn't read. The volumes contained profound thoughts from hundreds of *shqiperise* (Albania, as they word it over there) minds, none of which spoke to me. All the great and enduring literature of Albania was lost to me, a truly disconcerting failure in my education. Perhaps one of the books contained the secret of the universe or, even better, some of Snoopy's timeless wit and wisest sayings translated into Albanian. At least I could appreciate the cartoons. "Peanuts" is universal, even if English isn't, and a picture is worth a thousand words, many thousands if they're all in Albanian.

One day during the tour my hopes rose in Gjirokaster, a picturesque hill town in the south. Walking up a steep street paved with cobblestones I happened upon a small antique/knickknack/bric-a-brac shop whose proprietor greeted me with a cheery "*Bonjour*." Anticipating at long last a conversation with a local I dialed in my French but, *hélas.*, those two words had exhausted the man's vocabulary, so for me it proved to be a *mal jour* for communicating with an Albanian. I guess the shop owner had given up after his first day of French lessons, in which case at least he'd avoided the subjunctive.

Not only were foreign languages foreign to Albania, so too was electricity. Very alternating current from time to time sporadically and erratically energized the country, but the electrical system took a lot of time off. Around Africa power outages are common, both the electrical kind and those featuring revolutions which oust from power regimes such as the *coup d'état* which short-circuited the governing clique the day in October 1987 when Edith and I managed to escape from low-voltage Upper Volta to Mali. Albania, however, was the first European country I'd visited where the power functioned based on Heisenberg's Uncertainty Principle.

In Sarande by the coast in southwest Albania I encountered something of a shocking new electrical hazard. We stayed at a hotel with four or five stories—more stories than stars—in a building which in Albania passes for a skyscraper. The skies are low there. Our tour leader warned us that if we opted to take the elevator we might be stranded inside between floors if the power went off. We thus faced the choice of walking up and down the stairs to and from our fourth floor rooms or being trapped for hours in a stalled lift that didn't lift or descend. Elevator risk: that was a new one for me, a hazard I'd never before worried about. Around the world I'd encountered any number of other risks—hygiene, safety, security, officialdom, strikes, unrest, riots, bandits, pick-pockets and other such annoyances and dangers, but elevator risk? That wasn't on my watch-list, at least not until Albania. The electricity eccentricity typified travel in that weird land.

Then there was the odd power episode, electrical and political, we encountered in the provincial town of Korçe, south of Tirana. Walking to a restaurant one evening in the center of town, I noticed that in one area the streetlights were off while nearby they were lit. Puzzled by the discrepancy, I asked the tour leader for an explanation. He informed me that in the last election the dark section had voted against the winning party, which in retaliation frequently turned off the lights in that precinct, not a power shortage but a power struggle problem.

Albania was a collector's item of a place, one of many strange, unusual, odd lands with other-worldly customs, practices and traditions I visited over the years. I greatly enjoyed seeing such off-beat countries. Although nowhere else did I face elevator risk, I did encounter various other risks, inconveniences, curiosities, oddities, amusements and bemusements while on the road. Yemen presented some of those characteristics, the sharp-edged kind which lend adventure travel an edge which, like a Tuareg sword, can be cutting or merely fanciful and decorative. After flying in 1989 to San'a from Ethiopia, which we'd toured on our own, Edith and I proceeded to an interior ministry office to obtain a travel permit for areas outside the capital. *Laisser passer* described the document even if, like *Bon jour* in Albania, those were the only two French words known in Yemen. On the form you list each and every area you intend to visit and provide other details of your planned route. Once the officials approve your itinerary and collect the fee—there's usually a fee, hand-out, dash, baksheesh or other pocket-picking, often straight from the pocket, in Third World travel—you make enough copies of the form, 30 or so for us, to hand out along the way as you cross from one clan land into the next territory. Invisible but guarded boundaries crease the map all over Yemen. Every 20 or 30 miles tribesmen flagged down our car to demand a copy of the permit, but never any money. Surprisingly, it was only a hand-out and not a hold-up. We distributed the *laisser-passer* forms as we followed a circuit out to Marib in the eastern desert, down to Hodeida on the coast, on along the Red Sea and then inland to Ta'izz, north through Ibb and finally back to San'a.

In Yemen each clan, band, tribe, mob, militia, sect was determined to show that they, not the central government back in San'a, controlled the local territory and governed who was allowed to enter their domain. Competing groups divided the land into small fiefdoms, each jealously guarded. Compared to the tribal rivalries in Yemen, the Hatfield-McCoy feud back in the States seemed tame. Yemen's turf wars represented the story of the world writ small. Over the centuries since the earliest days humanimals had partitioned the lonely planet into any number of protected and defended enclaves. The earth's surface was cut up by countless borders, barriers, boundaries, frontiers, lines of demarcation, carve-outs and shut-ins, road blocks, control points, city limits, wards, precincts, counties, states, provinces, parishes, territories, countries. Exclusionary parcels divide the cosmic pebble's terrain into domains occupied and controlled by countless humanimal groupings. Territorial imperatives animated those creatures to shred their little terrestrial home into fragmented parcels, each with its own rules or governance. Such is life on earth, not the best of all possible worlds—things are no doubt better elsewhere—but that's how earthlings have treated their only homeland.

Our stack of travel permits gradually dwindled as we made our way around Yemen. Along the coast beyond Hodeida as we headed south we came across in the countryside a camel herd which included a new-born—womb wealth for the owner. He milked the mother and offered us an ultra-fresh straight-from-the-tap sample of the still warm maternal elixir. As with the toxic glasses at the Tuareg camp near Timbuktu, Edith and I were once again faced with hygiene risk, much more common in the Third World than elevator risk of the sort I'd encountered in Albania. Prudence vied with a probable once-in-a-lifetime opportunity to enjoy the unique experience of sampling camel milk on the spot, sharing it with the infant rightfully entitled to the nourishment. I knew that drinking the fresh milk ran the risk of brucellosis, but sampling the beverage offered a chance Borden's back home couldn't match. We decided to take the risk and go with the flow. We drank the foamy beverage, thick and creamy and almost Dairy Queeny in its rich texture. No undulant fever undulated in our system, so the experiment proved harmless. Next time we visit Yemen, we'll stop at the same drive-in.

We continued through the Tihama, the coastal plain along and near the Red Sea, not red in tooth and claw but red only in name. Blood-thirsty tribesmen, not content with camel milk to assuage their thirst, at times reddened the inland sands with crimson tooth and claw drawn stains. We stopped in Zabid, a sleepy town 10 miles inland, then drove on to Mocha, a coastal settlement which from the fifteenth century on was Arabia's main coffee-exporting center. Although Arabica coffee still grows in the Yemen highlands, the crop and Mocha lost their importance after coffee plantations were later developed in Java and South America.

Leaving the coast, we turned inland to reach Ta'izz. In the crowded marketplace there someone began pelting us with rocks, a reception less friendly than offerings of camel milk or the traditional three cups of tea, appreciated even if served in filthy glasses. Only in three places, far removed from each other by distance and in time, have I been a target of stoning: there in Ta'izz; in a back street in Kandahar, southern Afghanistan, where boys pelted Edith and me with rocks; and as I stood on the small balcony of my hotel room in Santiago watching street demonstrations against the Allende government in Chile. Although only a stone's throw away from rabble-rousers in each of those three places, I managed to escape the hail of rocks without injury. The rebels' aims were more focused than their aim.

Back in San'a we faced a more rocky, sinister and unsettling situation. One day Edith and I visited the small national museum to view the history and archaeology exhibits. Curious about the only other visitor, a Westerner, I struck up a conversation with the man, a middle-aged American who I supposed worked for the U.S.-based Hunt Oil Company, which explored and drilled in the Marib area. But when I inquired about his mission in the country he replied that he had no connection with Hunt. Now even more curious, I asked if he was with one of the international organizations which operate in many underdeveloped places. He again rejected my supposition. I then asked if he was, like us, an American tourist in that out-of-the-way and volatile corner of the world. This, too, he denied. No business in Yemen, no organizational connection, no touring—what was left? Persisting, I now gave up guessing and asked him directly why he was in Yemen.

"I'm a government worker," he said.

"U.S. Aid?" I asked. "Agricultural Department attache?"

"No," he replied. "I'm the American ambassador." By chance we were in good company at the museum.

Ambassador Charles Dunbar was as curious about us, a stray American couple traveling alone in Yemen, as I'd been about him. He didn't encounter very many free-lance independent U.S. adventurers in that unruly land. He seemed rather surprised and even impressed that we were braving the perils in Yemen, a country seething with danger, by visiting the place on our own without the protective benefit and comfort of a tour. He also noted with admiration that we'd just finished a trip through Ethiopia in the same way. Endorsing our adventurous spirit and daring travel, Ambassador Dunbar invited us to lunch at the embassy the next day. At the time, the embassy occupied a splendid old palace built in the distinctive style of traditional Yemeni architecture. By chance we'd managed to gain access to the diplomatic enclave, U.S. territory beyond control of the local chieftains. We didn't need a *laisser-passer* to enter the embassy. The ambassador was the key to our access.

Arriving at the embassy as scheduled, we enjoyed there a pleasant tasty meal—for once no worries about hygiene, we were in America—garnished with a *tour d'horizon* conversation with the ambassador which gave an overview of the political lay of the land in the Middle East. The regional situation was as complicated as Yemen's physical lay of its land, with the country's territory broken into small sections, each dominated by a different clan armed to the *qat*-stained teeth to prevent rival tribes from intruding onto fiercely guarded enclaves. Ambassador Dunbar's comments offered some inside insights into the current state of play in the region, where many finite and militant forms agitated the political scene. Things in the U.S. Middle West were simpler than in the Middle East.

When Edith and I returned to our hotel after lunch the desk clerk informed us that an agent from one of the security services had come by to ask about us and to examine our passports, held at the reception during our stay. It seemed that I wasn't the only curious person in town. The interior minister was apparently quite curious about us following our visit to the embassy. I immediately realized that the authorities kept the embassy under surveillance and that by going to see the ambassador there we'd become suspicious characters, not a comfortable status in a country like Yemen. Would we now be detained, arrested, jailed, deported, fined or face some other sort of sanction or punishment? Possible scenarios, none of them appealing, played through my mind. Not wanting to alarm Edith, I said nothing, and for all I knew she was contemplating the same sort of threats. I consoled myself with the fact that at least we knew the ambassador, who could probably bail us out of trouble. Meanwhile we continued on our visit, taking care to give no further cause for suspicion.

In repressive countries such as Yemen, where authority exerted itself in arbitrary ways, it can be dangerous to come to the attention of the authorities. Dictatorships, strong-man regimes, authoritarian governments, guerrilla-installed rulers, revolutionaries, despots, autocrats, tyrants, thuggish overlords and their underlings, and other such controlling characters who manage to grab power employ secret police, security services, undercover agents, over-armed thugs, gangs, the army, militias, torture squads, enforcers to retain power. Hard-won power gained by hard men is carefully guarded. Countries run by such regimes present a traveler with potentially serious challenges. The suspicion which permeates those lands induces a vaguely sinister tone. One misstep or misspoken word and your troubles begin. You behave in a guarded and—if under surveillance—a regarded way. You're careful where you go, what you do, what you photograph, who you talk to, what you say. I've traveled in all too many of those repressive and unsettling places, ones where you constantly suffer from a certain anxious nervousness until finally you're across the border, leaving the spooky country behind unless you've just entered the next one, which was the case when we flew from Ethiopia to Yemen.

Those two back-to-back spooklands on opposite sides of the Red Sea kept us on edge for a few weeks. It was high risk but hopefully high reward travel in highly uncertain lands. At the time, strongman Mengitsu Haile Mariam, who ruled Ethiopia from 1974 to 1991 with an iron fist as intimidating as the Iron Curtain, controlled the country. Although aware of the main risks, Edith and I nonetheless decided to venture into Ethiopia, hoping that we wouldn't fall foul of the ever-present ever-watching security apparatus. Once its agents decide to sequester your corpus, no habeas corpus is available to free you. You're in for the duration, however long it might be. As we expected, an undercurrent of hidden menace unnerved us as we traveled around the country. But we stuck to our plans and followed the itinerary—at times followed by an agent—that we'd prepared.

By the time we reached Harar in the lowlands off to the east we felt somewhat more comfortable. We'd so far encountered no difficulties in Ethiopia, and if we had been under observation we weren't aware of it. In the old walled city the surveillance, more Keystone Kops than sinister operatives scrutinizing our every move, was quite apparent. The plainclothesman who followed us everywhere might just as well have been wearing full military regalia, so obvious was his presence and surveillance. No secret agent, the man was only too evident and persistent, a character who shadowed us as closely as our own shadows. We gave him a good tour of his city. I considered charging the un-secret agent for my guide services but thought better of it.

We arrived in Harar after flying from Addis Ababa, the capital in the highlands, to Dire Dawa, like Bobo-Dioulasso in Upper Volta on the other side of Africa a town with a name not to be ignored. It would be remiss to travel in Ethiopia and fail to visit a place named Dire Dawa, just as a tourist in the U.S. neglecting to see Dollywood or Graceland would miss a prime attraction. From the Dire Dawa airport we took a car to Harar, 30 miles farther east. A remarkable and heart-warming display welcoming us greeted our arrival. Maybe the warm welcome was because we were perhaps the only foreigners in months to reach that part of Ethiopia. For whatever reason, along the road to Harar stood crowds of people waving to us and shouting out greetings and applauding us as we passed by. This reception certainly topped a few glasses of tea or fresh camel's milk as hospitality offerings. The crowd's enthusiastic reception mitigated my concerns about traveling in a repressive country like Ethiopia. Suddenly it seemed a much more friendly place. The ardent welcome greatly flattered us. Acknowledging the energetic greetings from the throngs lining the road, we waved back. Only later in Harar did we learn that Ethiopia's prime minister had been expected to travel along the same route about the same time as we did. I guess not too many chauffeured cars passed that way.

About half way between Dire Dawa and Harar, beyond where the crowds had gathered to greet us or someone, the driver stopped at a lively roadside open air *qat* market to buy some of the traditional chew, a mildly narcotic leaf locals in Ethiopia, Yemen and parts of East Africa like to masticate. After buying a stalk of the regional version of chewing gum, gum with a punch in a flavor Wrigley will never produce, the driver offered me a sample. I chewed the leaf briefly but it produced no effect on me, other than lodging a small piece between two teeth. Otherwise, nothing. I'd stick with Wrigley chewing gum, a one-time sample of fresh camel's milk, some Tuareg tea, along with a few other substances, mainly Dairy Queen and M&M's, which served to satisfy my tastes for and curiosity about special treats. I never had the slightest interest in stimulants. Tobacco, alcohol, downers, uppers, narcotics, mood enhancers and whatever else humanimals devised to get a kick didn't appeal to me. I needed no mind-altering substances. My travels provided me with more than enough such stimulation. Every trip I took altered my mind in one way or another. Reality proved more interesting than any artificial visions or pipe dreams produced by adventitious substances. I felt no need to augment real-world experiences with induced ones. There was no reason to add to the remarkable and dazzling play of infinite forms any enhanced versions. For me empirical proofs and high spirits were more satisfying than alcoholic ones. Artificial intelligence and virtual reality seemed to me inferior to natural intelligence and actual reality, attributes which endowed humanimals with enough resources and diversions for a lifetime. But our driver seemed to greatly enjoy his *qat*, the thick wad of leaves puffing his cheek into a bulge as he chewed the narcotic. If he went off on a trip before delivering us to Harar our trip would be delayed, which would be quite a catastrophe.

Pleasant random roadside surprises represent one of the most characteristic and delightful kinds of experiences for an independent traveler. The *qat* market in Ethiopia, the camel's milk in Yemen and many other such unplanned and unexpected episodes greatly enhanced my travels. Going it on your own exposes you to serendipitous opportunities, such as meeting an American ambassador, a Guatemalan president, strangers who become friends, a TWA stewardess with whom you share a pseudo-honeymoon, and other such memorable unexpected encounters and happenings. But traveling independently in places like Yemen and Ethiopia also brings special concerns, for authoritarian forces of a capricious nature govern such countries, as our stay in Harar evidenced. Each time we emerged from the hotel our friendly but vigilant plainclothes escort began to tail us. Casual but purposeful, he followed us everywhere. When detained in Córdoba, Argentina, years before I didn't see myself as someone who looked like a threat to civil order. Similarly, I didn't think that either Edith or I looked like a terrorist, agitator, guerrilla, revolutionary, secret agent, spy, trouble maker, agent provocateur. But somehow, in the eyes of the Ethiopian authorities, we seemed suspicious. The security services were taking no chances with us. As two foreigners traveling independently around the country we might well have been cutthroats, even though I wasn't carrying my Tuareg sword with me, or assassins or undercover operatives, agents tasked with organizing strikes and demonstrations or stirring up dissidents aiming to overthrow the government, or saboteurs sent to poison the *qat* plants. In a way it was rather flattering that we aroused so much attention. It was like the unexpected enthusiastic greeting we'd received on the road from Dire Dawa to Harar. It's nice to be noticed—sometimes. In fact we were really only a pair of mild-mannered middle-America travelers interested not in undermining the government but only in minding our own business, which was to visit the country's curiosities. For us, Taking Care of Business in Ethiopia was simply to see the country as best we could in our allotted time before flying on to Yemen.

One evening in Harar we arranged to see the well-known local attraction featuring the hyena man, not a man who looks like a hyena but a humanimal who at dusk attracts the animals by feeding the beasts large chunks of raw meat, a play of finite and rather gruesome forms. Out at the feeding area the impresario illuminated some spotlights to begin the performance. As if on cue, the hyenas appeared and began to devour the slabs of meat, clawing and biting and chewing the flesh. Truly ugly looking creatures (and no, there were no mirrors around), the scavengers grabbed and snatched large portions, ripping chunks from the bloody carcasses, an animalistic thought-provoking spectacle. The strange scene reminded me of the paintings of the aptly named Francis Bacon, whose works depicted human bodies as hunks of raw meat, fleshy blobs which resembled the chunks of animal meat the hyenas devoured. The grotesquely deformed human forms Bacon painted appear to be fresh flesh, the same as the hyena food and, brothers under the skin, made of the same substances as the hyenas themselves.

The grisly analogue between humanimals and animal flesh appears strikingly in Rembrandt's portrayal of "The Anatomy Lesson" at the Rijksmuseum in Amsterdam. With the possible exception of some of Lorenzo Lotto's portraits, no more haunting face in all of painting exists than the one depicted in the right-center of the dissection scene. The man's large white ruff circles his neck like a delicate noose as he gazes, as if dazed, looking out directly at the viewer. His somber expression and fixed stare portray the anatomy lesson observer as greatly sobered by the blood-red arm of the unhealthily pale cadaver. The specimen reclines as an inert body of flesh dissected by Dr. Tulp to present his anatomy lesson. The seven on-lookers observing the procedure all belonged to Amsterdam's Guild of Surgeons, which commissioned the painting. There's nothing like a hands-on dissection to reveal the inner truths of a humanimal. You see all sorts of things in there, each of them part of a system fated eventually to fail. An even more shadowy and spooky inside view—the inner workings beneath the skin—is displayed by an X-ray image which shows the skeletal structure underlying the vulnerable and evanescent fleshy soft superstructure. Observing a dissection is more instructive than reading about one. Learning by looking and doing beats books. The sixteenth century medical renegade Paracelsus expressed his disdain for academic theories and books which substituted for observation and experiments: "not even a dog-killer can learn his trade from books."

The dead body on the dissection slab is believed to have belonged to the corpus once occupied by a man named Adriaen Adriaensz, a 28-year-old arrow maker hung as a criminal on January 31, 1632. Seeing a humanimal's body parts being dismantled or diswomantled dehumanizes the creature. In *Reminiscences* Rabindranath Tagore describes his visit to a medical school in India where in the dissecting room he saw an old woman's body "stretched on the table. This did not disturb me so much. But an amputated leg which was lying on the floor upset me altogether. To view man in this fragmentary way seemed to me so horrid, so absurd, that I could not get rid of the impression of that dark, unmeaning leg for many a day." To disaggregate a humanimal somehow suggests that the being is less than the sum of its parts, an unsettling realization. Watching the hyenas feeding on meat slabs in Harar 357 years after young Adriaen was hanged in Amsterdam to become a cadaver cut into pieces gave me much food for thought. One day before long it might be my remains on a cadaver table or, if not, for sure boxed up like cuts of meat for consumption by hyenas or worms.

Apart from the gorging hyenas, a twilight far from home spectacle I'll never forget, Harar offered some less sanguinary sights. We made the rounds through the squares and elsewhere, followed always by our shadow. The city itself was a featured attraction, as colors, scenes, sounds, people, places, activities presented exotic sensory perceptions, all rather different from my quiet little suburb back home. (I should note that my village did away with the hyena show some years ago, a great loss to the cultural scene there.) In addition to Harar's general ambiance, a major local feature was the house once occupied by French poet Arthur Rimbaud. As usual—except at the hyena spectacle where we were on our own, just the hyena man, the hyenas and us—the agent followed us, by now looking more bored than vigilant, while we visited the dilapidated mansion where Rimbaud had lived a century before. The vintage wooden structure with its elaborate frets and trim offered fanciful decorative touches, the décor reminding me of similar embellishments at the Hotel Oloffson in Port-au-Prince, Haiti I'd seen with Carol years before. As time went by and my travels added ever more sensory impressions, often one experience reminded me of another. The connections which happened to come to mind seemed to form "the great chain of being"—as Arthur O. Lovejoy titled his book—which linked many episodes of my life. But unlike the chain Lovejoy described—which was interconnected with continuity and gradations linking the plenitude of forms into a coherent system—the random incidents I experienced came to me with no plan or structure. No pattern endowed my adventures around the lonely planet with theme or meaning. Those I arbitrarily imposed in an attempt to grasp some sort of minor significance from my brief earthly existence.

After abruptly abandoning poetry, or maybe poetry abandoned him, in 1880 young Rimbaud moved to Harar where he became an arms dealer and a coffee merchant. The 26-year-old Frenchman occupied Jugol Harar, an Indian merchant's mansion, until returning to France in 1891 for treatment of his leg, which had to be amputated—a stray limb of the sort which so upset Tagore and one which the hyenas would no doubt have savored. Although Rimbaud hoped to return to Harar, he died in Marseille in November, age 37. When we visited his Harar house an impoverished family occupied the dilapidated building. There wasn't much to see at the shabby residence, no traces of Rimbaud, not even a few coffee beans or bullets to evidence his long ago professional activities in the now crumbling house. Only his phantom long gone one-time presence haunted the house.

The hues of Rimbaud's colorful local life had long since faded by the time of our visit. Still a young man when he moved there, the prodigy poet had already seen life in Technicolor, a view of the play of infinite forms I gained in earnest only in early middle age when I began my far-flung travels. He was very young when his poetic visions began. Not long after Rimbaud at age 16 wrote his soon famous "*Le Bateau Ivre*" ("The Drunken Boat") the poet produced his sonnet on the colors of each vowel: "*A noir, E blanc, I rouge, U vert, O bleu.*" Just as letters represented colors to Rimbaud, so did the outside world strike me as a wonderfully colorful place, its rich tones and hues presenting endless chromatic visions without the benefit of *qat* or other such mind-altering substances. The lonely planet was a colorful place, even if half the time, as the earth spun in the firmament, a realm of darkness. It's a chiaroscuro kind of place, with both bright colors and tenebrous forces. From what I saw, nature's most beautiful color was the seemingly silky soft yellow-gold-orange hair tuft parts of the King Penguins in Antarctica. The brilliant hue offered some relief to the blankness of the white continent, a desert of ice. Like Rimbaud's vowels, the penguin feathers spelled colors, so in that and many other ways nature had cast a spell over the enchanted but doomed ghost planet.

Rimbaud's time in the house was revived and reviewed when the mansion was restored some years after our visit. In 2000 opened the Arthur Rimbaud Cultural Center, which ranks right up there with the hyena spectacle as a must-see Harar attraction. Rimbaud would no doubt be pleased to join the hyenas as one of the leading tourist spots in the city. I suppose our shadow, by now an old man, still follows foreigners around as they make their rounds through old Harar to see its sights. Maybe he's even been promoted to the lucrative position of hyena man. We'll find out if he's made the grade on our next visit.

From Harar we returned to Dire Dawa, this time with no *qat* stop and no crowds lining the road applauding us. From there we took the early twentieth century single-meter gauge train—no Twentieth Century Limited, only limited—back to Addis Ababa, some 270 miles away in the highlands at an altitude of about 7,500 feet. The day-long trip takes you through a complete change of scenery, from arid countryside up to the much cooler and greener area around the capital. The lower stretch crosses a desolate and largely empty landscape with only a few stops at hamlets along the way. We didn't lack for company, however, as a government agent accompanied us on the entire trip. I couldn't quite understand why he rode with us, other than perhaps to collect some frequent traveler miles. The train ran on a fixed route and for most of the way, until we approached Addis, operated far from any signs of life, apart from the few scattered tiny settlements we passed through. There was no risk that we'd leave the car to venture into the forbidding countryside. We were prisoners of the rails, iron bars which locked us onto the line and barred our way to free movement. There was no escape out into the landscape. A railed road reaches the terminal point on a predetermined route, a confining way to get from beginning to end. I didn't want to end up as a cadaver never having jumped the rails.

~ 8 ~

Of various means of ground transportation, track-trapped trains, trolleys and trams are the least favorable for detours and diversions which might lead to memorable pleasant roadside surprises. A line's rigid metal ridges and wooden ties tie you to a fixed route, predetermined and invariable, which limits freedom and flexibility. Every so often something like murder on the Orient Express enlivens the trip, but most of the time you ride the rails from beginning to end to arrive at the terminal destination on a fixed course. For free-spirited travelers conveyances like planes, ships, cars, camels, donkeys, feet, horses and hearses (even corpses, truly free-spirited, have to be moved) offer more freedom to carry you all over the map. Those modes of movement allow random roamings, while trains remain stuck on their rails. In my early thirties I found myself being carried along on a trolley headed straight on to the end. I could see it all coming—the passing years with their repetitions, routines and rituals like the invariable scenery along the same familiar route the trolley traveled. I sought a change of scene—other lines, different stops and go's. Although not a Brooklyn resident, I was the ultimate trolley-dodger. To find the diversity of experiences I sought I transferred from my regular line leading straight on to the future to more varied itineraries. The new routes took me on unpredictable adventures characterized by erratic random happenings in far-flung places around the world. Some of the adventures included trips by rail. Although confined to fixed lines, those journeys by train through a wide variety of scenery offered me many new perspectives on the play of infinite forms.

I trained myself all around the world. On the railroad from Rangoon to Mandalay in Burma we passed through colorful typical local rural scenes. In Alaska the ride from Anchorage to Denali took us through thick forests, the narrow line cut between vast stands of trees. The *Autoferro* from Quito to Guayaquil descended to the coast, the Dire Dawa to Addis train ascended to the highlands. All around India I traveled by train, often in hard class and occasionally a little softer on ultra-slow creeping "express" runs, more like walks. In India we lacked the time to take the adorable 54-mile-long Darjeeling Himalayan Railway, known as the "Toy Train," but we saw the vintage cars in action as we traveled by road up to Darjeeling. Built about the time Rimbaud lived in Harar (he arrived there in 1880), the narrow-gauge track train is in a way poetry (even if not Rimbaud quality) in motion as the car convoy chugs along to connect Siligari near sea level with Darjeeling some 7000 feet high. The train's antique steam engine and quaint old cars, delightful relics of yesteryear, rattle their way up on tracks which crisscross the road between the two towns. On the faster road route we played tag with the train, crossing its path as the very senior citizen engine huffed and puffed its way on. When we later saw the cute cars and Raj era locomotive stationed at the town terminal in Darjeeling we toyed with taking the Toy Train back down, but our time in the area was too short to spend traveling in Nineteenth Century Limited style. Like the old-fashioned train, time itself advanced with its slow and steady moment-by-moment progression as the even more archaic wingèd chariot drew ever nearer. We could skip the Toy Train but time, ever on the wing, was unavoidable. Time was nineteenth and twentieth and every century limited so we needed to spend it carefully. Investing in a Toy Train trip would limit our available time to visit Darjeeling.

In Darjeeling we treasured our stay at another local anachronism—the old-fashioned nineteenth century Windamere Hotel, one of those legendary British Empire era establishments like Dean's in Peshawar, the Strand in Rangoon, the Taj Mahal in Bombay, the Eastern & Oriental in Penang, the Peninsula in Hong Kong, the Connaught in New Delhi, Raffles in Singapore, the Galle Face in Columbo, all of which I saw and some of which I stayed at. Of those famous and historic vintage hotels the Windamere, redolent of the long gone days of Empire, most evoked the ambiance of a classic colonial era hostelry. A true relic, the Windamere hosts history as a permanent guest. For me the Windamere is the dean or queen or maharajah of hotels, my very favorite anywhere (the grossly rank Sultan Mohammed in Ghazni, Afghanistan ranks as my favorite awful hotel).

Although by no means the fanciest or most famous, the most esteemed, or the best for service, food and amenities, still and all the old Windamere is the most delightful and charming place I ever stayed, apart from my own house, even if the service and (especially) the food were better at the hotel than at home. My seeming hyperbolic opinion I propose based on such intangible delightful factors as the establishment's ambiance, its feel and flavor, its history and folklore, the lingering sense of the old days and ways. The hotel is a time machine which, in a way, years ago operated to stop the wingèd chariot in its tracks. The chariot continued to run on for all of the hotel's guests but not for the Windamere itself, which has been delightfully left behind. The Windamere has character—is a character, hosted characters—and possesses its own distinct personality. The hotel maybe even possesses a sort of soul if that's taken to mean something of an essence, a kind of transcendental quality transcending its visible and tangible traits. Over the more than half a century of travel during which I've stayed at countless hotels the Windamere, along with a few others, remains in my memory with the strongest staying power. The Windamere and the miserable Sultan Mohammed Edith and I stayed at in Ghazni represent the for-better-or-for-worse extremes among all the hundreds of hotels which sheltered me during my twilights and nights far from home. The entire range of infinite forms which came to me over the years I treasure. Whatever the quality of accommodations and food, however the vexations and hassles of travel challenged or discouraged me, I valued everything which comprised my experiment in living on the open road. It was more energizing and life-enhancing than remaining on the fixed-line trolley at home. Even the lowest category no-star hotel brought me a memorable experience of the kind unavailable on the predictable rail-straight life I'd led in my native habitat until my mid-thirties when I transferred to new lines of endeavour.

A long pedigree endows the Windamere with a distinguished and colorful heritage. In the late 1880s the Ada Villa, the original structure, became a boarding house for bachelor tea planters and British Raj officials. In the late 1930s the Tenduf-la family acquired the property which they converted just before World War II into a hotel. Still alive in her nineties at the time of our visit, Mrs. Tenduf-la occasionally observed that the only older fixture on the premises was the vintage red British pillar (letter) box from which mail was regularly—a flexible term in India—collected by the postal service. Some of the hotel's venerable traditions continued on over the years, even after the old practices had become archaic or unnecessary. In 1959 Hope Cooke met at the Windamere her future husband, the crown prince of Sikkim, who she married in 1962 after he'd become the king. In *Time Change: An Autobiography* Cooke noted that at the "warm, sedate, predictable Windamere there remained unneeded old-time water-filled ashtrays used down on the plains where overhead fans blew ashes around the room." Although I was a fan of the hotel, no overhead fans were necessary up in the mountains, but nonetheless the Windamere still retained the special ashtrays. Hope Cooke underwent a "Time Change," but never did the venerable predictable Windamere change much. Stranded in time, the hotel offered a predictability more pleasing than the certainty of a streetcar on its invariable route.

Our accommodations at the hotel seemed a century behind the times. With the delightful yesteryear atmosphere we reverted for a time to the old days. For once we could live as in another era. We occupied a bungalow called a "snuggery." Instead of an overhead fan which scattered tobacco ashes, the ashes of a fire remained, unscattered, to evidence the vanished fuel—not logs but coal—of a fire lit in the cottage to chase away the chill of cool Darjeeling nights. Every evening an attendant came to the snuggery to make it even more snug with a fire, its coal gradually melting away to disappear into smoke as the flames temporarily warmed the room.

After the fire expired we went to high tea—quite high there in Darjeeling in the Himalaya country—served from 4:00 to 6:00. Later, at dinner, turban-topped waiters gave us attentive and at times over-attentive service. They watched our every move, bite, swallow, gulp, burp, waiting to pounce to give attention to our perceived needs. We were treated like royalty, but wished we could abdicate. A library at the hotel contained an eccentric selection of books which matched the Windamere's eccentric personality. Can a hotel be adorable? That's the word I'd use to describe the Windamere, an expression I wouldn't choose to describe any other hotel, nor to designate my own cozy little snuggery at home, a pleasant and livable but not lovable place. My residence is functional, comfortable, convenient, a facility to like but not a place to adore. However, in one way my homestead recalls the hotel, as these days the house shelters a relic whose antiquity evokes that of the Windamere.

Beyond the Windamere's delightfully antiquated premises, picturesque Darjeeling, which sprawls irregularly across hills and hollows, offers another upscale attraction—the Himalayan Mountaineering Institute museum with its displays on Mt. Everest and other expeditions. In a small cemetery nearby reposes Tenzing Norgay, Edmund Hillary's climbing partner for the 1953 conquest of Everest. In his memoir *View from the Summit*, Hillary (who stayed at the Windamere, "a comfortable but elderly relic of the Raji," when in Darjeeling in May 1986 for his colleague's funeral) recalled the "dull crack when Tenzing's skull split open" as his body, reduced to smoke and ashes, was burned on the funeral pyre.

There out at the edge of Darjeeling we gained a view of the nearby countryside where tea estates spread over the steep slopes. Women plucked the tea leaves and tossed them into large wicker baskets strapped to each workers back. The back-breaking work performed by the pluckers to gather the leaves enabled drinkers in distant foreign lands to enjoy cups of the renowned Darjeeling tea. The laborious plucking reminded me of the chopping operation workers at Tiquisate in Guatemala carried out to amputate banana stalks from the trees. It took a lot of work to tame nature and wrest from it resources to serve humanimals. It took a lot of effort to capture, process, package, handle, truck and then ship the tea to the far corners of the earth. Whenever and wherever I sipped that beverage—by a Tuareg campfire in the desert, at a moment-in-time Japanese tea ceremony, elsewhere—I always spared a thought and some appreciation for the women I'd seen laboring so hard in the Himalaya hills to pick the leaves whose ambrosia I so casually consumed.

Thanks to good Fortune, the tea plantings took root in Darjeeling after Robert Fortune, a Scottish plant hunter, stole and smuggled out of China in 1848 tea plants for the British East India Company. Fortune disguised himself as a Chinese merchant and used portable boxes like mini-greenhouses to transfer 20,000 plants and seedlings to the Darjeeling area, the entourage accompanied by a crew of experienced Chinese tea workers. Like all flora and fauna, the plants faced mortality and most of them didn't survive. But eventually the Darjeeling tea industry flourished, and that region's good Fortune became China's misfortune as its monopoly for the crop ended.

The hilly tea estates and pluckers in Darjeeling recalled similar scenes I'd seen years earlier, in January 1974, on the road between Kandy and Nuwara Eliya in the Ceylon highlands. The route passes through splendid scenery which includes some of the most striking and memorable panoramas I ever came across. On the hillsides stretching into the distance were fields of tea plants clinging to the slopes. Just as in Darjeeling, women saddled with wicker baskets bent down to pluck the leaves. A high tea requires lots of low stooping. A chiropractor would enjoy a robust practice in that area. I never saw any men in the fields. The females were more plucky.

On Laden La Road in Darjeeling we visited Nathmulls, the well known tea emporium which sells some of the area's choice pluckings. The sales clerk gave us a few samples which we took back to our snuggery at the Windamere where an attendant brought us a pot of steaming water so we could taste the various vintages in our vintage cottage. We selected a few types which we later purchased at Nathmulls. When I drank the pungent tea back home I fondly recalled my time at the old Windamere, Proustian sort of tea-stimulated memories even without a *madelaine* to accompany my sippings. It was both spooky and enormously pleasing how at random moments stray fragments from my past suddenly happened to surface in my mind to remind me of a particular place, a particular time. In some cases I never quite understood what activated those capricious and unexpected prompts. In any case, however, at random times a specific stimulus like stirring or sipping tea served to stir memories of past experiences. When I happened to hear certain songs, long-dormant memories came back like a song. Refrains reprised years after the original version recalled the same noteworthy notes I'd heard long before in another place, another time, remnants of the past echoing across the years to revive the moments and sites where I'd previously heard the music. Tea, notes, notations, passing momentary impressions, whatever, enlivened some moribund memories, born again only soon to re-die.

I remember many of the vintage Raj era hotels East of Suez like the Windamere. The ones I stayed at over the years each offered a distinctive flavor. My twilights far from home in those colonial establishments recalled the vanished world when the sun set on the British Empire—a dusk more momentous than my passing twilights. No matter how well kept or modernized the old hotels, they continued to represent relics of a bygone era. One true relic, the old Strand Hotel in Rangoon, remained unchanged from its original state at the time I stayed there autumn 1973. Built in 1901, the Strand was stranded in time. Its well-worn furniture, fixtures and floors creaked with age. By now I know the feeling. The public spaces and guest rooms were true period pieces, as am I now. The Strand has since been restored, which is more than I can say: by now I'm beyond any structural or decorative improvements.

Back when I stayed at the Strand the hotel exuded not only more atmosphere than I could absorb but more than I wanted to absorb. The past smothered the present with an almost tangible presence. Our shabby room, the very image of genteel poverty, reeked and creaked with its overlay of dust and a venerable bare floor underfoot which sighed with every step. A ferocious looking large lizard clung to the ceiling as if designed to decorate it. Downstairs at the bar a mosquito garnished the insipid warm beer; no extra charge for the insect. Overhead ceiling fans squeaked slow wobbly revolutions, while the teak floor squeaked protests as guests trod upon it. The lizard eventually checked out, the dead mosquito I removed from its beer/bier bath, the fans whirred on: just another passing slow lazy day at the old Strand in Rangoon.

I stayed at the Strand and traveled through Burma with a Dutch diplomat I'd met by chance on the flight from Bangkok to Rangoon. It was pure luck that we'd randomly been assigned adjacent seats on the plane. B was at the time posted to the Dutch embassy in the Thai capital and I was in the early stage of my five-month-long Southeast Asia and India trip. By an odd coincidence B started to read on the plane the very same book on Asian art I'd not long before consulted at home to prepare for my trip. This *déjà lu* prompted me to strike up a conversation with the stranger, and before long we agreed to travel together in Burma. During our chat it had quickly become obvious that we had similar interests, traveled in the same independent style and, most important, both suffered and benefited from keen curiosity.

It was of course necessary for me to evaluate on the fly this stranger who by chance had been seated next to me on the flight. Although a few facts and some information help you to formulate a quick judgment, in the main it's instinct rather than evidence which determines on-the-spot decisions. Travel sharpens your skills for such speedy responses, as on the road or in the air you often confront situations which demand immediate choices based on only fragmentary information. My judgment in this case proved sound, as B and I became lifelong friends, a rare example of an unexpected travel encounter which happens to result in a continuing and rewarding relationship. Seldom do passing chance meetings during a trip lead to an ongoing friendship. Years later after B retired and settled in The Hague I occasionally visited him there. One day I happened to notice in his library the Asia art book he'd been reading on the Rangoon flight, the very volume which had evidenced to me a commonalty between us. It was strange, decades after I'd met B, to hold that book, an object which now transcended the years to link past with present and which long before had served as an intermediary to introduce B and me.

When I bought a second-hand book I often speculated on the type of people who'd once read the handed-down volume. I wondered who had touched the pages and how their text had touched the reader. Based on the nature of the work, I envisioned who might have originally bought the volume or later read its contents. But such precursors remain unknown, and even a bookplate offers only a name and, based on the design, a slight hint about the reader who once possessed the volume in which he placed his mark (for some reason few she's had plated the books I came across). Previous readers otherwise remain anonymous, as I shall be to those who in the future happen to acquire the books which will one day be dis-aggregated from my library and scattered to the four winds, all gone with the wind. Readers (and their authors) come and go but the books remain, changing hands and shelves, maybe even changing minds. Books provide tangible evidence that once upon a time someone thought and wrote about a subject which somehow interested him or her, even if only to earn some fees or un-majestic royalties. Written works bring to a transient world some continuity and—in contrast to the screen-based but largely unscreened writings in the digital world—some tangibility. Staying power is a comforting thought as the ceaseless whirligig of time and the slow creaky Strand hotel fans and low-land Indian overhead fans, blowing cigarette ashes away, spin away the days. It's comforting as well to happen upon in another person's library a book identical to a volume you also own. This both serves to confirm your own taste—not that it needs confirmation: the fact you're reading this book evidences excellent taste—and also to connect you with another like-minded reader. You feel less removed from the world as you engage in the isolated and intensely personal activity of reading, which involves not social but unsocial media. Books held in common with other readers in a way socialize reading, even though that activity is performed alone. Awareness of others who fancy the same works you value makes the lonely planet a little less lonely.

Unlike the Strand back in my time at the dilapidated facility, most of the other Raj era colonial hotels were well-maintained and upgraded to modern standards but, fortunately, without eliminating the old-time charm. Thankfully and memorably, the Windamere retains its early-day unmodernized persona. Although I stayed in some of the old Empire hotels, I usually preferred less elegant and less expensive accommodations. I never sunk quite so low again down to the no-bargain-basement sub-human level of Ghazni's Sultan Mohammed, but I did favor lower-end places rather than up-scale establishments. I enjoyed small locally-run hotels offering historic, cultural or architectural features which reflected the region's typical way of life. Although such extras as ceiling lizards and mosquito swimming pools in glasses of beer—as at the Strand—seldom featured at places where I stayed or which I saw, other hostelries often offered some special touches.

I always tried to include visits—even if not residence at—hotels with special characteristics. Some of the establishments were like mini-museums, with artful touches or historic backgrounds or other unusual elements. At the old-fashioned Alfonso XIII in Seville the bellboys standing at attention by the entrance saluted me as I entered and left, even though I wasn't a guest there. The *botones* attendants attention made me feel like King Alfonso himself. When in the Alfonso neighborhood I went in and out just to get the salute, a service more respectful than any offered at my rather modest hotel. I enjoyed the salutes, but didn't want to pay for them. Truth in advertising prevails at the truly splendid Hotel Splendido, perched above the picture-perfect jet-setty sea-side village of Portofino in northern Italy. Farther north in Italy at Stresa stands the Grand Hotel Borromeo overlooking Lake Maggiore. The grand old place is one of the famous hotels which appear as a character in a work of fiction. Real hotels in made-up stories make imaginative use of places which actually exist; invented hotels in fiction, lacking a true history, are only impostors.

The Borromeo makes its appearance in Chapter xxiv of Hemingway's *A Farewell to Arms*. Arriving there from Milan, Frederic Henry takes at the Grand-Hôtel des Iles Borromées "a good room" in the "very luxurious" establishment where "I went down the long halls, down the wide stairs, through the rooms to the bar." These days the bar serves a Hemingway Cocktail, a confection in memory of the author whose alter ego once stayed at the grand hotel. The morning after a rainy night Henry looks out the window of his room: "Down below were the gardens, bare now but beautifully regular, the gravel paths, the trees, the stone wall by the lake, and the lake in the sunlight with the mountains behind," all quite civilized and orderly and removed from the disorderly war which raged farther south and which in Stresa Henry left behind. He rents a boat to row across to Isola Bella, an accurately named beautiful island in Lago Maggiore. During the annual August Stresa music festival I've on occasion attended concerts held in the Borromeo palace on the island. Even Elvis Presley's songs and Graceland can't match the civilized refinements of the classical music and incomparable setting on the enchanting island.

Hotels such as the Grand Borromeo in Stresa which feature in fiction help make the story come to life. Graham Greene liked to include a hotel setting in his novels. In *The Comedians* the "Trianon" serves as the fictional version of Haiti's Hotel Oloffson, its "gables and balconies and towers, the fantastic nineteenth-century architecture [typical] of Port-au-Prince." At night the "towers and balconies and wooden fretwork decorations" remind the narrator of a Charles Addams-type *New Yorker* cartoon spook house. For my part, the Oloffson's decorated trim—whenever years later I thought about it—would remind me of the similar gingerbready touches which embellish the Rimbaud house in Harar I saw some 15 years after the Haiti building, and my recollection of the Oloffson in turn cast my memories back to the time when I saw the Haitian hotel with Carol, the well-trimmed and architecturally impressive well-built TWA stewardess I'd by chance met in Quito. Experiences as they were lived often reminded me of earlier images and sensory impressions, just as the random connections and recollections which ricochet through my mind as I record in this reminiscence long-gone adventures recall to me many past happenings. Many memories seem to connect, while others remain isolated and—like the Strand Hotel—stranded in time.

More spooky than the haunted house Oloffson, Greened into the Trianon in the novel, was the City Hotel in Freetown, Sierra Leone, an establishment the novelist renamed the Bedford, which features in the opening lines of Greene's *The Heart of the Matter*. When Edith and I were in Freetown (a misnomer) in early October 1987 I was curious to see the setting the novelist chose to begin his story. No budget buster, the "Bedford" was an oppressive place, unkempt, unmaintained, unsavory, uninhabitable. As the novel opens, "Wilson sat on the balcony of the Bedford Hotel with his bald pink knees thrust against the ironwork." Modesty prompted me to keep my bald pink knees covered as I entered the lobby where lingered and loitered sinister looking locals who stared at me as I arrived. Languid ladies of the night and of the day, ready and willing to work at any time, lounged about, obviously paid rather than paying guests. In December 1967 Greene returned to Freetown some 25 years after his mid-1940s interlude when he collected there the experiences he imagined into the novel. In a brief 1968 reminiscence Greene wrote that the City Hotel "had not altered at all. A white man looked down from the balcony where my character Wilson sat watching Scobie passing in the street below." But in late 1967 the prostitutes weren't as well-placed as in my day, 20 years later, as outside the hotel, Greene noted, "a tart in a scarlet dress danced to attract attention (tarts were not allowed inside)." By 1987 the girls had managed to enter the hotel, all the closer to the Bedford's beds where the action took place.

After my fill of the seamy scene in the lobby I emerged to encounter out front a thuggish character who claimed to be a police spy. Detaining me, the alleged agent started to question me: Why had I visited the City Hotel, what was I doing in Freetown? "Isola Bella" quite accurately described the picture-perfect Lake Maggiore island, but I feared that "Freetown" disguised the reality of the city where suddenly I was being interrogated by a claimed security agent. A cell there, depriving me of my freedom, would no doubt provide very unaccommodating accommodations even below the standards of Ghazni's Sultan Mohammed. Better to stay at the Sultan Mohammed-class City Hotel than in the hellhole where the agent might take me. I intuited that the self-described spy was in fact a shake-down artist, a hoaxer who had indeed shaken me up but who lacked any authority to detain me. Adopting an assertive attitude, I demanded that he show me his credentials. I stated that I'd been in contact with government officials, a perfectly true claim as immigration and customs officers had processed us through on our arrival at the airport. His bluff now revealed, the thug renounced his efforts at extortion and, once again free in Freetown, I continued on my way.

In Freetown we bedded down at a decent hotel which greatly outclassed the City Hotel, not a difficult accomplishment. At our hotel I encountered a bureaucratic procedure I'd never before and never since experienced. When we checked in the desk clerk stamped the hotel's name in our passports, the oddest entry ever put into the document. All sorts of colorful visas and entry and exit stamps filled the pages of my passports, but never a hotel's name. I could pay the hotel with a Visa but apparently I needed a visa-like stamp to check in. The desk clerk explained that visitors exiting Sierra Leone were required to show that they'd stayed at the hotel specified on the visa application. For sure the City Hotel hadn't tempted me to move from the hotel I'd noted on the application. Visa forms often vary inversely with the importance of a country. The less prominent the place, the more detailed the form. The Sierra Leone application included an item which required you to state that you wouldn't seek residence or employment in the country. After only a few minutes in Freetown I knew that I'd honor my commitment not to live or work in Sierra Leone. Never would I apply to manage the City Hotel and to live there as a permanent resident. Somehow Freetown failed to appeal to me as a place where I might settle, but maybe I'm too particular.

For the City Hotel/Bedford one quick visit was enough for me. I didn't desire to join the hotel's frequent guest club which would entitled me to an upgrade to a room with clean dirt. The same was true for most of the hundreds of other places which housed me over the long years. The hotels were for the most part un-notable and unmemorable. In fact, I remember few of them. But some standouts like the delightful Windamere I'd very much enjoy staying at again. I realize, however, that never shall I return to that delightful corner of the world—the world as it once was. My memory of the place, not misty but unfortunately un-refreshable, will have to suffice for a lifetime, for I won't go back—not to the Windamere, not to the City Hotel in Freetown, not to the Sultan Mohammed in Ghazni and not to most of the places I saw over a long lifetime of travel. Almost everywhere I visited, and especially in remote areas, my working assumption was that I'd be there only once. I operated with the belief that my first visit would be the last. I found this a useful attitude, as I could never be sure I'd ever return. My outlook was somewhat oriental in its stoic belief that every experience, happening, impression represented only a fleeting ghostly passing interlude, an *ichigo, ichie* "one time, one meeting" once-in-a-lifetime occasion as for a Japanese tea ceremony. Even accessible cities like London, Paris, Rome, Vienna I might not visit again, so while there I tried to see and experience as much as I could. It happened that over the years I returned many times to those cities and to some other places, but you never know. I took the precaution of trying to extract from any visit, whether the first or repeats, as many of the ambient infinite forms I could collect, leaving undone as little as possible. As you grow older the "one time" or last time attitude represents an ever more realistic and likely possibility. All too many random and uncontrollable forces can at any time intervene to prevent you from returning to any place, no matter how close or accessible. The same applies to people. You may never again see your family and friends. They can suddenly disappear on you, or you from them. Anything can happen to eliminate further contact. The ability or opportunity to spend time with others is finite. The play of infinite forms doesn't include infinite chances for fellowship and relationships. They end. Meaningful relationships are given to us with the understanding that at any time they can cease to exist. Treasured friendships and close family connections remain ever vulnerable to disappearance. Those are the terms of engagement. That's the deal, take it or leave it.

Based on the above considerations, I aimed to create experiences and relationships without delay. Any omissions might never be remedied, and all intentions remained subject to capricious intervening forces. Attempts to tie up loose ends offered no guarantee that they'd be neatly knotted. Especially now in my later years, I've assumed that each interaction with family, friends, places, activities, events might prove to be the very last one. A seemingly endless series of pleasurable experiences eventually finally comes to an end. Even the most charmed life is bewitched. At a certain unknown time you enjoy your favorites for the last time. People, places, plays, paintings, museums, music, books, relationships, Dairy Queen cones, Snoopy's wit and wisdom and all the rest terminate. Future springs will bloom and autumns wither without you. You're anyway just a bystander for the changing seasons, and someday you'll no longer witness them. Nature doesn't care: the cycle continues, even though you're out of the loop. In *Prospects of Provence* James Pope-Hennessy tells of visiting Avignon with a French friend and his aged mother. When they arrived at Villeneuve across the Rhone River from the old town the old lady says, "I should like to stop here, for this is the best view of Avignon of all; and I have reached an age when I realize that I may be doing something for the last time. In all probability I shall never see Avignon again!" Pope-Hennessy adds: "Pausing there since, I have sometimes thought how this is what we should all, at any age, feel about a given place." The Pope may believe in an eternal life beyond the grave but Pope-Hennessy is more realistic. Don't wait for the next life, live in this one. Already as a young man I realized

Section V: The Misty Chords of Memory

that places, people and all life activities and experiences were extremely perishable and would soon pass their "do by" date. In all probability I shall never again see some of my friends or favorite places or enjoy various pleasurable experiences and activities. I am now at the "for the last time" stage and age.

Because I knew that never again would I return to Darjeeling and the dear old Windamere Hotel or to many other places, I always made a special effort wherever I was to see and do as much as possible. While in the Himalaya region in 1999, traveling with a friend from New York, we also visited the remote mountain kingdom of Bhutan. Flying from New Delhi to Paro in Bhutan we passed near Mt. Everest. Just as for my Angel Falls fly-by more than 25 years earlier, the unusually clear conditions afforded outstanding panoramas of the mountain's famous form and its jagged peak jabbing a sharp point into the blue sky. Everest's snowy flanks glistened brightly in the sun as we passed nearby. Because no haze or clouds veiled the view neither my sightings nor my memory of the spectacular panorama are misty. Back on the ground in Bhutan we crossed the mountain range-bound little land from west to east via Thimpu, the capital, then on over the nearly 11,000 foot Pele La Pass and continued on along a narrow curving road clinging to the cliff-side high above the valley. We suddenly arrived at a magnificent viewpoint looking out to one of the world's great vistas. Down the valley nine miles away appeared Trongsa Dzong, there in a magical setting among the mountains. Founded in 1647, the complex includes a monastery, the seat of the district government, and a fortress with a commanding view over the valley. Trongsa Dzong's 23 temples, maze of corridors and courtyards and rooms, hidden corners and higgledy-piggledy erratic architectural accretions comprise the ancestral home of the royal family which reigns over Bhutan from the Dzong, 141 miles east of Thimpu.

From Trongsa we continued east, ever farther into the remote reaches of the mountain kingdom. After another 42 miles we reached Bunthang Valley where by chance our visit coincided with a colorful and spirited local festival. The celebration was for sure no Chamber of Commerce event contrived to attract tourists. We enjoyed an unexpected opportunity to view an authentic folklore event, an example of the chance happenings which endow independent travel with memorable unplanned experiences. One day during our Bunthang stay our driver took me (my friend stayed behind) onward farther east over the nearly 12,000 foot Sheltrang Pass into the valley beyond to visit Ura, believed to be Bhutan's first settlement. Ceramic shingle roofs, cobblestone streets, sheep and yak herds, an absence of shops and signs and other archaic touches lent a medieval feel to the village. In a way it reminded me of Sucre in Bolivia, also a remote picturesque town well out of the mainstream. Both places seemed somewhat other worldly, as if removed from the scrum of the modern world. Ura's aura made it seem distant in place and time, lost in the mountains and in another era. I was seemingly the only foreigner there, and apart from the driver and guide probably no other Bhutanese outsiders were in Ura just then. Mass tourism hadn't yet reached the town, which in fact received almost no tourists at all. Mallory climbed Everest "because it was there." I went to Ura for the same reason. Although my friend was content to remain in Bunthang, itself extremely remote, I wanted to go on to the end to the even more remote village of Ura. Lost in the far reaches of a valley in a kingdom locked away in the Himalayas, the village seemed a Shangri-la kind of place far removed from the turmoils of the outside world. Land-locked Ura was as isolated as sea-locked Easter Island. Beyond the mountains to the northwest lay Tibet, off to the east was Assam. But those areas were inaccessible by road, which ended not far beyond Ura. As at *finisterre* points and capes I'd visited, I felt at the edge of the world there in the isolated corner of far eastern Bhutan. The valley was an unearthly other-worldly off the map cul-de-sac, a lonely outpost of the lonely planet spinning in the lonely emptiness of space.

Ura differed from the rest of Bhutan not only by its isolated location but also because I found no one in the village who spoke English. Elsewhere in the country many people knew the language, as English was taught in the schools. In the far eastern areas of Bhutan the locals used Sharchop. Tongue-tied in that tongue, Albanian to me, I couldn't communicate directly with anyone, but my guide interpreted my conversations with the residents. Not only their language but also fatal circles of the locals differed completely from mine. A vast gulf separated our lives. But at the same time we were all of us humanimals subject to the cupidity of chance and to the ultimate fate of all beings. The only basic difference was that the Uras conducted their experiment in living in Sharchop, with its words for "chance" and "luck," while the terms I used for those fateful factors were in English. In whatever language, chance and luck ruled the day. Every tongue has to have terms for those two words, as without ways to express those controlling forces you're tongue-tied and unable to be a true-story teller.

It's rare to find areas where English isn't spoken. Apart from Albania and a few other remote or isolated corners of the world, my native language has become the common means of communication. Over the more than half century I traveled the globe one of the most pronounced developments has been the spread of English. Were, perchance, an Albanian happen to met someone from Ura they couldn't use English to bridge the linguistic gap. The world desperately needs an Albanian-Sharchop dictionary and interpreter. These days at least an elementary form of English prevails almost everywhere. Many versions of my native tongue are spoken, some comical, a few quite innovative, most comprehensible. While sitting one morning in August 1985 in the lobby near the Intourist travel office at Moscow's Kosmos Hotel, no Windamere but fit for purpose, I overheard every customer of whatever nationality speaking English with the attendant. They'd all appropriated my language. By then a *lingua angla* had become the world's *lingua franca*. Since 1985 foreign English of the kind I overheard at the Kosmos has become even more dominant in the travel cosmos. World travelers from Ura or Tirana who speak only Sharchop or Albanian are at a considerable disadvantage. Having traveled in Albania and to Ura I know the feeling. Everyone can communicate except you. In such a place you're incommunicado.

Although Ura was one of the most remote corners of the earth I visited, I also occasionally found myself in other "off the map" kinds of places, exotic corners of the lonely planet where the whirligig of time had stopped spinning and the locale was behind the times. Just as the Tuareg camel driver in Timbuktu had never heard of aspirin, so in other remote places the locals were unfamiliar with everyday products and conveniences common in developed lands. In October 1987 Edith and I crossed by land from Togo into Benin in West Africa. Rummaging in my suitcase, the border control inspector retrieved a strange (to him) object he closely inspected. Turning the mysterious implement back and forth, the official carefully examined the device. Finally giving up his attempt to discern its function, the functionary asked me what the mystifying thing was for. I demonstrated the shaver's use by switching it on and then handed it back to him to look at more closely. Alarmed by the electrical buzz, as if it was some sort of activated evil spirit, he quickly returned the razor to me. He was obviously a Gillette rather than a Norelco guy. I could well understand why the official might never have heard of Occam's Razor, but I was surprised he'd never seen an electric razor. After some searching questions and searching hands rummaging in our luggage, he finally let us cross the border into Benin. It was a close shave.

In his own way Henry David Thoreau was as provincial as the Benin border official. Just as Thoreau disdained a journey to Zanzibar off the coast of East Africa to count cats, so he turned his nose up at Benin (formerly the kingdom of Dahomey) on the west coast: "A single farmhouse I had not seen before is sometimes as good as the dominions of the King of Dahomey." But what did Thoreau know about felines or house lines in Africa? He never went there. It's easy to sit around Walden Pond making up stuff about places you've never seen. For those of us interested in traveling the world, Thoreau seemed a spoil sport. I bestirred myself to visit both his sylvan Pond habitat and, less convenient to reach, disadvantaged Third World Dahomey along with hundreds of other places scattered around the globe. Ura, Timbuktu.

Easter Island, Swat Valley and many other hard-to-reach places I roused myself to see, while Thoreau was content to remain close to home to count the local cats and look at nearby farm houses. The very finite play of infinite forms the sage saw at home inspired only invented notions of the outside world, not imagined ones based on experience.

Nearly 40 years ago in China a modern-day device more sophisticated than an electric razor aroused great interest in a provincial city we visited. In 1981, not long after China opened to foreign travelers, I joined a tour to the previously long-closed country. Our group comprised, so we were told, the first outsiders in years to visit Zhenjiang, the city where Nobel Prize winner Pearl S. Buck had once lived with her parents. Born in the United States in 1892, Buck was taken to China at age five months by her missionary parents. In Zhenjiang the family occupied a hilltop house overlooking the river; in 1916 Pearl left her family in China to attend college in Virginia. We drove by the house, at the time unmarked and nondescript. The city later developed the property and the adjacent former factory site to establish the Pearl S. Buck Museum and Philanthropy Center which opened in October 2008. Like the Rimbaud Cultural Center installed in the poet's one-time residence in Harar, the Buck memorial—more bang for the Buck—honors a Western author in an exotic foreign land: East meets West.

The arrival of Westerners in Zhenjiang caused a great stir among the Easterners there. By the time our bus reached the center of the city, news of our arrival had spread and hundreds of people had gathered to observe us. Locals crowded around our group to stare at the strange other-worldly creatures who'd suddenly appeared as if from outer space. We looked and dressed funny. There in the inscrutable East we were completely scrutable. Women held babies and little children high so they could see the curious odd-looking foreigners, weird humanimals from somewhere in the outside world far, far away.

Now things became even stranger when the strangers presented to the crowd a wondrous magical trick. What next developed, instant photos, no doubt persuaded the locals that we were wizards. Someone in our group used a Polaroid camera to take at random pictures of a few of the folks in the crowd. What a sensation! Hey, presto, you were no longer just another face in the crowd but now by magic a someone instantly captured on film to have and to hold. It was a miracle, right up there with the Holy Ghost, Norelco razors, Dairy Queen and all that. In a minute or so the picture appeared out of nowhere, just as had our group, just as new-borns did, just like Thoreau's invented travel notions—all from nothing to something. The picture was like a reverse ghostly image, one which starts as vague shadowy streaks and develops into a form, the opposite of how a humanimal's life dissolves from form to ghost. An X-ray film reveals the ghost which lies beneath the skin, the fleshless skeletal reality of the phantom behind the facade of existence. Contrary to an X-ray, the Polaroid film starts from nothing and, a reverse ghost, slowly evolves into a full-fleshed human image. As the Polaroid portraits gradually emerged from the magical photo box, the astonished subjects there in Zhenjiang who saw themselves pictured were at first speechless, then speech-full as they Mandarinized excitedly about the miracle. A near riot broke out as bystanders jostled to push to the front so they could get an instant-gratification photo. Suddenly the residents had been exposed to the wonders of the West. I wondered if I should awe the Zhenjiangers even more by demonstrating to them my electric shaver, but no—one cutting edge invention was enough for the locals. They had enough to ponder with the Polaroid. I didn't want to give them sensory overload.

By his brilliant feat of creativity, Dr. Edwin Land had managed to reach the promised land of instant photography. The inventor's own image appeared on the cover of *Time*, and time was when even in developed lands the Polaroid was truly a sensation. As time went by, however, digital technology superseded the old methods and Land's ingenious invention became a kind of photographic land's-end. Polaroid got Kodaked, as both firms were fatally over-exposed to the new digital developments. The company went bankrupt in 2001 and a successor went bankrupt in 2008. In 2017 a revived version, a mere shadow of the original Polaroid, started up, so proving that resurrection is possible, at least for defunct brands if not for humanimals. One day years after my Zhenjiang visit I was riding with a Bostonian along Route 128, the beltway around the city along which a number of early-day technology companies, many of them now defunct, were located. As we passed a partly demolished large structure, like a half-castle on the Rhine, I asked what the building used to be. The structure being dismantled was none other than the once-upon-a-time world headquarters of Polaroid. Just as *Time* and *Life* and other print media had suffered from the digital revolution, so too did the new technology obsolete Polaroid, its magic no longer enchanting. Digital brought a refutation of *Time* and destroyed *Life*. Remnants of Polaroid's one-time headquarters presented a picture as evocative of time's power to destroy as the other ruins, ancient or recent, I'd seen all around the world.

I'm no stranger to ruins. I need only look in the mirror to see one. No matter how many antiquities I visited, remnants of once-proud and by now long-vanished civilizations, it was hard for me to imagine that the familiar fixtures of my time will, in their turn, one day crumble into ruins. Seeing the abandoned forlorn Polaroid headquarters on Route 128 reminded me that no institution, construction or civilization created by humanimals remains immune to eventual disappearance. No generation believes that its certainties and seemingly permanent surroundings and culture, its entire way of life, will cease to exist. Until digital shuttered them, Polaroid and Eastman Kodak seemed invulnerable monarchs ruling the world of photography. But they were dethroned. What appears to be invincible is in fact highly vincible, ever subject to decay, decline, deterioration, destruction and disappearance. As I saw around the world, once lively cities filled with earthlings, the New York and London and Paris of their time, became phantom ghost towns haunted by their departed residents and lost in the mists of time.

~ 9 ~

Ancient ruins often exert a morose effect on visitors to the tumbled stones of one-time lively but now vanished human habitats. All around the lonely planet I wandered through age-old remnants of defunct civilizations. Societies and cultures once dominant and seemingly enduring no longer existed. Conditions on the lonely planet didn't favor durability. It was sobering to realize that the scattered old stones and time-haunted relics of past eras represented previews of what one day our own ambient world would look like. In James Joyce's *Ulysses* Stephen Dedalus said that "History is a nightmare from which I am trying to awake." For me, however, the past was so interesting it woke my interest in what existed before my time. Nightmares I reserved for truly serious concerns, such as the possibility— admittedly unlikely but always a worry—that Dairy Queen, as had Polaroid, might fade away, or that the World Trade Center in New York might collapse (not possible), or that the immortal Snoopy would one day grow old and go to dog heaven, his biting comments and dogmatic but always welcome wisdom no more to be heard. Compared to those truly alarming possibilities, history was hardly a nightmare. Quite the opposite: the past was a dreamy sort of thing, a misty haze of long-vanished happenings which endowed the present with time deposits later generations could excavate. At least the past was settled, its terrors known; the future, with its unknown and unknowable emptiness, presented more of a nightmare. In its day, Kodak sold lots of film used for photos of photogenic ruins which caught the passing interest of humanimals before they passed on to their next sensory impressions from the never-ending play of infinite forms. The mutating forms never ended, even if Kodak and the observers did.

Among the most evocative excavations I saw which unearthed time's remnants were the remains in a remote and hard-to-reach corner of Libya. Although it's not easy to enter Libya, once you manage to get in it's fairly easy to reach the ruins at the ancient sea-side cities of Sabratha to the west of Tripoli and renowned Leptus Magnus to the east, but to access the isolated age-old settlement of Cyrene requires long travel. First you fly or drive to Benghazi, on the coast some 400 air and 620 road miles east of Tripoli, then continue on through the countryside to Cyrene, slightly inland in a little-visited area far from the mainstream. Few visitors animated the ancient ruins as our group made its way in November 2010 through the decades-decayed city. The eminent Libyan archaeologist Dr. Mohammed Fadel led our tour of Cyrene, a site he'd excavated and worked at for years. I didn't hold it against the distinguished scholar that the execrable Sultan Mohammed Hotel in Ghazni might have been named after him. Over the hills spread remnants of the extensive Greco-Roman city settled in 631 B.C. by Greeks sent there on recommendation of the Oracle of Delphi. But according to Dr. Fadel, the Greeks were rather late arrivals, as traces of a civilization dating back 80,000 years have been found in the area. Maybe an earlier Oracle had already delivered oracular pronouncements to prehistoric beings who sought a soothsayer for guidance. Such sages were early-day versions of financial advisers, wealth managers, know-it-all "experts," social scientists, economists, pundits and other current day oracles who profess to give authoritative guidance. The lonely planet can be confusing. All through the ages humanimals in any era have searched for meaning and looked for answers, a spirited search but one without any ultimate findings.

We started at the imposing Temple of Zeus, where back in the day Cyrenes gathered in their attempts to practice their belief system. The temple's massive fluted columns still standing—or perhaps by archaeologist slight-of-hand again standing—were a refutation of time, which had tumbled the stones. Farther on, a colonnade around the Ptolemaion enclosed a large forum, now empty, where the townspeople once gathered. Other typical classical structures evidenced the once bustling but now moribund metropolis: Roman baths, an amphitheater, mosaic pavements, temples, the agora, a necropolis, statues, all remnants of the ancient once lively and now lifeless city.

The past seemed to perfume ruined Cyrene with the scent of human history, an essence distilled from old time. Or perhaps the aroma stemmed from the wild flowers which brightened the site. Maybe they'd grown there year after year over the centuries since even before the first settlers had arrived. Perhaps the flowers and their descendants had predated and outlasted the people. Some of the relics salvaged from Cyrene are displayed at the small on-site museum, established and curated by Dr. Fadel. He lovingly described some of his favorite pieces, rescued from the past so that evidences of bygone times would not all simply vanish forever.

Something of a similar notion, a desire to preserve a few fragments of a personal past, inspired me to spend some of my now little remaining time to produce this reminiscence. It's not really a memento of my life but more of what might be called a "mento" as I've omitted the me from my account, which—as I've explained—deals not with my life but only one element of it: how one earthling's compelling curiosity encountered the play of infinite forms as they existed on the lonely planet during a particular era. I didn't want my travel experiences of more than half a century and various other adventures over a lifetime to vanish like the smoke of the campfires at Phantom Lake which once burned bright before turning to ashes. Why bother to light the fire if nothing of it remains but ashes? I once came across a child's epitaph which lamented, "What was I begun for if I was so soon done for?" Why live, only to have everything disappear with the person? Just a few selected examples of my memories—fond or foul, pleasant or oppressive, misty or clear—have I recorded in my account. But that's better than nothing. Space restraints and the even more restrictive limitations of time—for me by now extremely limiting and a factor which imposes on my enterprise a great urgency—preclude inclusion of most of my recollections. Everything unrecorded in this reminiscence will all soon vanish. The memories I recalled but omitted haunt my final years like ghostly phantoms risen to bid me farewell. It took me years to create my past, which before long will self-destruct. Like the ruins I saw at Cyrene and in ghost towns around the world I, too, will soon be empty of life.

I remember those forlorn ruins: Cyrene; Hampi in South India; Balkh in north Afghanistan; leavings of vanished empires in the jungles at Angkor Wat in Cambodia and Tikal, Copan and Palenque at the disembodied old worlds in the New World; at Palmyra in the Syrian desert; ancient Holy Land religious sights and wholly secular Greek and Roman sights around the Mediterranean. In Bamian, central Afghanistan, I saw the huge cliff-niche Buddhist statues before the Taliban destroyed them, and in March 1998 visited Cuba to see that relic of a place, an entire society fossilized, before the island (no doubt) modernizes after the Castros' time. Countries like Cuba, Albania, New Guinea (not visited) and remote habitations like Ura in Bhutan and hamlets deep in the Amazon exist in a time warp which leaves each such place behind the times like inhabited temporal ruins. Venerable places with odd, fanciful and fabled names drew me to them: Machu Picchu, Marrakesh, Kom Ombo, the pampas, Braggadocio in Missouri and Cappadocia in Turkey; Byblos and Baalbek, Tegucigalpa, Belmopan, in India Fatehpur Sikri, Ooty, Pondicherry, and elsewhere such lands as Swaziland, Iceland, Finland, Thailand, and voweled places like Lome in Togo and Salala in Oman, Ushuaia at the ends of the earth on the south edge of South America. A mouthful and eye-full of Ceylon sights I saw: Poloannaruwa, Sigiriya, Anuradhapura. At Ganvie in Benin stands a stilt village raised above the water, at Cuchi northwest of Saigon a maze of tunnels form an underground town where Vietnamese sheltered during the war. Prison cells on Robben, Devil's and Gorée islands recall once caged humanimals, gown-shrouded mummy-like sleeping bodies in Indian train stations at night previewed future corpses. A wide spectrum of chromatic sights with colorful names brightened my travels: Capri's Blue Grotto, Iceland's Blue Lagoon, the Blue Danube, the Black Forest, the Black Hills, the Black Hole of Calcutta marker, the White Mountains, Dover's White Cliffs, the Yellow River, the Red Sea, Red Bay in Labrador, Delhi's Red Fort, Orangemen in Belfast, the Orangerie in Paris, on Parris island green Marine recruits, the Green Mountains, and many other hues and views. Concentration camps of a kind Camp Ironwood in Maine little prepared me for like Dachau and Auschwitz darkened my travels. I went to an alphabet of grade A quality rhymers: Albania, Algeria, Arabia, Australia, Austria, Bolivia, Bosnia, Bulgaria, Burma, Cambodia, Canada, Colombia, Cuba, Estonia, India, Indonesia, Latvia, Lithuania, Malaysia, Molvania, Syria, Yugoslavia. Remote picturesque towns I can still picture in my mind: Sucre in Bolivia, Cuenca in Ecuador, Antigua in Guatemala, Congonhas and Ouro Preto in Brazil, Dawson City in the Yukon, Luang Prabang in Laos, Suzdal in Russia. Monuments from the past seemed lonely in their isolated locales: Borobudor and Prambanan in Indonesia, deserted desert temples in Egypt, the Krak des Chevaliers fortress out in the Syrian wilderness, the remote monasteries at Mount Athos in northern Greece, Santa Caterina monetary in the Sinai desert. In Uzbekistan Bukhara, Samarkand and Khiva represented Central Asia exotica; Xian, Souchow, Guilin, Chengdu in China, Tibet (Chinese by possession if not in spirit) symbolized the mysterious East, in Japan Hiroshima and Nagasaki represented cities destroyed—not by time but by man—and revived. Each of those places and so many others I saw played a part in the play of infinite forms which for me comprised the passing views of my life's content.

All of those travel experiences have by now passed into the past—all gone. As I climbed the Apadama Stairway in the ruined city of Persepolis the lines of Marlowe's *Tamberlane the Great* played through my mind: "Is it not passing brave to be a king and ride in triumph through Persepolis?" There you have it: whatever brave new world you happen to experience—new to you even if age-old like Persepolis— you're always and ever just passing through, a bird of passage, a transient, a vagabonder ultimately un-bonded from existence, here today and gone tomorrow with no staying power. You're like time, always on the wing, but unlike time you're finite. Each and every away place you see is a way-station, a momentary stopping point on the way to the end of the road. Even with a sweeping perspective affording me an overview of my entire odd odyssey over my long past, as I near the end it all seems a rather curious journey— "a long strnge trip," as the Grateful Dead said. Should the deads, beyond plots and counter-plots, be more grateful than the alives still able to experience the play of infinite forms? The experiment was a confusion of random disjointed happenings, a jumble of passing episodes here, there, everywhere scattered through time and all over the map. A blur of people, places, scenes, sights, a hazy maze of amazements, a crazy-quilt patchwork of visions revolve through my mind as if seen through a mist-veiled slowly turning kaleidoscope.

My pointillistic text attempts to reproduce the disjointed, impressionistic sensory impressions of the kind I experienced. It all just keeps coming at you, a ceaseless play of forms, always coming at you like you're speeding around Route 128 outside Boston past the ruined Polaroid building, or hurtling down Highway 58 toward Mason City, hurrying as if trying to outrun the wingèd chariot, rushing ahead as described in the opening lines of *All the King's Men*, with the highway "straight for miles, coming at you, with the black line down the center coming at you and at you...coming at you with the whine of the tires." Not that I didn't occasionally stop to smell the flowers, but often they gave me hay fever so I kept going as the forms kept coming at me and at me. You try to slow it down to piece it all together, somehow making some sense from the jigsaw puzzle's puzzling pieces, but they never hold still long enough for you to form the forms into an integral pattern so that they make some sense, as something's always missing from the picture.

It's only by chance that you might be able to tease some meaning out of the random flow of happenings. In 1865 English naturalist Thomas Littleton Powys, the fourth Lord Lilford, climbed a tree in central Spain to examine the nest of a red kite, a bird which embellishes its twiggy home with random scraps. Lilford found there a fragment of a Spanish newspaper which gave him news of Lincoln's assassination. Bird-like tweets spread the news faster these days. My reminiscence consists of a magpie-like kite nest mix of memory scraps chosen from the confusion of random original sightings which were coming at me, coming at me over many years. My monkey mind, as the Chinese describe a failure to focus, was constantly attracted and distracted by the ceaseless play of infinite forms which kept coming to me. By design my eclectic and ever-mutating text zigs and zags in order to suggest how my random earthly experiment lacked a design. Any theme which happens to appear in my account is only arbitrary, imposed by me as a stop-gap to fill the otherwise empty and meaningless nature of my existence. The absence of a pattern is the same for Everyman, many of whom disappear without ever managing to derive some sort of meaning from what happened to them. Those souls just vanish, their lives ended with no rhyme or reason or discernible motif. No narrative order is definitive, nor is any life transcendental or thematic but only transient and episodic with no existential theme.

I always realized I was just passing through. Each place, each visit, each encounter, each experience represented only a fleeting play of forms, like the "one meeting, one time" Japanese tea ceremony never again to recur. Like my very existence on the lonely planet, everything was a once-in-a-lifetime phenomenon, a passing form. On an excursion out to an isolated corner of the Japanese countryside Kazu, in Yukio Mishima's *After the Banquet*, noticed how a cloud cluster hung "glowing along its twisted upper edges, but receding sculptured shadows below." Just as the cloud was suspended in the sky, so the moment was suspended in time: "Thus a moment's light on a late spring afternoon created a strangely delicate yet certain landscape, never to be seen again." So it was for me: a play of infinite forms—and me, a passing form who saw some of the play—never to be seen again.

Section VI
SUBJECTIVE OBJECTIVES

~ 1 ~

Only in passing did I puzzle over my decision to trade comfort for the rigors of the open road. I gave up a cozy enclosed existence at my homey provincial suburban subdivision for open-ended open road adventures which removed me from relative security and threw me out into the world, there to face uncertainties and challenges. While carrying out my new agenda I diversified my life and assuaged my curiosity at the cost of losing my comforts. Whether or not this would turn out to be a good trade remained to be seen. Exchanges—financial and otherwise—meant to improve your position often damage it. But for me, leaving well enough alone didn't conform with my revised agenda. The inevitable discomforts of adventure travel were the price to be paid for engaging with the great wide world beyond my tiny corner of the lonely planet. My fatal circle had determined my original conditions. Subsequent circumstances and choices influenced how my life had developed. Fate, luck and chance would largely govern how my experiment worked out as from my mid-thirties when I changed course to set forth on my adventures based on my new format. Where this would lead I didn't then know; where it did lead I now know and, in small part, relate in this reminiscence. Both the living and the writing could all have been different.

Although I greatly enjoyed my roaming travel adventures in far lands and exotic areas as described in previous Sections, I also wanted to settle for a brief time in a few choice places to connect with daily life there. I'd still just be passing through, but at least I could linger for a time to savor the scene. I opted to focus on the Old World where the fatal circles of my forebearers had originated and which eventually determined my heritage. The much older worlds like Egypt, India and other ancient civilizations and more challenging destinations I reserved for later trips, such as the trio of five-month-long journeys I took in the early 1970s after my year-long stay in Europe. At the outset of my new way of life, in January 1970 I moved to Europe to enroll in the Sorbonne and then in the University of Vienna. I was curious about the French and German methods of education. In addition, being a student in Paris and in Vienna rather than simply a long-term visitor would in a limited way attach me to the local setting. My decision somewhat contradicted my general intention to pursue a hands-on real-life post-graduate education out in the real world rather than classroom lessons. By my thirties I'd already been there and done that, probably to excess. I'd become not so much over-educated as mis-educated, with all too many courses and professorial discourses and too much book learning. Rebalancing that wordy education by examining the outside world rather than taking exams represented my true new agenda. But the mind-expanding activity of free-form travel to far lands I deferred in favor of returning to the classroom in Europe in 1970. Enrolling in foreign universities would at least give me a change of scene and serve to start me off with some travel experiences, an easier entryway into my new way of life than simply roaming about and just passing through without any substantial staying power in places I visited.

I realized that studying at the Sorbonne and then over the summer at the University of Vienna would add yet another overlay of academic-based strata to my already rather too many layers of book-based academic learning. But connecting with a university represented the best way to become temporarily in a small way part of a city. The designation "student" in Paris and then in Vienna made me an official resident and gave me an identity. With the status of student I was part of something, not just a passing sightseer. Yet a third Old World university had served to give me an official status and some standing when in 1957 I studied briefly at Oxford. I experienced each of the three places both as a purposeful resident, even if quite temporary, and as a visitor. In that way I enjoyed the best of both worlds as a student and as an active and avid visitor. Apart from getting to know Paris and Vienna, as facilitated through my university connection, I also came to know well two other European capital cities, London and Rome, thanks to good friends in each place who shared with me their lives and contacts and who frequently hosted me. Their network of relationships widened my circle and introduced me to experiences in London and in Rome unavailable to tourists and casual visitors. So it was that I enjoyed the opportunity to gain inside views of the four great European capital cities—London, Paris, Rome and Vienna.

Living the life of a student in two such civilized and cultured cities as Paris and Vienna wasn't particularly adventurous, and such a privileged academic life postponed my more daring world travels. But I still had world enough and time for those wide-ranging journeys, and later in the 1970s I made up for time spent at schools with my three solo five-month-long adventures to South America, the Middle East and the Far East, trips which gave me experiences far beyond the halls of the academy. For openers, however, I reverted to student status, a rather more cloistered and mild-mannered way of life than the hard knocks of the open road. Later I could follow the advice of Wilfred Grenfell, the ultra-adventurous renowned doctor who in the late nineteenth and early twentieth century braved rough seas to establish and deliver medical services in Labrador and Newfoundland. Dr. Grenfell once advised a young American, "When two courses are open, follow the most venturesome." Of course, university courses in a venerable institution like the Sorbonne didn't represent the most venturesome kind of course but, as described in previous Sections, to my academic activities I later added post-graduate studies out in the real world beyond the ivory tower. I wanted the whole range: worldly First, Second and Third World adventures, Fourth and Fifth ones if available, and also contemplative ruminations about the nations I'd seen to add some mental or sentimental substance to the raw sensory impressions I'd experienced. Descartes said, "I think, therefore I am." My motto was (as I've already noted and now, in the spirit of my time in Paris, repeat), "I am, therefore I think." As a living humanimal for a brief time I wanted to form from the play of infinite forms as complete a picture as possible of the lonely planet as I happened to encounter that strange place embedded in time and space.

Whether in the First World with its amenities and its regularities, or the Second where conditions are somewhat less favorable but still functional, or the Third where you're guaranteed to find dysfunction, discomforts and confusion, I enjoyed it all. You never know what lies ahead on the *rue* or the *Strasse* or the *strada* around the next corner or what the next day will bring. You can only set in motion a general agenda, after which the specific events which occur take place largely by chance. Although First World lands offer somewhat more predictability than do less-developed areas, uncertainties can arise anywhere based not on place but from the capricious nature of existence itself, a truly problematic and perilous status for anything alive. At any time your life could be diminished by setbacks, accident or disease or you could suddenly be not alive. Before embarking on my European experiment I knew only that spending a year on the Continent would in principle further my goal of widening my experiment in living. This hoped for result I based purely on subjective objectives, ones without any general practical or functional purposes but meant only to enable me to live in Europe without roots but with routines. It would be something of a half-way house, with furnishings and surroundings, customs and practices somewhat different from the familiar ones at home but also not completely exotic as they originated in cultures not altogether unfamiliar to me. My heritage was rooted in Europe.

Although my aspirations and intentions were clear, how they happened actually to work out remained subject to many contingencies. But it was this very uncertainty which lent my enterprise some excitement, precisely one of the elements which I sought in order to enliven my existence. I wondered who I'd meet, what experiences I'd find—or maybe somehow they'd find me—and what lessons I'd learn, both in the Sorbonne classrooms and, more alluring, outside in cosmopolitan metropolitan Paris. There was a big blackboard waiting to be inscribed with what in the coming months I'd chalk up to experience. The text which would eventually appear on the slate would be written in indelible letters, never to be erased. No swipe of the eraser could ever wipe out and no revisions could ever amend the one and only version of the story which would be told.

My new way of life was by no means an attempt to "find myself." I wasn't at all lost and certainly not like members of a "Lost Generation," those expatriate creative characters chit-chatting and scribbling away in Left Bank cafes who haunted Paris in the1920s and 1930s. Once asked if he'd ever been lost, American frontiersman Daniel Boone replied, "No, I ain't never been lost, but once I was befuddled for a few days." My subjective objectives well in mind, I wasn't at all befuddled. I suffered no mid-life—or early or late life—crisis. I simply wanted to live life in my own way, which now included a desire to see as wide a variety as possible of the lonely planet's infinite forms. Before the planet became even more lonely without my presence—the loss of any earthling accentuates the loneliness for those left behind—I hoped to see what the place was like and how different parts were unalike. Pursuing novel experiences and real-life contexts rather than reading novels or textbooks represented my goal.

Starting January 1, 1969 I faced the new year with a new life. I'd cut my professional ties the day before, my last day at the office, and embarked into the *terra incognita* of an unstructured future rich with potential and fraught with uncertainty. I went skiing, took a few short domestic trips, then in the summer traveled with a friend to Morocco. After she returned to New York I proceeded on my own to Palma de Mallorca to meet there an American couple, old friends from Pittsburgh then traveling in Spain. As I sat in the hotel lobby awaiting their arrival Juan Carlos, the future king of Spain, passed by. That was my closest brush with royalty, a category of majestic figures or figureheads for which I had at most only a passing interest. My only connection with royalty at home was to listen to the music of Duke Ellington and Nat "King" Cole, read Prince Valiant and the Little King in the Sunday comics, consume the cuisine at Dairy Queen and Burger King, and sip Royal Crown Cola. White Castle offered less regal fare, but even that eatery appealed to me more than European royals.

After Mallorca my friends and I traveled to Barcelona where at the American Express *poste restante* I found a letter from Henry, my good friend in London, requesting me to come there while in Europe to meet his fiancée, Kathrine. Back in that prehistoric era people on the move were incommunicado. You could reach them only by a prearranged time for a (land line) phone call or by mail sent to a designated address along the way, often at American Express offices. While studying in Paris in 1970 I picked up my mail at the ornate and grand American Express office at 11 rue Scribe around the corner from the Opéra. On a recent visit to the city I found that the firm's Paris presence had been reduced almost to the vanishing point, as Amex now occupied a single slot at a counter inside a nearby Western Union office. The Express had been derailed by the internet. The once mighty Paris institution was a mere shadow of its former prominent and dominant place in the travel lives of many Americans who went to Paris. The company's one-time services there now belonged to a pre-digital lost generation from a lost world.

In compliance with Henry's request, I rearranged my itinerary to include London. From Barcelona I continued up the coast toward France, stopping in Cadaqúes on the Costa Brava. The main reason for going there was to try to meet Salvador Dalí, who lived with his Russian wife Gala in the house he'd bought in 1930 in Portlligat just outside Cadaqúes. I walked out to Dalí's house and knocked on the door. No response. I knocked again. No answer. I had some more knocks in stock so I banged again. Presently a rather formidable looking lady opened the door and glared at me. I told her I wanted to meet Dalí. She gave me an unfriendly "who are you?" stare and asked me in French what my business with him was. I couldn't say it was just curiosity, so I told her what a great admirer of the painter I was and that I'd be honored to meet him. She invited me in and after some brief *commerages*—chit-chat, a French word I learned only during my Sorbonne days, still some months in the future—I politely hinted that it would be nice if Gala produced Dalí so he could sign *Salvador Dalí: Visto Por Su Hermana* by his sister Ana María Dalí, the only volume by or about him I could find in the Cadaqúes book store. Dalí's autograph was as eccentric as the painter himself. Rather than writing in the book he wrote on it, painting his name on the cover in thick black strokes with the brush he was just then using. Usually you can't tell a book by its cover, but with the volume Dalí autographed for me the cover clearly evokes the work's contents and evidences the artistry of the person described in the text. Along with meeting Philomé Obin in his Cap Haitian studio a few years later, that was my closest brush with a famous painter.

I finally reached London, where Henry introduced me to his fiancée and asked me to be his best man for his mid-December wedding with Kathrine. This surprise request immediately suggested to me a plan to return to Europe in three months for the marriage ceremony and then stay on in 1970 to realize my ambition of living on the Continent for an extended period. Suddenly my generalized subjective objectives had become crystallized into a specific program. I returned home in September to arrange for my year-long absence there, then flew back to London in mid-December for my part in the marriage ceremony. After celebrating the new year and new decade in London, in January I traveled on to enroll in the Sorbonne. Thus began the "gap year" I'd arranged to fill in some gaps in my life experiences.

By the time I became an official Sorbonner I was somewhat older than the other Sorbonne students, but not yet so senior that a generational gap would prevent me during my gap year from being part of student life. My time at the Sorbonne in 1970 was more a senior than a junior year abroad, but nonetheless I blended in well with my classmates. I had not quite yet reached the age of an older generation lost to the ways the younger students—themselves soon, sooner than they realized, to become oldsters—pursued during their fleeting care-free university days. My seniority in fact served to give me a more seasoned, mature and nuanced perspective on the experience of studying at the Sorbonne than how the junior students experienced their time at the university. How we viewed the experience depended in part on what we brought to it. Behind me stretched a hinterland my classmates lacked, as they hadn't yet lived long enough to develop enough of a background to fully appreciate their brief time at the Sorbonne. My seniority gave me something of an advantage, and although it distanced me from some student activities my age in some ways brought me a richer experience than that accessible to younger classmates. One of my professors there said in a lecture that no one under age 30 could comprehend Proust. I might have been the only student at that lecture who truly understood the point.

Although I well realized it was getting a little late in the day to pose as a student, enrolling in the Sorbonne connected me with the city, enabling me to some degree, even if not a regular academic degree, to integrate myself into a minor but official phase of Paris life. As a Sorbonne student I belonged somewhere in the scheme of things. I was now identified as a student, both by status and with an official identification card which evidenced my part in the Parisian play of infinite forms and, even better, allowed me to get discounts on many purchases, mainly for cultural attractions. By chance, Henry's wedding had given me an opportunity to return just in time to student life in Europe before it was too late for me. Time was on the wing, I was growing older, college students seemed ever younger, and before long my opportunity to fit in at the school would be over. It would have been embarrassing for me to be the age of a distinguished professor with a chair and still sitting in a student's seat. As the new year and the new decade arrived I grasped my chance before any new new years further eroded my allotted time, and so—there in Paris, early 1970—began my new life.

Yet unknown to me at the time was the shape my new life would take. The future lay completely open, desert-like in its emptiness. I had no idea if or how I'd continue on with that new agenda or, if I managed to proceed, how it would all work out. My European university year represented a start, but what came next was still unsure. It's all clear now, but back then nothing could be or was known. Everything lay in the realm of potential, an amorphous prospect lacking specificity. Still ahead were those three grueling globe-girdling five-month-long trips and many other travels to exotic places, some of them discussed in Section V, and then in the late 1970s and through the 1980s my ten books and some two hundred travel articles. As of 1970 my hinterland I knew, and from that land which represented my past at my back I always heard as the days went by time's wingèd chariot drawing near, while looking forward I saw looming in the far distance the desert of vast eternity. How I'd proceed through the dark woods to span the years to reach that eventual wasteland remained hidden in the future.

The year 1970 thus represented for me the great pivot point which served as the transition between my former settled provincial suburban existence, pursuing a rather modest professional career, and my new amateur life out in the world. To me the designation "amateur"—one who follows a particular pursuit, study or science without proficiency or a professional status—described the essence of a curious person. I was an amateur in the best sense of the word. The only other "best" I'd manage to attain was serving as Henry's best man. As for the rest, I could only hope for the best in my new life. I figured that even if the worst should happen to occur—whatever that might be—I could always retreat to my little house and try to revise my already once revised thwarted agenda, then start over with another plan. Failing to take the risk of a major change in 1970 would for sure leave me with regrets. It was better to seize the opportunity then, not later or never, to pursue life-enriching new and different experiences which in the end would leave me with less remorse over might-have-beens. Inaction was sure to produce an unfavorable outcome, while by venturing forth into the world of travel I at least gained the possibility of living in my own way to create something of a charmed life. It was worth a try. I decided to suck it and see.

As a Sorbonne student I was entitled to get inexpensive accommodations at the *Cité Universitaire* out by the Périférique ring road to the south, but that remote inconvenient location was not the *cité* I'd come so far to experience. Living so far away from the center would require commuting back and forth by *métro*. I didn't want to live a mole-like life during my limited days in Paris, tunneling through the city. Living at the *Cité* would exile me to a location far removed from the cultural, intellectual, culinary and historic neighborhoods of the city, the very areas I came to Paris to enjoy. Those legendary center-city quarters were not simply various fourth-rate attractions but represented the heart and soul and mind of Paris. Being stuck out in a fringe area would defeat the very purpose of my enterprise. The *Cité Universitaire* was far less appealing than places like the Ile de la Cité in the Seine, the river whose ever-flowing currents were the lifeblood of the city's vital center. I paid considerably more to live at a pension not far from the Seine and within walking distance of many attractions and distractions and of the Sorbonne. I started at the top in Paris with a comfortable if cramped room on the fourth floor walk-up atop Pension Orfila at 62 rue d'Assas in the upscale VIth *arrondissement*. The nearby Luxembourg Gardens west gate led into the park which I crossed every morning to reach the Sorbonne. The long climb up to my room and the walks through Luxembourg to and from my classes offered commutes much preferable to traveling by subway between the *Cité Universitaire* and the center.

I wondered how the un-French name "Orfila" came to designate my home away from home at the pension. What or who was "Orfila"? I thought the name might refer to some obscure battle, or perhaps to one of Napoleon's mistresses, to a character in one of Balzac's novels, or to a savory regional dish. Did any restaurants in the *quartier* serve a tasty orfila? But none of the above. It seems that Joseph Orfila (1787-1853) was a physician who'd emigrated from Minorca to Paris where he held chairs in legal medicine and then medical chemistry, specializing in toxicology. I didn't know if Dr. Orfila had lived in the building which bore his name. Maybe he just fiddled with his toxic compounds there and lived elsewhere. I hoped no poisonous residues of his toxicology experiments lingered in my room. Residing there could prove to be a breath-taking (other than the climb up) experience for me.

Although Dr. Orfila may have lent only his name and not his presence to Pension Orfila, another famous foreign figure had once lived there. On the facade of the building a plaque noted that August Strindberg had resided at Hotel Orfila from February to July 1896, "a decisive phase of his life." In a similar but far less consequential way, my Paris interlude at Orfila also represented a decisive phase of my life as that period was the first major experiment for my new way of life. Without the slightest shred of evidence, I liked to pretend that the famous Swedish playwright had lived and written in my little garret-like room and that his spirit still lingered there. The play of infinite forms around me might well include phantoms of the characters Strindberg imagined for his plays there in my room. But maybe what I took to be emanations of his spirit were only bits of dust—as was by then Strindberg himself—which wafted around the small space. In any case, I imagined the author's ghostly presence hovering over me as I bent over textbooks to master the French subjunctive, one of the less delightful elements of my subjective objective in studying at the Sorbonne.

While Pension Orfila boasted a connection with the good doctor and with Strindberg, many other defunct characters haunted Paris, a lively ghost town bustling with inhabitants surrounded by places named for departed countrymen. Most of the names I didn't recognize, but the French seemed to fancy living among streets and squares which evoked the dead. Over the months I systematically got around to see most of the city and its network of streets and squares commemorating once prominent but by now long-gone obscure figures. The measure of a great city is its delightful unpredictability. I lived in a good city—my provincial life there suited me—but it wasn't a great one. Back home my daily activities were characterized by repetition and regularity, which for the most part lacked surprises. That predictable way of life offers many advantages, but at a certain point I decided to seek some surprises by restructuring how I lived. I of course hoped they'd be pleasant surprises, but anything could happen. In a kinetic metropolis like Paris all sorts of new and unexpected happenings and connections might develop. Never had I imagined that by chance I'd be living in the same building, maybe even in the very same room, a famous playwright once occupied. Coincidences and curiosities abound in great cities like Paris, London, Vienna, Rome. In *Pantomime: A Journal of Rehearsals* Wallace Fowlie tells of his stay at a Paris pension in the summer of 1928. One day at Montparnasse Cemetery Fowlie saw on the grave of poet Charles Baudelaire an odd head sculpted by an artist named Charnoy. Back at his pension Fowlie asked the owner, Madame Yvet, if she knew anything about Charnoy. She replied that the sculptor had occupied the same room Fowlie was in, and that after executing Baudelaire's tomb statuary the sculptor had fallen ill and was nursed by Madame Yvet until he died. Like my pension, Fowlie's came with a story. Paris is full of stories—fourth stories, where I lived, and quarters rich with *histoires* from history and current daily life

I was happily, conveniently and comfortably, even if crampedly, settled there at Pension Orfila with its excellent location and pleasant ambiance until suddenly, as if a plot twist in a Strindberg drama, the story took a turn for the worse. One evening at dinner just before the Sorbonne spring break the propriatoress announced that the building had been sold for conversion into condos and that all residents had to vacate the premises within 30 days. My spring vacation would indeed be a vacation. All tenancies eventually end, for both Fowlie and Charnoy at Madame Yvet's, for Dr. Orfila, for Strindberg and for me at Pension Orfila, and for each and every resident in every other building in Paris and everywhere else on the lonely planet. Every lease must at the end be released, as no permanent tenancies exist for anyone. I'd hoped to remain at Orfila, not forever but just for a few months longer, but time had run out for me there. I now needed to find new quarters in the *quartier*. I resolved to stay in the same area there in the VIth, a convenient neighborhood I'd gotten to know and where the tradesmen knew me, at least by appearance if not by name. The shopkeepers of course cared less about my name than about my purchasing power. In any case, I was more than just a passing stranger in the VIth. In a small way I belonged.

I knew that my limited time in Paris was a fleeting "one meeting, one time" one-off interlude, a long coveted and never-to-be-repeated once-in-a-lifetime experience. I wanted to extract from it as much as I could, and living there by the Luxembourg Gardens and near the Sorbonne I was in the center of things. For me Paris was less a movable feast, as I knew it wasn't a portable or exportable experience, than a moving feast, not my move to a new pension but moving in two ways: emotionally and also because the days moved on quickly. My tenure in the city seemed to shrink at a rapid pace. Time was on the wing and my Paris days, flying past, would soon be only a misty memory.

Before long I managed to find another place at an even better located pension up the street at the corner of rue d'Assas and rue Vavin directly across the street from the Luxembourg Gardens gate. I could see the entrance into the park from the little balcony of my corner room on the fourth floor of the building. It seemed that I was destined to live in Paris four stories up. Madame Pedron, who ran the pension, was pleasant and efficient and the tenants, many of them foreigners, were more varied than the mostly French Pension Orfila residents. Two meals were included in the rent, a skimpy breakfast consisting of *café au lait* and a croissant (no seconds) and either lunch or dinner. I almost always chose dinner as I didn't want to interrupt the day by returning to eat. Mornings after breakfast I'd walk across the street to enter the Luxembourg Gardens, a park rather than a garden. Sometimes in the afternoon after classes I'd return to study there in the French-style regular formal park irregular French verbs and other eccentricities of the language. I wished the verbs had been as regular as the park. The many maddening colloquial expressions, unreasonable pronunciation and baffling irregularities of English, all so simple and natural for me, I never noticed or had to learn. Studying foreign languages sensitized me to the kinds of problems faced by students trying to learn mine.

Over the months I'd see familiar faces in the Luxembourg Gardens, an oasis which attracted many regulars. The French seemed to want to be left alone so I never talked to many of them. I was anxious to use my hard-won subjunctives, but the locals were indifferent to my struggles to learn their language. From time to time I'd spot Samuel Beckett ambling through the park, always preoccupied and often looking down or gazing straight ahead. Maybe Godot was waiting for him at a cafe on the nearby Boul' Mich'. Because Beckett seemed lost in reverie I suppressed my curiosity and never interrupted his meditations by approaching him for conversation. He, in turn, was kind enough never to intrude on my study of the vexing irregular verbs. What I lost in refraining from introducing myself to Beckett I gained in an enhanced French vocabulary.

Back in my time in Paris you had to pay a few cents to occupy one of the green metal chairs in the park. As the ticket ladies made the rounds to collect the fee I kept tabs on the pesky assessors so I could abandon my chair before they arrived. I felt it was entirely inappropriate for me to have to pay for the privilege of sitting in the Luxembourg to learn French verbs, even if I occasionally enjoyed a Beckett bonus by sighting the author there. One of my proudest life accomplishments is that never did I pay for a Luxembourg chair, an admittedly rather modest feat I mention because I really don't have all that many achievements to brag about.

Although Beckett, deep in thought, seemed lost to the world as he walked through the park at the time he was quite alive. But the ghost of his Nobel Prize predecessor William Faulkner haunted the Luxembourg then, as by 1970 the author had been dead for eight years. Like so many Paris residents, foreign and local alike, those who came before and passing visitors like me who came after, Faulkner was enchanted by the park, a quintessential Parisian space where daily life unfolds as the generations come and go. The Parisians called the parks "lungs," not because they're breathtaking but because the green spaces offer some breathing room in the urban jungle. So taken with the Luxembourg Gardens was Faulkner he set the final scene in *Sanctuary* there. As Temple Drake and her father wander through the park they felt the sad gloom "of autumn, gallant and evanescent and forlorn." Temple notices the stone statues of "the dead tranquil queens in stained marble," then she looks up to the sky in the autumnal "season of rain and death." So ends the novel. When I later returned to Paris in the fall of 1970, after my time there, I'd also see the Luxembourg in a new light, with its forlorn autumnal feel as dead leaves littered the ground.

After studying in the park in the morning I'd proceed on to the Sorbonne to attend lectures, many of them held in the spacious Richelieu Amphitheater. The amphitheater was decorated with rather unattractive murals by Puvis de Chavannes, but what can you expect from an artist with a name like that? It was hard to avoid looking at the distracting murals, which dominated the lecture hall. It's also hard to believe that back in his day de Chavannes was viewed as one of the great artists of his era, along with Manet, Renoir, Degas and Cézanne. Art connoisseur and collector Leo Stein, Gertrude Stein's older brother, referred in a letter (early 1905?), quoted in a chapter on Leo by Irene Gordon in *Four Americans in Paris*, to "the men of '70 whom the Big Four and Puvis de Chavannes are the great men and the inspirers in the main of the vital art of today." It's far too late to retouch the painter's images to refresh his now faded reputation, but never too late for an image makeover. Puvis might well benefit these days from a new name-brand, such as Snoopy de Chavannes. Although I viewed the "great man's" paintings as an unpromising product for export, much to my surprise I later learned that the artist's murals decorate the McKim Building of the Boston Public Library. Monsieur de Chavannes produced the panels—the artist's only murals outside France—in his Neuilly studio in 1895-6. In my opinion, Elvis more than Puvis has benefited American culture. The Boston Library would have done better to mount in the McKim Building a photo montage featuring photos of Presley along with background music of his songs. Such enhancements would have enlivened the Richelieu Amphitheater as well: Elvis at the Sorbonne to match Puvis at the library. Although the Graceland singer was absent from the Sorbonne, another famous American once graced the halls of the ancient university. It was there in the Grand Amphitheater—very similar to the Richelieu and even with its own large unsightly Puvis de Chavannes mural above the podium—where on April 23, 1910, Teddy Roosevelt delivered his "Man in the Arena" remarks, with a soon famous passage which endorses "the doer of deeds ... who is actually in the arena, whose face is marred by dust and sweat and blood" as contrasted "with those cold and timid souls who know neither victory or defeat." It seemed to me passing strange that there in the sheltered rarified Cartesian precincts of the Old World university Roosevelt advocated a vigorous worldly hands-on outdoor encounter with life, while the academics represented bookish minds-on thought experiments.

The very first day of classes I took a place in the front row near the center as I wanted to be as close as possible to the lecturer. I hoped to improve my chances of comprehending the foreign language which was about to envelop me. Unsure if my French was then strong enough to allow me to survive as a Sorbonne student, I was delighted and a little surprised to learn that I understood almost everything. From then on I sat farther back rather than directly below the lecturer, one advantage of that more distant position being that I was somewhat removed from the unsightly murals.

An illustrious cadre of professors taught the courses I'd selected. The eminent scholar Antoine Adam lectured on seventeenth and eighteenth century French literature; Raymond Piccard, a renowned academic, spoke about nineteenth century literary works; Georges Matoré, a celebrated philologist, lectured on and also in the French language. My fourth course, modern French philosophy, was taught by René Poirier, "pear tree." He in fact was somewhat shaped like a pear, with something of a rotund belly atop his spindly legs. Or maybe an overdose of literature lectures induced me to see in him a bodily simile, or is it a metaphor?—I can never keep them straight—which in fact didn't exist. As Professor Poirier continued his course I noticed that he never referred to Albert Camus, an omission I mentioned one day to someone at the Sorbonne. My informant explained that when Poirier had supervised Camus' studies years before at Algiers University the teacher had failed to recognize his student's potential. Perhaps Camus was a late bloomer, but in any case Professor Peartree seemed not to notice the orchard's most fertile mind.

In 1931 Camus had enrolled in the Faculty of Letters at the University in Algiers under the direction of Professor Poirier. After he'd become a philosophy professor years later at the Sorbonne, Poirier had described young Camus (as noted in *Pour Connaître la Pensée de Camus* by Paul Ginestier) as "an exceptionally gifted student." But Poirier seemingly didn't hold that opinion back at the time he was supervising Camus. In *Camus* Patrick McCarthy relates that when Camus consulted the professor about doing a thesis on Indian philosophy "Poirier ridiculed the notion." He "thought little about the thesis ... and Camus was not a good enough philosopher to write anything outstanding." When Camus finally delivered his thesis (on another topic) Poirier "was mildly surprised when he read the completed work. Camus had been conscientious and the thesis was, if not outstanding, then at least competent." My Sorbonne source claimed that Poirier's oversight in failing to spot Camus' potential had later damaged the professor's academic reputation. This seems unlikely, as after Algiers Poirier did become a well-known Sorbonne professor. But his lapse in recognizing Camus' ability seemed to explain why Poirier never referred in his lectures to the renowned Nobel Prize winner. As always, what seems clear in retrospect is never apparent looking forward, so how was Poirier to know that Camus would eventually star? I imagine that even Elvis's mother never imagined that she birthed a future super-star performer. Nor, no doubt, did Puvis's mother.

By the time I reached Paris in January 1970 Camus had been dead for exactly 10 years, but he was still alive as a major cultural figure. He'd been killed at age 46 in January 1960 in an auto accident while speeding back to Paris from his house in Courmarin in Provence. Unlike Edith Piaf—"*Je ne regrette rien*"—I regretted that I was too late to be able to meet or at least to see in person the great philosopher-writer. The closest I came was my Sorbonne seminar at which we spent the entire semester studying Camus' *La Peste* (*The Plague*). Although Camus was gone, his intellectual rival Jean-Paul Sartre was still around and I hoped to find him in one of his Left Bank haunts. If I couldn't meet the late humanist I might at least encounter the still lively and active communist. Apart from the slight disadvantage of being dead, Camus seemed much more vital and contemporary than did Sartre, who clung to his Left Bank left-wing doctrines which by 1970 were getting somewhat stale. The famous "events of 1968," riots and strikes touched off by student demonstrations, had occurred two years before and by 1970 things had quieted down even at the Sorbonne, to which the rebellion had spread after starting on the Nanterre campus in an unfashionable suburb west of Paris.

The contrasts between the two philosophers, both in form and content, were extreme. Sartre was much more bookish and wordy; Camus more worldly and human. In a way Camus represented the imaginative and Sartre the inventive mind-set. Sartre entitled his mini-autobiography *Les Mots* (*Words*), which reflected his literary way of looking at the world. Although he used words to craft his written works, Camus engaged more with the emotional side of life rather than the severely cerebral dimension favored by Sartre. Sartre was something of a theoretician, Camus more a practitioner. Each of the sages "had a quite different experience of reading," observed Paul Viallaneix in his introductory essay in *Youthful Writings* by Camus (translated by Ellen Conroy Kennedy). "The divorce between literature and living that Sartre became aware of very late, Camus on the other hand discovered even as he was learning how to read." My sympathies lay with Camus, and I would have preferred to meet him rather than his rival, but even for a ghost writer it's difficult to arrange an encounter with a phantom so I concentrated on meeting a pre-ghost in the person of Sartre.

Meeting or even just seeing Sartre presented a challenge. I knew that he frequented various places around the St. Germain and Montparnasse neighborhoods where I spent much of my time. But to rely on the possibility of a chance sighting didn't represent a realistic expectation. If I depended entirely on luck I'd most likely end my Paris experience without ever seeing the famous intellectual. It happened that an opportunity arose for me to view Sartre from a distance when I noticed a newspaper item announcing that he was to speak one evening later that week at the Maubert-Mutualité, a cavernous meeting hall located a few *métro* stops from mine at Notre-Dame-des-Champs on Boulevard Raspail. I assumed that Sartre's appearance would be my only chance to see him, an opportunity which would give me only a distant impression without being able to meet him. But seeing Sartre from afar was better than nothing, and by then Camus was nothing so I sought to see Sartre. The title of one of his books, *Being and Nothingness* (that covers all the possibilities), described the situation: to perceive him would bring into being a memorable Paris experience as compared with the nothingness of never even being in his presence. Sartre had it right: there's no middle ground between existing and the nothing which defines the prenatal and postmortem states. You either are or you are not, and most of eternity you're not.

The evening of Sartre's scheduled lecture was rainy, dreary and gloomy. I was tempted to renounce my excursion to Maubert-Mutualité in favor of remaining in my cozy little Pension Orfila garret to study French subjunctives. But I finally persuaded myself that Sartre was worth giving up a few arcane refinements of the French language. They'd be there forever, those maddening verbs, but Sartre wouldn't and neither would I. The wingèd chariot continued to gain on me and time was running out for me in Paris, so this might be my only chance to add Sartre to memories of my stay in the magnificent city, one of humanity's greatest constructs. I emerged into the rain, made my wet way to the *métro*, then proceeded on to the auditorium where acrid smoke from the pungent cigarettes favored by the French hazed the hall. Sartre wasn't present. I hoped that by the time he arrived the thickening haze wouldn't veil him from view. The place was full. Sartre was a big name, a big draw. People wanted to hear the Left Bank leftist's pronouncements on various topics of the day. For sure Sartre was never at a loss for words. He lived for and through words.

There was no sign of Sartre, but still plenty of people patiently and puffingly persisted to await the great man's entrance. Smoke continued to stench the hall. Unlike campfires, from which smoke curled into the open air and disappeared, indoors the fumes accumulated, the nothingness of smoke for once becoming something. The thick haze seemed almost tangible. But like all mists, vapors, clouds, airy nothings and much else the smoke was really no thing at all. Shortly after I arrived someone came out to announce that Sartre was running late. Well, at least he was running, hopefully toward the auditorium. Time moved on but the smoke didn't. It hung heavy in the crowded hall. Now another announcement: Sartre delayed, would arrive in another half hour or so. The "or so" prevailed, as another 30 or so minutes passed and still no Sartre. The audience waited patiently, fuming not with anger but only with cigarettes. By now it was about 10:30. My sunk costs of time, effort and energy factored into my decision whether to stay or to leave. If I left, my entire investment would be written off; if I stayed, Sartre might never appear and I'd lose even more time and inhale even more fumes. I invested another half hour or so before finally deciding to depart. Returning Sartre-less through the rain to my room I resigned myself to the probability that my Paris experiences wouldn't include Sartre. I'd have to content myself with an occasional glimpse of Beckett in the Luxembourg Gardens, with frequent sightings of the Puvis de Chavannes murals, and with irregular French verbs and the subjunctive. But at least I was getting some stimulating irregularity into my life and, even without Sartre, actually being in Paris was better than the nothingness of never living there or of only reading about the city.

The next morning before my first lecture at the Sorbonne I by chance happened to encounter Claudia, a classmate from Sincelejo, Colombia, who observed that I looked a bit tired. I recounted the events of the previous evening, noting that I didn't get to bed until midnight. Greatly to my surprise Claudia then told me that she knew Sartre and would introduce me to him. In the person of Claudia Lady Luck smiled on me as if to make up for the frowns I'd faced in her countenance the night before. Claudia first swore me to secrecy: I had to promise never to reveal where Sartre could be found almost every afternoon. The Paris intellectual community and Left Bank insiders, but not outsiders or tourists, knew what Claudia revealed to me about Sartre's whereabouts. The famous figure's desire to keep his location secret from the general public arose because he didn't want to be interrupted by strangers while working at his regular table at La Coupole, the renowned Montparnasse restaurant where the writer-philosopher sat with his papers—and his comely helper—most afternoons. Claudia told me to meet her at the restaurant the next day at 2:30. Arriving early, I spotted Sartre at his table but waited until Claudia showed up to give me a proper introduction. Next thing I knew, there I was face to face with Sartre and chatting with him. I enjoyed not only a close but also a clear view, as this time no smoke filled the air.

Sartre was different than I'd imagined—or, rather, invented as until I met him I had no actual experience on which to base my impressions of the celebrated writer. My bookish invented impression was based on his rather forbidding turgid and convoluted writings. Compared to Camus, so lucid and direct in thought and style, Sartre's works seemed wordy and often opaque, more philosophical and theoretical than humanistic. In person, however, the brainy philosopher was pleasant and friendly and not at all abstract or obscure. He wasn't airy or foggy but quite down to earth. We spoke in French, my irregular verbs and subjunctives now coming in handy. To Sartre's inquiry as to my nationality, I replied that I came from a country I knew he disliked. Ever the philosopher, he made a distinction between the American government and its policies, which indeed he abhorred, and individual Americans, who he quite liked. He found us open, friendly, informal, unstuffy. I did my best to uphold our reputation. At a later session with Sartre, toward the end of my tenure in Paris, I asked him to sign a copy of *La Mur* (*The Wall*). He wrote: *Á Thomas Weil trés cordialment*, adding his signature and 22 *juin* '70. I really liked that "*trés.*" Sartre's wording expressed that not only was he cordial towards me but very much so. His cordiality in fact extended beyond his inscription, because when I'd asked Sartre the afternoon we'd met if I could occasionally drop by his table at La Coupole for a brief chat he agreed. He told me that he rather enjoyed a little attention from people he knew. He explained that although he was a famous figure, there in his own neighborhood familiarity bred indifference and the locals completely ignored him. Although he fancied some occasional recognition from residents of the *quartier*, most of the time he remained rather invisible to the Montparnasse regulars.

Even when Sartre sought publicity, sometimes he was ignored. He related that not long before, when he was out on the street hawking the extreme left wing *La Cause du Peuple* newspaper, he hoped to get arrested to attract publicity for his extremist "cause." The police never even looked at him. Sartre was disappointed he didn't land in jail and in the headlines. At least he tried, in his way, to promote his ideology, unlike "the kind of French intellectual cleverness that has done so much for literature and so little for people," as Romain Gary wrote in *White Dog*. Although a wordy fellow, Sartre occasionally emerged from his thinker-writer role to operate as an activist involved in "the cause of the people."

I didn't often intrude on Sartre, and whenever we spoke I continued to marvel at the fact that I was engaging in a dialogue with the great man. Little me, a hayseed, a hick from the sticks and not even French ones but from the far off U.S. Midwest, a provincial there at an archetypical Left Bank eatery chatting with the world-famous Jean-Paul Sartre. Young Yankee meets Paris intellectual, the stuff that dreams are made of. He was a big cheese, a *gros légume*—a big vegetable, so say the French—and I was small potatoes, a mere student, the lowest of the low in the egghead pecking order, not even a *petit légume* or a small cheese. I wasn't on the menu at all. I was just a temporarily uprooted stray American, a passing visitor briefly a part of the Paris scene and soon to be gone. Sartre had staying power there in Paris, but I had none. The Sorbonne, Luxembourg Gardens, La Coupole and all the rest would before long fade away into my past, leaving me with memories as hazy and evanescent as the smoke which filled Maubert-Mutualité before gradually wafting away into nothing.

Sartre may have been a man of vision but he suffered from poor eyesight. So wall-eyed was his right eye he disconcertingly looked way off to one side rather than directly at you. Sartre saw you in only half of his visual field but for me half a look from him was better than none. At Sartre's table sat an attractive young woman who read to him various clippings and articles from the press. I imagined that his good-looking helper—she enjoyed better looks than wall-eyed Sartre did—performed other less cerebral and more physical services for him. As permitted by his "arrangement" with his long-time companion, Simone de Beauvoir, Sartre enjoyed many female pleasurings. In fact, de Beauvoir could observe Sartre's interaction with his nubile assistant there at La Coupole, as often Simone sat at a table at the back of the restaurant. I noticed her there a few times but somehow never asked Sartre to introduce me to her. To my Paris regrets, which include arriving there far too late to see Camus and being forced to look far too often at the Puvis de Chavannes murals, I added my failure to seize the opportunity to meet de Beauvoir. That omission I regretted even more years later when I came across her remarkable *A Transatlantic Love Affair: Letters to Nelson Algren*. Seldom have I read a more human and humane account of an intimate personal relationship. I couldn't put the book down—the highest compliment a reader can pay to a writer, unless somehow the binding glue has kept the book stuck to the reader's hand. I fear that all too many would-be readers will say about my reminiscence—I couldn't pick the book up. Their loss, as this weighty tome can at least be used for an exercise workout.

Simone de Beauvoir's revealing letters depict a heart-felt account of her romance with the hard-bitten Chicago tough-guy writer Nelson Algren The intercontinental relationship was a one-sided affair and the letters equally so, as only Simone's to Algren are included in the book. He knew no French so she wrote him in excellent English, with only a few charming and understandable mistakes. I imagine neither was the case for the French as they heard my errors in their language. Her command of the language allowed de Beauvoir to express her emotions, feelings, passions and intimate thoughts in a self-revealing way which compels the reader to sympathize with her. It's a truly touching and moving love story, not one invented by a shut-in novelist but a lived and felt life experience. For me it was both enlightening and heartening to read how such a cerebral, rational, logical French intellectual could at the same time be such a loving, emotional, committed romantic being. With Algren, de Beauvoir's mind took second place to her heart, a moving anatomical re-ranking. Although de Beauvoir writes Algren (October 30, 1947) that "love is not something you can tell," in her letters to him it's precisely a great love that she eloquently in a foreign language describes. She refers to a "honey meeting" with her "Division Street Dostoyevsky"; says she kept hoping "you had preserved some little bit of the old love in some dark, far away corner of your heart."; and in the end she sadly concludes that "it just could not go on."

The Atlantic Ocean and their respective backgrounds formed a true "Division Street" between the two of them, separating the American and the Parisienne. Each of them was stranded in a completely different world incompatible one with the other, two contrasting his and her settings. The cultural and intellectual hinterland which nourished each writer comprised two vastly different formats. Cut off from Chicago or from Paris, Algren or de Beauvoir would become rootless and wither. Her last letter to him (November 1964), long after the transatlantic affair had ended, concludes: "Tell me something about you, dear old beast. Or are you too busy? As ever, with much love, your own Simone." Perhaps too busy, apparently Algren never replied. Distance, disinterest or death eventually separates all couples, even the most devoted ones. All relationships, all stories of love—ghosts in waiting—sooner or later come to an end. But while human connections last they make the lonely planet less lonely.

~ 2 ~

My own localized transatlantic-type Paris relationship spring 1970 was much more brief and developed with considerably less emotional intensity than the de Beauvoir-Algren love affair. I wish it had begun earlier and lasted longer, but in any case a spring fling flung me into a new and pleasurable connection during my quickly passing Paris days. As was the case for my unexpected meeting with Sartre, my romantic adventure occurred by chance. I sometimes studied in my room and at other times, in the tradition of Left Bank intellectual life, in a cafe, and when spring arrived I often read in the Luxembourg Gardens, which after the moribund days of winter bloomed into a garden-like park. Every so often I studied at the Sorbonne library, a setting which made me feel more like a student, even if not more studious. The library offered the additional advantage that there on the "campus"—no greenery, just a bunch of urban buildings, a streetscape rather than a landscape—I could observe a number of attractive *mademoiselle* students. This attraction offered a pleasant diversion as well as a welcome distraction from studying irregular verbs, obscure words and the subjunctive. On one occasion I enjoyed at the library brief relief from my language studies by peeking at the notebook of the cute coed seated next to me to get a clue what she was working on. I thought that information might help me strike up a conversation with her. Although my Peeping Tom glance didn't advance my social life, it otherwise proved very enlightening as her notebook entry gave me a profound lesson in European history. Written in large letters and underlined was the phrase "*roi contre Pape*." To my mind that ultra succinct bumper sticker length formula—"king against Pope"—explained all I needed to know about some 2,000 years of Europe's past. Since those three words so completely encompassed and encapsulated two millennia of developments it was no longer necessary for me to pursue the topic. Suddenly the tangled and confused Old World past had become clear and I knew everything important about Europe's history.

A later visit to the university library proved less educational but more stimulating. Ignoring her notebook, I noticed seated near me the notable face and figure of a striking statuesque student, her dark hair cascading to her waist. This descent was in its way more impressive, even if much shorter, than Angel Falls. At some 1.85 meters the beauty overshadowed me by two or so inches, my five foot ten falling short of her towering stature. Using my best subjunctives to impress her—they now came in handy—I struck up a conversation with Dominique. As things developed we spoke the same language, both to communicate and emotionally. We began to hang out together and before long, *voilà*, I had a Paris girlfriend that spring. Former University of California president Clark Kerr deemed that the school's main problems were "sex for the students, athletics for the alumni, and parking for the faculty." At the Sorbonne teams and parking presented no problems, nor did student sex as in springtime Paris nature took its course without preoccupying (but who knows—perhaps occupying) the university rector.

With Dominique in my life my time in the city now seemed to be depleting even faster than before. It doesn't take an Einstein to figure out that when you've added to the formula of your life a form and figure like Dominique's the days fly by. Time of course wings on at the same pace with or without a Dominique, but the clock and the calendar seemed faster with her around. Our relationship reflected in an abbreviated version the differences faced by de Beauvoir and Algren. There I was, a provincial Midwesterner involved with a *Parisienne*, two people whose language, nationality, hometowns, culture, upbringing and much else differed. We even measured things differently, she in meters and me in feet and inches. But it so happened that—unlike de Beauvoir and Algren—for a time Dominique and I lived in the same city, sharing in common being at the same place at the same time studying at the same university, and as is often the case propinquity trumps differences if you hold some things in common.

Dominique seemed to find me *amusant*, not amusing—although I did my best in her tongue to be entertaining—but diverting, different, a novelty. In the fatal circle of her Paris-centric life I seemed something of an exotic creature. I was more than just another American in Paris, a common bird of passage flying through town like a tourist. Those types were a franc a dozen. Like Dominique, I had a local identity as a Sorbonne student. But unlike Dominique, I was a foreigner from the remote reaches of the U.S. Midwest, out where the cowboys and the *peaux rouges* roamed the prairies, or whatever kinds of creatures lived in the middle of the distant New World beyond St. Germain and the "Boul' Mich' " student quarter. France is a highly centralized country and the Parisians, convinced of their superiority, disdained provincials stranded out in the hinterlands. Residents of the capital—not only the political center of the *hexagon* (as the French sometimes called their country) but also the main financial, cultural, social, intellectual and almost everything else headquarters—referred to the rest of France as "*le desert*," not a desert of vast eternity so much as a wasteland populated by peasants. I hoped my status as an American provincial wouldn't cause Dominique to look down on me, which she did anyway because of her height.

After seeing one another in the French way in public places—the Sorbonne, cafes, walks in Luxembourg, at *spectacles* (performances)—there arrived the truly meaningful moment when Dominique decided to invite me to meet her family. This is a big thing in France. The French like to keep their distance. It didn't have to be an Atlantic Ocean-size distance, but they kept to themselves. Getting access *en famille* represents a rarity, especially for foreigners. A certain barrier exists between French families and the outside world, and also even within families. The 30-something daughter of Paris friends once told me that she continued to use the *vous* rather than the more familiar and familial *tu* form to address her parents. When I observed that foreigners never knew just when it was appropriate to advance from the *vous* to the *tu* she replied that the French didn't know either. It's an invisible line, impossible to perceive but one which exists somewhere along the spectrum between formality and familiarity.

With Dominique's parents the *vous-tu* dilemma wasn't an issue. I immediately realized that they would not be "*tu*-able." As is common between students, Dominique and I had "*tu*-ed" almost immediately after meeting, but she certainly didn't inherit her friendliness from her parents. Extremely reserved, they were very wary of me, both in the typical French stand-offish way and because they no doubt feared that a transatlantic love affair might inspire Dominique to run off with me to live forevermore in a land infested by cowboys and Redskins, by provincial Midwesterners and other such barbarians. Although both Dominique and I realized that this was not a possibility, her parents remained cautious and even sullen. I was obviously an intrusive threat to the family circle, part of Dominique's fatal circle, which I'd suddenly managed to enter, invited there by Dominique but otherwise unwelcome. She continued to bring me into the closed circle at the family's upscale rue du Bac apartment in the fancy VIIth Quarter within walking distance of my modest pension across from the Luxembourg Gardens. Although her parents never warmed up to me, Dominique did. Fortunately the daughter didn't mimic her parents' dispositions and hadn't adopted their predispositions regarding foreigners. She was stand-onish rather than stand-offish, and we enjoyed a cozy relationship. I used to call Dominique *La Grande*, a designation referring both to her stature and to the grand times we enjoyed with each other. I appreciated that she didn't call me *Le Petit*. *La Grande* was in her way more fun than the Grand Tour through Europe I'd taken in 1955 when, as a teenager on my own, I'd visited Paris for the first time. Back then Dominique was still very much *La Petite*. The chances then that we'd ever meet were less than slim but, somehow, in the course of time chance happened to bring us together.

Our different backgrounds seemed to be an attraction for her. Spring 1970 was apparently the time for foreign romances, as a Sorbonne friend of Dominique started to date an Englishman who taught English at the university. The two girlfriends referred to their boyfriends as *les Anglo-Saxons*. I rather liked that description. It was certainly better than *Le Petit*, which had had too many unfavorable connotations. Dominique's label for me somehow sounded more exclusive and dignified than simply "an American in Paris." Lots of American frequented Paris. They were all over the place, a veritable invasion of characters from the New World. "Anglo-Saxons," however, were rare beasts. No one had ever called me that. I realized that for the French the term wasn't so much an ethnic or nationality designation as an omnibus catch-all phrase encompassing attitudes contrary to the usual French political, economic and communal beliefs. To the French "Anglo-Saxon" represented someone who followed the Protestant ethic precepts and other doctrines based on individualism, market-oriented economies, self-reliance, decentralized political power, public-spirited private charity rather than a Paris-centric know-all, do-all state. An Anglo-Saxon was a completely different animal from the home-grown Gallic version of the humanimal. I nonetheless, perhaps mistakenly, took Dominique's label for me as a term of endearment rather than as socio-political shorthand.

I'd never thought of myself in the way Dominique did, but that's how Dominique saw me. Her term for me brought me a new self-view, one which enlightened me and served to give me a different perspective on myself. Such revised perceptions, not only about myself but about anything and everything, were one reason why I wanted to travel. All through the years I'd viewed myself as a plain vanilla provincial Midwestern American guy, as ordinary and as vanilla as a Dairy Queen cone, but with Dominique I'd become an "Anglo-Saxon." I knew what "Anglo" referred to but, in truth, I wasn't exactly sure what a Saxon was—maybe something to do with the *roi contre Pape* stuff. Perhaps the Saxons, one group among all the king's men, served the monarch as he fought against the Pope. In any case, the word sounded more dignified and weighty than being described as a Yankee. Anyway, I was a Cardinal fan but that word seemed to imply that I was on the Pope's team. Could I be both a Saxon and a Papist? Even with the "King against Pope" bumper sticker-size formula, European history was a tangle. Things got confusing if you thought about them too much. For sure, though, was that in Paris that spring I'd gained an entirely new persona. No longer an American in Paris or an Anglo-Yank, I was now an Anglo-Saxon.

Travel enlarged my perspectives, a desired result for my experiment in experiencing the big wide world. Perceiving things from an Anglo-Saxon angle brought me an a-centric and maybe even at times eccentric new perceptions. Dominique liberated me from seeing myself only as an American. A person can be identified in many different ways. It's all a learning experience. Some of the altered perceptions can lead to new behavior. Foreign travel refracts your vision into new alignments, so enhancing your insights, hindsights and foresights. Revised perceptions facilitate your ability in some ways, even if not fundamentally, to transform your life, improving it with previously unseen perspectives and opportunities. Making a life out of the raw material at your disposal, your fatal circle and all the rest, is a do-it-yourself enterprise. I once heard about a foreigner settled in America who announced that the buffet dinner she was hosting was "self-serving." My adventures around the world were both self-serving and self-service. For me, Paris and travel in general served as a workshop where I could assemble some of the elements of my experiment in living.

Sartre was the great advocate of creating your own meaning in life. Since there exists no intrinsic transcendental significance to the phenomenon of human existence, each individual must fabricate in his own workshop the earthling's own meaning. How a humanimal shapes its life always depends on the confines of the creature's fatal circle and then on responses to the random operation of chance and luck. With those considerations in mind, I set out in my mid-thirties to endow my brief experiment in living with as many experiences, good ones I hoped, as possible. During my limited time in Paris I followed the same precept. I didn't confine my activities to the Vth where the Sorbonne was located or the VIth where I lived, as delightful as those *arrondissements* were. Remaining there in the heart of the refined and civilized cultured Left Bank—with its bookstores, galleries, schools, intellectuals—without seeing the rest of Paris would prevent me from getting a complete picture of the city. Transcending my own quarter was the equivalent of my practice of leaving capital cities in every country I visited to get the flavor of provincial areas. Unlike the Parisians, I never referred to the provinces as *le desert*. Not only was I myself a provincial, I enjoyed escaping urban jungles to see provincial "deserts," even if—or perhaps because—such "wastelands" reminded me of my ultimate destination. That reality kept me ever aware of time's wingèd chariot hurrying near. Paris added up to more than just an XI based on the V and the VI, so I tried to make every week count by adding to my experiences in the city as much Paris as could fit into the brief and quickly passing available hours. Just as for life in general, the possibilities on offer greatly outnumbered my constrained capacity to participate in it all. The city was too oceanic, more vast and in its way deep than the Atlantic which separated de Beauvoir and Algren. Endless attractions and distractions lured and tempted me. Countless alluring diversions beckoned—museums, cafes, street scenes, performances, spectacles, galleries, book stores and stalls, flea markets, exhibits, architecture, culture, characters, monuments, cemeteries, theaters, concerts, lectures, classes, seminars, shops, street scenes, oddities, curiosities, *roi contre Pape* history and all the rest. I could never see or experience it all.

In the busy workshop that was Paris, where I tried to fabricate and structure the content of the fleeting days, I did my best to build what would be my experiences and then my memories. Some of my time I devoted to my Sorbonne classes and to *La Grande*. Much of the rest I allocated to sightseeing. I methodically organized and carried out my Paris explorations by following each of the 38 detailed walking itineraries in the 207-page 1968 French edition (the 16th) of the Michelin *Guide Vert*, more complete and informative than the English language version. After finishing all the main walks I then visited all the places listed in the *Autres Curiosités* section as well as the excursions to nearby areas included in the Guide's 26-page *La Banlieue Parisienne* section. By the time I completed everything detailed in the *Guide Michelin* I felt as if I knew the city well—at least its externals. For the inner workings and hidden dynamics of a metropolis like Paris, no guidebooks existed.

Even for tourists and other visitors, guidebooks can only tell you so much. Chance finds and events, little-known corners and unexpected happenings greatly augment travel experiences. The guides can't guide you to those kinds of lucky discoveries enabled by serendipity. Guidebooks back in my day also chronically suffered from a time-lag. The books were always behind the times. By the time the guides saw print the world had moved on. But whether up-to-date on-line digital versions or lagging indicators on paper, guides only reflect the interests, tastes, beliefs, attitudes, values, viewpoints, experiences and opinions of the author(s). In any case, if you follow only what others write about a place those sources lead you around by the nose—a common problem in any circumstance for people who accept pronouncements and opinions as authoritative. Although at times it can be convenient to be told where to go, what to see, what to do (and, in some problematical places, what not to do), the ideal way to travel is for the traveler himself there on the ground to determine—based on preparation, curiosity and a sense of adventure—how to arrange his stay to create the most enriching experience possible. In *Redburn* Herman Melville noted that "This world, my box, is a moving world...it never stands still, and its sands are forever shifting." He cautioned that guidebooks "are the least reliable books in all literature; and nearly all literature, in one sense, is made up of guide-books. Old ones tell us the ways our fathers went, through the thoroughfares and courts of old; but few of these former places can posterity trace, amid avenues of modern erections; to how few is the old guide-book now a clew [sic]! Every age makes its own guide-books, and the old ones are used for waste paper."

Not every guidebook deserves that fate. The Michelin *Guide Vert* for Paris and the Green Guides for other regions and countries served as valuable resources to guide my visits in many areas. I still retain my old companion, the 1968 Michelin Paris Green Guide which nearly half a century ago I used to see the city in detail. Melville's derisive opinion of guidebooks for sure doesn't apply to a masterpiece of the genre, *Molvania: A Land Untouched by Modern Dentistry* by Santo Cilauro, Tom Gleisner and Rob Sitch and published by Jetlag. The book is truly an outstanding and necessary comrade for travelers who happen to find themselves in that benighted dentistry-deprived land. The Molvanians don't fear the dentist's drill—only the militia's drill. For a writer, Molvania is a very appealing place, as books that sell only two or three copies get on the best-seller list. The guide contains much useful information and many valuable suggestions to enable visitors to cope with travel in the eccentric country. Odd facts and curious comments about Molvania pepper the piquant text. It would be quite difficult and probably even impossible to visit that Fourth and maybe even Fifth World country—a land empty of dental fillings but filled with weird incidental but not transcendental experiences— without the guidebook in hand. Like many toothless mouths and cavity-hollowed teeth in the country, the dental school at Molvania University was also empty, but other faculties functioned in a normal way. MU is internationally famous for its Institute for Retarded Study, modeled on Princeton's Institute for Advanced Study. At the Institute in Molvania renowned scholars labor night and day to understand the theories behind such advanced technologies as radio, horseless carriages and cutting edge novelties such as can openers, electric razors and nail clippers.

In the *Molvania* guidebook we learn that because of a power shortage the revolving restaurant atop a skyscraper takes six months to make a complete rotation; that waiters at another restaurant spray insect repellent on the food to improve its taste; that a nearby cafe catering to academics and intellectuals is (unlike the busy Paris Left Bank cafes) always empty; that Molvania's space program boasts the singular accomplishment of being the first country to land a man on Poland; that a visitor to Molvania who asks where to find a no smoking area is told "Austria"; and that Molvania exacts a departure tax of $3000 which, understandably, most visitors deem worth the price. Such odd countries and areas like (fictional) Molvania in fact actually exist. I've been to many of them, thankfully suffering during my visit no dental problems. A place like Albania seems to be quite a normal country compared to Molvania. E.M Forster described such a comic opera kind of place he visited in India 1912-13 as an "amazing little state, which can have no parallel, except in a Gilbert and Sullivan opera." But light opera wouldn't do it for a land like Molvania, much more suitable for a darker and heavier phantom of the opera-type production based on the dramatic story of a befanged Dracula-like character who goes berserk due to his lack of access to modern dentistry.

Melville's comments in *Redburn* didn't apply to Michelin's evergreen Green Guides, which contain so much history and cultural information they're never completely out of date. Just as the obese rotund Michelin Man never seems to age, so do the Guides remain un-retired and serviceable. Although lacking any references to modern dentistry, the Michelin Paris Guide served me well to fill my days (if not my teeth) and guided me to places renowned and obscure. As a curious man I was especially attracted to the *Autres Curiosités*, but even sights not considered *curiosités* interested me. I enjoyed teasing Parisians I encountered by asking them if they'd visited or even heard of some of the more obscure corners I'd discovered. As Sartre knew, guides can only take you so far. Of whatever shade, green or blue or otherwise, no book—whether the Good Book or any other sort of guide—pointed the way to how to live. Each humanimal had to find its own way. Even in Molvania, a true *curiosité*, a visitor was obliged to cope on his own with the lack of dentistry and other local challenges.

The Michelin Guide I used spring 1970 didn't provide my introduction to the attractions of Paris. Many I already knew, as by 1970 my stay in the city represented something of an exercise in nostalgia (a re-visit to Molvania would be an exercise in neuralgia). By then I'd visited Paris twice before, the first time as a teenager traveling on my own in the summer of 1955 and then again in 1957. During my Guide-led walks around the city in 1970 I recalled how 15 and 13 years before, fully a decade and more after the war, there still lingered a kind of old-time pre-war ambiance. Back in the mid-1950s the city seemed somewhat behind the times—not as retarded as a place like Molvania (among other features, France did have dentists), but as if at least a few elements of the old Paris of the 1920s and 1930s still remained. The Lost Generation, by then truly lost, of course no longer existed, and in the 1950s and later members of new generations had found their own Paris. For a brief while I was one of them. In the summer of 1955 Paris was a laid-back and rather low-key place, far more livable than the bustling metropolis of 1970. On my two earlier visits in 1955 and 1957 I stayed at the Hotel Pas de Calais on rue des Saints-Pères not far from Boulevard St. Germain in the heart of the VIth. More than half a century later I stopped in at the former zero star hotel, now remodeled and modernized to three star standards. A desk clerk there told me that in the 1950s Sartre, then apparently with no fixed address, occasionally roomed at the Pas de Calais. Maybe we crossed paths there, but it took me another decade and a half to encounter him at La Coupole, not far from his and my mid-1950s Paris accommodations.

Often I left the Michelin Guide back in my room at the pension and just wandered around, almost always by chance coming across in the great city something new, different, odd, endearing. Street scenes, window displays, cafe life, food shops, building details, passing characters, fleeting vignettes, curiosities, shifting patterns of light and shapes, an endless play of infinite forms enlivened the city. I'd stroll along the Seine, browsing among the bookstalls as I slowly made my way upstream toward the west, the river flowing on behind me and on and on as it continued on its way down to the distant sea. At twilight I'd reach the Pont Alexandre III, my favorite Paris bridge, there lingering as the vintage globes came to life with light. Below on the Seine *bateaux mouches* glided by silently, their lights casting moving reflections on the river until the boat passed by, darkening the restless ever flowing waterway. Everything kept moving, flowing on and on, time on the wing much as Vladimir Nabokov described in *The Real Life of Sebastian Knight* how birds wheeled across the sky and then settled "among the pearl-gray and black frieze of the Arc de Triomphe and when some of them fluttered off again it seemed as if bits of the carved entablature turned into flaky life." I, too, watched those birdy bits flake into life, rising triumphantly above the Arch into the twilight sky over Paris, winging west to chase a setting sun they'd never catch. Then, all gone. As did all too many passing events and forms, the fleeting flying birds—poetry in motion—brought to mind Andrew Marvell's lines:

> But at my back I always hear
> Time's wingèd chariot hurrying near;
> And yonder all before us lie
> Deserts of vast eternity.

I could never get that image out of my mind nor the true-story reality it represented out of my life.

I saw it all once upon a time—the river, the birds, the Sorbonne coeds, the Parisians and all the rest, all ever moving onward in time to the end. Birds, those time-on-the-wingers, had a habit of flying off, off into the sky as they winged their way through the emptiness—on the move as if imitating the wingèd chariot—with a ceaseless motion of wings scything the air. Birds show you how things are—transient fly-away moments ever on the wing. In an essay on Istanbul in *Destinations* Jan Morris described how she saw at Galeta Bridge "clouds of pigeons, suddenly emerging from their roost in the dusty facade at the end of the bridge, sweep across the scene like huge gray raindrops." The rains, the snows and the days of yesterday—all gone, vanished into nothing.

It came to pass that my brief time as a bird of passage in Paris that spring of 1970 came to an end. I knew at the beginning that the end would soon arrive. From my first moments in the city that January I realized that before long my time there would run out. But while it lasted I savored each moment. I was happy to have been part of the city during my brief interlude in Garlandia, "a lively section on the other side of the Seine, opposite of Notre Dame," as the early-day Latin Quarter was called. The name came from John of Garland, who studied and taught in the Quarter in the early thirteenth century. "For a long time the memory of *Garlandia* remained alive in student traditions," noted Astrick L. Gabriel in *Garlandia: Studies in the History of the Mediaeval University*. "Those who left remembered the area with nostalgia and wrote, 'We too were once in *Garlandia*' *(fuimus simul in Garlandia)*." So, too, have I with nostalgia written here about my long-gone Sorbonne days.

Long after my Sorbonne spring I occasionally returned to Paris for a visit. I enjoyed seeing some of my old haunts, now animated with new generations as the people I once knew back in my time were by now only phantom memories, ghostly figures from my past. All the familiar faces were gone. I was a member of a lost generation, one lost in elapsed time. On those return visits I'd of course cross the Luxembourg Gardens, passing the chairs and benches where I used to sit. At the Sorbonne in the Richelieu Amphitheater remained the unsightly Puvis de Chavannes murals, but long gone were the distinguished now extinguished eminent professors who in the lecture hall had professed their knowledge about old French literature. By now their learned lectures would fall on deaf and maybe dead ears as no students from my time remained in the auditorium.

I'd continue on upstairs to the library where one fine spring day I'd met *La Grande*, by now an old or at least older lady somewhere—if anywhere, for perhaps Dominique no longer existed. *La Grande* was probably a *Grande-mère*, her grandchildren sailing their toy boats in the Luxembourg Gardens round pond just as children in the VIth have done for ages. When strolling through the park I'd linger a time, perhaps sitting a brief while in one of the now fee-free metal chairs to *pique-nique* on my favorite local lunch fare, a baguette and a wheel of Camenbert. I'd then continue on past the the little marionette theater where generations of children squealed with delight at the Punch and Judy show, the same I once watched there with children long since grown. Through the familiar west gate I'd pass, past the corner where I lived at Madame Pedron's pension and then on down the street to the former Pension Orfila where I paused to read the Strindberg plaque on the facade. Looking up at the window of my long ago room, I remembered my long gone days there and wondered how much the place was worth now. Long after my day the VIth had become Paris's most expensive *arrondissement*, a chic and probably sheikh-filled upscale quarter with no affordable student housing as in my time.

I'd then continue on along rue Vavin to Boulevard Raspail and past the Rodin statue depicting a brooding Balzac—he probably resented the pigeons who perched and pooped on him—and on up to Boulevard Montparnasse and to La Coupole where I'd momentarily stand by the table where I used to chat with Sartre, now just a ghostly phantom haunting his long-time regular place. One time I walked on to Montparnasse Cemetery to visit the adjacent graves where Sartre and Simone de Beauvoir lie side by side, his "Cause du Peuple" and all his other causes and philosophical becauses now dead issues and her forever oblivious to her once impassioned but doomed transatlantic love affair with Nelson Algren. Now all three were *passé decomposé*, a fatal tense. De Beauvoir once wrote him: "How quickly the deads are forgotten."

Such are some of my memories from Paris where I fleetingly lived a long, long time ago.

~ 3 ~

In this reminiscence I can speak only for myself, not for anyway else. Any generalizations or pronouncements arise from my own experiences, which include the subjective objectives I initiated and managed to accomplish, and aren't meant to be authoritative or conclusive for other earthlings. The Paris experiences I recall relate to nothing beyond my own perceptions and impressions and my reactions to them. Each visitor will experience and perceive the city in his or her own way. I have nothing to add in the way of general comments or descriptions regarding Paris as I can think of nothing new to say about a place so often and extensively written about over many centuries. A reminiscence such as mine represents only one person's singular view. My account is simply a bridge, a kind of drawbridge to draw a subjective picture, which attempts to connect a reader with a few experiences which comprise part of a particular writer's life—its outward-looking component, not the personal side. In my case, only a few pedestrian recollections and observations manage to cross the bridge, as I'm unable to offer any more substantive and over-arching comments. No worthwhile authoritative statements, sweeping pronouncements or general interpretations about Paris or for anything else occur to me. My reminiscence obviously includes some general observations and various conclusions about what I learned from my experiment in living, but those views and reviews represent only my own opinions and reactions, without application to other earthlings, to the strange experience of briefly existing as a humanimal on the lonely planet.

Since time began for humanimals those earth-tethered creatures have imagined all sorts of reasons why the world is the way it is, how it functions, how to deal with it, what life means. I have no idea. Many earthlings believe that some sort of divinity or a master plan controls how things on the lonely planet go, or don't go. On a more individual basis, successful types will often attribute their accomplishments to native ability, a winning personality, high intelligence, hard work, whatever. Characters who've lived something of a charmed life might say that they enjoyed such a favored existence because they were especially savvy or smart and were shrewd enough to make the right decisions. I leave to others those kinds of explanations, which may well be correct. My life evolved in a different way—less by operation of the me element and much more by the effect of random extraneous factors. For my infinitesimally tiny and fleeting part in the play of infinite forms I know only that in my particular case—as illustrated by many of the examples I've included in my account—such factors as luck, chance, fate, destiny, happenstance, circumstances and serendipity represent the main controlling elements which influenced how my life played out. For other mortals it may have been different.

Not long after I finished at the Sorbonne in early summer 1970 I moved on to Vienna to attend summer school at the university there, my third and last European university. The first was Oxford, where I studied briefly 13 years before the Sorbonne. Although foreign travel represented my primary post-graduate education, I also hoped to learn how other nationalities thought and taught. This I accomplished by enrolling at different times in the three universities, each in its way a representative of the mindset and way of life in England, in France and in the Germanic lands. Studying at a foreign university enables you to learn not only academic subjects but also offers insights on the county's national character and the residents' way of thinking, their temperament and how they approached life. By way of example, for the Parisians "*le desert*" referred not to a wasteland of vast eternity but to the benighted, unfortunate French provinces. I found such lessons informative. They taught me some of the subtleties and nuances of how people in each country and region saw their world and also the world. Studying at each of the three European universities, I learned about various practices, attitudes and assumptions foreign to my own way of life back home. Familiar and settled ideas became unsettled as I discovered new ways of thinking and of doing things. Seen by Dominique from her perspective, I was not an American in Paris but an Anglo-Saxon. Oxford offered one set of principles, beliefs, procedures, attitudes, assumptions and views while at Molvania University there no doubt existed quite different ways of dealing with education and the world in general. I can speak from experience about Oxford, but unfortunately I never managed to study at M.U. Although I'd hoped to take a few courses at Molvania, the only faculty operating there was the School of Dentistry where no operations had taken place for more than half a century as no students had enrolled or procedures taught there since the school was established. Molvania was an eternal dental desert. Oxford, however, was still taking students (although not for dentistry) seven centuries on after the university had been founded, and so in 1957 I arrived there to study all too briefly but long enough, so I hoped, to learn something about how the Anglos, if not the Saxons, educated their elite.

As I realized 13 years later when at the Sorbonne, my experiences at the English and then at the French university were quite different. Apart from the fact that in England I didn't meet a local equivalent of *La Grande* Dominique—Great Britain was in that way not so great for me—the contrasts between the two universities were striking. The very setting of each institution differed substantially. A college town, Oxford seemed subordinated to and subsumed within the dominating university. Oxford was essentially the colleges. Dons, domes, deans presided over the city. In Paris the city surrounded and dwarfed the Sorbonne, just one of many educational, intellectual and cultural institutions in the vast metropolis. Oxford's Isis, a modest gently flowing stream, typified the university's dreaming spires and dreamy scholars and the serene green meadows and quiet quadrangles. In Paris the busy boulevards and the brawny (but, still pre-sea, not yet briney) waters of the Seine which crossed the huge urban area were in no way serene or quiet. Unlike the college town's tranquil lanes, the big city's roads overflowed with traffic; contrary to the backwater Isis byway, the Seine was a highway; Paris was a vast urban complex producing all manner of goods and services, while laid-back Oxford functioned as a compact center of thought engaged in manufacturing scholarly works and graduates.

Apart from the cities themselves, a more substantive difference contrasted how the Oxford colleges and the Sorbonne functioned. The teaching techniques at each university varied greatly. The French presented settled doctrines in a didactic way. You were lectured to. I sat passively in the Richelieu Amphitheater, looking at the unsightly Puvis de Chavannes murals, as the professors looked down on the students while telling them what they should know—no if, buts or maybes. The Sorbonne teachers admitted to no known or unknown unknowns—everything was a known known and presented authoritatively by the savants in their lectures. At Oxford the professors professed at lectures, but the presentations seemed more tentative, provisional, and suggestive than the certainties pronounced at the Sorbonne. In Paris the scholars spoke down to you; in England the dons discussed with you at your level subjects in discursive and nuanced ways. The one technique seemed akin to rather abstract invented thought systems, the other based on more worldly empirical imagined notions. With insights such as how higher education functioned in each culture, I learned something about the prevailing mentality and attitudes beyond the topics of the courses I took.

Prior to starting at Oxford I spent some time in London visiting friends there. Even as early as 1957, when I was still a very young man, I'd already managed to establish the beginnings of a network of local contacts. These connections originated in part thanks to Henry and Kathrine, for whose marriage in December 1969 I'd served as best man. This occasion in turn had served as the catalyst which offered me an opportunity to remain in Europe in 1970 once I was in London for the wedding. Some of my London circle came to me other than through Henry and Kathrine. Neil and Yvonne Salmon, of a generation senior to mine, I'd met through my father. In the late 1940s he'd hosted the Salmons when they stopped in our city during their long trip across the U.S. My father paid for the Salmon's hotel and other expenses, not because the visitors were short of money—anything but—but because at the time they were short of dollars as currency controls prevented the British from traveling abroad with more than very limited amounts of foreign funds. As they made their way across the country a network of contacts subsidized Neil and Yvonne's trip. The Salmons later graciously more than reciprocated by offering their various U.S. hosts visiting London outstanding hospitality. In fact, Neil and his brother Tony were both in the hospitality business together. The brothers were even more closely connected as Tony's wife, Valerie, was Yvonne's sister. Neil and Tony were members of the family which controlled and ran the J. Lyons Company, a large publicly-held corporation operating hotels and restaurants and producing a variety of other well-known goods and services, including the iconic Lyons ice cream and tea. During my London visits over the years I greatly benefited from my father's generosity in paying for Neil and Yvonne's stay in my home town. If you cast your bread on the right waters, time and tide bring you not only soggy baked goods but also lots of Lyons ice cream, meals, hotel stays and other special London hospitality.

My father always endorsed travel as a form of education. Although not much of a traveler himself, he encouraged me to go places and paid for my trips. I guess he created something of a monster, as eventually travel became an obsession. But it was a benign sort of monster, a toothless creature quite unlike Dracula and more like the dental-deprived denizens of Molvania. While still a teenager in the summer of 1955 I embarked on a Grand Tour of Europe, starting in Britain where in London the Salmons gave me royal treatment, which in England is saying a lot. Later, over on the Continent, I reverted to my normal status of commoner, staying in youth hostels and traveling second class and sometimes by hitchhiking. Back then you could get rides; these days you'd probably get arrested. Thanks to the Salmons, in London my life was rather majestic. They gave me complimentary accommodations at the Regent Palace, the Lyons-owned hotel at Piccadilly Circus. It wasn't a palatial place but the price was right. On a recent visit to London I found that the Palace was demoted to ordinary premises as the building had been converted from regal to retail and office space. The Salmons also hosted me for dinners at the Trocadero restaurant and at the Cumberland Hotel, both Lyons properties. Neil and Yvonne even lent me their limousine, complete with Street, the chauffeur, who addressed me as "sir," a heady experience for a young man. No one at home ever called me "sir." I knew that in England "sirs" ranked somewhere up there with lords, earls, knights, so I was pleased to join that exalted company, even if only on a temporary and brevet-like basis. Titles were important in England; they gave you status, like a student ID card in France or being named Duke Ellington, Nat "King" Cole, General Motors, Colonel Sanders or Captain Midnight.

Thanks to the Salmons in London, I was punching above my weight, which increased by the many meals the family fed me. So appreciative were the Salmons for my father's earlier hospitality, before my first London visit they'd even invited me as their guest to attend Princess Elizabeth's coronation in 1953. Back when I was a youngish teenager, Royal Crown Cola, Dairy Queen and Burger King, White Castle—or their then equivalents—interested me more than British royalty so I stayed at home. The coronation somehow managed to take place without my participation. I hoped the new Queen wasn't insulted by my failure to attend the ceremony.

On one of my later London visits I was finally able to render a service to the Salmons, the least I could do to reciprocate for their truly exceptional hospitality. My supposed hamburger expertise as an American induced Neil to seek my opinion about a new product Lyons planned to introduce in the company's nationwide chain of informal restaurants called Corner Houses. One day Yvonne took me to the flagship store on Shaftsbury Avenue to sample a thing called a Wimpy, the Lyons version of a hamburger. As all Americans were thought to be hamburger connoisseurs, Neil believed that I could serve as a one-mouth test market to evaluate the quality of the Wimpy. I ordered two Wimpys and a Whippy (or maybe it was called a Whipsy), a kind of milkshake. It's claimed that when future prime minister Margaret Thatcher, who'd studied chemistry at Oxford, was working in the research lab at Lyons she formed part of the team which invented the "Mr. Whippy" soft ice cream confection. Whatever its origins, I sampled the mushy concoction, sipping it along with a taste test of the equally insipid burger. Eagerly awaiting my verdict, Yvonne gazed at me expectantly, closely observing my reaction as I chewed, or tried to. The sandwich tasted like the cardboard box which contains a White Castle hamburger, hardly regal fare but much better than the Wimpy which was indeed a wimp of a burger. I'd never tasted a worse version of the all-American sandwich.

Yvonne kept staring at me, awaiting my pronouncement. I forced myself to take another bite, as if savoring the treat. I wondered what the patty was made of. Nothing could taste that bad without a lot of thought and special ingredients put into it. Yvonne waited silently for my judgment. What should I tell her? I was a young man beholden to the Salmons, who'd coddled me with their gracious hospitality. Hotels, meals, dinner parties at the family home, a chauffeured limousine, and now a Wimpy—all that I owed to my hosts. Should I reveal the truth to Yvonne, perhaps irritating Neal and evidencing ingratitude for all that the Salmons had done for me? I felt rather cornered there at the Corner House. I stalled by slowly eating the second Wimpy. At least the jejune Whippy served to suppress the hamburger's remarkable taste. I finally decided that it would be impolite not to endorse the product, so I said, "It's good—I like it." By my calculated judgment I managed to refrain from putting my Salmon relationship in jeopardy, but at the same time I put the digestive systems of the British into great jeopardy. My decision meant that I'd lost the chance to change the course of British culinary history for the better. I'd instead inflicted on the population the truly inedible Wimpy, as before long Corner Houses all across the land sold the product. To this day I greatly regret my part in introducing the wimpy Wimpy hamburger to the British.

Back in the 1960s and 1970s Neil and Yvonne lived in an elegant house in a park-like setting by Holland Park, an exclusive part of London. They later moved to a condo at Hyde Park Gate, near the Royal Albert Hall in the same enclave where Winston Churchill, my fellow "Sir," lived at Number 28 in his later years. When I was a boy I sent to the great man a photo of himself asking the statesman to autograph the picture for me. After a long delay his secretary, Jock Colville, eventually returned the photo to me (by sea mail) with a printed card stating that due to the large number of such requests Churchill received he was unable to comply with mine. I still have the photo and the card as mementos, and from my unsuccessful exercise retain a British postage stamp I added to my stamp collection.

Years later on a visit to Chartwell, Churchill's country estate in Kent, I finally managed to obtain a more substantial souvenir of the man. I visited Chartwell off-season when no other visitors were present. Churchill's former assistant, who worked at the house-museum, gave me an unusually complete and informative tour of the property. When we reached the library she asked me if I'd like to have a copy of one of the Churchill-authored books in the collection there. This question was considerably less challenging than Yvonne's inquiry as to how I liked the Wimpy burger. The woman at Chartwell gave me one of the duplicate copies of Churchill's soft cover 1931 *India* published by Thornton Butterworth. Although of little monetary value, the book is one of my great travel treasures. To it I much later added another Churchill memento I highly value. In 1980 I visited the Churchill Museum on the campus of Westminster College in Fulton, Missouri where on March 5, 1946 the statesman had delivered the soon famous "Iron Curtain" speech. On the original typescript, which I asked to see, Churchill had added in his shorthand some cryptic hand-written prompts. On my request. the museum copied the page with the Iron Curtain passage, a tangible fragment of modern history I still have. More recently, yet a third family item came into my possession when I met Edwina Sandys, Churchill's granddaughter. She signed for me a catalogue of the exhibition of her grandfather's paintings where I met her. After I told Edwina about my visit to the ancestral family estate at Chartwell she replied, in a somewhat less eloquent way than her renowned grandfather and with typical English non-commital reticence,"Lovely." Little did Edwina know that she was conversing with the culprit partly responsible for the not so lovely plague of Wimpys which once infested the U.K.

After my London visit in 1957 I went up to Oxford. The first fellow student I met there, Kakuichiro Fujiyama, became a lifelong friend. At the time Fuji—as he asked Westerners to call him—was studying business at MIT's Sloan School in Cambridge, a place name deemed "unlovely" and unwelcome in Oxford. But perhaps the Massachusetts version grated less harshly on Oxonian sensibilities than did references to the rival English university. Not long after the term began I noticed in the *London Times* a reference to a certain Aiichiro Fujiyama, the Japanese foreign minister. I jokingly asked Fuji if he was related to the cabinet minister. My new friend seemed reluctant to reply, but finally admitted that the mentioned Fujiyama was his father. Knowledge by Fuji's classmates of this relationship would have enabled him to gain face, but he really didn't want any more: he had quite enough face already. Fuji asked me not to reveal the family connection to our Oxford colleagues and I never did. A decade later when I visited Japan in December 1967 Fuji introduced me to his father, by then an elder statesman no longer foreign minister. The senior Fujiyama, a tall distinguished looking silver-haired gentleman, officed at a Tokyo hotel the family owned. The former cabinet member discussed with me developments in Vietnam and offered a number of insights on world affairs. Although never very diplomatic myself, I enjoyed getting an inside view of various diplomatic situations and matters of state as seen from the perspective of a Japanese insider.

Rather at random, at Oxford I was assigned to a tutor at All Souls, unusual as that college doesn't take undergraduates. I was placed there because at the college a young fellow, already a distinguished All Souls Fellow, specialized in seventeenth century English history, the field I'd chosen to study, and so by chance he became my tutor and All Souls my college. In truth, I really didn't much care where or what I studied at Oxford. The point of the exercise was simply the experience of being there. If my curriculum happened to include "king versus pope" stuff, that suited me okay. A subject like "British versus Wimpy" would have been more appropriate, even if—once the truth of my role in the disaster was revealed— rather embarrassing for me.

Being assigned to Keith Thomas at All Souls gave me by happenstance access to that hallowed institution's inner sanctum. Appropriately located on the High Street, All Souls stands at the apex of English academic and intellectual life. I was clearly not up to the usual heady intellectual standards of the college. I gained access to the hallowed precincts of All Souls by chance. I was essentially an outsider looking in, a passing Yankee who'd gotten his foot in the door even if not his mind and soul. I was no high-powered High Street academic but only a curious character seeking experiences. I in no way ranked as a dedicated scholar, nor was I bookish enough to merit the statues of being a true Fellow, even if I was a reasonably convivial and somewhat studious fellow. All Souls belongs to the category of those highly selective and exclusive educational institutions where only top students study and the most accomplished scholars profess and produce their learned academic tomes. In France it's the *École Nationale d'Administration* which trains the *Énarques* who to a large extent run the country. In Japan the leading school is Tokyo University; in Britain, a few of the colleges at Oxford and (excuse the reference) Cambridge; in Italy, perhaps Bocconi; in new Cambridge, Harvard Law School; in Molvania, the renowned Autonomous University of Higher Studies, out of sight and so high you need oxygen to attend classes there. The High Street in Oxford is more down to earth.

Of those and other exclusive institutions, perhaps All Souls tops them all. Founded in the fifteenth century, as were some of the other Oxford colleges, All Souls over the years evolved into a rarefied academic institution for post-graduate scholars in the humanities. Many members of the College aren't resident teaching Fellows but serve in high positions in government, politics, the arts, business, the professions, cultural life and real-world activities outside the ivory tower. In a certain way, some of the cloistered resident All Souls academics, lost in a wilderness of books and words, seem lost souls. Sheltered inside the tower, All Souls scholars typically produce weighty wordy works on subjects arcane or obscure, but occasionally even a few less esoteric tomes of practical value. The often voluminous volumes are not known for their brevity. With complete freedom to think and a lot of time on their hands to write, the erudite dons confect thought-full papers and books. The All Souls historian Rohan Butler concluded his 1078-page first (and, as it happened, last) volume of his intended multi-book biography of the French statesman Choiseul with the sentence: "The diplomatic and political career of the Duke of Choiseul had begun." Unfortunately, Butler died before completing the next 30 or however many volumes he'd planned.

An exception to the bulky books favored by All Soulers is the College's 2006 "Concise" history, which notes that "fellowships offered for competitive entry every September became Oxford's most sought-after prizes and attracted outstanding candidates." One of All Souls traditional examination papers, no longer required, asked candidates to compose an essay improvised without any prior knowledge of the assigned topic. The three-hour essay responded to a single-word subject, among them innocence, miracles, water, originality, bias, style, charity, corruptions, mercy, possessions. (I was disappointed to note that such topics as Elvis, Snoopy, Dairy Queen, the Wimpy were not included on the list.) The on-the-spot, if not effectively on-the-mark, essay was intended to test "exceptional analytical ability, breadth and depth of knowledge, independent-mindedness and clarity of thought and expression, " all with the aim of selecting for the "rarefied intellectual air in three Gothic quads ... the brainiest of brainiacs." The only part I—as a non-braniac—would have a chance to pass would be "independent-mindedness." In 2008 the fellowship selection process was somewhat dumbed down when Sir John Vickers, the College Warden (the head inmate), decided to eliminate the famous and fearsome "Essay," calling it outdated and less useful than other exams on general and on specialist subjects.

Whether soulful, soulless, solo, solipsistic or otherwise, All Souls scholars are often perceived as pretentious, eccentric, dreamy, ethereal, cloistered or donnish, or all of the above. These supposed traits have inspired observers to mock, satirize or caricature the College's brainiacs. For my own part, I admired their eccentricities, many of them much more original and fanciful than my own, and I even adopted a few. I also respected their learning, even if it was only ivory tower relevant. The dons' learned minds and bookish ways and tasty meals offered me much food for thought. When I returned to All Souls after my time, there to reabsorb the heady atmosphere in the three Gothic quads and to see Keith Thomas, the renowned Fellow would occasionally invite me to lunch in the Buttery where good fellowship and stimulating table talk garnished the meal. You never knew what would come out of a professor's mouth as he put the grub into it. Hopefully, the food wouldn't come back out. Maybe the savant would observe, "I say, the turnips seem rather overcooked—or is that just my imagination?" The reply: "What exactly do you mean by 'imagination'?" The counter-response: "What in fact is the meaning of 'turnips'?" Such exotic conversations occur of course not only among Oxford's dreaming spires but also in other ivory towers, but the truly brainy souls at All Souls discuss such momentous matters with more gravitas and panache than elsewhere. The convivial and conversational repartee at Buttery meals often inspires notions which eventually find their way into published works produced long after the food has been ingested and digested. In another 15 or 20 years, say, there might appear a 1078-page tome entitled *Turnips Unpeeled, Their Secrets Revealed*, Volume I in a projected series of X books on the topic, plus a 3000-page Volume XI containing bibliography, footnotes and footnotes to the footnotes. Over the centuries, college luminaries extant and extinct have illuminated all sorts of topics, greatly advancing human civilization. Now, finally, turnips have their turn as an All Souls savant presents the final word on the vegetable. Up next, kumquats.

Better qualified and more articulate observers than I have tried their hand at satirizing All Souls types. In *Truffle Hunt* Sacheverell Sitwell refers to All Souls in a passage relating to the fastidious and self-important head porter waiting on him at the Dijon train station. The top functionary, as shown by his "*numéro un*" badge, summoned the chef and the steward to orchestrate with great ceremony Sitwell's meal. The porter and his colleagues seemed "as though they were a body of dons, the French equivalent of the common room of All Souls." Such perceived pretentious and affected donnish behavior exists at other Oxford colleges as well. Enid Starkie, an academic at Somerville College, described "that special talent of dons of which they have made an art, fanciful speculation, but speculation based on logical arguments once the premises have been accepted...—brilliant conversation arabesques, growing more intricate and complicated." But All Souls seems in a class all its own. In Javier Marías's novel *All Souls*, the direct two-word title more clear and succinct than many long-winded don-delivered lectures or books, he includes a vignette which may in a way capture the very heart and soul of the College, describing "one of the grandiloquent suppers known as high tables" where the Oxford dons "are expert at simultaneously talking, eating, drinking" and on one "occasion I even heard someone gargling." The Warden really should impose a gag-rule to prevent such mouthing-off: an occasional guzzle or gurgle permitted, a gargle banned. These are the sort of distinctions All Souls academics enjoy parsing.

To his credit, my tutor Keith Thomas hardly fit the All Souls stereotype. Although a leading

English scholar at a young age, Keith was informal and down-to-earth, and although he may have guzzled and gurgled at the High Table I doubt that he ever gargled. And if he did, it would be with water and not wine. Convivial Keith was erudite without being pedantic, intellectually rigorous but flexible in his thinking, and at tutorials a mentor rather than a tormentor. He was a shrewd judge of his students' tutorial essays, as proved by the fact that he often complimented mine. Keith's academic achievements were later recognized when he became a "Sir." But unlike Neil and Yvonne Salmon's chauffeur Street, who "sir-ed me," never did I attach that honorific to Keith's name when referring to him. To me the eminent professor was plain ol' Thomas, a regular guy whose last name and my first name we shared. I always gave Thomases the benefit of the doubt, even if I myself was a Doubting Thomas. For me, the knighted historian remained "sir-less," as I always referred to him as Keith. My refusal to "Sir" Keith wasn't simply a Yankee's inbred republican reluctance to acknowledge English titles. Even renowned Oxford don A.L. Rowse was unwilling to "Sir" his fellow knighted All Souls Fellows or any other such anointed figures. In *A Cornish Childhood*, Rowse noted that he was not "in the habit of saying 'Sir' to anybody ... let alone say 'my Lord' to anyone enjoying that rank." At least once a week I'd visit Keith's rooms at the front of the college overlooking the High Street for a tutorial. Either I or another of the one or two other students in attendance would read a paper—not the *Times* of London, sometimes referred to in the college as "the All Souls parish gazette," but a paper we wrote. After the reading Keith and my fellow classmates would comment on the essay. Such intense personalized attention motivated careful research and composition which could be defended in dialogues with the interlocutors who critiqued the paper.

The famous Oxford tutorial system represents a special method of instruction, back then new to an American student like me at the university. The technique is meant not as a didactic process to indoctrinate students—as is the teaching at the Sorbonne—but to address the challenge of training young people how to think. Tutorials are designed not so much to teach a subject as to educate students how to learn. The one-on-one format also exists at "the other place" where "supervisors" oversee "supervisions," the Cambridge equivalents of Oxford's tutors and tutorials. Maybe Cambridge has the best of it in the wording, as process gives students a super vision on how to reason and clearly express ideas, thoughts and opinions. In his *The Idea of a University* (1852), Oxford's Cardinal Newman suggested that "the objects of a university" are to produce "a cultivated intellect, a delicate taste, a candid, equitable, dispassionate mind, a noble and courteous bearing on the conduct of life." This mouthful of a result Oxford attempts to accomplish in part by means of the tutorial. The personalized tutorial aims to give the student a rational and nuanced way of looking at the world.

The process practiced at Oxford and Cambridge continues the training found at the renowned English public schools. In *British Gentlemen in the Wild West*—the story of the refined and cultivated titled and entitled blue-bloods who in the late nineteenth century owned and operated cattle ranches in the U.S. West—Lawrence M. Wood describes the classical education of Eton and Harrow boys who "were being prepared for public leadership; the classics they read spoke of the duties of service and the fitness of patrician rule. Their readings from Horace indicated the feelings of detachment...and the virtues they learned to admire were order, symmetry, balance and restraint. The Latin and Greek studies called for a powerful memory and, more important, formed a common bond with other gentlemen that could not easily be acquired by outsiders." The preparatory education served to set in place a certain mind-set the Oxford tutorials and Cambridge supervisions continued to strengthen. The methodology was aimed at shaping modes of thought—not what you thought but how. Some English blue bloods didn't take much to this system, favoring instead practical experience out in the world. The *Who's Who* entry of Osbert Sitwell, Sacheverell's brother, stated, "Educated during holidays from Eton." In a similar way, Ernest Hemingway's 1930 U.S. *Who's Who* entry for "education" stated only "abroad." I sympathized with this view, as my real education took place beyond the academic classrooms and lecture halls of Oxford, the Sorbonne, the

University of Vienna, and the other universities I attended. But both formal and informal lessons are necessary to facilitate something of a charmed life. A synthesis of book learning and real-world experience represents the best sort of education. Just as I attempted to reach an equilibrium between my home and away lives, so I tried to keep a balance between the two modes of learning—the academic and the experiential.

The tutorial technique is in essence meant to prepare you to bring a certain attitude to how you think about a wide diversity of situations. The play of infinite forms presents you with all sorts of variables. Will G. Moor, Fellow and Tutor at Oxford's St. John's College, explained in his 1968 essay "The Tutorial System" (quoted in *The Oxford Tutorial*, edited by David Palfreyman) that "the roots of the tutorial method" involve "a skeptical method, a method that inquires, probes, scrutinizes. It is not at its best an _ex cathedra_ authoritative statement. It prefers the relative to the absolute, the tentative to the dogmatic, the essay to the treatise." Quoting an Oxford colleague, Moor continues: "Its function is _not_ to instruct: it is to set the student to the task of expressing his thought articulately, and then to assist him in subjecting his creation to critical examination and reconstructing it." In Keith Thomas, then, I encountered the Oxford don version of a professional Doubting Thomas. In that way, my tutor matched not only my name but also my own temperament. The tutorial process conformed to my personality and served to confirm, shape and enhance the way I thought and expressed myself. The tutorial approach fit to a T(om) my intrinsic skeptical mentality characterized by a tentative and provisional mind-set, which was in fact largely un-set. I tried to go with the flow, resetting my views as circumstances changed. I was wary of fixed positions lacking flexibility. As I later learned in Paris, the Oxford approach was the opposite of the didactic techniques imposed at the Sorbonne. That revelation was in itself part of my education at the French university.

Perhaps the closest analogue in the American higher education system to the Oxford tutorial is the Socratic method of instruction used at Harvard Law School. In 1991 former dean Erwin N. Griswold described the school's pedagogical goals in a way which resembles the aims of the tutorial:

> You go to a great School not so much for knowledge as for arts or habits, for the art of expression, for the art of entering quickly into another person's thoughts, for the art of assuming at a moment's notice a new intellectual position, for the habit of submitting to censure and refutation, for the art of indicating assent or dissent in graduated terms, for the habit of regarding minute points of accuracy, for the art of working out what is possible in the given time, for taste, for discrimination, for mental courage, and mental soberness.

The ultimate goal of both the Oxford tutorial discipline and the arts and habits taught at Harvard Law School as articulated by Dean Griswold is to enable graduates to engage with the world in open-minded, creative, flexible and functional ways in order to deal with the play of infinite forms which you happen to encounter. My worldly travels over the years gave me an on-going refresher course in those useful principles. My engagement with the world derived not from fixed ideas but from an open-minded attitude which would let new impressions and perspectives enter my consciousness.

Part of the Oxford experience involves the common post-study practice of returning to the university over the years to recall, if not relive, your vanished time there. When in London visiting my friends, I occasionally took time to go up to Oxford. I once intruded on a don who occupied Keith's old rooms, apologizing by explaining that having studied at Oxford the place continued to exert a magnetic attraction which drew me back. "Damn it all," he said, "it's the same for me. We always want to return to Oxford." As I strolled the streets of the old town my memories quickly revived. On some of my visits I'd see Keith, to his relief, no doubt (and to mine), finally without the need for me to produce and him to read a tutorial essay. In

May 2006, a year short of the half century since my brief 1957 interlude at All Souls, Keith signed for me at the College a copy of his *The Ends of Life*, that single letter "s" optimizing the otherwise morbid title. Keith wrote: "To Tom Weil, who came to me in this College for history tutorials in 1957 and appears to have survived the experience." The "appears" qualification appears to reflect the characteristic Doubting Thomas tutorial mentality, an attitude which refrains from drawing definitive conclusions.

On my return visits to Oxford over the years I saw that the ancient university town has, not unlike me, both changed and remained the same. Equally ancient as Oxford in terms of a humanimal's life span, I've in some ways changed while fundamentally remaining the same. The venerable colleges still stand like sentinels along the High Street and line the lanes and byways elsewhere in the vintage city. To the same historic pubs flock new thirsty students, gathering at the "K.A." (King's Arms), the Turf, and a veritable menagerie of other picturesque hangouts—the Bear, the Trout Inn, the White Horse, the Lamb and Flag, the Eagle and Child (known as the Bird and the Baby) where once gabbed and gulped The Inklings, among them J.R.R. Tolkien, C.S. Lewis and other dons who guzzled and perhaps gargled there. Many of the old second-hand bookshops—Thornton's on Broad Street, a splendid store with many unusual and esoteric volumes once located on The High near Queen's College, and also the small book stall in the Covered Market—have disappeared, driven out of their city center premises by both increased rents and digital formats which replace paper-based materials. High rents have in a way changed the nature of The High Street. The colleges remain but their commercial neighbors, now boutiques, fancy hotels and expensive restaurants serving "cuisine" rather than meals, have gone upscale by luxury, so to a certain extent degrading the city's old-time charm.

Of a crisp autumn eve when Evensong sounds from a college chapel, or on a balmy spring day with winter's chill just a memory, it's pleasant to stroll about the old town where now a new generation of students animates the ancient university's halls of learning. I'd leisurely amble down The High, past the tree which marks All Souls, then continue on to glance into the tranquil quadrangles where linger the ghosts of yesteryear students and defunct dons, now—like me—long gone. On through the medieval side streets I'd stroll, and then pass through picture-perfect Christ Church Meadow, a peaceful green park removed from the busy colleges, pubs and shops. The students and the professors change—they come and go—as do some of the business establishments, but little else. Over the years since my youthful Oxford days I too changed, for I became an old gentleman wandering about among the fresh young students. They, too, in their time, would soon leave the dreaming spires of old Oxford for the greater world beyond and, years on, would no doubt one day return, as did I, to look back on their flown vanished student years.

As ever, boaters would be poling their girlfriends as the punts glided along the tranquil Isis, there in Oxford a modest dreamy stream before the waterway joined the Thames on the river's way to London and then on down to the distant sea. That eccentric Christ Church College don Lewis Carroll began his tale of *Alice's Adventures in Wonderland* with a dreamy description of how "All in the golden afternoon/Full leisurely we glide"—

> Alice! A childish story take,
> And with a gentle hand
> Lay it where Childhood's dreams are twined
> In Memory's mystic band,
> Like pilgrim's withered wreath of flowers
> Plucked in a far-off land.

Such are some of my memories, less mystic than misty, I've plucked from my withered days a lifetime ago in a far-off land.

~ 4 ~

My days in Europe glided by, the golden afternoons of spring and summer flowing quickly on, on like the Isis, the Thames, the Seine, the Danube, and back home far away the Mississippi. For the year I wasn't there, the Mississippi rolled on without me: nature's systems indifferent to my presence. Autumn and spring, summer and winter would come and go oblivious to whether or not I was around to experience the changing seasons. As time winged on during my year in Europe, it somehow seemed to fly faster than at home. Time's wingèd chariot gained on me at the same measured pace always and everywhere, but how I experienced its tempo often depended on situational factors. For a child, a prisoner, a convalescent time seems to advance in slow motion. It takes forever to grow up, to get out, to get well. For me during that year 1970 in Europe time flew by, winging its way toward the end. By the time I finished at the Sorbonne half the year had passed. I realized that before long I'd be back in my familiar surroundings at home, my adventures in Europe ghosted into memories of bygone times. In the meantime, the second half of my Euroyear remained to be completed.

After my Sorbonne semester came to an end I continued on to Vienna to enroll in summer school classes at the university there. This completed the triad of European institutions of higher education I attended. Oxford and the Sorbonne presented two quite different ways how each nationality went about teaching upper level students. The English and the French methods reflected some of the national characteristics of each country. I was curious also about the Germanic techniques and mentality, which—as much as the specific subjects I'd study—I hoped to learn about at the University of Vienna. To understand those Teutonic-type traits I had something of a head start, as my heritage and character were Germanic. By both inheritance and temperament I was essentially German, as modified and diluted by being a fourth generation Midwestern American. Perhaps Dominique's description of me as "Anglo-Saxon" was close to the mark: a Saxony-like mentality Americanized with Anglo traits. To my many German characteristics I wanted to add the language to fill in that ethnic gap. Studying in Wien would enable me to get my German up to *Geschwindigkeit*. Germans unfortunately don't call it "speed," which would be much easier for all concerned, even for the Germans, as they could get across the idea much more speedily by using the simple single-syllable English word. But *nein*, you've got to learn all those strange fearsomely long words, tongue-twisters needed to produce the right sounds to speak that foreign tongue. In French, Spanish, Italian you could make educated guesses about many words; in German you had to memorize them. Without vocabulary you're shtum, a double-duty word used in German (spelled *stumm*) and also by the English but until now not by Americans.

Over the years I sometimes wondered why I devoted so much time and effort learning four foreign languages. My time might have been better spent in other activities, such rereading the speech balloons which contained Snoopy's wit and wisdom or gliding in golden afternoons along the Isis poling a cute coed along the way. Instead, I spent hours in the Luxembourg Gardens in Paris studying irregular verbs, the subjunctive and other French language nuances and nuisances, in Vienna memorizing multi-syllable words, and elsewhere learning how to vowel into voice Spanish and Italian. These linguistic exercises helped to inform me that my own so familiar tongue was nothing more than an arbitrary way to reduce the play of infinite forms to meaningful expressions. What I presumed was a canine was nothing of the sort. "Dog" represented only one way of looking at the animal, a creature people elsewhere described as a *chien*, *perro*, *cane* or *Hund* and, what amounts to the same thing, a man in other lands became an *homme*, *hombre*, *uomo*, *Mann*.

Like every language, German reeks and speaks with eccentricities. Those Germans—they capitalize nouns, an oddity except for the Krauts for whom it's natural. But they no doubt find

Section VI: Subjective Objectives

it strange that Anglo-Saxons capitalize "I" as if the IndIvIdual is especially important, even though an Englishman held that no Mann is an I-land. With the high "I," English speakers seemingly endow themselves with an exaggerated sense of the significance of each separate humanimal. Because this book isn't about me, i've suppressed the big "I" in this book. Although each person indeed—in word and in deed—exists as an I, his or her essence is in fact as an It. That's the word which should be capitalized. If "I" merits the capital, then why not "Dog"? That bIg "I" standing up so pretentiously prompted Émile Zola, when in London in 1893 in exile from France, to speculate on "the effect of the capital 'I' on English character." HIlIghtIng the "I" seems too self-referential, as does "me/moi" in ME-MOI-R. Too inward looking, the "I" focuses on the person rather than on the vast everything else. In that I'm not the featured character in my reminiscence, the capital "I" should be demoted to "i." Removing the teller from the tale prompts the question of who the main character is. You needn't ask for whom the tale is told. Turning the pages on and on one by one to the end—as time takes its toll—you'll eventually see that the story is told for thee as much as for me.

The four foreign languages I learned came with some useful general lessons beyond grammar and vocabulary. They taught me that all around the world people perceive things in terms other than how the play of infinite forms appeared to me. My own limited world view expressed in English and seen from my perspective was just one way of looking at the play of forms. Many other ways existed to designate in words the exact same sensory impressions, to which humanimals elsewhere might react in ways quite different than mine. Learning another language represents the same sort of exposure to other perspectives as does foreign travel. You can of course learn a language without giving up home comforts, but to see the world you need to move. Armchair travel can get you only so far, such as to the fridge for a snack. By way of contrast, these days English, no longer only a stay-at-home, goes a long way. The foreign languages I labored so long and hard to learn lost value. The supply-demand equation moved against me. With the widespread proliferation of English, its supply increased, so diminishing my need to speak other languages. Unfortunately, I can't trade in my French subjunctives and compound German words in exchange for the time I invested to learn those mind-bending and tongue-twisting expressions..

In addition to their once practical value, languages served to satisfy my curiosity about them. I wondered how they functioned and differed from English. Learning a language is like solving a jigsaw puzzle, as you gradually assemble the pieces until some sort of pattern originally unfamiliar to you emerges. I would have enjoyed learning Russian, Arabic and Chinese, but every experience and experiment comes at the cost of effort and time. The play of forms was so infinite that even a very long lifetime wouldn't suffice for me to pursue activities and topics which interested me. With the time at my disposal I did my best to cover as much ground, intellectual and geographical, as I could, but in the end I managed to accomplish relatively little, even if more than my stay-at-home contemporaries. I was nonetheless grateful for my opportunities to expand and enrich my experiment in living, an enterprise which for no mortal ever yields a proof of concept. The novelty of speaking and hearing a foreign language in a foreign setting served to revise and enlarge my perspectives, including my self-image. Different contexts, such as those in Oxford, Paris and Vienna, helped to put my "I" perceptions in a new light, as if reflecting back views which enabled me to see myself in ways other than the mirror image I perceived at home. Such additional perspectives were at times unsettling, disappointing, comforting, concerning, exhilarating but in any case always revealing. It's a new you, a variation on a theme which in part serves to redefine the old you formed in your native habitat. The new images induced me to reflect how I appeared not only to myself but also to others. Although I didn't much care how anyone else viewed me, I was in a minor way curious about what they saw in me. But all that was quite subordinate to my main agenda, which was not to focus on the "I" but to look outward at the world. The lonely planet—not my transient and quickly depleting self—was what interested me.

There in Vienna, where I gained a few new self-images, I became a kind of third Mann following the two earlier versions as, first, a young man at Oxford and, later, a second more seasoned *homme* at the Sorbonne. As in Paris, in Vienna I took a room located near the university and not far from the center. I wanted to be near Wien's beating *Herz*—*das*, neuter. It's kind of a sad commentary on the mentality of the Germans that to them heart-felt emotions stem from an organ which is neuter. You'd think that *Herz* would be either a he or a she. But you'd be wrong. For the genders of German nouns you'd often have to guess, with a one in three chance of being right. The odds were better in the romance languages—a one in two chance—and English presented no worries in regard to noun gender, but problems aplenty otherwise. At the university I enrolled in classes on German language and literature. A short walk took me down (*die*) Währingerstrasse, where my pension was located, past the Votivkirche (also a *die*), along the Ring and on to the nearby *Universität* (yet a third *die*). I wasn't deceived by those three-*die* nouns. All too many *ders* and *dases* lurked in other nouns, knowledge of their genders necessary to speak properly. Even more than when I began at the Sorbonne, I wondered if my language skills were sufficient to enable me to understand the lectures. German presented many challenges, among them learning words like *Herausforderung*.

Some of the language's difficulties Mark Twain detailed in "Learning the Awful German Language" in *A Tramp Abroad*. Referring to the gender of nouns, Twain notes that "In German, a young lady has no sex, while a turnip has." He illustrates this oddity with a dialogue. A girl asks a boy where a turnip is, to which he replies, "She has gone to the kitchen." She then asks, "Where is the accomplished and beautiful English maiden?" and he answers, "It has gone to the Opera." Twain concludes with suggestions for improving the language, one of which is: "I would recognize the sexes, and distribute them according to the will of the Creator." Young Winston Churchill responded with a similar sort of quizzical reaction to grammatical complications when a teacher at St. James' School in Ascot told the boy that a word like *mensa* required a case: "You use the vocative when you address the table," to which young Winston replied, "But I never speak to a table." Twain's imaginary dialogue of course finds its way (even if only in a footnote) into that All Souls don's 1078-page page-turner turnip tome, at least so they tell you at the college for the book has apparently not yet been completed. The professor is still sampling turnips at the High Table between gargles.

In chapter XI of *A Tramp Abroad* Twain describes his stay in Heidelberg the summer of 1878, noting how he'd made progress in learning the "awful" language. Visiting the city in the mid-1950s as a very young Mann, I tramped up the steep street to the castle, which Twain had painted to produce his first "great picture." It was no doubt a relief for him to take a break from *der*/*das*/*die*—and especially *die*—to work on paintings rather than genders. "Die" ("the" in German for feminine or plural nouns) was perhaps too suggestive to Twain of the English word.

At the castle entrance I encountered a grizzled old attendant, his bristly large white mustache contrasting with his neat black uniform. The Germans like uniforms, except for the un-uniform nouns whose random genders so vex foreigners. The old man asked where I was from, and upon hearing the word "Missouri"—not American English but Native American—he informed me that years before as a very young man he'd met my fellow Missourian Mark Twain there at the castle. The author had represented Missouri in Heidelberg 77 years before I got there. I was honored to follow in his footsteps as a latter-day representative of our state. At the time of my visit both the castle and even the ancient guard remained, but not Twain. His life may have been shortened by the strenuous efforts he exerted to learn the language he deemed awful. As related below in this Part 4, some three decades later a hands-on intergenerational encounter also put me in touch with Twain.

One of my German language instructors at the university claimed that English was even more awful to learn than his language. I wouldn't know, as I never studied English as a foreign language. As an example, the teacher wrote on the blackboard "police" and "lice" and challenged non-English speaking students to pronounce each word correctly. I think the teacher was right. Spelling, pronunciation and idioms all present major problems for foreigners who try to learn my language. In any case, and including grammatical cases, my studies of German went well and unlike Twain I found it hard to be cross toward a language with such words as *Auspuff* (masculine; car exhaust) and *Beisskorb* (masculine; "bite basket=muzzle"). My literature classes also proved informative. One paper I wrote which quoted Goethe yielded a high grade. In the course of my studies abroad I'd learned that if you included in your written work quotes from a national icon you'd get better marks. Professors liked to see references to the homeland's revered heroes. In England you won't go wrong quoting Shakespeare; in France any writer will do, as the French think all their authors are better than foreign writers. In the States—here's a tip to foreigners studying there—something from Snoopy always pleases.

In addition to improving my German I also managed to upgrade my Italian, thanks to Gregorio Piaia who roomed in the same pension where I lodged. Although studious and well educated, Gregorio spoke little German and less English so we communicated in Italian. We became good friends and have kept in touch over the years. Gregorio became a distinguished and an extremely productive scholar at the University of Padua where he served as professor of medieval philosophy. In 2017 his colleagues honored Gregorio with a *Festschrift* commemorating his many years of high-class written and class work in his field, including a number of books and articles. A few years after we'd met in Vienna I visited Gregorio and his wife Giuliana in Montebelluna, north of Padua. Some six or seven months later he wrote to tell me that they'd had a baby boy who they named Tommaso. Gregorio explained that he and Giuliana had chosen that name because during my visit with them I'd been such a great guest—so charming, so delightful, such a wonderful conversationalist, an all-around swell guy. He concluded by noting: "The fact that my father is named Tommaso had absolutely nothing to do with our decision to give the baby that name."

During my time in Vienna I followed my usual practice of assuming that I'd never return there for another chance to see the city. This motivated me to see as much as possible. As in Paris earlier that year, I used the Michelin Green Guide, German language edition, this one for the entire country rather than just the capital city. Guide in Hand—really an easy German word to learn, but what's the gender? Is "Hand" a he, a she or an it?—I saw much of Vienna and also some nearby areas in eastern Austria. I visited all of the places described in the Guide's dense-text 20-page Wien section, including the *Hauptsehenswürdigkeiten* (the main sights), *Weitere Baudenkmäler und Parks* (additional built monuments and parks) and the *Umgebung* (surroundings). I wonder what Twain would have said about the turgid words which designated those categories. At least "Park" would please him.

One *Hauptsehenswürdigkeit* understandably omitted didn't exist at the time—the Arnold Schoenberg Center, established in Vienna in 1998 after the family had removed the composer's collections from the Schoenberg Institute at the University of Southern California in Los Angeles where they'd been housed for 25 years and moved the materials to Vienna. Long after my student days in the city I visited the Center when I returned to Vienna to see friends there. A strong sense of *déjà vu* startled me when I saw a replica at the Center of the sitting room at the composer's house in Brentwood (Los Angeles) where he lived from 1936 to 1951. It was odd to see a room in Vienna identical to the actual parlor already familiar to me. In the late 1950s I'd by chance met Ron Schoenberg, Arnold's son, and we became good friends. Whenever I was in L.A. I used to visit Ron and his wife Barbara, who lived in his father's former house. We always sat in the sitting room, furnished with pictures, portraits, furniture and other objects unchanged from the time Arnold and his second wife Gertrude (Ron's mother) occupied the residence. The room in fact was an antiquated outpost of the Old World there in trendy Los Angeles, as the parlor seemed to be a typical *Mittel-Europa* enclave from an earlier time. Ron's claim to fame was not music but as a nationally ranked tennis player, while professionally he also served on a court—as a distinguished Los Angeles judge.

A certain symmetry characterized the life course of Ron's famous father. Born in Vienna in 1874, the composer moved to the U.S. in 1933. His archives and other personal items returned 65 years later to the city of his birth, thus completing a kind of round-trip. Seeing his original lived-in room at the Brentwood house and a duplicate displayed in distant Vienna also represented a symmetry, one which typified how as the years went by for me patterns seemed to form from the play of infinite forms which had by chance happened to come to my attention. The elapse of time tends to connect some of the otherwise unwieldy confusion of past sensory impressions. The longer you live the more the loose ends in your life knit into a kind of double-helix, entwining experiences and memories into an integrated format. If you lived forever everything would gradually eventually converge.

While I was a student in Vienna the university sponsored two weekend trips, one to Prague and the other to Budapest. At the time, of course, both capitals lay behind the Iron Curtain, Churchill's descriptive phrase included on the copy of the page from his 1946 Fulton, Missouri speech I obtained while at the Churchill Museum in that Midwestern college town. A chance to peek behind the Curtain piqued my curiosity so I joined both excursions. The two cities showed me at first hand the effects of a repressive communist regime on the inhabitants and on the economy. Just as for my on-the-ground experiences later in Santiago, Chile and elsewhere, the real-life lessons learned in Prague and Budapest during my short visits supplemented the scholarly subjects I studied in Vienna and at other universities. There's nothing like seeing in practice that which professors and professional pundits learn and profess based only on academic theories and bookish notions. Those authorities were print-educated paper tigers. A few glimpses behind the Curtain revealed back stage views of how people actually lived in a Communist system. Fate had somehow decreed that my life originated on the right side of the Curtain. My fatal circle wasn't surrounded by a wall or curtained off from the outside world.

After the war Vienna suffered from a split personality. Divided into four zones—U.S., British, French and Soviet—the city lay one-quarter behind the Iron Curtain. The movie based on Graham Greene's novel *The Third Man* pictures Vienna as it was during the early post-war era. Along with the players, the city itself features as one of the movie's main characters. In some fictions the characters seem more real and are certainly more long-lived than living people. Just as Mark Twain's Tom and Huck and Becky appear to be actual Hannibal kids, Holly Martins, Anna Schmidt, Major Calloway and especially the elusive Harry Lime seem to haunt Wien. I could visualize all those characters as I made my way around the city to various settings which appear in the movie. The figures came to life as I visited Sacher's Hotel, the giant Riesenrad wheel in the Prater, the portal of the apartment building in Schryvogelstrasse near the university. There in the shadowed doorway Harry Lime's face suddenly becomes illuminated when a woman in a room across the way draws back a curtain, more easily lifted than the Iron one, and a shaft of light hits him. Suddenly the mysterious Lime is in the limelight. An Iron Curtain might have saved Lime from exposure, but now his secret was exposed. His funeral had been staged; he was still alive. Before long Lime meets his definitive end in the underworld labyrinth of sewers beneath the manhole covers in the Viennese streets. Lime's remains rose from the underworld to be returned to it in his grave at Vienna's Central Cemetery (1874) where some three million defunct Wieners—about twice as many as the city's living population—repose. Vienna was more dead than alive, as are all earthlings—mere ghosts passing in the form of a humanimal. Running from the old Stock Exchange building in the center out to the entrance of the cemetery, tram 71 connects the two realms—the land of the living and the dominion of the dead. It was precisely that sort of fixed-rail trolley route to the end of the line I wanted to avoid by varying my trip to the terminal stop. Although the destination couldn't be changed, I could alter the route. Harry's second funeral was his last. But Lime still lives on as he enjoys the sublime status of being a fictional character. His alter ego, Orson Welles died; Lime lives.

As Andrew Sinclair wrote in the Introduction to the Modern Film Scripts edition of *The Third Man*, "the wet brooding, defeated presence of ruined Vienna is ... dominant." He relates how film impresario Alexander Korda sent Greene to Vienna to find a contemporary story, "which Greene grafted onto an old idea about meeting 'a dead man'." Lime feigns his death and a fake funeral to escape pursuit for a racket he ran in shattered post-war Vienna. The Greene-guided plot involving a supposedly dead character still alive resembled a similar resurrection in Eric Ambler's novel *A Coffin for Dimitrious* and also the real life case of one Ye Qinghe, also known as Paul Yip. After Fujian Province on the Chinese coast north of Hong Kong seceded from Chang Kai-Shek's Nationalist government, the new regime recruited Yip to run the region's opium monopoly whose profits financed the Fujian People's Movement. By 1937 the wheel of fortune—for Yip misfortune—had turned and the drug wheeler-dealer was arrested on narcotics smuggling charges and (seemingly) executed in Nanjing. But in a Lime-like caper Yip was in fact still alive and had escaped to Hong Kong. Death later got a second chance, a major disadvantage for those characters actually alive rather than existing as a fictional character, immune to mortality, like Lime or Dimitrious.

In 1947, the year before Carol Reed directed *The Third Man*, he'd directed *The Fallen Idol* (called *The Lost Illusion* in the United States) based on Greene's short story, "The Basement Room." After completing *The Fallen Idol* Reed planned to fly to Los Angeles to direct Joan Crawford in a movie, based on the play *Portrait in Black*, to be produced by Jack Skirball. A Cleveland rabbi before he moved to Los Angeles in 1938, Skirball ran the Educational Film Corporation of America and later became a multi-millionaire after he developed Vacation Village Hotel in Mission Bay, San Diego. It was Skirball who gave Bob Hope the comedian's first exposure in films. The first of Hope's 55 feature films and eight short subjects was *Going Spanish*, produced in 1934 by Skirball for Educational Pictures at Astoria Studios, predecessor to Paramount Pictures, in Queens, New York. After the short appeared Hope commented about the movie, as Walter Winchell reported in his column, "If they ever catch John Dillinger, they ought to make him sit through it twice." In *Have Tux Will Travel* Hope relates that after his suggestion on how to punish Dillinger hit the press the comedian's agent phoned him: "I've just had a call from Jack Skirball at Educational. That was a very funny crack of yours in Winchell's column, but Skirball doesn't think it's funny and they're dropping your option." Skirball's film business suddenly became Hopeless, but the comedian's career thrived. Released from any future Skirball-produced releases, Hope went on to star in *Big Broadcast of 1938*, featuring what became his famous theme song, "Thanks for the Memory," perhaps meant as a sardonic farewell to his first producer.

Although I never met Jack Skirball, who died in 1985, I encountered Hope at a private club in 1958 when he was in town for a performance (but not in the title role) of the musical *Roberta*. A group of waiters stood nearby as I chatted with the famous comedian. He was very matter-of-fact and even, it might be said, rather humorless. When he asked an attendant for a glass of milk, Hope's request elicited great laughter. With a slight shrug of his shoulders and a wry smile, Hope looked at me and said, "You know, everyone always thinks I'm trying to be funny, but I really just want a glass of milk." Before long one of the waiters brought Hope the requested beverage, which he drank without a wisecrack. Famous figures I've met often behaved offstage in ways differently than suggested by their public persona. I once conversed with Richard Nixon in his modest and rather cramped Wall Street law office where he worked during the interlude between his losses in campaigns for president and California governor and his later election as president. Nixon didn't strike me as a "Tricky Dick" type, and he didn't even have a swarthy five o'clock shadow complexion, although admittedly I saw him in the early afternoon and maybe the shadow came a few hours later. Nixon was cordial, straightforward, responsive and trick-less, all rather different from his public image. Like Hope, Nixon operated with two personas—one performing, the other when behind the scenes.

After Jack Skirball's death my mentor and good friend Bill Jacobs (discussed in Section III, Part 4) introduced me to Jack's widow, Audrey Marx Skirball, by then married (in 1987) to her second husband, Charles Kenis, a Los Angeles importer of French wines. Audrey and

Charles owned a condo villa adjacent to Bill's on the grounds of La Costa resort near La Jolla in southern California. I became a friend of the couple and continued my connection with them even after Bill died. On one occasion Audrey and Charles invited me to join them for an afternoon at Del Mar rack track where one of their horses was running. As a guest of the steed's owners I felt it would be impolite not to show some support, so I bet $10 to show, which gave me three chances to win something—odds better than trying to guess the gender of a German noun. Much to my surprise the horse finished first and I was in the money. Before proceeding to the winner's circle with Audrey and Charles I went to cash in my winning ticket, which had produced a tremendous return on investment—50 per cent in a matter of minutes. Annualized, this would yield a fortune beyond tally. I could could abandon reading and give up writing articles and books to become a bookie. But this would mar my 100 per cent success rate wagering on horses. I thought better of the idea on the way to collect my winnings when I saw littering the floor hundreds of discarded unlucky tickets. Only very few chancers held winning cards at the track; most of the betting tickets were discards. Luck determined a humanimal's track record not only for horse races but also for all of life. Each member of the human race faced the race to the end with luck and chance influencing the result. Those who managed to reach the winner's circle would be helped by a favorable fatal circle, other beneficial circumstances, and mainly luck. The only sure winner was the wingèd chariot, which sped ever onward.

Although I enjoyed seeing Audrey and Charles from time to time at their Century City condo in Los Angeles, my main reason for visiting the city was to be with other friends, including Ron and Barbara Schoenberg and, especially, Aleen Leslie who hosted me at her 1923 Beverly Hills mansion. Aleen had created and written the popular 1940s network radio show "A Date With Judy." She and her husband Jack (Jacques), a prominent Hollywood lawyer who'd died before I knew Aleen, lived at 1700 Lexington Road located not far behind the Beverly Hills Hotel. Marion Davies, William Randolph Hearst's paramour, once occupied the palatial residence he'd bought for her as did as a renter—Columbia Pictures much feared head, Harry "White Fang" Cohn. Fangless and without a resident courtesan, I didn't quite fit the cast of characters previously connected with the house. Bing Crosby rented the 1.5 acre estate in 1936, and such luminaries of "the business" as Charlie Chaplin, Rudolph Valentino, Norma Shearer, Harold Lloyd, Irving Thalberg and Clark Gable attended parties there over the years. In September 1927 Greta Garbo and John Gilbert were scheduled to be married at the mansion but, a show business no-show, she never showed up to tie the knot with the silent screen star. The even more silent and reticent Garbo supposedly once pushed the fully dressed Gilbert into the swimming pool at #1700. As a provincial Midwesterner living far from the silver screen celebrities, I certainly contributed nothing to the lore and folklore of the fabled mansion. I wasn't even a bit player but only a face in the crowd—a nondescript sometime member of the audience gazing passively at the ghostly flicker of forms which danced across the screen in the darkness of the theater. Out at the big white house in Beverly Hills, I greatly enjoyed Aleen's sparkling wit and her inside tales of Hollywood doings. She knew a lot of the ins and outs of the leading men and the misleading casting directors with starlets and otherwise, and secrets of many other members of the vast cast and outcasts who populated the Hollywood scene and the formerly seen.

In her novel *Fleur de Leigh's Life of Crime* Diane Leslie, Aleen's daughter, who I also knew, describes her childhood at the house and her parents' star-studded social circle, part of Diane's inherited fatal circle. Until nearly the end when she died in February 2010 at age 101, Aleen remained her usual witty, wacky, funny and delightful self. In July 2011 #1700 was sold for more than $9 million, but for me the mansion was priceless for all the fun times and treasure trove of experiences I enjoyed there, thanks to Aleen's gracious hospitality and her willingness to share parts of her life with me.

By now, for me, "the deads," as Simone de Beauvoir called them, outnumber the alives who still people my life. Many of the friends and acquaintances who meant something to me are by now gone, leaving me with fewer alives. My ever-diminishing number of relationships due to mortality is a function of my advanced age, which by now encompasses enough time to have killed off many friends and family members, and which before long will finish off the job by ending my own existence. All those lost mortals who once populated my life haunt my memories like ghosts from the misty past. They seem like the shadowy characters from *The Third Man*, the phantom figures I imagined seeing in Vienna that long ago summer of 1970. Those fictional residual characters in a way seem more substantial and real and remain more alive than their real-life analogues—Orson Welles, Joseph Cotton, Alida Valli, Trevor Howard, Anton Karas and the other now dead players. Harry Lime is as vivid and as vital a figure as Jack Lemmon. Those fictional *Third Man* figures continue to cast their shadows around Vienna, haunting the city in a realistic way found only in a few other scenarios. It's rare that make-believe people seem to people a place with a life-like presence even more lively than that of the actual inhabitants.

In the United States two such places where literary characters seem as real as residents and visitors are William Faulkner's Oxford and Mark Twain's Hannibal. In both small towns, phantom figures created by a native son haunt the scene. Although the Mississippi Oxford boasts a university, the town differs greatly from the Oxford I knew as a student in England. Sharing only a name, the two Oxfords are otherwise worlds apart. On my first visit there in 1980 Faulkner's niece, Dean Faulkner Wells, signed for me her *The Ghosts of Rowan Oak*, a collection of ghost stories the novelist told his daughter Jill and granddaughter, along with Dean, at Rowan Oak, his antebellum Oxford homestead. The house-museum remains as if its author-owner had just stepped out briefly to walk to the town square before returning to his familiar setting, still today unchanged from his time. In his introduction to Wells' book, Willie Morris evokes the mansion as the house was when the children were little, a place "beautiful and serene in the daylight hours, full of dancing shadows and ghostly imaginings at night." He noted that "In memory there resides a beauty, a tenacity, a splendor which enriches and gives meaning to our lives and to those we love who will follow us. The gift of memory is a haunted blessing." As a pre-ghost soon to be phantomized, I treasure ever more the haunted blessing of my memories as they recede into the distant past like a campfire's smoke coiling into the air before disappearing forever.

Upstream from Oxford, which lies not far east of the Mississippi, Twain's Hannibal perches by the legendary river, his home town a "white town drowsing in the sunshine" with "the great Mississippi, the majestic, the magnificent Mississippi, rolling its mile wide tide along, shining in the sun." There at Hannibal where the waterway braids its way between a scattering of small islands once frolicked and played their pranks Tom Sawyer, Huck Finn, Becky Thatcher and the rest of the Twain gang as the river town drowsed the days away. Those characters seem as real as the boys and girls you see these days on the streets of Hannibal, but unlike the current kids now frolicking there by the river's on-flowing currents, Twain's creations will live on. Twain himself remains a phantom presence in Hannibal, haunting the old river town where he spent his long ago soon spent youth. Even more than 80 years ago so many local commercial, cultural and other establishments featured the names of Twain and his created characters a 1937 visitor, Major R. Raven-Hart, an Englishman on a kayak trip down the Mississippi, commented that he found in the town too much "Mark Twainery."

After Sam Clemens left Hannibal in 1853 at age 17 he occasionally returned to his drowsy home town, the native son transformed by then into the world-famous author Mark Twain. During his last visit in 1902 Twain addressed some 300 guests from the steps in the entrance hall of the Rockcliffe mansion, built at the turn of the century by lumber magnate John J. Cruikshank. Present on that occasion was Helen, one of four young Cruikshank daughters. At the time I met Helen Cruikshank Knighton at the mansion in March 1980, she was believed to be the last surviving Hannibal resident to have met Twain. Relating memories of her encounter with the renowned author, Mrs. Knighton mentioned that she'd shaken hands with him. I immediately jumped from my chair and rushed over to the old lady to shake the hand which 78 years earlier Mark Twain had held. It was touching to think that handed down to me now was a human contact with the famous long-gone Missouri author, the same who the old attendant at Heidelberg castle had met in 1878, so he'd claimed in 1955 when I encountered the relic of a fellow there. Thus do the generations one by one touch one another in a great chain of being as the on-going links are forged. Occasionally I relate how Helen Knighton handed off to me Twain's 1902 handshake with her. My mentioned memory often inspires listeners to ask to shake my hand, so enabling them to acquire their own derivative connection with Mark Twain. Before long my magic touch will soon expire, for no longer will I be able to continue to hand down what came to me from Twain through Helen Cruikshank when he visited his old hometown for the last time.

Twain's final visit to Hannibal in 1902 suffused him with sad nostalgia. "I was profoundly moved and saddened to think that this was the last time, perhaps, that I would ever behold those kind old faces and dear old scenes of childhood," he said at the time. Late in his life, at age 61, Twain recalled the words of an old time-haunted song, a boyhood memory from his vanished youth in Hannibal: "The day goes by like a shadow on the wall, with sorrow where all was delight." That sort of bittersweet sense of the slow flow of time, like the Mississippi slowly, steadily rolling its mile-wide tide along there at Hannibal, evokes how the days slip by like shadows as day by day they disappear. And so the days went by for Helen and Huck, for Mark and his Tom and—by now, not for much longer—for the Tom making his mark on these pages as he recalls in this reminiscence the phantom memories of his vanished days. Now near its end, my life, as I see it in retrospect, seems in some ways fictive as if I were a character in a film or a novel, a shadowy Harry Lime in Greene's Wien or an imagined Tom in Twain's Hannibal. Long before I became a ghost writer who produced this reminiscence, I was already a ghostly kind of character, as are all earthlings. Although real and vivid at the time they happened, the experiences which defined my earthly existence are now phantom memories doomed soon to disappear as I reach my ghostly end.

~ 5 ~

In Vienna in 1970 the summer moved on, the days one by one advancing like twilight shadows creeping across a wall until before long my time at the university came to an end. From Vienna I moved on to resume my travels, making my way to Venice by a roundabout route through Yugoslavia, then across the Adriatic by ferry to Ancona and up the west coast of Italy via Ravenna to Venice. From Italy I continued on to Paris and then back to London, there to end in December 1970 my European "gap year," or *Wanderjahr*, where it had started the previous December with Henry and Kathrine's wedding. The Brits gap; the Germans wander; and the Americans—how do they describe the way I lived that year 1970? Was I goofing off, a slacker, off my rocker or on a sabbatical, or what? The Yanks don't seem to have an expression to describe my foot-loose and fancy-free break from my home-based routines with overseas studying and roaming adventures. But words describing action aren't necessary to endow it with meaning. The action speaks for itself.

As usual I traveled with the attitude that since I might never return to an area I should try to see as much as possible within the limits of the available time. I followed that precept when in Eastern Europe, so before heading back west to conclude my gappings and wanderings— "wanderlust," that was the English word to describe my gap year—as 1970 aged into autumn, I started my return travels through Yugoslavia. Back then the country was still unitary, but— frontiers eventually scoring and scarring the previously unmarked terrain—Yugoslavia later balcanized into mini-states. As if echoing the noteworthy melodic musical sounds which resounded around Vienna, the symphony of foreign names lent an exotic note to the places in

Section VI: Subjective Objectives

Yugoslavia I hoped to and did see: Zagreb, Belgrade, Sveti-Stefan, Dubrovnic, Sarajevo and, somewhat gory, Split and Bled. Curiosity-teasers all, they irresistibly drew me to them. Such nominal curiosities tempted me the world over. Knowing only a name was often enough to motivate me to discover the place. In *The Asiatics* by Frederic Prokosch the narrator recalls traveling in Vietnam to "Plan-Thiet, Phan Rang, Nha Trang, Qui N'hon, Binh-Dinh, Quang-Ngai—names that sounded like a parrot calling, or the beating of a tin gong." The names in Yugoslavia sounded more like a siren song than a tin gong. I satisfied my curiosity about them by converting the abstract nomenclatures into their terrestrial realities, then crossed the Adriatic and "V-ed" my way onward via Ravenna, Padova, Venice, Vicenza, Verona. Along the way I made a "v-tour" to visit Gregorio Piaia, who I'd met in Vienna, at his hometown, Montebelluna north of Padova.

Before leaving northeastern Italy I spent a few days in Venice, still flooded with tourists during the late August high-season high-water mark of vacation time. Since I'd systematically already seen on previous visits most of the main sights, this time I wandered at random through the labyrinth where I came upon hidden corners, picturesque street scenes and typical lively local vignettes. It was pleasant to leave the busy tourist areas and lose myself in the maze-like quiet back streets before I found my way back to St. Mark's Square named, for all I knew, in honor of Mark Twain. At least it should have been, as my fellow Missourian was a square deal true-story teller kind of guy: he knew that the days went by like a shadow on the wall, and that earthlings were similarly just passing shadows. There in the square I'd sit for a time on the steps by Florian's to listen to the violin music at the famous cafe. By remaining outside I avoided paying the exorbitant price Florian's charged for a cup of coffee. Based on economic principles, you were really paying not for the beverage but to rent with a brief lease prime real estate. To a certain limited extent you could extend the lease, lingering as long as you wanted as the waiters never evicted you but waited patiently for your departure. At my age now, they wouldn't have to wait long. I nonetheless preferred the steps where I could enjoy the music for nothing.

A certain melancholy tone suffused the music with bittersweet evocations of passing time and bygone times. Although not as high-strung as the violins, I fell into a pensive mood as the mournful music played on. "Play Misty For Me" played through my mind as I recalled my already many misty memories and thought of all the people who'd passed through St. Mark's Square, few of them leaving a trace—other than memorable characters in the play of infinite forms like Mark Twain—and all as transient as the pigeons which, like time on the wing, fluttered and flew around the *piazza*. The violins mourned on as the people and the birds passed through the Square, then passed on.

Sitting there by Florian's in a music-induced contemplative mood brought to mind *in futuro*—or would have reminded me, had the order of happenings been different, but random disjointed timings can on paper be rearranged into meaning—another memorable musical interlude, one which occurred four years later miles away in Indonesia. The misty chords of musical memories—Venice's vanished violin vibes, the *gamelan* tones I heard at a *wayang* performance— often echoed in my mind. The *gamelan* music in Jakarta early 1974 seemed even more haunting than the melancholy Florian violins. At the national museum in the old part of town, a remnant of the Dutch colonial days in Java, I watched and heard the shadow play as the dreamy sounds of the *gamelan* orchestra accompanied the moving puppets. "*Wayang*" stems from the Indonesian word *bayang*—shadow. The earliest references to *wayang* puppets, colorful flat figures cut from water buffalo hide, date from the ninth century. As noted in *On Thrones of Gold: Three Javanese Shadow Plays* by James Brandon, a court poet in the reign of King Airlangga (reigned 1035-1049) wrote: "There are people who weep, who are sad and aroused watching the puppets, though they know they are merely carved pieces of leather manipulated and made to speak." The spectators "do not realize the magic hallucinations they see are not real."

The show may seem magical, but in fact the performance consists of stagecraft. Like the play of infinite forms and the earthlings who experience them, the puppet show is illusionary. On a screen lit from behind, the puppets form shadowy moving figures which create the deceiving "magic hallucinations." The play of puppet forms form the cast of a magic show which in a way is quite true to life. The moving figures slide across the screen, now here, now there, now gone as fleeting as the notes of the subdued wistful tones of the *gamelan* orchestra. The soft sad rhythms play along as the play of shadowy forms flit and flutter around the screen before disappearing, their performance at an end. The play of insubstantial forms reflects how earthlings may already be the posthumous ghosts of once living predecessors and, as such, only phantom characters of a shadow play which is the earthly play of infinite forms. No wonder the lonely planet was so lonely: the inhabitants who briefly haunted the place were all simply transient and fleeting pre-ghosts.

Time was running out for me in Europe. I moved on from Venice, and by the time I reached Paris September was getting on and autumn was beginning to chill the air. My year in Europe was moving toward a close. The Luxembourg Gardens trees, green the previous spring during my time there, were now starting to brown and their leaves would soon disappear, those leavings fated to vanish into the earth. Not long after my autumnal visit snow would blanket the winter-dead park and its decaying leaves. I'm quite familiar with leaves, as lots of trees garnish the subdivision where I live. In the fall I've watched fall lots of defunct leaves—looks, rakes, leaves—leaving their airy perches to descend to earth. It's a real come-down for the leaves after they've enjoyed a privileged life up above. Before long, I'll find out for myself how such a decline feels.

There in Paris that fall I remembered April but, as the song sings, by autumn the fire had dwindled into lonely ashes. Memories of my time at the Sorbonne and of Dominique haunted all our familiar places and I wondered who was kissing her now. You know how autumn comes on in Paris and in the U.S. Midwest and elsewhere as the days gradually cool and the shadows lengthen until finally all traces of summer have passed away. Fall has pinched the last warmth from the vanished final balmy Indian Summer days, and before long the crisp days of autumn will give way to winter's cold. That's how it goes. After my winter and spring time in the French mega-city I'd moved on, but I was lucky enough to return for a farewell visit. For me "the Paris Review" designated not the house organ of the American literary establishment but my passing autumnal stay in the city when I reviewed my vanished days there. Although never a part of *The Paris Review's* bookish in-crowd, for a brief time I had been part of Paris.

I managed to see Dominique a few times, but it wasn't the same as back in the spring. April's warmth had by now been replaced by cool September. She was still a Sorbonne co-ed, but no longer was I a student or Dominique's Anglo-Saxon boyfriend. By fall I was just another American passing through Paris, a bird of passage winging my way through town and not part of the city as during my Sorbonne time. Almost all the familiar faces from those days were now gone, my classmates scattered to the four winds like soon would be the fallen leaves of the Luxembourg trees. Only my memories survived, "Memory Green" recollections as the autumn wind stripped "branches of their dying leaves" and you along Friedrichstrasse at dusk "Or you in Paris" walking along the windy quay shuffling "the fallen leaves before you" as "sweetness suddenly fills your heart " and "tears come to your eyes." Where and when had I remembered those words before, "Ah where?"

My autumnal September visit in Paris brought me not only twilights far from home but also a crepuscular feeling that, as James Joyce lamented in his poem "Bahnhofstrasse," "High hearted youth comes not again." What is it about those *Strassen, Friedrich* and *Bahnhof*, which so inspires Hibernian-Anglo-Saxon poets to wax so eloquently about dying leaves and vanished youth? You don't need a Germanic Fred Street or a Station Street to suggest that nothing remains stationary, that all is in motion, and that all roads, rail and other, lead on and on to a point of no return, a one-way track or route to the terminal point where the trip ends. Paris that fall was for me a way-station leading to the end of my year in Europe and then, beyond, onward to the final stop back in London. Then what? My three five-month-long early 1970s travel adventures were as yet unformulated and in the future. The near past—my time in Europe—was set and fixed for all time, but ahead loomed desert-empty days I'd have to fill with contents that would become my life.

My gap year in Europe fulfilled my expectations but, like the Japanese tea ceremony, it was a one time, one meeting interlude. Paris was over, and by then so was my formal education at the three European universities and elsewhere. My classroom student days had reached their end. But another more classy type of education out in the wide world continued over the next four decades as I experienced life outside the cloistered halls of ivy. At times I encountered out in the world poison ivy, plots and counter-plots and other such threats and hazards but those, too, contributed to my education. First World, Second World, Third World, Third Man, Thomas Mann, Thomas Sawyer, Dylan Thomas, Keith Thomas, whatever or whomever, they all contributed to my experiment in living. The worldly studies didn't represent junior year abroad but more mature senior years afoot with twilights far from home giving me a post-graduate course from which I've not yet graduated and never will. Life is too short to earn a diploma. You learn by degrees but don't receive one.

~ 6 ~

As an important part of my past, Europe brought me many fond memories, and over the years I hoped by occasional visits to deposit in my memory bank further treasured recollections of my time there. During my senior years abroad I from time to time returned to Paris and to the three other great European capitals I'd come to know well. Vienna and Paris I'd experienced primarily as a resident student, while London and Rome I knew through English and Italian friends who graciously shared with me the inner workings and playings of their cities. Rome I'd first visited as a tourist in 1955 when still a very old boy or a very young man, and since then I'd often returned. These repeated visits I attributed to a reason other than casting a coin into the Trevi Fountain to honor the tradition that this ritual ensures you'll return to the Eternal City. Rather, I went back to Rome over the years to see my long-time good friend G, a classmate who I'd met at an American university in the fall of 1959. But to be on the safe side it didn't hurt to toss a coin into the fountain just in case the juju procedure somehow actually worked to guarantee you'd again return to Rome. Given the role of chance and luck in life, you might as well try to coddle those random forces by following a ritual humanimals seemed to find useful, even though I suspected that it was the city of Rome collecting the Trevi's proceeds—rather than the tourists who contributed the coins—which found the practice most beneficial.

On my first visit to Rome the summer of 1955 I looked on as a Scotsman, obvious by his pronounced burr, tossed a coin into the Trevi. After he'd parted with the piece, making it truly a liquid asset by throwing the coin into the water, I remarked to the man that never had I supposed that I'd see a Scotsman so casually throwing money away. He immediately replied, "I d-d-drilled a small hole and put a st-r-r-ring through it so I can pull the coin back and keep it." This explanation restored my faith in the admirable Scottish trait of frugality, and it served an an example to me never to overlook ways to minimize expenses. However, I never adopted the Scotsman's fanciful Trevi Fountain technique. I felt that retrieving the coin might jeopardize my chances of returning to the city.

Would that we could so easily withdraw the spent and, even better, the misspent hours and revise the cast-away past to recall—other than only by memory—ill-fated decisions or unlucky happenings, misbegotten choices, unfortunate events, undesirable outcomes, opportunities tossed away which, unlike the Scotsman's coin, are irretrievable. On the time-worn 1410 *orloj* (*horloge*) Old Town Hall astronomical clock in Prague a little window opens on the hour and small figures of the Apostles and Christ file by, a European version of a *wayang* shadow show.

As Death tolls a bell, an old man shakes his head to indicate he's not yet ready for death—will he ever be?—and Death then reverses the hourglass. Backward running sands of time to foil deserts of vast eternity exist only in the imaginative minds of characters like a medieval clockmaker or a thrifty Scotsman. Old men shaking their heads won't shake off the years gone by and Death to come. Clocks may stop, but Time doesn't. Tock-tick, tick-tock, on and on clicks the clock of Time, marking the seconds as has the *orloj* for more than six centuries. On runs Time until what seems forever becomes forenever and the clock finally winds down, the little figures no longer circling and Death no longer reversing the hour glass. Time's up.

On my most recent Rome visit, sixty years on from my first time there in 1955, the Trevi Fountain was waterless, perhaps due to global warming but more likely shut down for restoration. I wish I could say as much but, just as for the vanished years, I'm beyond restoration. Although the fountain's spouts didn't spurt, tourists nonetheless tossed coins into the water-less bare basin, the little metal discs paving the floor like an ancient Roman mosaic but without a pattern. The litter of coins reminded me of the discarded losing betting tickets which carpeted the floors at the Del Mar race track in southern California I'd visited with Charles and Audrey Skirball Kenis. Both the coins and the tickets represented hope, which all too often was misplaced. Once the race was over you can't revert to the starting gate or reverse the hourglass or re-run the course. Renaming the Trevi Ivert wouldn't allow the fountain to "rivert"—like Death on the *orloj*—to the time when flowing water enlivened the now moribund monument. With the fountain now dry, perhaps the Trevi coin tradition represented a losing bet. Nevertheless I decided not to tempt fate by omitting the ritual, so I relinquished my coin, painful as that was, and added it to the others. In the spirit of the Scotsman I'd seen there six decades earlier I used a very small denomination coin, which I deemed to be acceptable as a senior discount. But I felt as if I was throwing money away, as somehow I doubted that I'd ever again return to Rome. So far I haven't. Maybe I should request a refund. I don't like to think that I won't see Rome and my old friend there again, but there comes a last time for everything, a final moment when the hourglass sands have run out. The only remaining sands are those in the desert of vast eternity.

During more than half a century my investments in the Trevi Fountain pay-to-play system produced results by bringing me back to Rome where G's generous hospitality enabled me to enter into the city's social and cultural life in ways which gave me an inside view of the ancient metropolis. This privileged access let me see and learn about all manner of typical day-to-day ways of being as lived by Romans. I thus came to know the city well. Like many places, Rome holds its secrets closely. Hidden worlds lie behind closed doors which conceal inner sanctums inaccessible to casual visitors. As did my London friends, my Rome friend opened some of those doors for me so I could look inside. A few of the doors I managed to pry open myself. On one visit to a fabled local establishment, not far from the Trevi Fountain, I observed another Rome ritual by getting a *gelato* at Giolitti, the too-famous ice cream emporium in the center of Rome. "Too" renowned, I say, because so well known is the shop tourists and Romans alike cram into the cramped space to line up for the renowned frozen confection, noticeably better than the mushy and tasteless Lyons Mr. Whippy I'd sampled in London nearly 60 years before. Giolitti is one Roman attraction I wish would have remained more obscure and operate behind closed doors. On my 2015 visit I struck up a conversation with a friendly fellow working at Giolitti who I assumed was a store employee but who turned out to be none other than Nazareno Giolitti, the current generation's owner of the enterprise founded in 1900 by his ancestor. Encountering the proprietor of Giolitti sure beat meeting the owner of a local Dairy Queen franchise back home. Nazareno shared with me some of the history and lore of the firm, which in 1930 moved to its present location in the heart of Rome a few blocks from the Pantheon. Among other comments, he noted that customers from different countries favored different flavors. For my part, I always chose a two-scoop cone of tiramisu and chocolate topped with *panna*, whipped cream which crowns the ice cream with a foamy white topping. If it so happens that I never return to Rome, Giolitti *gelato* is one of the city's traditional pleasures I'd most miss. But at least I'd save myself the expense of throwing yet another coin away into the Trevi Fountain.

On a previous visit, long before I met Signor Giolitti, I encountered another famous local personage, a "Ro-man" whose name was a brand which evoked the city. Alberto Moravia set many of his works in Rome, including the *romanzo La Romana*, a novel which recounts the seedy story of "a woman of Rome." While working as a prostitute she manages to retain some sense of dignity and grit as *la Romana* plies her trade in Mussolini's time. That combination of decay and dignity seemed to me emblematic of the Eternal City's chiaroscuro tones, the age-old settlement shadowed by a long and tumultuous past and brightened by a cultural heritage as rich as Giolitti's *gelato*.

Moravia and I met briefly one day when he attended a rehearsal of his play *La Vita è Gioco* (*Life's A Game*) at a theater on Belsiana Street in the center of town. When I'd by chance passed by the theater earlier I noticed that a rehearsal would take place that afternoon. At a nearby book store I bought Moravia's *Il Disprezzo* and later returned to the theater where I sat a seat away from the director, Dacia Maraini, at the time the author's *convivente* (living with) companion. She asked me to give her the wrapping paper around the book to use to write notes during the rehearsal. Following the performance and her comments to the players, Maraini discarded the pages which I retrieved and have retained. After Moravia arrived at the theater we chatted a while and he autographed his novel for me. Called *A Ghost At Noon* in English, the plot of *Il Disprezzo* (contempt, scorn) oddly mirrored Moravia's own story: brainy writer loses beautiful wife. Although only a live-in ghost, probably not very spooky at midday, and not married to Moravia, Dacia eventually left him, perhaps finding the intellectual author insufficiently Romantic. Or maybe the novel's plot inspired her counter-plot. As Moravia and I talked I noted his trademark thick bristling eyebrows like horizontal parenthesis and his prominent nose centered in his rather patrician face. The very image of a distinguished author, Moravia seemed straight from central casting there at the theater. But he was no actor—it was the real Moravia.

On my last visit to Rome—will it, alas, be my last? I threw the coin—my friend and I visited Moravia's apartment, opened in 2010 as a house museum. His residence has been preserved almost exactly as it was when Moravia lived there, just as have many other such literary shrines. Faulkner's Rowan Oak in Oxford, Hemingway's Finca Vigia in Cuba, Snoopy's kennel, and in Rome the apartments of Mario Praz in the center and of Luigi Pirandello located in an obscure residential area all remain unchanged from the time they were occupied. So untouched did Moravia's apartment appear we felt like intruders. It seemed as if he'd return before long to find there two uninvited strangers in his residence. Maybe he'd offer us three cups of tea like the Tuaregs in Timbuktu. More likely, he'd ask us to leave. If the author happened to return to his old haunts, a ghostly *revenant*, Moravia would find his familiar surroundings unchanged, even if he himself was greatly changed into a phantom. No tenancy is permanent. Furniture, photos, paintings, items of everyday use and other objects remained to evidence the writer's daily life. A large writing desk and a 12,000 book library evoked Moravia's professional life. I looked at a few of the books, including an Albert Camus novel inscribed (in French) "from Albert to Alberto." I sat at the desk momentarily hoping for some inspiration but it never arrived. Must have been on strike. (For foreigners traveling in Italy, *sciopero* is an essential word to know.) An outside terrace overlooks the languid Tiber, a river which in that quiet corner of Rome lacks the energy and presence of such robust waterways as the Thames, the Seine, the Danube and especially the dynamic Mississippi. But like those more vitalized rivers, even the sluggish Tiber eventually would find its way out to the sea, there to disappear.

Moravia's apartment on Lungotevere della Vittoria by the Tiber was located in a peaceful neighborhood outside the center of Rome. He'd moved there in 1963 from his more central place on Via del Oca, and had lived at the Lungotevere apartment with Dacia Maraini until she left him. He then married Carmen Llera in 1986, and the couple remained together at the apartment until he died there on September 26, 1990. In *The Time of Indifference*, Moravia's first novel published in 1929 when he was 22 and set in 1920s Rome, Michele tells (in Angus Davidson's translation) his paramour Lisa: "I can't love you. It's always the same reason—indifference; it's always indifference. And so...rather than to pretend to fall into your arms, to be dying of passion for you, rather than make declarations of love to you—seeing that I can't do it successfully—I prefer not to do it at all." So did life imitate art, and Dacia departed. Michele seemed like a cut-out character in a *wayang* puppet play: "across the flat, blank screen of his indifference sorrows and joys passed like shadows, leaving no trace...everything around him was without weight and without value, fleeting as a play of light and shade."

If a humanimal is merely a plaything in a plot-less shadow show, the creature is bound to view its brief earthly existence as a "time of indifference." How to convert indifference into some sort of transient meaning is the essential challenge for an earthling. This represents a difficult enterprise, as the system is rigged against you. It's impossible to boycott or girlcott the rigid realities, such as time, mortality, sleep, decay, the operation of chance, and all too much more. The way things are irrevocably arranged imposes on humans existential infelicity. The very nature of being entails non-being. I could do nothing about that, but rather than succumbing to indifference I could try to avoid the double jeopardy of both cosmic and personal meaninglessness. This I attempted to accomplish by imposing a small overlay of meaning upon the fundamental and ineluctable oppressive realities on the lonely planet. The foundation for my not very super superstructure was quicksand, for I could find no *terra firma* on which to built a more substantial construct. I knew that at bottom everything stood on the sands of a desert of vast eternity. My flimsy little stage on which to play out my life was the best I could manage to build. Part of a ghostly shadow play, I played my part as best I could. I tried to experience as wide a variety of the infinite forms as possible within the limitations and opportunities of my situation and the time available. Although in the end my experiment in living would make no difference, in the meantime I tried to avoid indifference.

Since that first visit to Italy in 1955 as an old boy or a young man, I've managed on many visits to see most of the country's main areas and attractions, both on and off the beaten track. Over the years and kilometers I traveled from the lakes in the far north edging Switzerland to Sicily not far from Africa, covering many places in between. Each city and site comprised a piece of the mosaic, residues of time which pattern the vast past of that ancient land, historically and culturally perhaps the richest in Europe. The strata of time which characterized the tiny mid-Italy village I came to know especially well—Civita di Bagnoregio—was in its way a microcosm of the many layers of past eras which defined Italy's history. The mini-village encapsulated the country.

The villagette of Civita (as locals commonly called the tiny town) served as the setting for a Pirandello-type play peopled by archetypical characters who made their entrances, performed their parts and then departed, sometimes never to return. Civita was in that way a miniature stage set where typical Italian dramas played out on a human scale. With fewer than 10 permanent residents and largely empty most of the year of the seasonal outside homeowners, the isolated diminutive hill town, some 90 miles northwest of Rome and about 13 miles south of Orvieto, comes alive in the summer with visitors—too many these days—and with residents who occupy their own houses. Especially during the August *Ferragosto* holiday, the tiny town brims over with tourists swarming through the cramped streets and byways and also the regulars in their restored residences.

Especially memorable is a visitor's first view of Civita. Perched atop the squared-off free-standing hill, the hamlet seems to grow from the high rock formation above the valley as an organic settlement grounded in the tiny terrain it occupies. This granular connection with the earth was seemingly more natural than the artificial constructions imposed on the lonely planet by the hand of man. Civita, so it seemed, belonged to the earth, forming part of it rather than just existing as an induced intrusion—or extrusion—fabricated by humanimals.

My old friend G, a prominent Rome lawyer, became the first person to restore a house in the village, which she first saw in the summer of 1961. In a charming privately published memoir of her long history in Civita, G recalls her first visit. "I didn't think that in the twentieth century there still existed spots like Civita, a timeless place."

Time was when Civita remained almost completely unknown and undiscovered, even in the modern age. In *Forty Plus and Fancy Free* (1954) by Emily Kimbrough, she quotes from an obscure letter sent by Bernard Berenson to some New York City friends describing the panorama onto Civita from below:

> Between Montefiesconi and Viterbo you will see on the right hand side of the highway a little sign pointing to Bagnoreggio [so spelled]. Take this road and when you reach Bagnoreggio, which is a little village, motor slowly down the main street until you come to a sort of fork. The left hand side of this fork should be a small road diminishing into a country lane. You walk down this on foot and when you get to the very end—as far as you can physically walk, going always straight ahead of you—you stop and look. What you see will be of large dimensions, so you need not worry about looking for something tiny and obscure. But you will have to keep walking right to the end.

Berenson was correct to keep his description nondescript, for one of the great pleasures of discovering Civita involves suddenly coming upon the hilltop village as a kind of magical apparition lost in the sky. Following Berenson's advice, Emily Kimbrough told her readers, "You will be rewarded. And I doubt that you will see elsewhere anything like this 'Mystery View.' It is not recorded on a tourist map." Like Afghanistan in the early twentieth century, Civita was at the time "off the map."

Civita still remained as obscure as when Berenson had happened upon the village at the time, some years later, G first saw the "timeless" town in 1961. In 1963 the distinguished "*Romana*" lawyer bought in the hamlet a dilapidated stone house which in 1966-7 she restored. For the restoration, the first in Civita, G engaged American architect Astra Zarina. The pioneering project was the forerunner of many similar restorations which over time revived the previously moribund village. Not only Italians but also Americans, French and other foreigners followed G's trend-setting example. Some of the Americans were former students of Astra, who until her death in 2004 ran in cooperation with her husband Tony Costa Heywood a University of Washington program which brought American students to Civita to study Italian architecture and city planning. By now dozens of outsiders spend August, and shorter stays at other times of the year, at their restored houses in the village, which in season bustles with lively social activity for the residents. The busy behind-the-scenes scenarios—morning coffee gatherings, afternoon chit-chat and gossip sessions, convivial evening social occasions, cocktail and dinner parties, private get-togethers—day visitors passing through aren't invited to join. Even in tiny Civita, as at Rome, much of local life takes place behind closed doors.

Set out of harm's way atop the steep-sided hill, the high ground was first settled in pre-Etruscan times, and then in the sixth century B.C. the Etruscans took over. Some 200 years later the hamlet became a Roman settlement, with the typical *decumanus maximus* Roman route, more *minimus* than *maximus* in little Civita, cutting through the center of the village. Over the centuries the hill's soft volcanic stone and clay have greatly eroded, wind and rain clawing away the fragile terrain. As do all earthly settlements and institutions, Civita stands upon unstable and vulnerable ground. A severe 1695 earthquake destabilized the settlement, further reducing its limited space, by now confined to an area smaller than a football field. Engineering works, supporting pylons and braces installed in recent years serve as a holding action to keep the cliff-sides from further deterioration. As the years go by, the historic hamlet will no doubt continue to decay away piece by piece and rock by rock until finally the town crumbles into dust. This ineluctable fate has given to Civita the designation *La città che muere*—the dying city.

Not yet completely crumbled into dust, Civita for now still crowns the crest of the hill like an island enclave isolated in the sky. In his 2015 reminiscence *Le mie crete* (*My Clays*) Professor Giuseppe Medori describes his native village—to which he often returns to stay in his ancestral home—as the "enchanting island of Civita." The village indeed continues to enchant the estimated half a million yearly visitors who trek up the steep bridge to reach the ancient hilltop town. Time was when a picturesque rickety old wooden bridge connected the island hamlet with the mainland below. Donkeys carried provisions and other supplies up to the village. Retreating to the north in 1944, the Germans destroyed the vintage span, replaced in 1965 by an unsightly even if more functional modern bridge. A golf cart-like vehicle known as *Il Tractore*, motorized horse power rather than donkey power, now carries supplies up to the town. Only very rarely are cars permitted on the bridge, so Civita remains one of the few places in Italy without traffic. Nonetheless, time's wingèd chariot still rolls on, even up there in traffic-free Civita. But at least you can stroll around the village without ever encountering a car, even if occasionally a rogue motorcycle roars through the town.

During my visits over the years I came to know many of the regulars, both the few permanent locals and the seasonal residents in town during *Ferragosto*. The community accepted me as one of its own, and as time went by I became a part of that remote tiny Italian hill town. As an insider I heard the town chit-chat, chatter, gossip and rumors, all of which provided a running commentary on the various dramas which played out on the little stage that was Civita. Plots and counter-plots included the full range of humanimal behavior—jealousies, family feuds, laughter, tears, comedies, tragedies, farces, rituals, passing disputes and refutes, passing visitors, disappointments and joys, births, deaths and all the rest of what earthlings on the lonely planet experience. The title of Pirandello's *Six Characters In Search of An Author* set the scene. Little more than that number comprised the repertory company of permanent residents, while to the cast were added some transient seasonal performers like G and me.

My role seemed to be *il visitatore Americano*, a slightly buffoonish and befuddled foreign figure, a character who mixed Spanish words into his Italian; who drank espresso at the wrong time of day, and sitting down rather than standing at the bar; who thought that Manzoni was a vegetable, Cardinals a baseball team, *Vietato Fumare*—his name written all over Italy—a really popular politician or pop music star; who put Parmesan on pastas suitable only for other cheeses; who supposed *gelato* was the Italian version of lotto; and who, when trying to be humorous in the local language, in fact spoke only laughable Italian. For my performance as a bit player on the Civita stage I received no Tony, not even from my local friend Tony Heywood, nor have any offers come my way to audition for other roles in the hamlet's dramas.

In recent years the village has much changed, and not for the better. Too many souvenir stores, upscale restaurants, fancy bed and breakfast facilities, and other commercial establishments have robbed the village of its former unspoiled charm. During summer days visitors clog the few narrow streets, turning the town into something like a mini-amusement park. Later in the day, however, after the transients depart, Civita's essential character, with its communal togetherness and underlying medieval atmosphere, revive. The time-haunted old stones once again come to life along with the residents as they resume their social rounds. Once again Civita belongs to those who live there. Old-style lanterns cast patterns of light onto the shadowy lanes, now empty and quiet. At dusk the town reverts to the age-old ambiance which still prevailed in recent times before the crush of daily tourists and the latter-day modernisms changed the hamlet. Civita is one place where, as twilight comes on, the clock turns back—back to an era when the remote village was lost in place and in time.

At twilight we'd walk out to the front terrace just outside the Porta Maria, the portal at the top of the ramp above the bridge. As dusk falls, one by one lights in Bagnoregio across the way blink on. From the isolated little island in the sky you can see a few stragglers descending the bridge, gradually fading away into the gathering dusk, shadowy wraiths lost in the darkness of the dying day as once again—as it has over the centuries—night falls on "the dying city."

Section VII
SHADOWS AND SMOKE

~ 1 ~

Long have I anticipated the end and now, nearing my ending, I return to the beginning—my early days—to start this final Section, my last chance, before I give up the ghost, to haunt the future by leaving behind my ghost-written reminiscence. Only toward the ending page and age can you gain something of a meaningful perspective on what by chance happened to comprise your earthly existence. In conformance with the year of publication, I can now view the past with 2020 hindsight. The structure of a backward-looking book and of the lived doings and then dones which comprise its content both depend on arbitrary constructs—the original random happenings and then an artificial pattern later imposed on what occurred. The concluded overview is not conclusive but only one of many possible arrangements. Even at the end, with an apparent order, the tangle of past events which form a life defy any ultimate meaning. An earthling's terrestrial existence is only an experiment, one without a defined end except for the creature's end. The largely random happenings which shape an individual life as well as most of the basic prevailing conditions which set the terms of existence remain beyond explanation.

Starting at the very beginning with your fatal circle—the genetic endowment and the when, where and into what circumstances a newly hatched humanimal emerges incarnated into its earthly form and format—most everything depends on the capricious operation of random factors. In the pre-earthling's first living room, the womb-room, the person-to-be enjoys, oblivious to the world beyond, cozy and coddling comfort with all the necessary conveniences supplied by an accommodating hostess. Unaware of what's to come, the creature begins its existence there in that protected environment. Then, as if leaving home for adventure travel in the outside world, the sheltered being is expelled from that hospitable space into less accommodating surroundings, thrown from a nurturing habitat out into the lonely planet, yet another character, one of billions, joining the play of infinite forms. Just as travel exiles you from home-based security by cutting the umbilical cord which ties you to your motherland, so does parturition distance you from your pleasant comfortable and familiar space. You're on your way, your earthly existence a brief way-station en route to the terminal.

The contrast between home and away I've so often mentioned in my account refers not only to the two elements of my own life, as from my thirties, but also serves to describe existence in general. For an individual life, as mine played out, home represented my normal, familiar and regular state of being, while away was an alien and foreign realm. From an existential standpoint, however, the opposite is the case. Being alive—a strange and exotic phase—is foreign to the natural order of things, as non-existence is an earthling's true home. The norm is not to exist: oblivion is the rule; being, an exception. Prenatal and postmortem non-existence is much more common, comfortable and non-threatening than is life as an animate form, an aberration which defies the normal state of affairs. A humanimal's natural habitat is the desert of vast eternity, a no-place where the creature doesn't exist. Being incarnated into an earthly form was like my visit to Ura in the remote eastern reaches of Bhutan where I knew no one, didn't speak one word of the language, had no knowledge of local customs or beliefs or habits. I was a complete stranger, far from home, a momentary passing presence soon to be absent. For earthlings, their true home home is Erewhon—nowhere, in Samuel Butler's term. Humanimals belong nowhere; during their brief earthly existence, those creatures are exiled from their natural habitat. When, where and how I will revert to the normal state of non-existence—transitioning from ghost writer to ghost (my second such non-being interlude, after my long pre-birth oblivion)—I do not know. The suspense is killing me, even if not yet nature.

Only chance determined when and where I happened to emerge from oblivion into being, by now a long lifetime ago. Born into an incurious family, for some unknown reason

I was endowed and cursed with compulsive curiosity, an innate force which over the years operated to shape my life. For me, curiosity determined one of the most compelling terms of my existence. An advantaged fatal circle, along with other fortunate circumstances and favorable chance happenings, allowed me to enjoy something of a charmed life (if a favored earthling like me doesn't realize his life is charmed, then it isn't), one I was able to live in my own way. But even with my head-start nothing which subsequently occurred was inevitable; it was all quite evitable. What happened to take place to format my life mostly depended on luck, chance and fate. In my attempts to create a satisfying outcome I did the best I could with the time I had. From the very beginning time preoccupied me. Even in my early years I understood that time quickly decayed and soon vanished, a realization which produced in me an acute temporal pressure bringing with it a sense of urgency. Nonetheless, although I was aware of the fleeting and limited nature of time, ever on the wing no matter how you filled it, when I was young the years ahead seemed long. Not only do the living and the dead occupy two different realms but so do the young and the old. When young, you really can't imagine time's velocity and how it is to be old, but oldsters with almost all of life behind them are well aware of time's slippery and fleeting nature. A senior sees the years with a perspective quite different than when he was young and empty-headed. Time fills heads while at the same time it gradually empties the hourglass of allotted time as the sands, evoking the desert of vast eternity, drop away second by second grain by grain.

The young and the old live concurrently but in two different time-frames. By old or even just older age, it makes little difference what you've come to understand about time because by then it's already too late. The sooner a person realizes the temporal constraints, the better he can attempt to order his life—always within the framework of his fatal circle and other circumstances—in the way deemed desirable. In *The Tartar Steppe* (translated by Stuart C. Hood) Dino Buzzati described the difference between how we experience time as a young person and then as an older and hopefully wiser humanimal: For children the road into the future "seems infinite, where the years slip by slowly and with quiet pace so that no one notices them go." As time goes by, however, your perspective changes for "at a certain point we turn round, almost instinctively, and see that a gate has been bolted behind us, barring our way back. Then we feel that something has changed, the sun no longer seems to be motionless but moves quickly across the sky...clouds no longer lie motionless in the blue gulfs of the sky but flee, piled one above the other, such is their haste. Then we understand that time is passing and that one day or another the road must come to an end."

Insubstantial as smoke, clouds puff into being, scud away, evaporate into thin air. Those insubstantial fleece-like errant forms glide and slide slowly and steadily across the empty skies, the random fleeting puffy patterns drifting away with no *deus ex machina* guiding them. In Aristophanes' *The Clouds* Strepsiadas asks Socrates what force sends rain. "Not Zeus, but the clouds," replies Socrates. Strepsiadas persists, asking "Who but Zeus makes the clouds sweep along?" Socrates answers, not a divinity but an "atmospheric whirligig." "Whirligig?" Strepsiadas echoes, observing that he never thought of that explanation— "Whirligig rules now."

As a young man, long before the road approached its end, I started to think about the future and about the big wide world where the great play of infinite forms spun by the ceaseless whirligig of time and chance played out. The forms were infinite but not my time to experience them. I thought about the track ahead, the seemingly direct track forward and the end toward which my fatal circle, setting and circumstances led. Casting my thoughts ahead, in an anticipatory way I began to regret all the roads not taken. I tried to imagine looking back, years later, after the gate had been bolted behind me, at how "time passed, time passed flowing in a dark stream, staunchlessly, as though from some profound mysterious wound in the world's side, bleeding, bleeding forever," as Aldous Huxley put it in *Antic Hay*. Early on in my experiment in being I realized that I didn't want to stand by passively as time drained away, flowing on and on like the Mississippi or a blood-red artery like the Red River, sapping me of the very lifeblood of existence without attempting to use the precious life-giving substance to infuse my years with some new streams of sensory impressions. On and on time flew whatever I did, so I contemplated how best to grasp the fleeting minutes and years to fill them with content meaningful to me. Even the most skilled hematologist couldn't replace a drop of the precious insubstantial substance which dripped away day by day. Only transformation and not transfusion was available to me, as any effort to slow the flow of time as it pumped on as if through a vein would be in vain. Time hemorrhaged like an untreated wound. Once time bled away it was gone and nothing could revive it. Its passage was like the Phantom Lake campfires at summer camp in Maine a lifetime ago when time melted away just as the flames' heat ashed the logs into nothingness, the smoke curling languidly into the air like clouds before disappearing forever. As I gazed pensively at those smoking campfires at Ironwood I thought it was passing strange how the wood which seemed so solid soon vanished. Even then, as a young boy, I realized how things really were in life—were, and then weren't.

Although somewhat contemplative as a boy, I was by no means morose, introverted, withdrawn, reclusive, dreamy, unsocial or anti-social. For the most part I was a regular kind of kid, a boy interested in sports to watch and to play, comic books, radio serials and the breakfast cereals advertised on those programs, pinball machines, fun with chums. When I was young there was no television, but later when it came in I enjoyed watching Howdy Doody, Hopalong Cassidy, Dave Garroway—popular back then but by now all gone and forgotten—and many other shows. Away from the screen, out in the real world, I laughed a lot, mainly at my own jokes just to be sure at least someone did, and at a certain age I started looking at girls even if they seldom looked back at me. I enjoyed hanging out with my friends but—just as I liked my own jokes—I was also quite content with my own company. I passed a lot of time by myself reading, although whatever my pastimes—then or later—time passed quite quickly on its own. I had no problem keeping busy, for as a self-sufficient child I could entertain myself without the need for outside diversions. From early days and all through the years on up to the present day one of my challenges was how to keep unbusy part of the time. Because all too much interested me I never had time enough to satisfy my curiosity. Almost everything both attracted and distracted me. The play of infinite forms intrigued me, and I found something worth learning about in almost everything which happened to come to my attention. As time went by, however, I knew that it would be necessary to limit my curiosity to exclude vast realms, as no human life lasts long enough to grasp other than a very small part of the available forms.

Rarely did I suffer a dull moment as boy growing up. Even the most mundane matters somehow fascinated me. I wondered why the tongue of Teddy, my childhood chow pet dog, was purple. I still don't know. I glued myself to the radio to keep up with Captain Midnight (I was overjoyed to get my hands on the secret code ring offered to listeners, even though I had no secrets to code), the Green Hornet, the Lone Ranger, Jack Armstrong the All American Boy, the Shadow—he had the power to cloud men's minds. I tried to keep my mind unclouded to avoid an "atmospheric whirligig," as Aristophanes wrote. It wasn't all a free ride—getting to hear about the daring exploits of Midnight, Hornet, Ranger, Armstrong or the phantom-like Shadow. To earn that pleasure you first had to listen to the commercial. From that I learned that even for a ghostly world such as the Phantom's you had to pay one way or another to participate. In the winter when the ponds froze over our gang organized hockey games, while in other seasons we played softball or touch football. I even earned a letter in high school freshman basketball, my only sports letter, although a quarter of a century and more later thousands of eventually published letters clicked through my early 1950s Royal Portable Quiet Deluxe typewriter. Neither particularly quiet and for sure not deluxe, the machine is an antique—just like its present-day user. The vintage metal chunk of now primitive technology, a once state-of-the-art device which produced the words contained in ten books and some 200 articles, still serves as my word processor, which I occasionally upgrade with a sophisticated

"NR" attachment—a New Ribbon. One of my great anxieties has been the nagging fear that one day typewriter ribbons will no longer be available. He who worries about such minor matters indeed enjoys a charmed life. It's important to keep things in perspective, so the threat of typewriter ribbons becoming obsolete represents a more significant and concerning matter for me than global warming, atomic warfare, world-wide epidemics, a Third World War, an invasion by Martians, cyber attacks and even Dairy Queen going out of business.

Whatever the fate of my typewriter, my time as a writer now reaches the finish line with my book's final lines. In writing this reminiscence the lettered phase of my existence nears its end, just as does my life. The alphabet of my being has by now nearly reached the "Z." Unlike Percy Fawcett, who unsuccessfully searched the Amazon jungle for the lost city of Z, my life, like this book, will soon inevitably get to Z—The End. Where Rimbaud saw a color in each vowel, I saw a stage of life in each letter. In getting nearly to Z I skipped Y as "Why?" didn't feature in my life. X, however, marked the spot as that letter highlighted how I tried before I disappeared to see below the surface of things to X-ray in this reminiscence the inner workings of my life, the skeletal framework beneath the superficial surface of passing happenings.

My retrograde typewriter typifies my disconnect from modern technology and from much else in the contemporary world. I'm not only old but also rather old-fashioned. Just about my only digital resources are my fingers. I type hunting and pecking, rather like the way I hunted for experiences the world over while trying to keep my place in the regular and orderly pecking order at home. Happily, I escaped getting entangled in the world-wide web. I've never owned an iPhone, a computer or any other such digital device. I never chained myself to LinkedIn, tweeted on Twitter or booked myself into Facebook. That was one book which didn't interest me, and anyway—like a proper ghost-writer—I preferred to remain faceless. "Virtual reality" I deemed unnecessary, as for me the real rather than an artificial virtual world completely engaged and satisfied me. Artificial intelligence is certainly more high-powered than mine, but at least my natural intelligence belongs to and is controlled by me. Although I occasionally access the internet for email and some research on a public library desk-top computer, I've for the most part remained off-line in a way similar to how I also opted to go off-line from the tracks which extended ahead of me on and on along the same route into the future.

For me social media seemed anti-social as I relate better to people in person. For reading, Kindle kindled no interest as I prefer tangible and tactile paper-based materials such as books, newspapers, periodicals, texts on Cracker Jack prizes or Chinese fortune cookie strips, bumper stickers and other such inky writings. My surge protector device for my reading is the hand I use to close a book or to put the paper down. Because written works are already once removed from the real world, I prefer not to insert yet another layer in the form of artificial digital images which screen me from the real world by intervening between me and life in its original sensory richness. I'm in many ways still living in the twentieth century, but that feels comfortable and suits me as it's the time-frame in which I spent most of my life.

I've always tried to keep things simple. As best I could I kept my time free of extraneous distractions, and few gadgets or unnecessary things clutter my place in space. The relatively few possessions I own are for the most part functional and without pretension. My car is old enough to vote. My pared-down and tidy regular domestic life contrasts with my somewhat disorderly and irregular on-the-go foreign life, which entailed various complications and challenges. But those sorts of jumbles and muddles represented the kind of experiences I sought. My vintage typewriter, my archaic values and interests, my somewhat outdated way of life all represent a mode of existence largely overtaken by the modern era. I have always to a degree been behind the times. At the same time, however, I was in some ways far ahead of my time. Although some of my practices, habits and attitudes remain in retroversion, I based many decisions on forward-looking and far-seeing perspectives on how the world and my place in it might look years ahead. More often than not my view proved sufficiently clear to give me a workable picture of how things things were eventually likely to develop. By now my exercise in futurology has been overtaken by events and these days any forward-looking thoughts apply only to a limited life-span. Looking ahead has thus become quite simple, as before long for me all will be settled. I will soon be beyond all earthly plots and counter-plots once I begin my endless repose in a tiny and lonesome plot—my final destination on the lonely planet.

By merest chance I was born into an area and an era which completely suited my temperament, even if I was often out of phase with the times. The world seemed slower back in my day, but that was just an illusion as time advances at the same relentless pace at any and all times no matter how it seems to move. For the first three decades of my existence I for the most part accepted the settled and sedentary setting and conventional life path which by chance comprised from the beginning my fatal circle and other circumstances. I entered the world, grew up and still live within the same very small circumference. My fatal circle wasn't particularly limiting, but by choice the place where I lived was geographically limited. On the tiny stretch of terrain where I spent my life my roots ran deep, and I saw no reason to uproot myself to abandon my familiar surroundings. It was more convenient and comfortable just to stay put. Only a few walkable miles encompass the circumscribed area where I passed my entire—at least to date—existence. My reserved cemetery plot lies a few miles beyond my life-long habitat. It will seem strange—a new adventure—occupying a space somewhat removed from the familiar little area where I lived during my terrestrial presence. Remaining in the provincial Midwestern city of my origin largely satisfied me. That is, such an existence satisfied up to a point. But then in my mid-thirties I decided that it wasn't enough just to keep going along on the same limited track to the end. I needed to diversify the routes.

I realized at that point that I didn't want to live my entire life as a one-trick pony or a one kennel dog, a creature restricted to a home-based one-ring circus. I resolved to induce into my existence some irregularity, a kind of "limited sloppiness" as described by biologist James D. Watson quoting a fellow Nobel Prize winner in his memoir *Avoid Boring People*. He noted that "if you are just a little sloppy, you have a good chance of introducing an unexpected variable and possibly nailing down an important new phenomenon," whereas "always doing an experiment in precisely the same way limits you." For my experiment in living as a humanimal I introduced some uncertainty and irregularity by foreign travel, an activity which by its very nature entailed sloppiness. Chance happenings while away would bring me unsuspected and unexpected variables—some favorable and others not. I preferred the thrills of a three-ring circus to the less exciting—even if more comfortable— limited play of forms found at home.

A few days after my involuntary pre-dawn appearance—unrequested by me, but suddenly there I was—one hot mid-summer morning I was carried to the big white stucco house where I spent my early years. Never get stuck with a stucco house my father later warned, explaining that the brittle material—perhaps something like the soft volcanic stone I was to see years later crumbling away at Civita in far off Italy—was difficult to maintain and vulnerable to disappearance. So are people. Humanimals are ever vulnerable both to appearance and disappearance without willing either. They come, they go. I was thrown into life—a new little form joining the play of infinite forms—without being asked, stuck for a brief time on the lonely planet and stuck into my room in the stucco house. Years later at age 60 my father disappeared as suddenly as I'd appeared on the scene 29 years before. Earthlings arrive and depart in a whirligig of random happenings.

After my father died I realized that maintenance problems presented challenges for living creatures as well as for their habitations. Whether stucco, stone, steel, brick, wood or whatever material, houses— like flesh—all eventually fail. That was the nature of things. The controls, foundation, plumbing, nails, the nuts and studs and screws, bolts and volts, wiring and electrical current, connectors, walls and windows, pipes and drains, chimneys and all the rest rest on a foundation of quicksand and eventually wear out. You can repair and sometimes replace the parts but in the end they no longer work. We inhabit what finally becomes a ruin, the fixer-upper deteriorating over time into a tear-down. But back at the very beginning, when I was carried into the big white house one hot summer day, a fresh bundle of joy—joy for my progenitors but would my own existence be a joy?—I knew nothing of such hazards, bodily or structural. At the time I didn't even realize that I was alive or what it meant, and in truth I still don't know what it means. My brand new earthly existence was an amazing novelty, quite a change from the previous billions and trillions of years when I didn't exist. One day there was no me and then, all of a sudden, the me appeared. It was a new experience for me. And before long I'll undergo another new experience—post-existence non-being, which I expect will resemble my prenatal non-existence. During the eternities when you are not, the days and the years are all the same.

In my early years the big white house represented my entire world. Only later when I saw the wider world beyond did I realize that the big house was really not so big, even if the boxy building had seemed that way when I was little. As I grew up and expanded my perspectives the structure seemed to shrink. Although the house grew smaller as I became bigger and traveled more widely my homestead, fixed in its place, remained a mooring in a restless world. Even after I no longer lived in the residence, my parents gone and Teddy the dog with his purple tongue six feet under, the house continued to represent for me a familiar and comforting kind of continuity. Other dogs and residents occupied the residence, but still today the house stands to remind me of my long-gone boyhood days at home there.

Although I traveled far and wide I always returned to my roots in the Midwest where I'd originated, and I always lived within a few miles of the big white house. My life was by choice both unsettled, with trips all around the world, and quite settled there in my home town where I began and will no doubt end. The big white house stood as a symbol of the past, my earliest years when my fatal circle and initial circumstances operated to shape what came later. Wherever I roamed I knew that back at home the old house, its stucco walls bright white, remained to anchor my memories. My trips entailed an enterprise whirligigging with spinning change, flux, comings and goings, here-today-and-gone-tomorrow movement, a dynamic and kinetic on-the-move hustle-bustle rush to see and experience as much as I could within the available limited time. Meanwhile, back home the big white house stood patiently like a sentinel on the watch to welcome back an errant former resident, missing in action until one day finally he reappeared, home again. Just looking, even as only an outsider, at the old homestead—long after it had left the family—was somehow comforting.

The big white stucco house, a large residence in a good neighborhood, stands only a few miles from my little red brick house, also in an upscale area. As I've made clear, my life started with advantaged circumstances. As I got older I tried to keep my situation that way. I couldn't help it that I was born into a fatal circle endowed with various benefits. Unlike Daniel Webster, I saw no need to apologize for my good fortune. At a large public gathering in Saratoga, New York in 1840 he declaimed: "It did not happen to me to be born in a log cabin." Webster couldn't claim modest origins, nor could I—but they weren't hugely immodest either. The log cabin the politician wasn't born in could have fit into the kitchen of the big white house, but other houses in the neighborhood were even bigger. A German proverb holds that all beginnings are difficult. Mine weren't. What was hard pertained to transcending the disadvantages of my advantages, then retaining and using them to attain a life lived in my own way. What you get for openers doesn't determine how favorable will be closure toward the end, which this reminiscence evidences. Although my duration was unknown at the outset, for me a long road stretched ahead from babe to bier, crib to crypt, from the first toy rattle to the last death rattle. I can ever more clearly hear the rattle of time's wingèd chariot as it draws near.

I well realize that this is the end—the end of my reminiscence and, for me, everything else. How will it all end? I'm dying to know. Curiosity killed the cat; had curiosity been fatal for me, I would have been dead years ago. I hope it doesn't kill me now. But something will. The final curtain will soon shroud the three-ring circus. And so—cherished reader—the time has come for us to part. Your ghost writer will soon be a ghost. By the time you happen to read this, I'll most likely have crossed the frontier between the land of the living and the realm of the departed. Of all my journeys, that's the one which leads to a dead-end, a terminal from where no return ticket will allow me to revert to my familiar and comfortable home-based life. You, however, are still endowed—for now at least—with sensory receptors which can capture some of the play of infinite forms playing on and on after my time. For a short while longer—whatever your age—you're still able to participate as an earthly form to receive and to emanate for other humanimals sensory impressions. But don't count on any staying power: at any moment you, too—however old you happen to be—could reach your end just as (no doubt) have I as you read my ghost-written writings. While I lie lifeless in my final resting place, before long you'll put my book aside, parting company with what I left behind, my reminiscence soon to be only a few ghostly shadowy fragments deposited in your memory. Our brief time together is over, and before long—hard as it may be for you to believe—your own fleeting earthly time will end. I'm only a few short steps ahead of you.

Now at the end, I like to recall my beginnings and to remember the long years between the then and the now. At times I drive past the family homestead where I began life and where strangers now live. I pause to glance up at the upstairs windows of the bedroom where I was conceived one cold winter night, or day, long ago. My occasional passing views of the windows behind which I was created prompt me to recall my very beginning. The happenstance of my origins—that zygote, that embryo, that fetus, that new-born, that me and no one else—resulted from mysterious and unknowable fateful factors, ones which by chance somehow all happened to combine to incarnate into existence just then and just there a certain humanimal. Thrown into life at that particular time and place, I began my experiment as the years started spinning by one by one, gradually filling with content, until by now my end is near and shortly the play of infinite forms will play on in my absence. They won't miss me. Now an old gentleman—my white hair a visual echo of the big white house where I spent my childhood—I'm recalling in my waning days in this reminiscence the odd happening a lifetime ago which engendered my very being. My conception and gestation into a specific humanimal was inconceivable until I actually appeared, a random congeries of improbabilities incarnate, to begin the strange experiment of briefly being an earthling on the lonely planet.

No one ever took the trouble to ask me if I wanted to be originated. It just happened—one moment I didn't exist and then the big bang to start things off and then, *voilà*, there I was, suddenly alive after an infinity of non-being. It's a misconception that a humanimal wouldn't want the opportunity to be asked if it would enjoy the privilege of engaging in the experiment of being alive. If, perchance, some celestial manager (vice-president in charge of procreation) had inquired whether or not I wanted to become a character in the the great earthly spectacle known as life, what would my answer have been? Wow—let's go for it. Don't bother. Hey, give me a chance. Forget about it. I'm curious—let me have a look. I'm not that curious—drop the idea. Just leave well enough alone. I'll have a go. Save your energy—play golf, fill those holes. Give it a try, why not? I'd rather remain in Neverland. I'm game—it could be a great adventure. Don't you have anything better to do? I'll suck it and see. It's up to you—I've got better things to think about. What's the point of becoming a something if it all ends in nothing? All those notions represent some of my possible answers. I don't know which one I'd choose. Maybe I'd decide to keep my mouth shut and let nature take its course. *Que será será.* In any case, by now it's far too late to consider the question: the corporeality of my incarnation and the reality of my years-long earthly existence are facts. My being has already taken place, and before long I will lose my place and revert to the near eternity during which I didn't exist.

The interlude between my pre-birth and my post-death is so short it hardly seems worth all the bother. What did it all mean? Now that I'm almost done, for what was I begun for? Stay tuned for my answer. You're almost at the end, and so am I. Keep reading.

~ 2 ~

Verily, nothing in my original setting or fatal circle suggested how I'd spend so many twilights far from home. Just as it was impossible before the fact to visualize my very existence—and, once alive, its content and duration—so my future wide-ranging travels to lands far beyond the big white house were unpredictable. Chance might have made me a completely different person inserted into the same setting. Maybe I'd then never have experienced a single foreign twilight, in which case you'd be spared the incubus of reading this book. Given my fatal circle and the other conditions which defined my situation, I was in fact on track to live a regular settled and comfortable provincial suburban existence. That was the prevailing way of life in my milieu at the time I arrived there. While growing up I wasn't inspired to pursue my worldly adventures by any home-based influences. I encountered no role models to emulate, no bold and adventurous chancers, no colorful figures, no curiosity-driven types, no eccentrics but only centrics. Down-home conformist provincial bourgeois suburbanites peopled my early life. Those conventional characters did in fact serve as "roll" models but in a negative way when at a certain point I decided to roll out a new agenda which didn't follow their examples of how to live.

As for my fatal fetal circle, perhaps some unknown long dead ancestor had bequeathed me a few rogue regressive genes which served to activate my progressively compulsive wanderlust. Should I bless or curse that forebearer? Twilights entirely at home would have offered a quieter and more settled existence, even if a less colorful one. That sedentary way of being would have also brought the great advantage of saving time, yours and mine, by eliminating this reminiscence, for almost nothing of interest could I say about a life consisting mostly of dusks in an upscale Midwestern suburb. Such local twilights offer few highlights.

My earliest travels were limited to an occasional tram trip on the now long-gone "04" streetcar (oh-four, never zero-four) which used to rattle along on tracks passing right by the big white house. The old-fashioned car rumbled past as regular as clockwork day after day along the now grass-covered Wydown median, the rails and ties having been removed some 70 years ago. The 04 is now but a ghost train from the phantom yesteryears when I was still a boy. How I enjoyed riding that Toonerville Trolley-type contraption of a streetcar, operated by motorman Red who never, for the passengers at least, had a last name. Ding, ding, ding went the bell; chug, chug, chug went the motor; bump, bump, bump went the brake—just like in "The Trolley Song" from *Meet Me in St. Louis*. When we reached the end of the line out in the country, now city, Red would crank a rope to lower the rear roof pole connecting to the electric line above and then raise the pole at the other end, then he'd flip all the seats back to reverse them for the inward-bound passengers to ride forward rather than backward. Back and forth Red rode, a daily trip to nowhere except to and fro.

Although I greatly enjoyed watching all the maneuvers and hearing the clang of the bell Red rang along the way—I can hear it still, as in my memory the long-gone gong has never been stilled—I didn't aspire to become a trolley car motorman. But I greatly admired and even coveted Red's cap, its shiny rounded bill attached to a circular black head-piece. Years later, on a visit to the offices of the *Neue Züricher Zeitung* newspaper in Zürich I was given one of the motorman-like caps worn by the *NZZ*'s newsboys, most of them old men, who sold the paper on the streets while shouting the Swiss-German version of "wuxtra, wuxtra, read all about it." With the internet, those colorful old cap-clad vendors may no longer exist, and for sure Red doesn't. Back in the Cold War days of my youth when atomic war threatened, people used to say, "Better Red than dead.," a sentiment with which Red would no doubt agree. But Red is dead, gone to trolleyman's heaven, and the iron rails he rode day after day no longer line the Wydown median. All traces of Red and his trolley have vanished, as have my days riding the rails.

Back in the early years of my Wydown childhood Tom Williams lived in the neighborhood, over on Arundel Place a few blocks south of the big white house. In the mid-1930s Tom briefly attended Washington University, a few blocks north of the Wydown house where at the time I was an infant. Tom probably walked right past our corner residence by University Lane to reach the cut-through to the nearby campus. Tom and I were different ages but we shared the same first name, initials and neighborhood. The Williams family lived at various times in two different Wydown area houses. Tom's memoir includes a picture of the first house, on Arundel, and in her 1963 *Remember Me to Tom* his mother Edwina relates how "We made nine moves in St. Louis...before I bought the house where I now live on Wydown Boulevard in Clayton, a few houses beyond the city [of St. Louis] line, just west of Forest Park."

Back in the mid-1930s Tom published in the first issue of *Flair* magazine a short story with a glass menagerie-like description (quoted in Edwina's book) of his time as a boy growing up with his sister in St. Louis. Looking back at his youth Tom wrote, "I saw that it was all over, put away in a box like a doll no longer cared for, the magical intimacy of our childhood together, the soap-bubble afternoons and the games with the paper dolls cut out of dress catalogues." The narrator recalls "that I began to find life unsatisfactory" and so Tom started "putting the blocks together in some other way that seems more significant to him. Which is a rather fancy way of saying that I started writing."

Although Tennessee Williams and his mother and I for a time lived a few blocks from one another, his game playing with his sister and his plays never motivated me by way of example to rearrange my blocks in the same tormented way that inspired that other Tom. We shared a few things in common, including the same Clayton neighborhood at the same time and a first name and initials, but our similarities ended there. While the youth Tom W. agonized over his unsatisfactory life, the child Tom W. was preoccupied by such matters as Teddy the dog's purple tongue. I never suffered traumas and dramas and anxieties of the kind which possessed Tennessee, although I was later to worry about such serious problems as the continued availability of typewriter ribbons. Unlike Tom, I never needed drugs or alcohol to induce visions of virtual reality, as I found the wonders of the real world and the play of infinite forms more than enough to enrich my life and stimulate my imagination. Poor Tom suffered to create his plays of finite forms, but for this book only readers, not its author, might suffer. While Tom was brooding over his troubling and troubled life I occasionally suffered from sniffles, skinned knees, pimples and a few times Teddy tried to nip me (bad dog). My dramas couldn't compare to the ones the other Tom experienced and wrote about.

Back when Tennessee Williams was starting to write I couldn't even read. My past was so short I never thought about it, and the future was so long I had no idea what to think about it. Looking both backward and forward I drew a blank. My youthful existence back then was rather devoid of content. My life lacked interest for anyone except me and my parents and maybe Teddy, who liked a good nip every now and then. My years in the big white house gave me few experiences worth remembering or recording, other than the delights offered by my 04 excursions. My other trips in my younger years were confined to invented journeys inspired by my collection of stamps from around the world. The stamps made me curious about the countries the little pieces of paper represented, and they stimulated me to envision what those far places were like. Perhaps those colorful and exotic postage fragments served in part to inspire my later travels, the early-day hobby pointing the way to my later interest in seeing the world which the stamps evoked.

Even then oddly named places I knew from their stamps intrigued me. Tana Tuva, Bechuanaland, Mongolia, Molvania and many other such exotic words prompted me to find where the lands were located on the map. I categorized and put into albums the serrated-edged rectangles and the more fanciful triangles and circles so that the stamps were all neatly organized. With my neat collection, the world seemed orderly and subject to being arranged and rearranged as I decided. I controlled all the countries, each one under my thumb, the world cut into little paper pieces easily moved around and then held in an album until I decided to change the arrangement. But as I later learned, this invented world was entirely different from the real world I encountered far from home a long way from the big white house. Out there life stamps its mark on you much more than you on it.

The ease of shuffling the stamps around to put them into different categories illustrated to me that no arrangement was definitive and that an alternate order always existed. Sorted by color, country, shape, size, year, value, denomination, subject or any other theme, no selection was final, and however I grouped the pieces some seemed more suitable for another category. I faced the same problem as the character in Georges Perec's *Life A User's Manual* sorting hotel labels into order. Hoping "to link each label to the next...based on some detail in common," he tried chronological order (the date collected), alphabetical order, by continents, by country. But finding the process too challenging, he abandoned the project. I enjoyed organizing and then reorganizing my stamp collection, which gave me a touch of the outside world beyond the big white house. The stamps helped to introduce me to the play of infinite forms which comprised the lonely planet.

So dense with references to the plenitude of the vast range of forms is Perec's novel, his book could well have been titled *A Play of Infinite Forms*. Unlike Perec, however, my reminiscence couldn't bear the title of his book, as my intention has not been to write about life or even a life but only to present a few elements—mainly, travels—of a particular life. I would not be so presumptuous as to describe my offering as "a user's manual" for life, for my agenda is more modest—simply to recount the encounter of a compelling curiosity with the wide world and some of its infinite forms. Perec's title is not fit for purpose to designate my book, which is nothing but a travel reminiscence with some marginalia in the form of side observations on what I happened to observe—and my responses to the experiences—over many years. I've included what I learned from my real-world classes—based on travel in first, second, third or lesser classes—with no claim to offering the reader a manual for general use. Each earthling has to format his or her own user's manual. My purpose is not didactic or aimed to present generalized pronouncements or nostrums as I don't know of any, other than very few, such as: brush your teeth regularly, don't pick your nose in public, don't forget to cut your toenails, and remember to take out the garbage. Otherwise, I can think of nothing particularly helpful to tell you, other than the momentous thought I mention in passing in Part 6 below. Apart from the title of Perec's book, an especially significant and crucial element which differentiates my work from his pertains to how each narrative was created. Mine originates with images derived from real life experience as supplemented with an imagined arrangement and various comments, all based on what actually happened.. Perec's fiction features long passages with invented inventories of things, made-up events, fantasized episodes, artificial characters and other mind-game creations fabricated not from real world forms which exist but from inventions confined within the author's consciousness.

As much as I enjoyed my postage stamps, I later discovered that they didn't at all accord with the way things really were on the lonely planet. Just as Tennessee Williams described his childhood with his sister as "magical," so for me the stamps were like mini magic carpets carrying me to foreign lands. But as I eventually learned, things weren't so easy out in the magic lantern show which was the worldly play of infinite forms. It was hard to move from land to land, getting around to see some of the nearly 200 countries which checker the earth's surface. Stamps didn't represent the real world. The U.S. "forever" stamps, introduced long after I collected and curated my specimens, bear a legend which makes a truly audacious and fallacious claim to longevity. It was beyond belief that the post office had somehow managed to devise a human artifact valid forever. When Howard Carter broke through a door closed "for ever" in 1923 to discover inside the "wonderful things" contained in King Tut's tomb, the "for ever" came to an end. Even ancient Egypt or the Eternal City, Rome, aren't forever. But—who knows?—maybe the forever stamps will enjoy better luck and last so that I can can use them to post messages from the desert of vast eternity. If, perchance, I'm able to continue in the grave to function as a ghost writer, after now being a lively one writing this reminiscence, the forever stamps would serve to send writings from my very own dead letter office. There in my posthumous post office box I'd scribble away as I wasted away, a phantom author from the netherworld. Existing as an earthling in fact leads to an organic version of a forever stamp because although a humanimal doesn't continue on, its death survives forever. Once the Grim Reaper puts His stamp on the creature, it's forever canceled without a return address.

Inklings of my later post-childhood pre-mortem writings surfaced as I was growing up, for I worked on the high school student newspaper and yearbook. Those wordy activities I chose in part because I was quite unqualified for more vigorous exertions, such as sports or chasing coquettish coeds. I'd never be able to catch them anyway, as surely they'd outrun me. I somewhat envied the jocks who in high school enjoyed greater prestige than the scribes, the grinds, the nerds, the geeks. The guys with the balls, basket- or base- or foot-, got the girls. Back in my time, so I was told—alas, I had no personal experience—guys could only neck with girls, their below-the-neck attractions (a frontier area as inpenetrable as the Iron Curtain) remaining off-limits. Writing for the paper and the yearbook wasn't a real-world contact sport, so being a wordsman rather than a sportsman gave me little opportunity for contact with the cuties. They preferred the luminous athletic stars, not those of us off-stage writing about things rather than living them. Writing and reading represented only derivative activities removed from hands-on encounters with the coeds and the outside world. Only much later would I engage with the wide world far beyond the big white house and my boyhood stamp collection, student publications, adventure radio serials and other such confined and localized unworldly boyhood trifles. And that's about all I can say about my early years, which admittedly merit little discussion—and probably already too much, as almost nothing memorable occurred back when I was a boy. Thanks to my stamp collection, my time at the big white house gave me something to dream about—travels to see the places the stamps represented—but little to write about. Unlike for my literary neighbor Tom Williams, my soap-bubbles were confined to the bathtub.

My boyhood was for the most part as regular and routine as Red's life on the railed 04 line. Every day Red piloted the trolley as it growled its way out and back, up and down Wydown and beyond on a fixed line, Red's hand grasping the wheel except that it wasn't a wheel and he didn't have to steer as the car was on autopilot, an early-day autonomous vehicle. Red never chose where to go. He and the passengers were conveyed only where the tracks took them. Riding with Red, I had no choice for my route. Red turned not a steering wheel but a knob handle he moved to switch the current on to go and off to stop, and that's about all Red did every day on the 04. That was his life day after day, turning the metal knob back and forth and forth and back to complete or to cut the electrical circuit, and then at the end of the line Red changed the poles and turned the seats back to begin another circuit back along the same route in the opposite direction. He saw the same scenes every day, never once changing course. The motorman of course probably had a home life, complete with last name—Red Greene? Maybe Red was Graham's tenth cousin. And perhaps Red's family included a dog, a wife, a red-headed kid named Tom whose fatal circle was completely different than mine, and possibly they all lived in a cottage-sized white house on the other side of the tracks, a place much smaller and more down-market than my family's big white Wydown house.

There in the big white house I somehow escaped the sort of childhood dramas which characterized our neighbor Tom Williams' scenario. While he suffered anxieties galore I was playing with purple-tongued Teddy and popping soap bubbles in my bath. Nor was I sensitized to the kinds of poetic impressions and perceptions of our near neighbor, another Tom. As I told Jorge Luis Borges in Buenos Aires years later, I never met Tom Eliot in his childhood lair (or anywhere else) in the Central West End, not far east of Wydown, where the pre-poet lived before moving to London, there to become the genteel Englishman transmuted from plain old Tom into the more literary T.S. Eliot. It was disconcerting to me that my namesakes had dropped my name in favor of other labels—Tom W. to Tennessee, Tom E. to T.S. I would no doubt have enjoyed a much more distinguished authorial career had I styled myself Missouri Weil or T.P. Weil. But in truth, my life wasn't tormented enough and my thoughts not literary enough to yield the sort of writings the other two local Toms produced. My existence was simpler and less anguished than theirs. No angst, dramas, turmoil, wastelandish or outlandish notions occurred to me. There were no early evidences, or for that matter any late ones, of exceptional literary talent or of psychological disturbances, and no epiphanies, no precocious writing efforts, no wanderlust, except for rides on the 04, or any other kinds of lust other than reading, enjoying and rearranging my stamp collection, listening to the adventures of the radio heroes, and ice cream. Teddy I also used to like until he started to take delight in nipping me—a precursor to the dangers and plots and counter-plots of the kind which threatened beyond the sheltered precincts of the big white house.

A few irregularities—less vexing than the irregular French verbs I years later confronted when at the Sorbonne—shadowed my youth, but in general my childhood, boyhood, teenage years all passed without any noteworthy episodes or experiences. Like my long gone summer days at Phantom Lake in Maine with the wood smoke in the air, my pre-adult years seem phantom-like ashes of dead time disconnected from me, much as I now remain a ghost writer removed from this reminiscence. My account thus lacks any discussion of the kind typical of memoirs and self-referential personal narratives. I have nothing to say about such matters as family crises, psychological "disorders," confessional revelations, life-changing mental or sentimental episodes, emotional turmoil and similar. In any case, even if those sorts of problems had bedeviled me I wouldn't mention them in print. Personal matters are irrelevant to my purpose. Dead men tell no tales—which is why I'm rushing to tell mine before it's too late—and neither do ghost writers, for they don't discuss private concerns, even those which might persist in haunting the story teller. Such tales should rest in peace.

The only significant remnant of my early days which can explain why in my mid-thirties I decided to change my life is, as I've discussed, my compelling curiosity. As a child I was somewhat curious, and as time went by and I learned more about the outside world my curiosity became even more acute. It finally so dominated me that I was no longer willing to continue on with my linear existence which stretched ahead into the future like the fixed 04 tracks. Both private and public transit lead on—like the wingèd chariot—to a terminal destination, but if your trip is only via streetcar you're locked into a route lacking variety. I decided to get off the trolley before it reached the final stop. Unlike Red, who never had to steer, when I gave up my settled and comfortable existence to include twilights far from home I had to pilot my own way. No one encouraged, influenced, inspired or guided me. No family members, no role models, no mentors, no exemplars or anyone else showed the way. I followed my own star, an indistinct marker unlike Sirius the Dog Star, the heaven's brightest—to earthlings at least. The Dog may be less distinct viewed from other planets and perspectives. No starlight or other light, other than twilights far from home, illuminated my path through the dark woods. Once I cut loose from my previous structured life, I was on my own. I could rely on no adviser or divine force to shape or form my new way of being, which might prove anything but divine. As the twig is bent so grows the tree, but in my mid-thirties it was too late to bend my twig and in any case, as Kant suggested, twig-bending doesn't turn out so well for "from such crooked timber as a human is made nothing quite straight can ever be fashioned."

My new life as from my mid-thirties suited me by catering to my dominant trait, curiosity, and also by diversifying my life to bring some adventure into it. New faces and places seen on my travels out in the hurly-burly far from home contrasted with my local roots and continuity. Unlike for Red, for me the world was an open book to be read as I transferred from my confined rail-limited existence to the open road. Red, less well read, was stuck with the same book, chapter and verse every day. He had no way of getting a volume rebate to subsidize a revised edition. The motorman's fatal circle led, Kant not withstanding, not to a bent form (even if the 04 trolley did pass a street called Big Bend, not far west of the big white house) but to a straight-line travel format determined by the rails. By way of contrast, my circle by chance let me free myself from the iron-clad restraints of the trolley route so I could realign my life.

Apart from the design of a humanimal's fatal circle—when, where and in what context the creature originates—how the individual's personality evolves also influences what later happens. Mine tended to ignore speculative considerations in favor of practical ones. Theory divorced from reality never much interested me. I sought not to analyze how my personality had developed but only to understand it. This process resembled something of the difference between invention and imagination. Practical and pragmatic, I conformed with my temperament by focusing on what I imagined to be realistic perceptions and assumptions based on worldly rather than wordy sources. In Petronias's *Satyricon* a one-time slave named Trimalchio, who somehow managed to become a wealthy free man, discusses in detail his tomb-to-be, requesting the epitaph, "He never once listened to a philosopher." Having just quoted Kant, I can't claim the same. But I didn't listen too often to those pure-reason thinkers. *Esar's Comic Dictionary* by Evan Esar defines "philosopher" as "A person who always knows what to do until it happens to him." Esar riffs on "curiosity," which happened to define my character, by noting "Bright eyes indicate curiosity, and black eyes indicate too much curiosity."

All of the sages and would-be wise men all through time have tried to figure things out and then proclaim existential certainties, but those propositions represent only just speculations, guesses, opinions, shots in the dark. The earthly play of infinite forms is far too varied and complex to reduce the spectacle to definitive philosophical doctrines. The great thinkers, writers, poets, painters, pundits, philosophers, the academics, experts, wise men and wise guys, all of the Toms—Williams, Eliot, Mann, Merton, Browne, Wolfe, Tom o' Bedlam (beggars and vagrants of the kind once confined to Bethlem Royal Hospital for the mentally ill), the ghost writer behind this text, and many others—who chatter about the meaning of life, the meaningless of life, whatever, are all just puffing smoke. Although it's bad for the Tom brand to call attention to the reality that the Toms are no more profound than the Dicks and the Harrys or anyone else, such is the case. To give just one example, Thomas Wolfe claimed that "you can't go home again," but I always did: every trip I took reached its final destination when I retreated from the world to return to my little suburban house on a dead-end street. Pen-ready pensive types can write whatever happens to come to mind. What each sage says may comfort him or her, but such speculations hold no transcendental meaning for anyone else. A lot of good those philosophers' certainties did them. All those thoughtful characters are now thoughtless, or soon will be, all their grave notions and big ideas reduced to the silent confines of the grave. Dental rather than transcendental knowledge is more useful, especially in a place like Molvania, "the land that dentistry forgot." When you have a toothache, the cant of philosophers like Kant and the meditations of other thinkers won't help you much.

Belief in a personal value system, or even a contrived system that's valueless, may serve to console the individual who espouses such doctrines and also help to structure the pundit's life, but such personal views don't offer an object lesson for other earthlings. My reminiscence, essentially a *memento mori*, makes no claim to be definitive for anyone but me. My funereal tone simply reflects how I saw the lonely planet. The experiences, perceptions and perspectives of other observers will differ from mine, just as have their lives. I've by design avoided philosophical systems and claims to authoritative views. I leave the philosophizing to others. Most of the philosophers and similar characters who profess wisdom operate by clothing mere opinion in seemingly elegant attire, as if the superficial image endows the concept with authority. In truth, the clothes are probably from Good Will and represent largely recycled ideas. Those who promote such well-worn views for self-serving pecuniary reasons operate with bad will. Other than in the sciences, originality is extremely scarce. I never sought—and have never propounded—any original notions. My originality was restricted only to the specific life experiences which happened to me.

Two of the most important factors in the process of changing my way of being were my own assessment of my strong and my weak points and also a realistic evaluation of the other elements of my situation. I tried to concentrate on my strengths, work around my weaknesses, and deal with the realities as they were for me, not as I wished them to be. Some sages opine that you should challenge yourself by venturing outside the box—try stuff beyond your comfort zone. It's claimed that this will stretch your mind, maybe even slightly straighten your crooked human timber Kant claimed was bent beyond redemption. For my part, curiosity stretched my mind quite enough, and my travels to far lands remote from the boxy form of my provincial suburban house presented more than enough challenges and experiences for a lifetime, and maybe even for two lifetimes. My travels in fact took me far beyond my comfort zone, for the big wide world beyond my regular and settled comfortable home-based existence was a vast discomfort zone. I didn't need any expert opinion or advice to tell me about the merits and attractions of extending my life beyond my upscale suburban residential neighborhood, zoned to preserve the comforts of home. Out in the far lands where my albumed postage stamps originated lay in store all sorts of new experiences, experiments and adventures which appealed to my curiosity. So I began my new life.

~ 3 ~

Few complications barred my way to transition from a purely settled and for the most part conventional way of being to my new double format. Because I intended to retain many existing elements, I needed only to rearrange rather then completely to change my arrangements. My plan was to stay and to go, not to leave and then return only for a passing visit. I wanted to retain what I had while adding to the set format an entirely new dimension. With this balancing act, I didn't pursue any escape-artist tricks. I sought only to diversify, not eliminate, what already existed. It was a line extension, not a complete abandonment of the existing regular and predictable line like the 04 line had once been. "Expatriate" sounded too definitive and irredeemable, like "ex-humanimal." I favored instead expatriating and then, at trip's end, repatriating. I assessed the advantages and disadvantages of my contemplated change, its risks and rewards, the plots and counter-plots which might arise. I considered my goals and the best way to reach them. In short, my enterprise involved how best to tailor my situation, as it limited and liberated me, to conform to Christopher Morley's observation that the only truly successful life is to live it in your own way. Taking into account the above factors, and various others, I tried to determine what my own way would be and how to produce the desired outcome.

Only by chance had so many random elements combined simultaneously rather than just seriatim to afford me the opportunity to restructure my life in an appropriate way. An extremely thin line separates a fortunate situation, which manages to continue and which serves to satisfy a chosen agenda, from the case in which unfavorable circumstances predominate so as to prevent such a beneficial result. It's rare that a cluster of conditions exists which permits choice and also an on-going ability to pursue what's chosen. Much more common are the less favorable situations degraded with constraints which block a person's freedom to try to create something of a charmed life. Given the fatal circle and other conditions, the operation of chance and luck, and other limiting factors, it's rare that someone can realistically hope to transcend their immediate circumstances. And even for the most advantaged and lucky earthlings who happen to be favored with a charmed life, their good fortune remains subject to bad luck and other capricious factors which in the end determine outcomes and incomes and whatever else is to come.

Random events can at any time damage or bring to an end an earthling's existence. That's the nature of the conditions which prevail on the lonely planet. Analyzing my own situation, I was somewhat surprised at its many enabling characteristics. I realized that it was extremely unusual to enjoy so many fortunate advantages. Luck had somehow operated to put me on the right side of the thin line between opportunity and the more common circumstances teeming with prohibitive negative factors. By chance my situation brimmed over with potential, an exception to how most lives were typically subject to insurmountable limitations. My circumstances conformed to Jacques Monod's belief that chance rather than necessity governed. I'd somehow escaped most of the constraints which restrained most humanimals from freedom to choose how to live. Luck had put me in the position of being able to reposition myself. Various concomitant factors existed to endow my experiment in living with opportunities to create my own way of life, an advantage I highly valued but also a challenge and a responsibility I took seriously. Chance had given me a chance. How I managed to grasp the opportunity depended on me. Not withstanding my view that chance and luck largely determined outcomes, I recognized that human agency also played a part. I didn't undertake my new agenda with the belief that everything depended on random and capricious factors. I knew that my own efforts could to some extent nurture and shape my way forward, even if in the end the main operative elements were the fatalistic and random influences which ultimately control the destiny of every earthling.

Although fortune happened to smile on me, it frowned on many other mortals who seemingly enjoyed some of the same kinds of advantages available to me. Ever aware of the capricious nature of luck and of nature itself, I treasured how they happened to benefit me, and I always realized that they might at any time change and go against me. One moment you're riding high, a king riding in triumph while passing through Persepolis. You're the toast of the town and then, suddenly, you're just toast. It can happen to anyone at any moment, a fatal instant which does you in, takes you out, works you over, puts you down. Seemingly small happenings can lead to big consequences. What happens in a matter of a few seconds can lead to endings which can last an eternity.

As a successful composer, Alban Berg was hitting all the high notes when an insect stung him in November 1935. By Christmas Eve Berg was dead, expiring in Vienna age 50 from blood poisoning resulting from the bite. This real life, or real death, happening is an analogue to Tolstoy's fictional Ivan Ilych, whose death occurred after a similar seemingly minor and benign bodily intrusion. Fact or fiction, the story is always the same: something happens. You're trundling along like a trolley on your track and then suddenly you're derailed. Tolstoy begins his tale at the end with Ilych's death in 1882. A minor mishap killed him, just as for Berg. One day while standing on a stepladder at home to arrange a window curtain the law court official slipped, hitting his left side. For a time the pain persisted, then it passed. Time went by, all was well, Ivan was well, his life continued normally. But then he started to feel some discomfort in the area of the bruise. As the pain worsened Ivan consulted doctors but none could help him and eventually "It" came: mortal Ivan died. At the end he thinks, "I lost my life over that curtain...Is that possible? It can't be real. It can't be true! It can't be, but it is." Like every humanimal Ivan, slipping from the ladder, was from beginning to end a fallen man, one thrown into being and fated to end in non-being. After Ivan's death one of his friends gave a brief passing thought to the fate of the late Ilych: "Why, that might suddenly, at any time, happen to me." But the meditative moment quickly passed. The friend's life went on undisturbed, at least for the moment.

In the play of infinite forms no scene which curtains lives remains unplayed. In his novel *Caprice* Ronald Firbank tells of Sarah Sinquier, a provincial clergyman's daughter, who runs away to the theatrical bohemia in London to become an actress. After her first stage appearance, as Juliet, she trips over a mousetrap and falls through a trap door to her death into a well beneath the stage. That performance became her last stage before eternity. Terminating in a well, Sarah's career and life didn't end well.

The hazards of misfortune scythe the great and the small. Kings lose it all, commoners give up less when they give up the ghost. The more you have the more you lose at the end. That's merely the part you leave behind. What's taken with you to the grave, however, is your most valuable possession, which is your very existence. Shakespeare's King Richard II feared the vulnerability of "this flesh which walls about our life," a fragile state subject to termination from the most minor happening, such as when "a little pin/Bores through his castle wall, and farewell king!" Small intrusions—a sting, a stumble, a stab, a stagger—and farewell, or fare not well. Richard Two, too, was subject like all his subjects to a boring death, his flesh breached. Like every earthling, wealthy and regal rich King Rich enjoyed his good fortune only in passing. As both an imagined and a real character, Richard II existed in double jeopardy, II mortalities, but he in fact suffered only a single death as, thanks to Shakespeare, the monarch lives on in the play, even following his demise as a living performer in the play of infinite forms. Like Ivan, but unlike Alban, Dick still exists. Although the composer decomposes, the fictional fellows continue on with different states and fates, all three dead but two still read. Alban, once alive, gone; Ivan, never alive, survives; Dick, alive and then dead and yet still alive. As fact, Richard existed; as fiction, he persisted. Harry Lime, Huck Finn, Hamlet, Snoopy remain while their creators Greene, Twain, Will and Schulz (the names by mere chance designating Molvania's leading, biggest and only law firm, specializing in dental malpractice cases: business is slow) are all has-beens. Every author (and every non-author) exists as a ghost-in-waiting, but he or she at least leaves behind characters who outlive the creator. They're durable, those fictions, even if their creator isn't. The poor thing, fighting time with words, remains ever vulnerable to the sting of outrageous fortune or, less dramatic, to the bite of an errant insect.

Only in a dentistry-deprived toothless land like Molvania are humanimals less threatened by bitings, even if the occasional Dracula-like figure leaves his mark on you. King Alexander of Greece suffered a fatal bite in 1920 at age 27 after trying to protect his German Shepherd Fritz when the dog was attacked by a monkey in a private park outside Athens. A second monkey intervened and bit the king on his leg and torso. The monarch's body became infected from the bites and twenty-three days later he died of blood poisoning. The Greek king met his match, or very close match, as the difference between chimp and champ, monkey and monarch is slight. It's all the same when it comes to the end: monarch butterflies, queen bees, king crabs, emperor penguins, royal palms, simians and kings all eventually give up the ghost. It's hard to imagine transmuting in little more than three weeks at age twenty-seven from king to corpse. Things aren't supposed to work like that: monkeys don't kill kings, it's the kings who do the killing. Perhaps the world, too, and all its infinite forms are like the king's monkey death, an aberration, a cosmic mistake that produced a corrupt version of what was meant to be a more ideal and hospitable and less lonely planet. But as things turned out, that was not to be.

The capricious operation of fateful happenings is unnerving. In *Angel in Armor* Ernest Becker laments that "what bothers us most about our strange career on this planet is that our lives are subject to complete catastrophe by the simplest accident, the merest chance occurrence. This is the thing we can't stand. The undoing of years of work, effort, good will, morality, patience, sacrifice—by one tiny random event." When the final "event" terminates existence, your last experience includes losing all your experiences, "for in one hour has been undone all in which my entire life was spent," as Garcilaso de la Vega put it in *Canción* (Song) III. Chance starts to operate in the prenatal stage: even the very process of being incarnated into a particular person who's thrown into existence is an accident. Once an earthling, the creature is an accident waiting to happen. The journey from gestation and into being, and then on to the end and once again into non-being, entails a fatalistic experiment on the lonely planet fraught with vulnerability to the devilish play of chance and fate.

As the common fate suffered by Alban and Ivan shows, "one tiny random event" can irrevocably change anything and everything. Contingency shadows what's to come while fixity haunts what has already occurred. Like the jungle, thick with the density of what irrevocably took place and which exists forever in time, the past is a tangle of once fluid and then congealed possibilities. Empty of growth and greenery, the desert evidences the blank openness and potential of the future. Major natural features such as jungles and deserts and rivers perhaps exist as emblematic visible analogues to invisible factors such as the past and the future and their joinder, time. As spooky as the unknown future is, with its threats of random minor but possible momentous events—like stings or falls—the absolute settled nature of the unalterable past is just as disconcerting. Both past and future shape "our strange career on this planet." Contrary to the flexible situation of a traveler like me with a return ticket home, life is a one-way street with no reverse lane. Unlike a trolley which runs on fixed tracks in both directions, a person's route may change along the way but can never be reversed. Nor, regardless of the route, can the ultimate destination be changed. The terminal is always terminal. An "Exit Only" sign points the direction ahead and there's no going back. Only a "No Entry" sign—not to enter life at all—can prevent a potential but not realized earthling from the inevitable threat of the wingèd chariot, which rolls on and eventually overtakes every humanimal. Just as the spectacular fair which is the play of infinite forms is non-repeatable, so is the fare you pay in time for the wingèd chariot non-refundable. Like the iron-clad fixity of rails, the past evidences your path. But although bygone times are completely gone, their effects and defects cling. "The persistence of time," to rephrase the title of Salvador Dali's famous 1931 painting, the one with the droopy watches entitled "The Persistence of Memory," suffuses the present and the future.

In *Pleasures and Days* Proust tells of a certain Honoré de Lenvres whose legs were suddenly broken by a runaway horse. The accident occurred on a Tuesday, by Sunday peritonitis set in, and on Monday de Lenvres died. Just before the accident the unlucky fellow had been walking along, relishing his existence and refreshed by his pleasant stroll, "profound joy embellishing life that morning." The doomed Parisien glanced at his watch, retraced his step and "then it had happened. In a second the horse he had not seen had broken both his legs. It did not seem to him that particular second must have inevitably been like that. At that very second he might have been a little farther away, a little nearer, or the horse might have deviated a little, or it might have rained and he would have gone home before, or if he had not looked at his watch he would not have retraced his step...But yet this thing which so easily might not have been, that he could for a moment pretend was nothing but a dream, this thing was something real, and nothing he could do would change it." Honoré might still be alive today, honored to be one of the oldest senior citizens on the planet, age 120 or thereabouts, and still enjoying his walks in Paris, if only he hadn't momentarily stopped to check the time on his watch. That was the wrong time to look at his timepiece. Timing is an important subset of luck.

Stuff happens. When, where, how you can never foresee. Cosmic or microcosmic, animal, vegetable, mineral, everything which exists, remains ever vulnerable to damage or disappearance. Random stings, falls, pin punctures, runaway horses, trap doors and other traps abound. Plots and counter-plots threaten. A symmetry characterizes the beginning and the end: each living thing becomes so by an improbable and unsought process, and by chance can end in the same way. The chance of that happening for me increased when I started off on my wide-ranging catch-as-catch-can adventures far from my native habitat. Life at home was less random. It resembled a utility, an always-on available service I could readily activate when needed. At home I was coddled; out in the world, harried. Home and away satisfied me in two quite different ways. Serving to stimulate me, my travels brought some excitement and variety to my life. When away, I fondly visualized how I lived at home, and when in my native habitat I dreamed of visiting far places I longed to see. This pleasant dichotomy didn't represent a split personality but one integrated into a single life enriched by two very different and satisfying elements. My balancing act afforded me the best of both worlds. The lonely planet is not the best of all possible worlds (much could be better), but I tried to make from the conditions I found in the world the best I could of what happened to exist.

By now a reader may be wondering (but possibly not) what my familiars thought about my dual existence, a double life which included both a comfortable home structure and, occasionally abandoning it, travels to far lands. Because of the conventional context in which I lived, I was the odd man out. Everyone else cherished and clung to the sort of life I in some ways valued but which I refused to accept as the be-all and end-all of my being. Being only a provincial suburbanite in the U.S. Midwest wasn't enough for me. Near as I could tell, those who surrounded me on my home turf viewed me as somewhat eccentric, which I took as a compliment. Eccentric as my way of life may have seemed, from my standpoint ignoring many of the rituals and practices common in my neck of the woods made sense. An a-centric life rather than one centered on conventions common at home suited me. Although I was a provincial at heart, I was also to a certain extent a dissenter, a rebel, a radical, while at the same time comfortably installed at home as a contented bourgeois suburbanite. I was no misfit as I fit neatly into both formats. Both modes conformed to the essence of my existential nature. Only by happenstance did two completely different ways of being define my personality and, fortunately, they both combined in a way which allowed me to live something of a charmed life.

As I've made clear from the very beginning, my personal and home life do not play a part in my account of how I encountered the play of infinite forms. Something of a counter-factual description would say that my here-today-gone tomorrow balancing act allowed me to calibrate a system satisfactory to—in order of importance—(a) dog, (b) wife, (c) son, (d) daughter, (e) cat, (f) parakeet, (g) hamster, (h) goldfish, (i) all of the above, (j) some of the above, (k) none of the above. Any and all of those earthly forms incarnate, such as they were, seemed to accept my compulsive wanderings. While I hoped the family was glad I could enjoy myself by pursuing my curiosity with visits to such distant strange lands as Albania and Molvania, I didn't expect those in my entourage to understand my compulsion. For all I knew, the various home-based creatures were, except for the dog, pleased that I occasionally got out of the house, gone to live under those twilights far from home. The homebodies would then be free, at least for a time, from my eccentricities, my humor, my cooking, my snoring and my other other problematic habits. In truth, the beings I left behind at home, if there were any, would probably think nothing at all—as long as they were fed—about my chronic absences, viewing them as unworthy of contemplation. Even a goldfish or a hamster has its own life to worry about without pondering how others lived. Or perhaps no goldfish, hamster, parakeet, cat, dog or humanimals lived at home, less a home than only a house empty of voices, barks, meows and whatever noise hamsters make (never checked it out, as I'm not curious about hamster sounds). Maybe no life other than mine, when in residence, animated the place, otherwise filled only with silence and lacking the joyful noise of yappy doggies or happy kitties and kiddies to greet me when I returned after a long trip. That's all irrelevant, for my reminiscence isn't about me or my personal life. Although my home base indeed comprised a good—a very good—part of my life, that's of interest only to me and to the very few beings who formed part of my domestic world. Some exist, or existed, but belong only in my life and not in my book. Like its ghost writer author, my familiars remain behind the scenes as phantom figures playing their parts off-stage rather than front and center. My narrative essentially deals with the encounter of a certain person's acute curiosity with the outside world and ignores the possible curiosity of readers regarding my inside world at home.

Experiencing life out in the world, seeing places both strange and rather ordinary, visiting regions far and near, encountering lands teeming with humanimals in all their dazzling variety and conditions and settings and upsettings, observing how those creatures dealt with the odd phenomenon of being alive, a condition they never requested—all that enabled me to enrich my earthly existence with as wide a range of experiences and experiments as I could create. Existence was a chance gift of uncertain value and duration formed out of the oblivion of non-being and destined to return to that state. I believed that the highest and best use of the years given to me should feature the intangibles of experiences and relationships, both as evanescent and as personal as was I. Although more durable, things held no attraction for me. I avoided all but the most necessary possessions. My time was better spent other than spending money on stuff. I valued things money could facilitate but not necessarily buy. You can't purchase durable life-enhancing relationships or create unusual memorable experiences without some effort which transcends buying-power.

One of the main ways I parted company from how people around me chose to arrange their lives was my refusal to join the consumer society. As for other bourgeois practices common in my circles, I didn't scorn consumerism or fail to see its attractions. I only rejected such activities for myself. I didn't think in terms of shopping, sales, buys, bargains, stores, stuff. I never gave a second thought, and not even a first one, to consumption of the kind material to most people in my milieu but completely immaterial to me. Malls greatly discomfited me. I felt uneasy in those temples of consumption, their shops bloated with stuff. Places out in the world—"wheres" rather than wares—interested me far more than merchandise. Like Borges' Funes the Memorious, who remembered everything in vertiginous detail, I suffered a kind of vertigo in malls. The surfeit of stuff made me dizzy. Enclosed in a mall I felt claustrophobic, confined to an oppressive space cram-jammed with things, stuff piled in stacks, arranged in rows, over-filling bins and crowding shelves and stuffed into cases. It was all too much. For me all those goods were bad. And yet I was pleased to see how so many people were engaged in an activity they enjoyed. Consumption for the shoppers represented what travel was for me. I didn't denigrate their habit. Just because shopping didn't interest me wasn't sufficient cause to criticize others who found pleasure and treasure buying stuff. I didn't disdain consumerism; I simply avoided it. I wanted to remodel my life, not my house. Those who bought into a consumerist way of life no doubt found malls just as exciting as was travel for me. I even appreciated that, unlike so many of the places I visited, malls offered many advantages and conveniences, such as air conditioning, heat, clean toilets, drinkable water, a secure environment, hygienic food, signs and conversation in my own language. There's a lot to be said for such benefits and amenities, but nothing to be said for making shopping centers the center of one's life.

Although I shopped at Good Will—even though it seemed a bit expensive—and was glad the well-meaning organization could process cast-off stuff to resell, with few exceptions I remained out of the consumption loop. What others deemed essential or desirable I viewed as clutter which would only complicate daily life and divert me from how I wished to live. I was truly and ever a world-class non-shopper. When a good friend of mine gave me a microwave oven I told her I thought the thing was a television set, wising up only when I couldn't get a picture on the glass door which I'd supposed was a screen. In truth, I didn't need, want or own a TV and still don't. This presented me with one major disadvantage, as an obscure provision of the MEAL Act (May Every Animal Live) makes it illegal to eat a TV dinner containing meat if you don't have a television set. This regulation prevented me from upgrading meals from my own cooking to the much higher quality offered by TV dinners.

Like the dogs in my neighborhood I ate okay—not as well as the canines, but palatable (barely). With the microwave I tried to upgrade my meals to the quality most of the dogs in the suburb enjoyed. But I didn't attempt to prepare "cuisine" as I was never a gourmet or a foodie. I wasn't picky. Like dogs at home, I ate what was put before me. In underdeveloped areas, however, I paid a lot of attention to food—not its taste but its hygiene. As every Third World traveler knows, safe rather than savory food represents the most important factor for what you eat. Although an adventurous traveler, I was a completely unadventurous eater. I always carefully contemplated the digestive consequences of what I consumed—for once I was a careful and finicky consumer—and when in doubt I avoided questionable food in favor of less problematic prosaic dishes. In difficult areas I traveled with a small supply of emergency provisions, including crackers, nuts, a few cans of tuna. These reserves could tide me over if, as occasionally happened, I couldn't find hygienic food. Although I usually tried to sample regional cuisine I remained wary of its possible toxic effects. Avoiding typical local dishes wasn't much of a privation as *basse cuisine* satisfied me just as well as did *haute cuisine*. Although I appreciated good cooking, at home or away I was content with simple and basic meals. Whatever the quality and wherever I happened to be I was always grateful to have clean food to eat, as I knew that out in the Third World many people remained undernourished and subject to unsanitary grub, insipid gruel and contaminated water. Warned that the tap water in St. Petersburg (Russia, not Florida) swarmed with giardiasis, we brushed our teeth with vodka, and in many Third World countries restricted our food to hard-boiled eggs, fruit we could peel and other mundane but safe items.

My indifference to food matched my lack of materialism. Both eats and things were extraneous to my interests. My life I reduced to basics to free up time and energy to pursue other modes of being which I valued more than possessions or quickly consumed highly transient food which passes in, through and then out of your system. I by no means lived a spartan existence at home, nor was my way of life as simple as a Trappist or a Benedictine monk. In any case, even though I have a monkey mind I'm not monk material as I know nothing about organized or disorganized religion. I don't know, for example, what a novena is but I like the sound of the word. I supposed the term had something to do with a bull fighting maneuver or perhaps it's some sort of car brand. A shiny new Novena would upgrade my tired and oft re-tired old wheels. Monks who vow silence nonetheless write books, so apparently the good fathers can engender words even if they can't speak them. Or perhaps Thomas Merton, who wrote many pages, got some sort of Papal dispensation exempting monks named Thomas from a vow of silence. If that special exception wasn't in fact available for Thomases, as a monk I'd have to silence my memory-based writings and instead of *Speak, Memory*, as Vladimir Nabokov titled his reminiscence, I'd order, "Shut up, Memory."

My lack of interest in possessions wasn't based on anti-materialism but only on un-materialism. I didn't view over-consumption with a moral judgment that deemed frequent or compulsive buying as something bad. Frequent and compulsive travel characterized a good part of my life so I couldn't criticize those kinds of habits as applied to other harmless activities. Travel and consumption were essentially similar: each diversion involved a somewhat analogous frivolous attempt by humanimals to experience and enjoy the play of infinite forms. My choice dealt with the intangible but considerable benefits of travel; consumers preferred more touchy-feely things. My attitude pertained not to how I thought others should behave but only how to organize my own life. My view conformed to the belief (as I've noted) William James expressed in *The Varieties of Religious Experience*, that "Lives based on having are less free than lives based on being."

I deemed objects as intrusive. Like dogs, possessions can take over, limiting and distorting your life but—unlike canines—at least you didn't need to stuff stuff with food every day. Things were not just inert useless objects: they were rather productive as they generated a lot of dust, which some souls might have viewed as a *memento mori*. But that thought was far from my mind whenever I contentedly looked at the various objects I bought and brought into my life, things of minimal monetary value acquired on my trips but mementos I greatly treasured. Some of the objects I didn't even buy but just plucked them from where they littered the earth. Those treasured travel trophies are literally worthless—they cost me nothing—but are to me priceless. They include: a leaf taken (and later framed) from a supposed descendant of the Bodhi Tree under which Buddha sat for enlightenment at Bodh Gaya in northern India; a rock from the very tip of Africa at Cape Agulhas and from the tip of India at Cape Comorin; a pine cone from atop Ernest Hemingway's grave in Ketchum, Idaho; a rusty nail from a Devil's Island prison cell; a stone from the Troy ramp up which the wooden Trojan Horse was pulled; a sprig of frankincense from Oman (perhaps soon to be renamed Woman); a shard from ancient ruins at Balkh in northern Afghanistan; a fragment of a blue tile from Mother of Shah Madrassa in Isfahan; a small composition rock from Mount Sinai; a shard (perhaps six thousand years old) from a Jerico *tel*; a small mosaic pavement stone from Carthage. I wanted a lock of Snoopy's hair and a few lab experiment notes from the guy who invented M&M's, but those souvenirs eluded me. Whether purchased or plucked, my travel mementos represented memories incarnate and in that way de-materialized, mutating from objects into intangibles which evoked doing more than having.

My travels served to respond to my curiosity and also corresponded to my preference for experiences rather than possessions. Yet a third factor operated to induce me occasionally to remove myself from my native turf. Although deeply rooted and comfortable in my normal surroundings, I didn't subscribe to many of the beliefs, interests, value systems and practices which defined and motivated the lives of my peers at home. The societal structure in the context where I originated and stayed didn't suit me. The conventional rituals, habits, behavior, expectations and standards which typified society in the provincial Midwestern suburban area where I lived failed in many ways to influence, engage or motivate me. Dissenting, I didn't conform to many of the ways people around me lived. At the same time, I rather envied their certainties. My fellow suburbanites seemed comfortably sure of their way of life, which was quite similar from one family to the next. It would be mere tomfoolery on my part to criticize such conventional and conformist behavior, but for me it was too hermetic and I needed some fresh air. I didn't rebel against the prevailing values but only quietly distanced myself from them both geographically and by how I functioned at home. By my globe-trotting and my practices at home I went my own way.

My attitude toward my friends and how they lived was acceptance but not acquiescence. I doubt if they completely reciprocated. They probably neither endorsed nor accepted my somewhat eccentric sort of life, which in a certain way implied a rejection of their conformist conventional ways. In *It All Started With Europe* humorist Richard Armour described a mindset similar to my skeptical views toward bourgeois normalcy: "One of the leading scholars was Tom S. Aquinas. Those who believed his theories were known as Tomists. Students, who weren't quite sure, were called Doubting Tomists. Superficial students, who only skimmed his works, were referred to as Peeping Tomists." As a Doubting Tom I questioned the merits of many beliefs and practices of my fellow humanimals, not only at home but everywhere. I was also a Peeping Tom as I poked and peeked into far corners to see how earthlings in distant lands responded to the problems and challenges which confronted the locals where they lived, conditions quite different from the more comfortable amenities which prevailed in my little corner of the world. But in a more existential way, people the world over faced the same confusions and bewildering and baffling mysteries intrinsic in being alive as temporary inhabitants of the lonely planet. A wide variety of responses reflected how each culture dealt with both worldly and existential problems. In response to the play of infinite forms, people in different areas attempted various ways to impose order and to find some meaning. Beliefs and rituals in my own culture and in most others represent attempts by humanimals to structure life, an unruly natural phenomenon, by introducing some sort of transcendental—or at least a more limited and transient but somehow consoling—earthly significance.

From what I saw, most of the support systems and social structures formulated over the centuries and around the world and which comprised the very fabric of human civilization or culture were largely meaningless. Some of the systems seemed only a shot in the dark, insubstantial phantom-like speculative specters with little basis in reality. From the grab-bag of beliefs and practices, I selected only the few which I found viable or beneficial. Although I didn't believe in the Father or the Son, as a ghost writer I perforce had faith in a ghost, even if not a holy one. For the most part, however, I ignored all the rest—religions, conventions, rituals, conformism and consumerism, what passes for success, the accretion of wealth, fame, prestige, power, standing, possessions, dominance, control, leadership and its charisma (for me, not a chance of that anyway). Such strivings and wantings—selfies on the hoof— didn't interest or motivate me. I left all those chimeras to my fellow lonely planet residents, all passing through momentarily as riders on the earth, that celestial trolley on its inexorable track toward oblivion.

The mystery of existence is both intriguing and disturbing. The puzzle attracts attention but repels solution. If the question of meaning could have been answered by discovering a missing "something"—still a known unknown—then mankind would have been deprived of all the somewhat consoling elements of religious belief and the related cultural forms, such as music, art, writings, philosophy, the architecture and décor of churches, temples and mosques, rituals and celebrations. Losing all those cultural riches might not compensate for gaining a definitive answer to the questions religion seeks to address. But if religion had become unnecessary by knowledge of the absent "something," many wars, crusades, conflicts, turf battles, be-headings and other dismemberings, and bloody "*roi contre Pape*" encounters could have been avoided.

Viewed from a cosmic perspective, far removed from simply how earthlings organized their lives, something seemed amiss. I sensed that in an existential way a certain elusive "something" was missing from the odd phenomenon of human existence. Life as the humanimal knows it struck me as strange and incomplete, as if an essential component which would bring some meaning and significance to the enterprise had been left out. In *A Writer's Diary* Virginia Woolf wondered "Why is there not a discovery in life? Something one can lay hands on and say 'This is it?' " There seems to be a "something" there, a kind of shadowy skeletal "it," but no one knows what "it" is and never will. But people searching for a meaning should be careful what they wish for. Clarity would eliminate much of the fascination of human existence. The missing element is precisely what makes life intriguing and worth contemplating, and perhaps even worth experiencing—but maybe not. If existence were clear and without mystery life would be a boring phenomenon and lacking its edge. Although discomfiting, the missing "something" serves to sharpen the life experiment. In section 82 of *Essays in Idleness* by Yoshida Kenkō (translation by Meredith McKinney) he observes that imperfection is more desirable than completeness: "only a boring man will always want things to match; real quality lies in irregularity." Kenkō continues: "In all things, perfect regularity is tasteless. Something left not quite finished is very appealing." This attitude reflects the Japanese notion of *wabi-sabi*, an expression which refers to the charms of imperfection, simplicity and transience. Imperfection can also be useful. In *Metals in the Service of Man*, William Alexander and Arthur Street note that "metal grains or crystals are far from perfect and that many of their attractive mechanical properties are due to imperfections." Perfection is vulnerable to degradation without allowing any potential for improvement. Apparent flaws in fact lend to a situation or an object a certain melancholy touch by suggesting impermanence and the reality that nothing which exists is complete or completely comprehensible. Mystery can lend enchantment, but for the phenomenon of human life and the worldly play of infinite forms perhaps the mystery is too unsettlingly enchanting for words. In any case, had Woolf finally found the elusive "something" she probably would have had nothing to think or to write about, and nor would I or any other author. Religions would be discarded, temples of faith razed, seminaries closed, the clergy discharged, university philosophy departments shuttered, many holidays eliminated, the Vatican turned into a amusement park—much of life turned upside down, once a "something" which explains everything was discovered. But that prospect is highly unlikely, unless Snoopy or some other sage somehow happens to come up with some sort of sensational original and definitive solution to the puzzle.

Humanimals struggle without success to contrive some sort of meaning for their existence. Maybe the meaning, such as it is, pertains simply to the process of searching for a meaning, in which case the unsatisfactory result is—seek but ye shall not find. In *Our Life with Mr. Gurdjieff* Thomas and Olga de Hartmann described how from 1917 to 1929 they followed the "enigmatic" Russian spiritualist around—like puppy dogs, although the authors didn't put it that way—to Moscow, the Caucasus, Istanbul, Berlin, Paris, all the while looking to the guru to reveal "something" to explain life. No Doubting Thomas, Thomas de Hartmann believed that a "something" existed. But like Virginia Woolf, and like every mortal who ever lived, lives, will live, he never managed to discover that ghostly "something." The much sought-after "something" is a chimera, a *fata morgana*, a longed-for vision never to be found. Even a Doubting Thomas doesn't doubt that reality. And if, perchance, the "something" happens to exist, such phantasmagoric will-o'-the-wisp insubstantial "thing" lies beyond human ken.

It was a waste of time for Tom and his wife to chase after the "something" and for Virginia to lament its absence, for in reality that known unknown is nothing. Some earthlings who long for meaning suffer from a reluctance to accept its non-existence. To ignore the extreme limitations on the ability of humanimals to access knowledge, not only for meaning but in many other categories, says more about the ignorer than it does about reality. A lot of problems in life arise from an unwillingness to recognize "epistemological limitations," an expression with a fancy 50 cent word plus a normal 25 cent one adding up to the 75 cent observation (on sale, marked down from $1) that many things are impossible to know. To pretend we know or can know something which is unknowable can lead to all sorts of complications. For earthlings, certainty is a scarce resource. To Montaigne's literary motto *"Que sais-je?"*—"What do I know?"—my answer would be, "almost nothing." One thing I do know, however, is that there is no "something."

Kierkegaard believed—but he couldn't and didn't know for sure—that although you live life forward it can only be understood backward. Like every wise guy who purports to present wise man existential certainties, the Danish sage was entitled to his own opinion. But he was only a philosopher, a contemplative type too little exposed to the rigors and hazards typical of such activists as Judah P. Benjamin and Percy Fawcett or seasoned savants such as Snoopy and Elvis Presley. Kierkegaard doesn't exemplify practical wisdom of the kind as proposed by Jacob Bronowski, who held that "The world can only be grasped by action, not by contemplation." Kierkegaard was in fact half right, a good batting average in the National League but not for big time major league philosophers. Life, as he claimed, can't be understood forward, but it can't be understood backward either. Human existence can't be understood at all, no matter how you look at it. Thinkers can make up what seems to be some sort of meaningful and comforting pattern or explanation—such as, for example, how I chose to organize this book—but the imposed apparent meaning represents only an arbitrary arrangement selected from the vast play of infinite forms. So much happens and exists by the operation of chance, luck, fate, circumstances, serendipity there's really no way to establish by reason, faith, artistic creation, procreation or any other means a definitive explanation for the way things are on the lonely planet or elsewhere in the cosmos. The apparent but misleading benefit of hindsight, which Kierkegaard believes would clarify a life, can deceive you into perceiving that what you believe to be some patterns or themes are in fact only artificial constructs without any transcendent value. Viewed backward or forward, there's nothing to understand. You can only pretend to understand. Woolf and other seekers long for a phantom which would tincture reality with substance, but there's nothing there. Where an animal casts its shadow you see shade, but there's nothing to it. An end-of-life retrospective perspective such as this reminiscence enables you to know what happened during your earthly tenure, but such an exercise is otherwise no more informative in an existential way than when as a youth you looked forward into un-elapsed time. Neither perspective yields any "something." All the years since my early days at Phantom Lake in Maine—where the campfires burned into smoke which vanished into the air—remain phantom-like insubstantial ghost-like yesteryears which haunt my existence but don't endow it with any meaning.

~ 4 ~

It was clear to me—as clear and as distant as the starry heavens seen in a pristine desert sky—that viewed from a cosmic perspective the lonely planet made no sense. That's one reason why it was so lonely here. Pebble earth, by happenstance populated by a horde of humanimals and spinning in timeless space, meant nothing to the cosmos or to the globe's inhabitants. In the big scheme of things the planet was irrelevant. With a closer look, however, perhaps a limited sort of meaning could be teased out of the random play of infinite forms. It wasn't much, but a more narrow earth-centric view was all anyone had to work with. A level-headed look, rather than one up in the clouds based on airy speculations, lets you see more clearly the conditions which form mankind's fatal circle as imposed by the nature of things. The down-to-earth realities explained some of the behavior favored by humanimals. They jostled to seek wealth, power, position, prestige, property, status, security and engaged in plots and counter-plots to those ends. Earthlings longed for belief, hope, love, compassion, caring, emotional sustenance. The seasons, the seas, the deserts and jungles and other natural features, the birds and bees and beasts—they all seemed natural and normal, as if they inevitably belonged. It all made a certain sort of sense. When viewed within the context of the planet, the world's good and evil and what exists can be understood. But that local provincial view was as limited as my own home-based suburban provincial existence. Such a narrow perspective didn't help to explain the meaning of why things, including the presence of humanimals, had been arranged as they are. People seemed to be one of the most puzzling parts. Earthlings really didn't fit well in the scheme of things.

Unfit for purpose, humanimals seemed unsuited to the conditions which defined and delimited their earthly presence. Just as I incompletely fitted into my own suburban natural habitat, so mankind itself was a cosmic misfit, a being thrown into a strange inexplicable existence, "a troubled guest/On the dark earth," as Goethe wrote in the poem "Holy Longing." Heidegger refers to existing creatures as having undergone *Geworfenheit*, "thrown in" to life—earthlings who happen to have been incarnated and tossed into life, there left to flounder in that alien and mysterious state. I must confess that I really don't understand all that much about Heidegger's beliefs. What little I know about the German philosopher's views could be written on a piece of paper the size of a bubble gum wrapper. But that would at least give me something to chew on, even if only a few of Heidegger's speculative bubbles. What he says is just one man's opinion. I mention him because it upgrades a text, probably much needed by now, to quote a guy like Heidegger. The very name suggests great gravitas and implies to a reader that an author is an erudite dude, maybe even a profound thinker. But nothing could be further from the truth. Never in my entire life did I have an original thought. I only stumbled upon—often only late in life (I was a slow, but steady, learner)—what others already knew. Whatever I perceived was original only for me and, so I supposed, applied to my particular situation without any more general meaning. Even if perhaps slightly chubby I'm no heavy thinker. I'm incapable of offering sweeping general definitive pronouncements dealing with the BIG nitty-gritty questions of existence. My meditative thoughts expressed in this reminiscence are simply stray notions which occurred to me and which framed my outlook. Those observations represent my reactions to and contemplations about the play of infinite forms I happened to experience. Other observers might well have drawn the opposite conclusions.

Existential questions remained far removed from my quest. I was more interested in the here and now, followed afterwards by memories of what I'd managed to experience. In *Cupid and the Jacaranda* Sacheverell Sitwell explained why he wrote by citing his "love of this world and of the wonders in it, which are enough for all the hours of every lifetime since the first human footprints upon the tawny sands." Unlike Sitwell, I didn't love the world, and if I did have a crush on the place my affection would have been unrequited. I only admired the world, even if at the end of my earthly existence I'd be forced to forsake the lonely planet when I reached the tawny sands at the desert of vast eternity. You really had to admire such a fascinating place—even if the earth is a meaningless corner of the cosmos—filled with such a stupendously imaginative play of infinite forms. Those spectacular sensory impression sources were indeed enough to fill a lifetime. The forms provided endless diversions right here on the lonely planet without any need for me to expand my horizons to consider cosmic matters. As riders together on the earth, that bright loveliness in the eternal cold, earthlings were for better or worse stuck down here below on their sub-lunar terrestrial home. I was content to remain earth-bound—where none of the big questions could ever be answered—and firmly rooted on the ground in which I was destined eventually to lie interred forever. I realized years ago that it would be more productive and realistic to focus on the world as it is rather than on why it exists and the meaning of life, topics I was completely incapable of addressing for lack of any ideas about them. I was around the place for a short time and wanted to devote my few fleeting years to the experience and the experiment rather than to theories about them. I was no philosopher but only a stray humanimal passing through the dark forest, groping my way along like everyone else. I never sought such existential answers because I knew I could never find them. Ultimate certainties—and even many transient ones—eluded me. Almost everything was tentative, provisional. My comments all through this reminiscence pertain only to the world as I experienced it and to life as I saw it. In that sense, my account presents a very narrow and limited story. Like billions of other earthlings I was by some mysterious force thrown into existence and forced to find my way onward. My path was both specific to me and also in some ways archetypical, for Everyman and Everywoman must navigate his or her way through the darkness, all fated to emerge in the end at the desert of vast eternity.

Apart from the personal and the archetypical format, in yet another sense my reminiscence presents a two-part account which includes both the play of infinite forms as I experienced them in my time and the pervasive controlling realities of life for all humanimals on the lonely planet during my era. In still a third way the book offers a double perspective by viewing the lonely planet's spectacle as both a play of infinite forms and as an infinite play of forms, as over time everything mutates to form and re-form on and on. This ceaseless process continues even after earthlings—one by one and eventually the entire species—have exited the theater and can no longer observe the performance. Thousands of my meditative predecessors—philosophers, professors, professed experts, pundits, sages, gurus, seers, writers, mystics, thinkers, dreamers—have proposed their views on the *Geworfenheit* reality. I could quote any number of them—Montaigne, Plato, Homer, Dante, Shakespeare, Goethe, Presley, Snoopy, and in fact I've already quoted some of them. They're all really high-class guys (You want a gal? Okay—the Oracle of Delphi or Lady Gaga) who help to endow my text with a high degree of authority. Who in their right mind would disagree with what Shakespeare or Snoopy says? My own view is that being thrown into existence is a poisoned chalice: although you can enjoy the opportunity to drink-in the bubbling, churning play of infinite forms to the point of pleasant inebriation, you won't survive the experience. But in truth, whatever anyone says about ultimate questions is just opinion, as is my book—one long opinion based on the particular happenings which by chance formed the content of my individual existence. From my life's hinterland—the past, my experiences, my reading, and my reactions to and thoughts about what happened—I record in my reminiscence how it all looked to me once I was thrown into existence.

Being tossed into existence suggests that you've been thrown into an alien setting, one you didn't expect or request. It's a world unsuitable for a humanimal, a being which somehow suddenly finds itself alive on the lonely planet. Something seems out of joint, as if the terms of engagement lack some essential elements. As anyone who thinks about the situation realizes, something's missing. But, unlike Woolf, I never bestirred myself with a longing to grasp the absent element. I devoted my time to more attainable goals. The place of humanimals in the scheme of things is inexplicable: the poor creature is suddenly expelled from nothing into being only to revert after a brief existence to nothing. Like slowly moving camels across the empty desert sands forming a passing pattern of dark-light-dark, so pattern for transient earthlings the dark never before followed by a brief bright spark of life and then once again a nevermore night of non-existence. Camp Fire Girls and Camp Ironwood boys appear out of nothing, live briefly like the fire's momentary sparks, then disappear into the darkness just as do the ashed logs, soon vanished into smoke. Earthlings don't seem to belong on earth. A certain unsettling disjunction between our limited perception of reality and the full measure of knowledge, information, intuition and feelings needed to understand our existence blocks comprehension. That never-to-be-closed gap between being and meaning leaves empty spaces which prevent a full picture from taking shape. Only a few scattered fragments can be perceived, and—like my stamp collection—they don't fit together very well. An Irish folklore saying claims that "What is strange is wonderful." On this view, human life is indeed wonderful. But in fact it's more wondered about than wonderful.

It all might have been more artfully arranged. Instead of leaving mortal souls stranded on the lonely planet, all doomed to disappear without ever knowing anything meaningful about their reason for being, some sort of revelation might have accompanied the *Geworfenheit* when the Big Bang threw into being the components which eventually comprised the infinite play of forms. But that comforting additional revealing element wasn't tossed into the mix. The terms of engagement of the earth as it exists within the cosmos—terms which include mankind in the world—seems nonsensical. The imposed random conditions are oppressive. Morbidity and mortality ever threaten the very existence of each humanimal. If pathogens or traumas don't do you in, time, the most toxic poison of all, eventually inevitably will finish you off. As much as you might enjoy your life, it's a journey from grin to chagrin until finally the age old old age realization that time is running out demoralizes you. Surely there somehow could have been a better arrangement. All the pathology and the inescapable finality, the disease and death which so bedevil living beings, is rather bothersome. Illness and non-being are intrusive and inconvenient. There must be a more favorable system, an arrangement without morbidity and mortality and all the other plagues which shadow the lonely planet. I admit this is hard to imagine, but so is the world as it actually is. No one in their right mind would have ever thought such a weird place could exist. Although life is created by natural forces, nature then perversely intrudes upon its own creation by reducing everything to nothing. It makes no sense to create only to destroy.

Many of the conditions imposed to allow earthlings to remain alive seem unreasonable. To continue to exist, an animal must spend much of its allocated time for the lifeless state of sleep. Although our little life is rounded with a sleep, it seems unfair that a person's earthly existence is made even more little by the need to spend one-third of the allotted years in slumber. It's hard to imagine why on earth a humanimal is subject to this tremendous depletion of the creature's limited time. Some strange force throws people into being only to remove them from consciousness, frequently and without respite, by forcing sleep. After gaining life we lose one-third of it in sleep. That's not a very favorable bargain. But of course existence is nothing that we bargained for. Sleep prolongs life but also eliminates it during the hours we're out. In *The Garden of Cyrus* Thomas Browne describes sleep as a time when a humanimal must "close the five ports of knowledge," an inescapable dark interlude, a life sentence to daily oblivion until eventually "sleep itself must end." As is almost always the case for a Thomas—long live the Thomases—Browne offers great wisdom.

Added to the eternal pre- and post-existence dreamless sleeps, the series of daily dozes which mimic the oblivion of non-being even further limits our brief time. This theft of wakefulness we wouldn't tolerate for other activities. No one would put up with a baseball game for six innings and you're out, a game with third base eliminated and triples counting only as doubles. A round of golf would be over after 12 holes; a gap year finished in eight months; a three act play curtained after two; a honeymoon interrupted two-thirds of the way in; a family picnic with seconds but no thirds; a vacation vacated early; cats limited to six lives; triangles reduced to elliptical shapes; three star hotels demoted to two; triplets aborted to twins. Mothers would give birth in the second rather than the third trimester, and summer vacations for the kids in school would be reduced from three to two months. Love triangles would simplify to duos, a *ménage á trois* to *á deux*, and the *Trois Mousquetaires* would also become a *deux*, and *adieu* as well to tricorns, tricotism, tricuspids, tricycles, tridents and any number of other trifles. Chinese triads would be diads, the Three Stooges would become a duo, and the Andrew Sisters the Drew Duet. One card would be discarded from three-card monte, and 3-In-One oil would be a third less potent. Believers would have to decide which one of the three members of the Trinity should be eliminated. We'd be buried not six feet under but only four. Minutes and hours would have 40 seconds and minutes, days limited to 16 hours and years to eight months. *The Third Man* would be cut by a third to become *The Second Man* and the Third World and all its miseries would disappear—not all bad, but if there remained only the First and Second Worlds billions of Third World beings would be omitted from the cast of the play of infinite forms and the lonely planet would be more lonely and less interesting. We'd miss all those truncated familiarities, but no one comments on or complains about one-third of life lost to sleep. As much as sleep seems natural and normal—an unquestioned necessity—that somnolent state robs us of a large part of our limited time on earth. Age 90 is really only 60, 60 only 40. Surely there must be a better way. Truly, the only advantage—apart from reducing German noun genders from three to two—of a one-third reduction for everything, as with sleep, would be that this book contains only some 260 rather than about 390 pages—probably much easier on the reader and for sure on the writer.

Lord Chesterfield educated his son about sex by telling the boy that the expense was damnable, the position ridiculous and the pleasure momentary. So it is with sleep. The ridiculous immobile and prone position mimics death and mocks life, and the temporal expense an oblivion accounting for one-third of our earthly existence ultra-ruinous as it greatly depletes our allocated time. As for pleasure, it's brief and fleeting because like sex sleep must before long be renewed with another interlude. Sex is at least discretionary, unlike sleep which forces you into a nyctitropic vegetative state whether you want to be knocked out or not.

Obviously, I could do nothing about sleep's incessant demands or any of the many other imposed terms of engagement with life in general. I was stuck with them like every other humanimal. That's life, take it or leave it. I needed my sleep, and with this book may be facilitating yours. Happy to be of service. However, within the unalterable terms decreed by the natural order of things I could make some changes by rearranging how I related to man-made conventions and practices which prevailed in my little corner of the world. A fateful triad of traits (maybe only a duo if sleep is taken into consideration) motivated my wish to revise my way of being. Added to curiosity and to my desire for experiences out in the world was my problematic relationship with society as structured in my home setting. In my own home place I felt somewhat out of place. I was in part alienated from the culture and society where I belonged and was most at home. I resolved to remain there to enjoy the benefits of my familiar setting and its conveniences while at times abandoning my home base for a more diverse existence. By the time I made the change in my mid-thirties, the timing seemed especially favorable. I was then still young enough to retain enough flexibility, resilience, energy and an adventurous spirit to pursue and enjoy far-flung travels. At the same time, I'd reached an age sufficient to bring to my enterprise well-grounded experience and a seasoned perspective. These mature enhancements enabled me better to comprehend and appreciate what I happened to encounter out in the world. Aldous Huxley described this fortunate combination of youth and age in a letter (June 22, 1955, quoted in Laura Archer Huxley's *This Timeless Moment*) to his brother Julian as "that mental openness and elasticity of youth, while being able to enjoy the fruits of an already long experience."

By my mid-thirties my probable earthly tenure was of course no longer all that long—it was possible I'd already lived half of my allotted span—and my longevity was in any case uncertain. Although I'd only been an adult for a relatively few years, by then I'd lived long enough and experienced enough to bring a certain maturity to my travel adventures. I managed to synthesize diverse components from my conventional upbringing, schooling, higher education, professional career and other factors to throw some light on what happened to befall me during my travels. The mix of my conventional past with my new adventures yielded a satisfying amalgam composed of disparate elements. My decision to diversify my life with adventure travel in effect represented my answer to the question posed (as already noted in Section V, Part 1) in André Malraux's *Days of Hope*: "how can one make the best of one's life?" To this query Garcia, a Spanish Civil War revolutionary, responds exactly as would I: "By converting as wide a range of experience as possible into conscious thought." I never lacked for conscious thought, but until I recast my somewhat narrow life to include wide-ranging travel I did suffer from a shortage of things to think about. I hoped that widening my limited range would endow my thought with granular content it would otherwise lack. My own "days of hope" would hopefully allow me to adopt Bob Hope's "Thanks for the Memories" refrain, a thankfulness which now, near the end, greatly gratifies and consoles me—a welcome consolation prize.

Although I hoped to enjoy a successful experiment, whether or not my experiences would prove useful or interesting remained subject to chance and luck. As Georg Lukacs observes in *The Historical Novel*, "it is a matter of chance whether the immediate material which life offers is suitable for revealing adequately the laws of life." The laws of life were the same everywhere. The celestial legislature mandated one-third lost to sleep, certain loss of family and friends and of life itself, disease and accidents, wars, disappointments and sorrows, and other such rather bothersome provisions. I ventured forth to learn how people around the world attempted in various ways to confront the unruly condition of being alive. Not only the laws of life but also laws of the land and other societal practices regulated and defined the context of earthlings' presence on the lonely planet. The play of infinite forms set the scene and brought a variety of responses to the conditions which imposed the terms of how people lived. My days of hope in broadening my perspectives were meant to bring me an assortive rather than just *A Sort of Life*, as Graham Greene entitled his memoir. I don't know if Greene traveled to write or wrote to travel, but for me neither process applied. I simply wanted to get around to see something of the world into which I had been thrown. I was here, so why not try to make the best of it, even if the best might ultimately not be all that good.

My quest didn't seek only novelties and curiosities. I was interested not just learning how people elsewhere responded to their local conditions but also how the eternal laws of life operated to shape how earthlings thought and acted. Transient trivial current happenings lacked the same educational value as more durable influences. It puzzled me why so many earthlings obsessed over the new when in fact it's the olds which most affect life on the lonely planet. The eternal verities matter more than passing events. Conditions as from the time of the Big Bang set the terms for what comprises the cosmos and what ultimately produced the age-old ineluctable laws of life on earth. Much of the news distracts people from those venerable permanent influences. Because my curiosity extended less to what was new than to what was old, my reminiscence contains little or nothing new except what was new to me. Although I recognize that one reason people read is that they hope to find new ideas, I have almost none to offer. All my observations are simply old hat upon a new head.

For some reason news, most of which may not be personally relevant, seems to preoccupy many people. Thoreau wrote that "Hardly a man takes a half hour's nap after dinner, but when he wakes he holds up his head and asks, 'What's the news?'" Humanimals go to a lot of trouble to get the news, distributed in each era by the contemporary intricate networks which move the words around. Auden's poem "Night Mail," written (1935) to commemorate the centennial of Britain's Traveling Post Office, describes how the train carrying letters from Euston station in London arrives in Scotland where "Men long for news." In remote areas where news is as scarce as water, new information serves as a kind of currency. In some places in the Middle East desert tribesmen greet newcomers with "*Wish al-barid*"—"What's the post, the news?" If you grant the "wish," the hospitality improves—you get four cups of tea instead of the usual three (which anyway would be two if sleep is factored in).

Although less curious about the news than the olds, for years I read four and sometimes five daily newspapers, always in their paper versions. Only one of them, however, truly brought me the feeling of belonging somewhere. The national or international papers reported on the outside world, while my hometown newspaper gave me news about my native habitat. Home is the place where you can almost entirely understand the local news, especially by reading between the lines. Only at home could I comprehend a newspaper with enough familiarity for me to grasp all the nuances, back-stories and implications. This comforted me because it meant that I knew I belonged somewhere, even if I traveled anywhere. At home the news came to life, so confirming that I wasn't just a rootless vagabond but that I remained bonded to a place, one where my roots ran deep. I valued both my hometown and its news as well as all the "news" my travels brought me—new places and faces, ideas, customs, experiences. But I also kept in mind the more permanent and ancient realities which characterize life as it has always been for earthlings on the lonely planet.

In recent years the first section I turn to in the local newspaper is the obituary page where the olds by their disappearance make news. For oldsters the occupation of reading who passed is in fact a preoccupation as, among other factors, we want to take the measure of our own actuarial possibilities. It's always comforting not to see your own name in the obituary listings, but in reading funeral news you're thinking about who's next. It could be you. I look first not at the date of death but at the year of birth to compare it with mine. Recently the deadly birth year is often the same as mine or even a later one. This sort of creepy mission-creep is unsettling. Death no longer creeps up on you but now wings its way toward you as the wingèd chariot seems to speed up. "Exit, pursued by a wingèd chariot" is the script for how the play of infinite forms will soon end for me. The chariot never runs out of gas but I am (except when I eat my own cooking). The final item on my bucket list is to kick the bucket. It'll be a real kick to satisfy my curiosity about what happens after the final curtain. My own instructions provide that no announcement or obituary be issued to provide news of my demise, and no memorial service take place to celebrate the occasion. My feeling is that once you advance from ghost writer to ghost there's no reason to announce or to commemorate the transition. This truly major accomplishment doesn't merit any kind of notice or ceremony. I'm content to let nature take its course without any discourse or news about what everybody already knows will soon occur, for me and for themselves as well.

My reminiscence is obituary enough. I wrote it in the spirit of the practice customary in the time of historian Jacob Burckhardt, who shortly before his death in 1897 prepared a brief summary of his life to be read at his funeral. My longer account recalls that custom in an extended format, but one which should be read only before or most likely after and not at my funeral (which anyway will be burial without a service, other than the grave practical services and rituals necessary to put me away). Reading my book at a memorial service might try the patience of the mourners, or mourner if only one shows up (thanks Fido), survivors who understandably want to get on with their lives. Omitting a service will not only spare those survivors from any readings, it'll also save the costs of a post-funeral reception. In his mini-memoir Burckhardt, referring to the death of his mother when he was 11, wrote (in the third person, which should be the second person because of what he lost to sleep): "Thus very early in life, not withstanding an otherwise cheerful temperament, in all likelihood inherited from his mother, he received an indelible impression of the great frailty and uncertainty of all earthly things, and this determined his view of life."

Section VII: Shadows and Smoke

Although I shared Burckhardt's sense of how uncertain everything is on the lonely planet, unlike him I wasn't completely certain how my uncertainty originated. I realized that it in part derived from and also influenced my temperament, but real life experience was a more substantial factor. The more I saw of the world the fewer my certainties, until finally I was left with almost none. What the whirligig of time happened to spin into being to produce Tagore's play of infinite forms arose from the random play of luck, chance and fate, all by definition sources of uncertainty. Similarly, a certain randomness determined which incidents and episodes I experienced and then later recalled in order to transition them onto the printed page of my reminiscence. My intention was to include not only some specific happenings but also to transcend them by adding some reflections relating to my experiment in living. I hoped in a small way to preserve from my experiences for a brief time after my time a few thoughts "so that time may not draw the color" from them, as Herodotus in a much more momentous and durable way attempted when writing 2,400 years ago about Egypt's early civilization, which predated the historian by some 2,600 years. By now however, twilight shadows darken the spectacular Technicolor play of infinite forms which comprised my life. The magic lantern is dimming.

In attempting to draw some brightly hued highlights from my life to illustrate matters beyond the events themselves, I've repeated some topics—a play and re-play of infinite forms—in order to view them at different times in different settings from various angles, so bringing to those subjects new perspectives, just as turning a kaleidoscope reveals changing patterns of shapes and colors. Perhaps my repetitions also represent a subliminal but ultimately unsuccessful attempt to counter the "one time, one meeting" evanescence which dooms each and every thing and happening to never again fleeting form or format. The grabagatelle of the play of infinite forms furnished me with a large stock of impressions, only a few of which I've recorded in these pages. My account thus contains only very limited withdrawals from my memory bank, a treasure trove for me but one which will soon become a buried treasure. Never fear—no further withdrawals will be made as no sequel threatens, no encores, no Volume II: this is it.

My reminiscence may seem a longish volume but I've had a long life and all these pages represent only a brief condensed vignette version extracted and organized from a convoluted and voluminous earthly existence. Reduced to the applicable metrics, the book is actually quite short. It amounts to only something like seven pages per year of my active traveling life, which equals less than 60 per cent of a page per month, and almost nothing for every week on the road. Even if you deduct one-third of my earthly time lost to sleep, then the book represents only about eleven pages per year of my waking hours. I'd like to think that my earthly adventures over more than half a century were worth more than these few pages, but now at the end I see that they aren't. Other lives merit longer treatment. Snoopy in his little dog house, where he blows up his speech balloons filled with wisdom, deserves more pages than my more wide-ranging adventures. One reason for my brevity is that I wrote with an ascetic aesthetic, the work's guiding consciousness remaining in the background as only a ghostly specter. In that the author omitted almost all personal references, he remains a phantom-like presence, or actually an absence, hidden in the shadows. Rather than talking about myself I preferred only to describe how the curiosity of one humanimal engaged with the curious and strange lonely planet. I wanted to put my two cents worth in, in fact a highly exaggerated amount overstated by 100 percent as my account is worth at most only one cent as I've produced only very few value-added perceptions. "A penny for your thoughts" more accurately represents an evaluation of my contribution to eventual waste paper.

Although my narrative isn't chronological I believe it's quite logical. I by design fractured time with episodes redistributed to connect with events which occurred at different times. This was meant to link some disassociated matters so as to associate them with a view to offering comparisons and contrasts in order to illuminate connections which might otherwise remain obscure. Whatever the arrangement, the text is but a pale copy of reality, for print is even more elusive and elliptical than the shadowy ghost world which hosts earthlings. Although new to me, my experiences and responses to them were in general already old to mankind. As Albert Camus realized as a young man, as he noted in *Youthful Writings*, "What I might say, everyone has said before. And there is not a single position with regard to life that has not already been taken." Who am I to pretend that I've said something new when even brainy deep-thinker Camus admitted that he couldn't? Even the braniac savants at All Souls College in Oxford largely only shuffle around old notions repackaged into new bindings, which aren't very binding as later generations review and remix the material. The experiment of being alive—a variation on a theme—involves how a humanimal will experience from among the old formats the mutations which for that particular creature are new. Each such being encounters by chance a different assortment of the infinite forms. So diffuse are the forms, as a whole they are formless. Although my reminiscence includes some thoughts on the few forms which happened to come to me, the book contains "not a single position with regard to life that has not already been taken."

Even though I greatly appreciate the passing interest of anyone who happens to read my reminiscence, in truth I'm indifferent as to whether or not readers come upon my book. Surprising as it may seem, I don't much care if anyone reads it. For me the only important thing is that I wrote it. It won't bother me if the world lacks interest in my terminal work, and in any case before long I'll be beyond caring. This may sound churlish, but it's how I feel. This sort of impolite attitude unfortunately seems to disqualify me for the Nobel Prize, whose "Code of Statues of the Nobel Foundation" provides in section 2 that the term "literature," for that prize, embraces "not only works falling under the category of polite literature [a polite way of saying *belles-lettres*], but also other writings which may claim to possess literary value by reason of their form or their mode of expression." Alas, my impolite attitude toward finding readers and my lack of pretension to "literary value" disqualify me on both counts. Neither my living nor my writing was noble or Nobel Prize worthy. My earthly existence and what I wrote were both just passing flickers in the darkness, like a campfire—burning logs into nothing—which momentarily illuminates the scene and the seen.

It's useful to retain a realistic sense of one's place in the world. The millions and billions who never read my account will far outnumber the very few, if any, who do. My primary goal has been not to attract readers or attention but to recycle my detritus so that it won't all disappear when I do. I wanted to avoid wasting the many privileged adventures, experiences and memories which, without this reminiscence, would simply vanish after my time. My text will for a time survive me in a ghostly way, like a shadow cast by the original now deceased form. It may be hard to believe that someone who goes to the trouble of writing a longish book doesn't care if anyone reads it, but that's truly my attitude. Put yourself in my shoes, or slippers—an old gentleman whose life has slip, slip, slipped away, almost all gone, a senior senior citizen beyond aspiration and very nearly beyond respiration, an old codger for whom all will soon be over. Writers don't publish or perish; they publish and perish. At this very late date I have no business dealing in ambition or expectation. My days of Taking Care of Business are over, and anyway I'm content that the business of life has by chance satisfactorily taken care of me. By now I want only to avoid waste with my responsible environmental practice of recycling some scraps from my life before they become landfill, which I myself will soon be. I'm neither a green nor a Greene, a worldly wordsman read all around the lonely planet. I realize that even if a humanimal has to some extent lived an active life, an obscure momentary earthly presence like mine may hold no particular interest for anyone other than the earthling who happened to live the experiences. I accept with equanimity that my book may remain unread. At my age, attracting readers hardly ranks as a matter which concerns me. By now I'm contemplating other topics, mainly ones of a terminal nature.

I hope that any reader who happens upon my book realizes that he or she stands in my shoes, by now for me slippery slippers. Time is slip slipping away—not only for me but also for you. Reader, you're just a few short slips behind me or, as fate will have it, maybe you're strides ahead of me after my time. You're somewhere in time, for now anyway, but not for much longer. Perhaps the unthinkable will happen and you'll get bitten by a bug like Alban Berg, or fall like Ivan Ilych, or be trampled by a horse or by an out-of-control horse-drawn wingèd chariot, or suffer some sort of sudden random assault on your being, an unlikely outlier until the event happens and then it becomes an inlier and does you in and wham-o, you're outta here, bye-bye, all over, nothing, dust, smoke even if you're not cremated. Please be careful, at least until you finish my book. After that you're on your own. In any case, even if you happen to survive me or read this in another era long after my decease then, over time, whether you manage to reach my advanced age or disappear earlier, your life will for sure gradually slip away and in fact it's slipping right now, as did mine, and eventually your very existence will end as you slip from life into an eternal nothing as soon will I. Toward your end would something like a lack of readers bother you or would, as for me now, other concerns preoccupy you? What about the threat of an undertakers' strike or a shortage of grave diggers or a runaway autonomous hearse carrying you away or the offal disposal vehicle running out of gas like me—those are the things, along with how, when and where you'll croak, which oldsters worry about, not whether their book will find readers. That's nothing compared to what matters to me now.

What I'm really concerned with at this point in my life pertains to matters more existential than mere readership. Every serious writer worth his salt—or pepper, Tabasco, curry or however he spices his life and words and works—truly contemplates at one time or another the same few mysteries: What does it mean to be a humanimal, what is the world about, what's the meaning of the play of infinite forms? As I've explained, I have no answers to these questions. I gave them a few passing thoughts, but such unanswerable transcendental matters didn't concern me. You can contemplate them from here to the desert of vast eternity and never will you reach a definitive answer. My concerns were much more limited and practical: I wanted to decide how best to pursue the experiment of being an earthling. As a young man I briefly contemplated those ultimate questions, not for airy speculative reasons—no soulless All Souls-type High Table high-powered intellectual repartee between gargles—but for the functional down-to-earth purpose of helping me define and create the good life within the limits of my circumstances. Rather than looking up toward the heavens with a trans-world search for transcendental truths, I tried to imagine looking down on the lonely planet observing from afar how things really were down here. This terrestrial rather than celestial orientation showed me a bright loveliness in the eternal cold inhabited by millions of condemned earthlings. The perspective revealed a spinning planet half-shadowed and half-sunlit, a chiaroscuro place of darkness and light. A glorious spectacle unfolded down there, an entertaining play of infinite forms. That was the bright side. But it all meant nothing, as in the end eternal darkness would shroud everything which had ever existed.

My contemplations included not only my own circumstances but also those imposed by the natural forces which prevailed on the lonely planet. A combination of me and the all-else governed my experiment in living, which involved a synthesis of world and mind—an individual thrown into and stuck within his own fatal circle and the systemic one which circumscribed earthlings. Tossed onto life, I had to deal with it on its terms, not mine. How I responded represented the experiment. From early days I realized that what confronted me wasn't a particularly good deal, but that's how the cards are dealt. They're stacked against you. You're faced with the prospect of a mighty effort to grope your way through the dark woods, only to emerge at the end into a desert of vast eternity. You'd think that all your exertions to find your way through the thicket would lead to something more promising. But, no—at the end there's only that empty eternal desert. What's promised is nothing, and that promise is always fulfilled.

When young I developed something of an antipathy toward the natural order of things. For one thing, the order wasn't very orderly as the infinite play of forms spun with ceaseless chaotic kinetic motion. For another, what existed wasn't really natural but only an arbitrary and in that way artificial format which represented just a single version of an infinite number of ways things might have been arranged. I rather resented the capricious phenomena and conditions which passed for natural and ordinary. I refused to take entirely seriously the infinite forms released originally from non-being by the Big Bang and then left on their own to proliferate and mutate. In my time I was part of them—an insignificant transient form—and I enjoyed the honor of participating in the spectacle, even if ultimately it all meant nothing. What chanced to be thrown into being and then somehow formed into particular components—water, oxygen, electricity, plants, animals, minerals, other available means—allocated to the lonely planet offered enough resources to make a world but not enough to make any sense. Earthlings find themselves stranded in a setting both accommodating and hostile. As far as I was concerned, all the infinite forms (including me) could revert to their pre-Big Bang non-existence. My fundamental indifference to the forms and the nature of things was in a minute way my revenge on the Big Bang for having started it all—the primeval soup and other stuff from which emerged the cosmos, the universe, the world, the forms, you, me, all the rest. I played along with the play of infinite forms—as an earthling I had no choice—but in a way I turned my back on them and also on the entire system. For all I cared, it could all have remained eternally Big Bang-less. The Bang seemingly created something from nothing but, in truth, the explosion simply set the stage for eventually turning something into nothing.

My story is now near its end. Only a few final years, months, weeks or days—which will it be?—remain for me to complete the terminating parts of my reminiscence which will determine how the story will end, an ending which is certain even if what happens in the meantime isn't. Writing a book by undertaking—as an undertaker before you need the services of that profession—and then completing any significant worthwhile enterprise seems like an endless task when you're doing it, but you just keep going. You keep going even as the ending continues to remain like a distant desert mirage in the remote future. It seems like the end will never come. But as time slowly slips away day by day you draw closer to the terminal point—the final page, the concluding stage, graduation, promotion, retirement, trip's end, the wingèd chariot's finish line. What appeared to be an endless process ends suddenly. Behind you bulks the past, your life overtaken by elapsed events. Such is the case for me as this book and its author approach their end.

Producing a book represents a truly unrewarding trade-off. What takes you a few hours to read took me a few years to write, an asymmetrical relationship between creator and consumer. You get three years (not including sleep) of my awake time while I get a mere three hours or less of yours. That reflects how the entire system is arranged, a deranged sort of scenario. A humanimal gets a mere blink of an eye of an earthly tenure, some 70 or 80 or so years, a nothing in the face of the eternity which precedes and follows the time during which the creature exists. Being incarnated into a human form is ultra-asymmetrical. Few, if asked, would agree to such a lop-sided bargain. If you think about it in those terms—as I did—remaining un-banged into being would be far more comfortable than facing the plots and counter-plots intrinsic in existence.

Writing a book of some length at an advanced age is a race against time. No member of the human race can ever win a race with time in the long run, but in a limited way you can outrun the clock and the chariot in a sprint, even if in the marathon event time always overtakes you and then takes you. Since everything eventually disappears, it's worth considering just what you want to create to contribute to the play of infinite forms which ceaselessly form and vanish. It all goes—books written, books read, travels, dogs, relationships, friendships, memories, lives, civilizations and civilization itself, planets, and even time, for time too must have a stop. Fortunately—for me, at least, if not the reader—I've been able to beat time in a limited way by completing my reminiscence. Even now as an old man I still, improbably, enjoy the ability to remember many adventures, and I also retain the time, the concentration and the facility as well as the will and energy to record some of those memories. To possess all those advantages at my age represents an unlikely and surprising nexus of capabilities for which I'm thankful.

Not only in writing this book but also for much of my life I was running a race against time. I tried to see, do, experience and experiment with as much as possible with the time at my disposal and within the limits of my fatal circle and other circumstances. Now near the finish line, I can see that Father Time, although much older than me, always wins. The old boy is indefatigable. I gave the oldster a good run for his money, coined not in gold but in the precious days I spent which endowed my earthly existence with treasured memories. Whether they realize it or not, all humanimals also run the same sort of race. Some seem oblivious to the match, and perhaps blissfully so. Those slow pokes are content to run out the clock with a jog. But after I realized in my early days that the race was on, whether I participated or not, I decided to accelerate my pace to cross the finish line and beat the deadline as best I could. I tried to use my allotted time so that by old age, if I managed to reach that last lap, I could look back at the empty track behind to recall some of the phantom memories of the run. This book represents such a backward look.

The major investment of time required to write my account was at my age a particularly expensive luxury. Choosing such a time-consuming activity in old age occasioned a careful cost-benefit analysis. Engaging in one project requires you to forsake others. When I was younger, the time I invested to produce a book or any other major endeavor represented deferred gratification—I gave up immediate diversions for the sake of a long-term result. In my old age that's no longer the case. My gratification now, with this reminiscence, comes not so much from the end result as from the pleasures connected with the process of remembering and recording my memories. At my stage of life deferred gratification or deferred anything else, except death, doesn't represent a realistic expectation. For me, now is it—it's now or never. The only delay which remains is for my earthly incarnation to mutate into remains, a transition soon to occur. Although I must defer to Father Time and Mother Nature, it's too late for me to depend on any sort of deferred gratification. This is what my life has come to: the end. No encore, no refund, no second chance, no deferred and very few undeferred pleasures. It's all over but the burying.

Writing this reminiscence depleted a large part of my remaining time. Trade-offs and opportunity costs represent key factors when considering any substantial undertaking or activity. When I measured the merits of writing a long book against suppressing such a major project, it was obvious to me that at my age a cost-benefit analysis based on temporal and practical factors indicated that such an enterprise would hardly be worthwhile. This conclusion holds for even the most favorable outcomes—the book gets published, people actually read it, some reviews appear in *The New York Times*, *The New York Review of Books* and, even better, in *Mad Magazine* and *The Onion* (or, if the rumored merger takes place, *The Mad Onion*). Thanks to the publicity, the book gets a few prizes—The Toenail Clipper Manufactures Association Award for Cutting Edge Writing, The National Association of Kumquat Growers medal, the May Every Animal Live (MEAL) trade group Top Dog Prize, or the Molvania University Distinguished Book Diploma for contribution, not to literature or humanity but to the university. Apart from the *Mad* or *The Onion* reviews, none of this matters to me, for the time, energy and thought expended to produce a work like this has already brought me a sufficient reward. To rely on any outside recognition of my efforts would be both unrealistic and for me useless as never did I depend on outside opinion to confirm my own. I'm certainly not interested in waiting around for outsiders to respond one way or another to my terminal work. As an elderly gentleman with very limited staying power, I won't be present to collect any kudos, and anyway, like me, they'd all soon be forgotten.

For an author with a short life expectancy to venture writing a long work involves an enterprise subject to incompletion. The report card which assesses the extent of my learning from the school of experience contains many "incomplete" entries—which I expected, as I never had time enough to finish the lessons. But by now I'm assured that the book will be finished before I am. I'm content with my decision to devote to the project what might well be my last weeks or months, an interlude during which I gave up many other activities in exchange for producing the reminiscence. Trading time for words, I expended days to earn pages. This required not a willing suspension of disbelief but suspending an outside life. I suppressed living for the sake of writing about living. Prior to the writing I lived; after it I won't, but the writing will for a time survive me.

Although a ghost writer, I'm not quite yet a ghost. That phantom phase will soon disembody me but for now I'm still here, which in a way surprises me. I thought I'd be long gone by now. I've been willing to live as a ghost to ghost-write my account because I knew that I'd greatly regret a failure to produce my reminiscence. Contrariwise, withdrawing to write by giving up the outside world would occasion very little remorse but afford me great pleasure. Even if the work is never noticed, my account at least preserves some of my adventures as encapsulated in my memories. Investing my near final earthly time allotment to produce a work which will for the most part be read, if at all, after my time, I've conformed to the spirit (a long-ghostly spirit) of Alfred de Vigny's view that creative effort represents an attempt to *"jeter l'ouevre à la mer"*—to throw the work into the sea of the future. While the French poet sees the sea as the future, I viewed it as a desert. For de Vigny, time's wingèd chariot would be a speedboat. However you view the future, it's a vast emptiness lurking ahead to await the random infinite forms—such as a poem or a book—which in the course of time will fill in the great empty blank. I've done my best to add for future generations one minute finite form as a hopefully worthy contribution to the ongoing play whose acts and scenes I'll never know.

The book has been a labor of like and, I hope, not a like's labor lost. I've enjoyed resurrecting my past before it's interred with me. No one encouraged me, so I was a self-starter and will before long be a self-stopper, done-in by some sort of organic systemic failure. In any case, no true writer needs outside encouragement, and a party of one can get the job done if it's the write one. Other book topics occurred to me over the years after I retired, but none motivated me to retreat from the outside world to invest the time to produce the volume. As for all unrealized potential, those possible books remain in the realm of might-have-been's but never-are's. What could come into being and doesn't far exceeds the very few things or events which happen to become part of the play of infinite forms. As my time slip slipped away I knew that my reminiscence was the one thing—among thousands of possibilities, many of them trivial— I wanted to transition from mere potential into being, and so now here it is. The book exists, and will after I don't.

~ 5 ~

From the outset of my new way of life, with its diverse duo, I realised that my desire to distance myself by travel and by attitude from the provincial bourgeois setting where I began represents something of a classic case. Guy who originates in a big white house in the provinces abandons his comforts and structure in his cozy suburban ranch house and his pleasant circle, fatal and otherwise, to seek adventures out in the big wide world. It's not an unusual story: Gauguin runs off to the Pacific isles, Rimbaud runs guns in Ethiopia, Lassie runs away. Toms follow the pattern: Tom Williams leaves Clayton for more dramatic settings (the Wydown 04 trolley wasn't exciting enough for him); Tom Eliot goes east to exchange his provincial midwest for a diverse life in far-off London; Tom Merton retreats into a Kentucky Gethsemani; and Tom W. runs off to brave the uncertainties and discomforts of adventure travel. But like Lassie, and unlike the others, I came home. Only passing brave, as a "Laddie Come Home" I always returned to my kennel. I didn't cut my ties by severing my roots from my native habitat and its familiar givens. From time to time I left, but only in temporary exile to escape from bourgeois suburban life at home. Wherever I roamed I always carried and used a return ticket to end up back on familiar ground where before long I'll remain at rest beneath the ground. Valuing both modes of being, I never sought to replant myself elsewhere or to travel in perpetual motion in a rootless free-floating way. Combining home and away suited my agenda.

Blended solutions, such as the quite different experiences offered in the suburbs and out in the world, always appealed to me more than binary exclusionary rather than inclusionary either/or formats. Opposites reconciled can combine two (or more) seemingly contradictory elements to produce a better outcome, but even at Best Buy there's no best because, as for life itself, something is always missing. Such a pleasing combination of opposites animated pleasure some years ago in Sun Valley where Jack Hemingway, an outdoorsman like his father Ernest, married his second wife Angela, reputed to be rather ardent. Someone who gave a toast at the wedding reception described the match (so a Sun Valley friend present at the celebration told me) as Orvis merges with Frederick's of Hollywood. Back then Victoria's Secret was probably too secret to serve as a comparator. In any case, the diverse pleasures of both outdoor and indoor sports no doubt appealed to Jack.

I met Jack Hemingway once in Sun Valley, just in time as less than a year later, in December 2000, he died in New York City on his way to Europe. Friendly and open, he discussed with me rather frankly his views on his three step-mothers. I also met at different times two of Jack's three daughters, Margaux and Mariel. In Jack and his girls alike, I could see some of Ernest's physical characteristics. Although Papa Hemingway's son and granddaughters were as close as I got to his once bodily presence, I saw Hemingway's houses in Cuba, Key West, and Ketchum at Sun Valley and also various sites in Paris and Oak Park, Illinois, connected with his passing role in the play of infinite forms.

Unlike more unsettled and ambitious types like Tom Williams and Tom Eliot, I wasn't motivated to abandon my native habitat by moving elsewhere. My more limited goal—analagous in a way to Jack Hemingway—was to enjoy both "outdoor," with far-ranging travels, and home-based pleasures. As for every earthling, I was faced with the challenge of reconciling varying values and interests in an attempt to facilitate my journey from non-being to non-being during a brief interlude of earthly consciousness. It was an individual experiment with no general application or import. Early on I realized that I was unable to add anything original to how humanimals viewed their existence. I could think of nothing new to enhance their perceptions, beliefs and conclusions. Back in the early days my new mind confronted an age-old world, one already experienced by countless other earthlings who had preceeded me and by the 7.6 billion now experiencing the lonely planet along with me—essentially lonely, even with all those fellow humanimals aboard. As a latter-day Tommy-come-lately, a latecomer soon to be the late Tom, I arrived on the scene long after it had been experienced, analyzed, and completely discussed and described by many other participants in the play of forms. Although the shadow play of a ghost world earthlings occupy was all new to me, other humanimals had already been there and done that and I was just another passing face in the crowd. In any case, the infinite forms didn't really yield for me any definitive conclusions, and the life lessons I've mentioned in my reminiscences are in no way meant to be didactic or proscriptive but serve only to illustrate what worked for me.

The globe, as I saw it, was like those cheap plastic globes filled with snow-like bits you agitate into motion and which then spin and whirl a few moments before falling back to earth in that tiny world which mimicked mankind's fatal circle. My failure to discern purpose didn't represent a view based on pessimism, fatalism, defeatism, negativism, nihilism, denialism, or any other "ism" except realism. For my own purposes, just as in this reminiscence, I was a true-story teller.

Realization of the truth that the elapse of time would doom not only my individual being but also eventually every earthly form greatly clarified my outlook. Like curiosity, which both motivated and victimized me, my awareness of time's finite nature was a mixed blessing. Such awareness both deepens and limits the experiment in living. Without mortality—personal and systemic—things would drag on forever, so depriving the passing scene of much of its meaning. Jonathan Swift wrote about the Struldbrugs, residents of the nation of Luggnagg, located near Molvania, beings who didn't die but continued to age. At eighty they're considered legally dead but live on forevermore. That's no fun, as even age eighty-plus zombies living forever face problems. It's more convenient for all concerned to keep the fatal circle fatal. A deathless life would rob the experiment of existing of its second most significant happening. It's really not worth living unless you die. In "For the Anniversary of My Death," W. S. Merwin wrote, "every year without knowing it I have passed the day" which will mark the anniversary of the poet's demise. I prefer to think about the anniversary of my birth, without which there could be no death, no play of infinite forms, nothing. Rather than a wish to versify about the end I opted to diversify by contemplating the other end, the time when my odd experiment began. By now it's nearly the end. No longer nurturing, Mother Nature will soon dispose of me, tossing my remains away as casually as I've discarded bare bones after passingly consuming their dead flesh. My soon lifeless bones are fated for the same end. The phantom-like X-ray image of the skeleton beneath the skin reveals the underlying reality. Before long I'll give up the ghost, becoming a late ghost writer who leaves the stage where new characters will briefly join the play of infinite forms.

My basic existential agenda dealt with the mini-momentous question—crucial for me, but meaningless for anyone else or for the cosmos in general—of how to arrange my life in the face of the true-story realities which prevailed. Although go-getter up-and-at'em types liked to say that they hit the ground running, from early days I realized that in fact you hit the ground moribund with all ambitions and accomplishments buried forever. Based on these and other considerations, I curated my life in order to decide how to live. On both a micro and a macro scale, ambition and many other values and practices favored by society shrunk into insignificance. But I didn't despair or resign myself to the futility of engaging in the experiment of being alive. Although my existence and this book are doomed—fated to disappear along with all the other works and ways of mankind—that reality didn't dissuade me from engaging in my experiment and then from writing about it. One way or another, I'd proceed to the desert of vast eternity. How I reached that final destination would represent the way my agenda and chance combined to form the content of my fleeting presence on the lonely planet, from which I will soon be eternally absent.

Thrown into existence, I had to do something between prenatal non-being and postmortem non-existence. As for everyone, that pesky trio of concentric fatal circles—cosmic, terrestrial, personal—delimited my choices. Having in this Part presented some of the fundamental beliefs which influenced my response to the challenge of how to live, I'll now phrase with specificity the question my life and my reminiscence attempts to answer: In the face of pervasive and certain universal mortality—individual, humankind, terrestrial, cosmic—and the unavoidable uncertainties of existing as an earthling, what kind of life represented for me the most appropriate way to carry out the experiment of being alive on the lonely planet, given my circumstances and the ineluctable realities which confront every humanimal? My reminiscence offers an extended commentary giving my own answer to that question. Each mortal soul thrown into life responds in his or her own way. How I decided to live, as related in this book, was my answer.

~ 6 ~

The world was not made for my delectation and delight, nor was I or is anyone made for the world. Finding myself in an alien place—a stranger in a strange land that grew ever stranger to me—I tried to make the best of the situation. Although not staged for my benefit, the play of infinite forms gave me the chance to create a life which kept me satisfactorily connected with a world I'll before long be forced to leave. I've by now nearly reached that forced exit. Memories of my twilights far from home are now shadowed by the twilight which darkens my earthly existence. Working on this reminiscence brought to mind many pleasant memories, while also serving to bring closure to my life experiment. Little remains to be done or said. Although I allotted my very scarce end-time to salvage for myself in this account a few echoes from the misty chords of memory before they fall silent forever, I also had in mind another purpose. Just as Mark Twain's handshake from 1902 was passed down to me through Helen Knighton when I visited Hannibal some eight decades later, so I thought it might be worthwhile to pass on to others a thought or two about the bare bones essence of what my experiment taught me. Long after my time perhaps you now (your now, not mine) reading my reminiscence will draw from it a few useful thoughts to help guide you through the dark woods. Like me, like Everyman, you were thrown into life, and like me you'll soon, whatever your age, be leaving it.

As I approach the end of both my account and my life I want to leave you with a notion which may seem obvious but perhaps too often remains unsaid. I do realize that some readers might claim that my entire book should have remained unsaid, and they may be right. In any case, I believe it might be helpful to bring out of the shadows an oft suppressed subject, one people don't like to hear about, in order to throw some light on the forbidden topic. I'm quite aware that you might prefer that both the subject and the object—you, the reader—of my comments here be kept in the dark. If so, skip the rest of this paragraph. What I want to pass on before I pass on is that no matter where or who you are, what your age and what your fatal circle and other circumstances may be, for you—as for every mortal, even those unlucky enough not to be alerted by reading these lines—time is on the wing, flying, fluttering, flitting, fleeting away, slipping away much faster than you might imagine. Before long the balance for you will shift and your passage through the dark woods will near its end and then you, too, as now have I, will come face to face with the desert of vast eternity which lies before you. Even if you're now young you really don't have that much time left, and if you're old, a status on which I'm now an expert, your days are very limited. It may be worth your while to contemplate your life, not only as it has been and as it's being lived day by day but also as it might appear to you from the more distant perspective of your ending vantage point. That terminal prospective retrospective perspective imagined now will offer you a previewed possible final view while you still have time to retouch the picture as it develops. Left untouched, the image may reach its final form with a negative content which will make you discontent. At my advanced age, the picture has by now very nearly reached its completed version, and is thus no longer subject to a new vision or any revision. For me my experiment is a done deal. Whether only a superficial selfie or an X-ray, the image will picture the same end result: a ghost writer soon to become a ghost. Years ago I hoped to reach the end with as little remorse and as few disappointments as possible. One way I looked at life was that the experiment represented an exercise in regret reduction, a belief which guided me along the way as I advanced to a time when it would be too late to make any changes. I tried to live so that I wouldn't look back wishing that things had been different for me.

By now, as I glimpse ahead with an uncomfortably close view of the desert, I can see looking back very nearly the entirety of my life. No one lives an entirely charmed life, and complete regret avoidance is impossible. In general, however, I lucked out. I can think of relatively few changes, ones realistically available, which would have enhanced my earthly existence with satisfaction even greater than I actually enjoyed. Perhaps I should have eaten more ice cream, guzzled more (root) beers (low calorie), petted more dogs, taken Snoopy's wisdom more to heart, sat out fewer dances and taken more chances for remembered romances, bought Microsoft on the underwriting. I probably should have operated in a more business casual way than in my rather buttoned-up, intense, Elvis-inspired Taking Care of Business mode. That focused and rather compulsive behavior typified how to good effect I arranged the rest of my life—organized, meditated, curated, planned, controlled, agenda-driven. I don't really regret how I operated; my *modus operandi* worked for me. The overly structured and detail-oriented format arose not only from my intrinsic character traits but also from my attempt to counter as best I could the capricious vagaries and hazards of fate, fortune, chance and luck. Because so much lay beyond my control, I tried to influence what little I could. Most of what happened was out of my hands. For those few elements which remained subject to my own choices for my experiment in living, I took a hands-on approach. Some of my interventions managed to help produce desired results, but in the end chance and luck determined the outcomes. A charmed life depends on incantations quite extraneous to an earthling who aspires to such a random and rare fortunate state.

Looking back, I see that a more frivolous and spontaneous attitude might have brought me some additional satisfactions. I could have left a few i's undotted (the pompous self-sufficient capital I needs no dotting) and t's uncrossed for a less tidy and letter-perfect "it." It wouldn't have diminished my life to have learned fewer irregular French verbs, read fewer books, sat through fewer classes, visited fewer museums, thought fewer thoughts. Better, most likely, to have enjoyed more purposeless frolics and pure fun, dilatory days, idle hours, down-time, aimless activities, to let it all hang out (some of it anyway) rather than keeping most of it in. Given another chance I'd trade the French subjunctive, some correct German genders, a few Italian idioms and a bit of Spanish vocabulary, maybe even visits to a small country or two (but not giving up Molvania, which everyone should see once for 20 minutes) for a little more *joie de vivre* or *dolce fa niente* or *Lust* (*die Lust* = pleasure), and also lust. But perhaps *Love's Labour Lost* rather than *As You Like It* was my fate.

I suppose my major regret is that never did I become a sugar daddy. I would have enjoyed being a father figure of sorts to a good-looking young woman. Although it's not too late, by now I'd be more of a sugar granddaddy than a sweet poppa. The main factor which deterred me from assuming such a role is that the sugar stuck to my fingers. As a frugal fellow, I couldn't shake enough of the grains loose to induce an attractive gal to become a sweetheart, and in any case perhaps none would have found me suitable as a source of sugar. But I noticed over the years that the more sugar an elderly gentleman dispenses the more attractive and desirable he becomes. Even Dracula would be popular if he sweetened his personality from bites to bucks so that his girlfriends could put the bite on him. Sugar or Sweeta makes the ladies sweet, so maybe there's still hope. I never understood why people referred to gigolos—perhaps the earliest examples of the gig(olo) economy—but never to gigolas, females who try to extract sweets from their adoptive daddy honeys. Like the diets of the most avid health fanatics, I lived a sugar-free love life featuring only low-cal offerings which failed to attract gigolas.

None of my omissions or commissions, however, blemished my life with serious regrets. Compared to many situations I saw around the world, including some close to home, I suffered almost no grave inconveniences, setbacks or misfortunes, although now I'm faced with a terminal grave condition. Even though my general mind-set might have been too firmly set and inflexible, my hard-headed and TCB attitude worked for me and, so I'd like to believe, operated to nurture good luck. I'm more concerned by what I failed to include, the omissions, than by what actually occurred to comprise the content of my life. But mild regrets over what didn't happen represent an unrealistic way of looking at things. For one thing, you can never be sure that a potential desired experience if realized would have worked out as you hoped. Even more crucial, to introduce another happening into the intricate skein of events which actually developed might have changed everything that followed, to my great disadvantage. If you've lived something of a charmed life it's folly to wish for an alternative version hoping that it might be even better.

An author and a reader are temporary partners. In this Part I've contributed to the enterprise some experiential and temporal capital. For your part, I encourage you, before it's too late for you, to take some time now (but only after you finish my book) to contemplate how your past will look to you from the perspective of the near end-point, one I've almost reached. You needn't try to kid yourself that your end lies in the distant future. It's closer than you think. You can observe the amazing never-stop shrinking machine depleting your life day by day right before your very eyes by thinking back to when you started this book and noting how time has elapsed from then to now as you read these words. Total up lots of brief passing thens to nows and, hey presto, you end up with a completed life. Don't say I didn't warn you, and—years hence when you're old, or at least much older (if not by then already ghosted)—of a crisp autumn eve or a balmy spring day, earthly delights no longer available to me after my time, spare a thought for Old Tom, who was for you a true-story teller. Look ahead now, before it's too late, to see where you're headed. Such a forward view will hopefully enable you, while some time still remains for you, to reset your agenda so as to add some charmed elements to your existence. It's of course easier to wish for than to live a charmed life, but at least you can try. As for everyone, your choices are limited by the constraints imposed by your fatal circle and circumstances, but within those restrictions perhaps you have some wiggle room. It's of course possible that you'll decide no change is necessary to conform your life with your long-term interests, goals, hopes, aspirations and expectations. If that's the case, you're already living a charmed life. In my case, a substantial change was indicated, one I've described in my reminiscence. Either way, regret reduction requires a timely consideration of possible alternatives to one's present way of life. Failure to engage in such an assessment leaves you in the default position of accepting your existing life without imagining and activating ways to improve it.

I now confess to you quite the oddest and most startling notion which has ever occurred to me, one so surprising to me and so nonsensical and preposterous that it completely contradicts my normal and, up to now, invariable life-long rational, logical and perhaps too cerebral way of looking at things. The thought suddenly came to me, while I was writing these passages, that perhaps I was incarnated—me, one infinitesimally tiny fragment of consciousness out of all the possible versions which might have gestated—for the very purpose of creating a certain specific mind, one out of a trillion, through which would pass a spin of random earthly impressions, ones which happened to come to me from the infinite play of forms, so as to allow me at the end to transmute my experiment, by the alchemy of writing about some of what I experienced, into the cautionary statements I've just mentioned to you. But on second thought I know that it's absurd to suppose that I was procreated to experience my particular life and now, at the end, to write about it. Try as I might, however, I really can't think of any other reason for my earthly existence, so if that explanation lacks validity then the obvious conclusion is that there was no particular purpose for my being. The truth is that like all humanimals I was born purely by chance, a random toss into existence with a short lease on life now about to expire. Meaningless, however, can in a small way be transmuted into belief in a meaningful existence by finding a way to believe and live <u>as if</u> things matter. Each person selects his or her own way, if any, of fabricating such an attitude. Although it's pretentious to believe that any particular life has meaning, we pretend.

With my life and this book now nearly complete, I've tried to convince myself that my life was indeed created and lived for the specific purpose of writing this reminiscence and, in particular, the last few passages. I'd like to think that my particular personality and fatal circle, the months of travel, the twilights far from home, the whirligig of experiments and experiences, my compelling curiosity, the ceaseless churn of thoughts, the sensory impressions, the infinite forms which happened to reach my consciousness, everything which furnished me with the threads to weave my tale, represent the concluding and conclusive reason for my being. Deem it if you will a tale told by an idiot, signifying nothing. But at least I told the tale and, however viewed, my account consoles me because I sense that it represents what I was born to create. Because my belief allows me to suppose that there was meaning to my life, I'm prepared for it to end. Meaning for me was thus not transcendental but simply incidental, limited only to the belief that by experiencing a wide variety of infinite forms I could at the end, with the time and leisure available to me, record my impressions of the great earthly pageant with a closing statement meaningful to me even if to no one else. That and nothing more is the only possible reason I was incarnated, and even assuming my supposition is invalid I was nevertheless by chance able to enjoy something of a charmed life, even if one completely without meaning. To suppose any more ultimate purpose would be nothing but a phantom fantasy belief. So there, Father Time, Mother Nature, Desert of Vast Eternity! Out of nothing I made something—a something which meant something to me—to counter how the three of you will soon out of me make a nothing. My low-order personal something gives me sufficient satisfaction so that I don't regret not having pursued the sort of transcendental "something" Virginia Woolf and so many other humanimals have sought—but never found— in order to explain the meaning of life.

It's really not like me to engage in rather mystical speculations about the purpose of my life or of life. From my early days, I realized that there was no essential reason for me to be alive. Although my being—like that of Everyman—was unendowed with any transcendental meaning, I could at least try to establish some sort of limited earthly purpose. I never knew what it was until now, at the end. My perception may be incorrect, in which case I'm back where I started—a meaningless life. Either way is okay, as before long I won't know the difference. In the meantime, pending further developments, I'll now abandon my strange passing aberrational behavior to revert to my usual less fanciful rationality. Going forward, I won't even pretend that my odd "as if" idea is sufficiently meaningful to offer a basis for believing that something matters. Nor will I bring any rational analysis to the rather mystical matter, for that would in fact operate to degrade my stated belief in what made my life meaningful. It's a notion best left unexamined. But, on second thought, perhaps it's only reasonable to allow myself at least that one comforting mystical vision in a long lifetime of rational thought and behavior.

By now time has nearly eclipsed my earthly existence. The penumbra of extinction shadows me as gradually dusk and soon darkness shroud my being. Memories of twilights far from home, the long-vanished into smoke campfires of my youth at Camp Ironwood, and all the rest will soon disappear into the eternal night. In a nostalgia-filled reminiscence about India before the country received independence, a one-time long-time English resident there recalled (in *Plain Tales From the Raj*, edited by Charles Allen), "I've only got to shut my eyes and picture an evening in a village in India, with the smell of wood smoke, which is the most gorgeous smell in the world. There was a very sudden twilight in India always and when the twilight came it was dark and mist would rise up in a sort of blue haze from the fields—and there would be this gorgeous smell drifting across the fields." Those traces—the vanished wood smoke, the twilights, the hazy past, misty memories—are all that remain for me now.

At this point the world has largely passed me by. But as I intended years ago, I didn't pass the world by. Many a happy hour did I spend out there far from home, and countless pleasant times in my native habitat as well. A lifetime ago I started off as a member of the new generation to pursue my experiment in living. Now the time has come for other new generations to discover in their turn in their own ways what the old once saw and learned. Although my temporal account is almost empty, my memory bank remains full, rich with recollections of the minutes and miles—those remaining now grown short—over the long years. As charmed as my existence was, I wouldn't want to live any or all of it over again. Once was enough. Never would I wish to roll back the years to return to Phantom Lake with the wood smoke in the air and start all over again from the beginning, even if I could be assured of a charmed outcome. For me it's all over, as soon I will revert to the phantom-like state of non-being which predated my youthful times at Phantom Lake. Father Time and his henchman the Grim Reaper are dogging my every move, and before long will catch up with me. I'm about to become a ghost writer reduced to a ghost.

~ 7 ~

I'm quite aware that a certain melancholy and morose tone suffuses my elegiac reminiscence. At least I hope it does, for the text is meant to reflect the chiaroscuro feeling which now at the end somewhat darkens my terminal mood. Although I enjoyed my exercise in reminiscing, perhaps my final substantive life experiment, contemplating my elapsed and soon to completely lapse existence and final matters has proved to be something of a bittersweet experience. Sweet were the many pleasant memories, but melancholy was the realization that my recollections are only shadowy phantom remnants of long-ago forever expired adventures, and that so many familiars who had so pleasantly populated my life were no more to be seen. Losing such intimates tends to de-charm a life. I felt like Mark Twain on his last visit to his hometown in 1902 when (as I mentioned earlier) he lamented in Hannibal how "I was profoundly moved and saddened to think that this was the last time, perhaps, that I would ever behold these kind old faces and dear old scenes of childhood." Although it's unlikely that I'll ever again visit many dear old scenes of yesteryear or see the kind old faces of my most distant close friends, I enjoyed remembering those bygones as I wrote this reminiscence. But those images are only pale copies of the originals as I experienced them. Even a work imagined from real episodes as derived from life—rather than invented fantasies based on abstract thought—scarcely resembles the original impressions. The scenes and vignettes arranged into neat paragraphs and now bound into this volume present only words which barely evoke the experiences. What happened through the years came and went, leaving only my memories and this book. All the rest is gone.

My enterprise in writing the book served as a constant reminder that the episodes which comprise the volume's contents represent the genuine but now insubstantial elements of my existence. A written version of the lived experience can never capture the reality of how things operated out in the world. As Jacob Bronowski said, "The world can only be grasped by action, not by contemplation." He himself was not just a contemplator, for the professor preached what he practiced. At the beginning of lecture three (of six) of his 1977 Yale Stillman Memorial Lecture Series entitled "The Origins of Knowledge and Imagination," Bronowski summarized his theme as "to construct a natural philosophy which was based on the physical ability of human beings to receive and translate their experience of the outside world." My post-graduate education was based not on my reading or the classroom sessions confined to the limitations of wordy lectures and discussions but from my adventures out and about around the world. Word sources supplemented but could not replace what actual actions gave me. As Lewis Mumford observed in *The Fulfillment of Man*, cultivating a granular rather than merely a virtual contact with the world yields a richer harvest: "The hot house fruits of life" offer only " a waxen beauty," but fruits grown in the wild have "the most interesting and significant marks of growth." It's the earthy hands-on interface with the rough-edged outside world which sparks the happenings that bring true substance to your life experience. Like Andrew Grove, Mumford believed in going "out doors."

Admittedly, my engagement with the world was somewhat superficial as I was ever only passing through, even though in some areas I lingered long enough to get the feel of the place. In *On War*, Clausewitz describes how "a kind of friction" accumulates to degrade your plans which then, buffeted by events, "always fall short of the intended goal … Friction is the only concept that more or less corresponds to the factors that distinguish real war from a war on paper." Friction-less (but not fictitious) paper trips as described in the pages of this reminiscence barely resemble what I actually experienced. My adventures now seem like ghost trips, *Flying Dutchman*-like phantoms from the past. It's impossible to reproduce on the printed page all the confusions, uncertainties, hazards and hassles, tedium, the waiting, the wrong turns and the dead ends, border barriers, bureaucratic controls and bandits uncontrolled, alien and friendless hotel rooms, bad water and good pickpockets, the low-giene lack of hygiene, the dust and dirt and delays and discomforts and dangers and discontinuities, and all the rest of the frictions on the open road. Print can't capture the irreproducible feel of the traveler's granular tactile on-the-ground abrasive contacts with the outside world. Nor can print present how all the positive and pleasurable play of infinite forms played out—the world's beauties and wonders, the delightful unexpected happenings, the glorious scenes and settings, the fun and fascination, the untold delights, the pleasant surprises, spectacles, marvels, enchantments, excitement, anticipation and satisfaction, the unforgettable oddities and curiosities and strangeness, the fellowship and camaraderie with fellow travelers, and all the other pleasurable experiences of my twilights far from home. None of those as I experienced them—the problems and the delights—translate well into print. Although I've tried my best to revive and to reproduce with my descriptions in this reminiscence *"wie es eigentlich gewesen"* (in German historian Leopold van Ranke's phrase)—how adventure travel actually was—I realize that the words only sketchily suggest the lived realities. These thin flimsy fragile paper sheets, eventually to crumble into dust as will I, contain only a shadowy ghost version of how one obscure Midwestern suburbanite, at heart a provincial, was motivated by his keen curiosity to engage with the play of infinite forms and how he carried out his experiment in living.

Every humanimal's life entails a story of one kind or another. My earthly existence was no more or less meaningful than the sort of lives my family, friends, acquaintances, neighbors and other contemporaries managed to create for themselves. Each had his or her own way of inventing some meaning. My way was both different and the same as that of others at home. I benefited from two formats. My home-based roots deep in my native turf as well as my rootless travels both meant a lot to me. Although I enjoyed something of a charmed life, I recognized that in some ways my existence was charm-deprived. Ceaseless restless compulsive curiosity bedeviled me, in a way contorting my life. This dominant trait operated to throw me from my comfortable home space out into the world in a way similar to how I was thrown into existence by being incarnated into an earthling. Travel's "chiaro" and "oscuro" also reflected the bright and dark sides of life itself. Both life and the world struck me as strange and fascinating, even if and perhaps because both were without any ultimate purpose. With a definitive meaning, earthly or personal existence would offer less fascination and probably less incentive to explore the lonely planet.

Many of my peers no doubt viewed their lives as charmed as I saw mine. Just as I treasured my distant twilights, my peers valued their home-based delights. The provincial suburbanites where I lived enjoyed their car pools and swimming pools, the Little League and the Junior League, the card games and the Cardinal games, and the many other benefits of a settled and normal conventional Midwest bourgeois existence, many of whose charms I too valued. For their part, my contemporaries might well have viewed my own treasured way of being, with its unsettled wide-ranging travels to far lands, as charmless. In fact, the charmed elements of each of my two lives were different. I enjoyed many of the same homey local advantages which so pleased people whose home life represented their entire life. I appreciated my little house and how my interludes at home afforded a typical bourgeois suburban existence. But I didn't want that way of life to comprise my only mode of being, nor did I make travel my sole activity. Because home meant a lot to me I clung to my native habitat, the familiar terrain where I'd begun and had remained my entire life (except for my studies and travels). All through the years I've lived within the radius of my fatal circle, just a few miles from the big white house where

I started and in the same area where I'll no doubt end. Need or ambition didn't compel me to move elsewhere nor, for once, did curiosity motivate me to change cities. I was a homebody quite content to stay put. My travels brought me quite enough change for a lifetime.

There, rooted in my native Midwestern soil where my ancestors had settled four generations back, I felt comfortable. I enjoyed the familiar setting, the sense of place and space, the cycle of the seasons which would come and go with a pleasing and predictable regularity. Just as I enjoyed twilights far from home, so I delighted in dusks while at home. Of a mellow summer's eve I'd sit on my screen porch as the day darkened. Flitting lightning bugs momentarily brightened the young night before their tiny glows disappeared, the fireflies as evanescent as a dying campfire's warm flickers of light which soon vanish into nothing. There at home in the Midwest I saw the seasons change year after year, and now for me the final cycle nears. Each passing season may be my last.

My residence by now seems an intrinsic part of the scene, as normal as the seasons. So long have I occupied my home, it's become a part of me. I can't live without it, although before long the homestead will be occupied by strangers and continue on without me. As I traveled over the years I always knew that my house stood waiting for me, ever accommodating to offer me a haven and a retreat from the outside world. It's hard for me to believe that the time will come when I'll no longer inhabit my comfortable welcoming space. Foreclosure will soon bring an end to my half-century-long tenancy, not by debtor's default but by elapsed time which will evict me and sever me from home and from everything I've known. A house cooling will mark my departure from the cozy comfy place where 50 years ago I began with a house warming. Like Wordsworth, I enjoyed the leisure and pleasure of sitting silently "In the loved presence of my cottage fire,/And listen to the flapping of the flame." But eventually the flames consume the wood and all warmth turns cold.

Although my home life conformed by appearances with the way other suburbanites lived, few of them had any idea how things were out in the wide world where, from my mid-thirties on, I passed a good part of my existence. No one at home seemed curious about my remote other life. I also sensed that almost none of my peers would want to emulate how I used my freedom to pursue peripatetic adventures in far lands. Nor did I follow many of the ways which typified how people around me lived. We occupied separate worlds, at least to the extent that my travels took me to distant foreign lands. But I also shared the world of the homebodies, who for their part seldom experienced the kind of life my ancillary activities abroad entailed. By keeping one foot on my home turf in the suburbs and the other foot-loose out in the world my stance was satisfyingly balanced and not at all awkward.

Although my equipoise required me to give up the structured work-a-day world, I wanted to keep connected as best I could with some sort of regular and continuing involvement which included productive activity. This presented a challenge, as I had no obvious reason to be anywhere in particular. No professional or pecuniary endeavor served to define me. At the dawn of every new day no fixed place or position beckoned me. I'd become a free agent, living an open-ended free-form free-fall existence without any of the usual labels by which others could identify or categorize me. I went from having a place in the scheme of things to displacing myself. In the local pecking order I was peckless and maybe also viewed as feckless: no standing, no profession, no job, no affiliations, no obvious handle for people to grasp in order to slot me into a set category. Never a joiner, I belonged to no clubs, societies, organizations or professional circles which might have given me an identity. I'd mutated not simply to be different but because of the format required by my new bifurcated existence. I became not by design but by redesign somethings of a Zelig-like figure, a character with a patchwork life typified by improvised activities lacking any discernible theme, even if the be-all and end-all was quite apparent to me. Although my agenda was invisible to my contemporaries, for me my plan was always clearly in view.

While my nondescript way of life set me somewhat apart from others, my double life didn't completely distance me from my suburban confreres. My irregular here-today-and-gone-tomorrow mode with its comings and goings seem to unsettle a few observers who didn't understand my experiment. But even lacking the usual characteristics which defined identity and evidenced a position in the regular working world, I managed to retain access to and interact with a fair number of active and accomplished participants in professional, business and civic activities. My ability to retain or to establish and maintain these connections somewhat surprised me. I'd supposed that after cutting myself loose from a structured and identifiable niche in the working world I'd in effect be relegated to an outsider status as a nonentity exiled from the system. But luckily I succeeded in keeping enough ties with the regular functional world so that I wasn't completely adrift. Eventually becoming a published writer served to restore an identity, permitting me once again to offer people a label which could be attached to me.

My withdrawal from the system changed how I operated but not with how the world around me functioned. Everything spun on as usual. Society, the suburbs, my friends, daily life and its routines and rituals all continued in their regular ways. Children were born, people died; kids grew up, adults grew old; dogs barked and begged, then didn't; autumn leaves littered lawns, winter snows blocked driveways, spring pollen agitated sneezes and wheezes, summer rays tanned skins, Indian Summer mellowed the new fall, and again ended one cycle and began the next. Money was earned, money was spent; birthdays and their celebrants came and went; holidays, sports events, performances, exhibits presented passing diversions and then vanished. People partied, at times as if there was no tomorrow—and they were right, because at a certain time there is no tomorrow: some unknown day will bring you your last earthly moments, and the next tomorrow will never dawn for you. There's never any certain tomorrow but only the evanescent present and all the yesteryears accumulated like discarded calendar leafs or autumn leaves scattered into crumbling piles soon blown away by the winds.

While my cohorts and contemporaries carried out their doing and having, I was often away in the big wide world. Although the permanent homebodies knew nothing of my travel life, I was familiar with how things operated back home, as that setting remained a part of my life. But it represented a way of being I couldn't accept as my only life experiment and experience. There in the suburbs and elsewhere in my circle were people walking dogs, dogs walking people, grilled hot dogs and bratwursts, treat-begging dogs, and undisciplined brats; kindergardens and subdivision gardens; playing—computer games, tennis, cards, the market; blingmobiles and shiny jewelry; big-screen TVs and phone, desktop, laptop, movie and window screens; humanimals falling in love, falling out of love; summering in Michigan, wintering in Florida, springing ahead, falling behind; gated communities and homecoming tailgate parties; church services, porcelain dinner services, service clubs, country clubs, golf clubs; perfecting lawns, remodeling kitchens; April showers, wedding showers, stormy divorces; baby-sitting and sitting babies; autumn leaves, spring arrives; Fourth of July blowout picnics, Fifth of July hangovers, back to the grind on the Sixth; Labor Day, women in labor, labor strikes, baseball strikes; Halloween tricks and treats, year-round plots and counter-plots; beers and biers, beginnings and endings; all that and much more back at home. While those sorts of activities and situations engaged my peers, I opted to disappear. I removed myself a long way from the conventional everyday provincial suburban way of life I knew so well. Pleased to be elsewhere, I treasured my twilights far from home. Pleased also to be at home, I valued the twilights there as well.

Although the doings, but not the don'tings, of my domestic life in some respects resembled how those around me at home lived and were clearly evident to my fellow suburbanites, my remote adventures remained invisible to anyone but me. Just as my neighbors treasured their tangible possessions, travel memories were among my most valuable assets, even if I could possess them only in my mind and not in the rooms of my house or in the driveway. Only a litotes phrase can describe the not inconsiderable effort my travels brought me and the not insignificant pleasures they gave me. To others, many of the places I visited were abstract—just names on a map. To me, they were memories. To understand worldly adventures you had to be there. Apart from me and anyone who happened to accompany me, no one else could grasp what happened or the event's setting, and in any case almost no one at home showed any interest in my travels. This suited me because I didn't particularly want to talk about my trips. Unique to me, my adventures became part of my earthly existence and, left unrecorded, would have vanished without a trace had I not tried to describe a few of the random happenings in my reminiscence, a now completed enterprise which for me has proved a life-saver.

~ 8 ~

I now bring my reminiscence to a close as my life draws to a close. After a lifetime of travels I'm approaching a dead end. I now soon face the ultimate twilight, one close to home. Night falls fast over the desert of vast eternity. Like the rivers flowing into the sea where they disappear, I'm nearly at *finisterre*, land's end for my adventures, for my memories, for me. The days have grown short, by now deep into winter and well past long-ago passed September when the autumn leaves turned to flame and withered. Just as for the wonderland Alice knew, the vanished years of the wonder-world I found on the lonely planet are "like pilgrim's withered wreath of flowers plucked in a far-off land." The once fresh bouquet of my springtime has wilted away, summer's come and gone, the leaves of autumn have flamed into ashes, and now winter chills the air. My reminiscence reviews the seasons of my nearly vanished life. My book isn't a travelogue or a log of my adventures but only an account containing a few residues of a long life nearly burnt like vaporized wood into nothing. By now my waning earthly existence smolders away, dying just as did the embers of the ghostly Phantom Lake campfire logs which a lifetime ago burned into smoke and disappeared into the air. Camp Ironwood's ironic name— the logs weren't durable like iron but only mortal deadwood which burned into nothing—exaggerated the wood's durability. Whether Dante's dark woods, Ironwood, Hollywood, Bollywood, Dollywood, Natalie Wood, Anthony Wood, Birnam Wood, all will over time ash away like the Phantom Lake Ironwood campfire wood, in the end nothing but smoke.

The most vivid and memorable moments slip into vanished time. In his *Autobiographical Novel*, Kenneth Rexroth recalls the time when as a teenager he met a woman some ten years older. An immediate epiphany about her suddenly possesses him. The young man's intuition proves valid, for the couple soon begins a romantic relationship. One day the two of them go to Washington Park in Chicago to picnic on an island in the boating lake. The scene, with "the blue air of the park, full of watery lights and log smoke and swooping purple martins, was charged with overwhelming power." The log smoke soon vanished, like time on the wing the birds whirred away, and Rexroth realized that "Nothing like this early autumn evening will ever happen to me again." Such was everything which happened in my life—an evanescent Japanese tea ceremony-like "one time, one meeting" experience.

The very being of every humanimal is doomed to be dismantled or diswomantled, as Rembrandt's "The Anatomy Lesson" illustrates. An X-ray reveals all the moist organic components and networks beneath the skin which will before long become dry and moribund. Their purpose ended, no longer will any of the long-useful—but in the end mere flesh-scrap—internal systems throb, pulsate, flow, secrete. They'll all finally rest in peace. Like Kleenex, a humanimal is disposible tissue. What once seemed second nature—the functioning body parts which meshed and operated without a second thought by their owner—will, by Mother Nature's ineluctable force, lapse. Hard to believe, but the so familiar extremely useful and even necessary handy-dandy conveniences belonging only to you—your personal format—which served you so well will soon lose their identity as parts of the form you assumed to play your part in the play of infinite forms. Before long you'll be *hors de combat*, disarmed and deprived of your arms, toes, tongue, ears, bladder, intestines—you name it: whatever the designation or function, the morphology ends in mortality—and all the other once functional but in the end inert bodily components which enabled you for a brief time to be a liver on the lonely planet. You'll for sure miss all those external and the inside X-ray visible ex-parts, an ex parte disintegration which will affect you and you alone. All the other earthly forms will whirl on without you. Reminding you of all this, I feel some remorse—but not enough to prevent me from being a true-story teller.

An autumnal feel now suffuses my being. As so many other authorial minds have noted, the twilight-like fall season, signaling the year approaching its dark December demise, denotes the sense of an ending. On the simple Hemingway memorial in the trees by Trail Creek in the Wood River Valley at Sun Valley the inscription taken from his words reads:

> Best of all he loved the fall
> The leaves yellow on the cottonwoods
> Leaves floating on the trout streams
> And above the hills the high blue windless skies
> Now he will be a part of them forever

So it is that I, too, will soon become a part of my native soil forever, planted in the same terrain where I began and where I shall end. Often I think back to many of the long-gone fleeting moments of my life when I realized at the time, as did Rexroth in Washington Park, that never again would the passing events recur. Those moments all vanished like the Phantom Lake campfire smoke, like passing lights reflected in water, birds on the wing, rivers into the sea, like people I knew, like all the twilights at home and far away, glimpses of earthly forms, and so many other momentary vignettes of "overwhelming power" but no staying power.

Well do I recall those long ago summer dusks around the Phantom Lake camp fires when I was a young boy. At the time, the years stretched ahead as if they'd continue forever. Back then a long and uncertain unmarked path through the dark woods led toward the unknowable future. As for every humanimal prowling its way through the thicket, lurking ahead in the woods were hazards and challenges, threats and dangers, plots and counter-plots. But as a carefree child at Ironwood, little did I know what lay ahead. Only now at the end can I see the route I took through the dense forest. My path was nothing like the linear 04 trolley which ran back and forth on the fixed iron rails, more durable than the Ironwood logs, as the streetcar regularly passed the big white Wydown house. As I arranged, my life went off the rails, my kinetic travels far and wide on journeys as predictably uncertain as the 04's route was predictable. Twists, turns, transfers, detours, side streets, off-the-beaten-track corners of the world, plots and counter-plots characterized my route, less straight-forward than the out-and-back and out again 04 line. In 1949 the 04 was reduced to 00 when the trolley stopped running, as soon shall I. All rails and trails and roads eventually reach the terminal.

Inexplicably, good luck companioned me along my long journey. Without the help of Lady Luck my road from the big white house to Ironwood's Phantom Lake in the Maine woods and on and on with my far-ranging travels and now to the far edge of the dark wood I've reached would have been completely different. But whatever the course, the road ends up overlooking the desert of vast eternity. By chance, both my fatal circle and good fortune favored me. Luck, chance, happenstance, fate, destiny, serendipity, coincidence—call those capricious elements what you will—happened to operate in unlikely beneficial ways to guide my way. The single word "luck—an 04-letter word—encapsulates the slippery, mercurial, uncontrollable, elusive, fickle will-o'-the-wisp intangible influences which facilitated my largely charmed life. Those random forces helped to make my tenure on the lonely planet less lonely. As a whole and against all odds my earthly experiment—all the plots and counter-plots, my fatal circle, a charmed life, my twilights far from home (and at home), the misty memories, my mostly fulfilled subjective objectives, and the shadows and smoke of my now vanished years—proved favorable, but before long a reversal of fortune is inevitable as I transition from existing back to nothing, an inescapable binary or, in Sartre's bumper-sticker-size tome title, *Being and Nothingness*. Looking back, I can now see how little my own efforts contributed to the journey. I was something like a hitch-hiker on time's wingèd chariot. Forces beyond me determined my fate. Ability, sensibility, persistence, volition, skill, will, purpose, energy, intelligence, shrewdness, judgment, awareness—what little of them I had—only took me so far. Those factors all played a minor role. Luck was the leading lady. My early-day realization of the great contingency faced by every humanimal, each possessing only limited means to influence the mortal creature's life, forced me to recognize that ultimately all depends on the uncertain kindness of strangers—Lady Luck and Father Time. Those two characters are by far the stars in the play of infinite forms. Without that leading lady and generous man, who gives you the temporal means to endure for some years, you end up with a luckless life of short duration, so limiting your chances for a charmed life. Many poor souls can never catch a break. At a pot-luck dinner those unfortunate guests at the feast suffer the bad luck of choosing to eat the one dish which produces indigestion. But somehow, against the odds, I happened to enjoy great good fortune. I inherited a fortunate fatal circle and managed to benefit from it; my natural habitat was favorable and pleasant; and I enjoyed world enough and time to live in my own way. The elements of a charmed life existed for me, and by chance those unlikely factors produced such a life. I can't explain it—I can only describe it.

Even for a charmed life, however, luck at times doesn't behave like a lady should. But with the right temperament, itself largely a matter of luck, you can sometimes react to unladylike behavior by the way you view her perverse acts. In *Somewhere Towards the End: A Memoir* Diana Athill reflects on "How successfully one manages to get through the present depends a good deal more on luck than it does on one's own efforts. If one has no money, ill health, a mind never sharpened by an interesting education or absorbing work, a childhood warped by cruel or mean parents, a sex life that betrayed one into disastrous relationships," then you suffer the disadvantage of "negative luck." She notes that "the greatest good luck of all is built-in resilience." Well, they all say that—respond to bad luck with a stoical attitude. It's advice easy to give and hard to follow, as you never know how you'll react to a setback until it happens to you. Only somewhere towards the end can you gain a clear-cut overview on how much negative luck you suffered and the mysterious capricious allocation of positive luck you happened to enjoy. But you needn't wait until the twilight years to realize that luck is a major determinant in what happens to you and that it comes and goes, and that luck always goes in the end because so does your life. You run out of time, of luck, of everything—it all goes. No matter how fortunate your own particular wheel of fortune has spun your fate, that luck-based whirligig of chance eventually turns into misfortune. Round and round she goes and where she stops everybody knows: *Rien ne va plus*. Nothing goes any more. The wheel stops spinning and then you're a luckless has-been. Even at the beginning of the game the finality is known, and at the end the view is clear: the desert of vast eternity. The only unknowns are when and how you'll transition from Phantom Lake to phantom.

The long road from Phantom to ghost through the dark woods took me to far lands where I experienced and now remember and in this reminiscence record some of my earthly adventures. I apparently emerged into being and then out into the world in order to pursue as many encounters as possible with the infinite play of forms. But I may be wrong: such a teleological view is illogical and contradicts my Doubting Thomas attitude that one of the intractable terms of existence is that being offers no meaning at all. Whatever the case, my enterprise served, by my worldly adventures, to convert the random into the routine by at times forcing myself to improvise my way through the lonely planet with free-form travels. Unstructured and open-ended as they occurred, my adventures—as I've recorded them now at the end—have served to bring my reminiscence to full closure as my earthly existence nears closure. *Rien ne va plus*. By chance I happened to be singled out to be the receptor whose life experiment involved the wide-ranging experiences which came to me, or I to them. I was really only a medium, perhaps something of a clairvoyant, a mechanism somehow fated to encounter a varied range of the forms which years later I'd use to transmute some of my memories, misty or vivid, into this reminiscence. Why me? That's just the way things happened to work out for my particular experiment in living. Back at home, other sorts of experiments defined the lives of my contemporaries. Their provincial existence centered on family, work, conventional suburban activities and practices, a life somewhat more certain than independent roamings out in the big wide world. Caring for a dog or producing and raising children gave the owner or parents a purpose, even if it didn't give the pet or the offspring a purpose. That they would have to find on their own, as did I and as does Everyman.

An outlier is by definition removed from the mainstream. Living in part in a non-conventional way unguided by the parameters of regular day-to-day work-a-day routines was in a way more challenging than simply keeping up with the Joneses at home by emulating their way of life. As Henry Miller noted in *Stand Still Like the Hummingbird*, "Everything is difficult, and everything becomes more difficult still when you choose to live your own life." Because I opted to live in my own way, which as Christopher Morley observed is the only successful life, the difficulties Miller referred to I soon recognized after changing my life in my mid-thirties. My previous arrangements needed to be rearranged. Long established guidelines, routines and rituals gave way to a blank slate. Societal patterns and practices no longer structured my existence. Habits and conventions help to tame some of the unruly play of forms and fortune. Remove the comforting scaffolding bourgeois life provides and the skeleton which frames society goes. I had to rebuild some sort of support to structure my new way of life.

Henry Miller also noted that the most important factor in living your own life is "the quality of your solitude." Distancing myself geographically with my far-flung travels, as well as partly remaining apart from the prevailing value system at home, was facilitated by my high quality aloneness. My new life, both away and at home, brought a certain degree of solitude new to me. My former regular life had been partly hollowed out, leaving me with some empty spaces. But the new format also gave me the inestimably valuable benefit of free time, an essential component for an independent life. Without access to unstructured time no amount of money can or will enable you to live in your own way. As May Sarton writes in *Journal of a Solitude*, "Open-ended time is the only luxury that really counts and I feel stupendously rich to have it." How you handle the priceless treasures of free time and solitude determines the degree to which those rare elements will benefit you. In Iris Murdoch's *The Nice and the Good* a character looking for the key to his companion's "interior castle...speculated often about the quality of his solitude." If Miller and Murdoch both happen to agree about the importance of solitude's quality, who am I to dispute their shared wisdom? And if Snoopy at one time or another had also offered the same thought, his contribution would for sure make the quality of solitude concept a truly irrefutable notion.

Although the fortunate few lucky enough to enjoy free time will opt to use that rich endowment in different ways, however the treasure is spent entails responsible budgeting. Because time is so scarce and erodes so quickly and relentlessly—the wingèd chariot ever hurrying near—it's rare to possess a surplus of it. Free time represents very much of a plus as you can use that benefaction for your own purposes rather than for what's imposed upon you. Because free time is a gift, wasting it represents an especially abhorrent practice. Time needs to be employed carefully and responsibly to fill the empty hours—before they close like a locked door behind you—with meaningful experiences and relationships. Time frittered away in aimless pursuits is an abuse of privilege. Unfortunately, those who squander time can't be sanctioned by having their time taken away for more productive uses. A freelance life based on freedom and independence isn't a license for self-indulgence. Some unfortunate fortunate characters of independent means come to depend on practices which lack substance, such as substance abuse, consumption, pretentious behavior and insubstantial frivolities. This mistreats Lady Luck and Father Time by failing to respect what they've given. She should be treated like a Lady and he like a *mensch* by giving the two of them some respect for what they did for you. They could have just as easily favored somebody else. It's advisable to manage free time and freedom with a realistic sense of their value. Being removed by luck from the practical exigencies of everyday life can distance you from the work-a-day world in which most humanimals are forced to function. The rare privilege of living an independent life comes with the possible disadvantage of distorting how the world actually operates. As artist Fernand Léger said, "To be free and yet not to lose touch with reality" represents a high order of being. Free time is not a license for fantasy play or for irresponsible behavior.

By now my use of time relates to allocating it at home as no longer do I travel as before. I've become a one-trick pony, the three-ring circus of the play of infinite forms out in the world now reduced to a single stage. My far-flung adventures are over, as I've for the most part resumed a settled and normal way of life which conforms to how I lived before my mid-thirties when I changed the format to include wide-ranging journeys to distant lands. My curiosity about the world, including its places and its play of forms, has by now to a large extent been satisfied. A dozen or so far lands remain of interest to me, but so remote and difficult of access are those areas it's unlikely I'll ever reach them.

Some die-hards never give up their travels. Although bed-ridden in his old age, Cardinal Richelieu refused to stop traveling. Six men holding leather harnesses attached to two beams beneath the Cardinal's bed carried him around on trips. Where a town gate proved too small for the mobile berth to enter, two masons who preceded the entourage dismantled the entryway to allow the bed to pass through. If the distinguished guest's bedroom was on an upper floor, the attendants would build a timber ramp up to windows enlarged to permit the bed to be placed inside. That would all be too much trouble for me, so these days I for the most part opt to stay at home in my sedentary bed. Every so often I still travel, but not with my own bed. Only a Cardinal like Richelieu—like a bird on the wing—enjoys that nocturnal nesting privilege.

Most of my trips now are to see familiar faces rather than unfamiliar places. Unfortunately for me and even more so for my old friends, ever fewer of them survive. Many of the people who populated my life I may never see again. So few Christmas cards do I now receive, I use a few from previous years to fill out the otherwise too empty space on the mantel above the fireplace where I display the holiday remembrances. In a way I was lucky to lose so many good friends, because that meant I had something to lose. It's not possible to live a charmed life empty of fulfilling relationships. The very absence of friends seems like a presence in my life, an existing nothing. Some of those lost were larger than life, and now much smaller. I well remember those vanished faces and the lives they fronted. Without my departed friends, nights on the lonely planet fall on an emptier, colder and more lonely world. At times melancholy suffuses my twilights close to home when I think of my late friends. Those who happen to remain I try occasionally to see. Rather than traveling to discover novelties and curiosities around the world, I leave home to visit friends and family while they, and I, are still able to connect. My hope has been that those in my circle could say that knowing me added something good to their lives. For sure they all in one way or another enriched mine. No matter how much my family and friends appreciated me, by now I'm fully depreciated. Now that my end-game writing is done I'm a nearly complete write-off with salvage value of zero, a discontinued operation soon to be removed from the books and from this book forever.

Apart from those occasional visits to see people, my enthusiasm for travel has waned. Never did I imagine in my younger days that I'd lose interest in worldly adventures, but time erodes even the most resolute and compelling aspirations. My seven league boots have been down-graded to minor league comfy slippers. Even an ever keenly curious character has finally seen and done enough. Museums, monuments, churches, street scenes, performances, sights and sites eventually fail to motivate travels. I managed to crowd into one lifetime enough such places and activities for two lifetimes. In any case, if I want to see a curiosity or a ruin these days I need only look into a mirror. That I can at this stage still travel and also, only from memory, write an extended work based on long-gone experiences somewhat surprises me. My sense of an ending and my awareness of finalities are for me based on existential realities rather than on the particulars of my own situation. Contrary to my funereal meditations and terminal thoughts in these late-in-the-day pages, I so far remain free of ailments common at my age and decay of the kind which plagues many of my contemporaries. This fortunate state of being prevails just now as I'm writing these words, but from moment to moment things can change. "Take one day at a time," oldsters are told, but at my age a full day is far too long. These days I take one hour at a time. By the time, if ever, anyone reads these words things will no doubt be different for me, most likely that a "me"—succeeded by the ghost which was once me—no longer exists. Ever the realist, I imagine that my reminiscence will soon represent the sole remaining residue of an expired life. My focus on mortality isn't a morbid or a fatalistic lament but only a matter of actuarial odds, which day by day move against me. Given those odds, there's nothing odd about my attitude. It would be odd if I wasn't thinking in those terminal terms. Much of my life was in fact defined by my recognition, even at a young age, of final matters.

The brief span of time remaining between my present age and when I become forever ageless represents an instant compared to the aeons which will follow my non-existence. Few if any readers will happen to come across these pages during my time, and even after I'm gone this book will become one of millions of shelved volumes gathering dust, as I disintegrate into dust, rather than gathering millions of readers. Should my reminiscence emerge from the shadows in the years to come, I will indeed be a ghost writer, a phantom speaking beyond the grave my grave thoughts to readers still alive. My only occupation will be the tomb I occupy, the place from which I will speak to you while decaying in my burial plot where no counter-plot can revive or vex me. There I will disintegrate as you read my distilled thoughts in the words I left behind after the motions and emotions and commotions of my life have been forever stilled. For me, as I speak to you from the grave, all is settled and fixed and no longer subject to revisions, uncertainties or change. As I conclude my reminiscence, my earthly existence is range-bound, for the accumulated years have crowded me into a corner. There's no way out, other than down. Looking for a victim to scythe, the Grim Reaper now shadows me wherever I am, and that devil of death will pursue me until I amn't. Soon I'll give up the ghost and receive nothing in exchange, a one-sided bargain. The clarity of my finality, like the dazzling sands of a sun-drenched desert, brings me a certain comforting acceptance, as—for once in my life—there's no uncertainty: death is for sure.

In the title and text of his poem "And Death Shall Have No Dominion" Dylan Thomas offers a hopeful thought. Although I shrink from disputing a fellow Thomas, as a Doubting Thomas I'd say that the phrase describes the exact opposite of reality. It's in fact life which has no dominion. Each individual living thing—dog, leaf, ant, humanimal, whatever form takes shape from the void—exists only provisionally. The chemical stew which creates and briefly maintains a fleeting corporeal existence soon degrades, resulting in the permanent dominion of oblivion. Although I'd like to believe along with my fellow Thomas that death shall have no dominion, to my mind William's Richard II presents the more realistic view when he intones, "Within the hollow crown/That rounds the mortal/temples of a king/keeps death his court." That sounds pretty dominiony to me.

Stephen Hawking didn't dignify the humanimal by calling the creature a chemical stew. He said that humanity is "just a chemical scum." This can't be right, as such a view diminishes animal life and no rational person would claim that dogs are only scum. Snoopy is in no way scummy. It's probably more accurate to see earthlings as the residue of a cosmic experiment which somehow went wrong. The failed effort produced a dazzling display of infinite forms, a great spectacle but one lacking a theme. Human beings brought to the scene wonderful works of the imagination, both creative arts and inventive scientific marvels, while at the same time those doomed creatures suffered disease, famine, wars, turf battles, accidents, defeats, intrusive plots and counter-plots, pestilence, natural and man-made disasters, mortality and oblivion, and all the rest which goes by the name of history, the scum of elapsed time. It was all a bold but ultimately failed experiment. Better luck next time.

Speaking to you—most likely from the grave—I don't mean to spook you, nor to be a ghost writer haunting your thoughts. In these closing passages I've simply tried to record a few of my final impressions about how it feels to reach a dead end after so many lively earthly adventures. It was all quite a show. The infinite play of forms, ranging from cellular to stellar, offered a great spectacle. My reminiscence attempts to capture a few of the performances I happened to witness, but attempting to grasp the essence of any earthling's existence on the lonely planet is like trying to hold smoke. In *Present Past Past Present: A Personal Memoir* (translated by Helen R. Lane), Eugène Ionesco describes "the unreality, the evanescence of the world, a fleeting image in the moving water, colored smoke." Time passes, the rivers flow on and on, the hazy bygone days recede and daze our memories as if veiled with smoke, shadows darken the past. The prism of print through which my impressions have been refracted contains only hints, clues, allusions, nuances, suggestions of the original experiences. Those now long-gone images remain only phantoms of what once upon a time happened to emerge from nothing to something. Richard Corbet (died 1635), Bishop of Oxford, spoke of "unspoken speeches...a thought that nevr [sic] was thought upon...like the shadow when the sun is gone." Left unthought and unspoken are many of the essential elements which by chance defined my experiment. All left unsaid will perish with me when I before long become like a shadow after the sun is gone.

Time ever on the wing, I now fleetly and fleetingly fly toward eternity. Even more than before I notice how nature hurries everything along to its end. Sometimes I espy high in the late autumn sky birds winging their way south along the Mississippi Flyway, their wings scything the shadowy twilight like the cuttings of the Grim Reaper Himself. Below the passing birds flows the great river as the Mississippi—like the Ganga, the Seine, the Tiber, the Rhine, the Danube, the Nile—winds its way down to the distant sea, there to vanish as all traces of the waterway disappear into the watery depths. Borne into the world on the wings of time, a humanimal exists in time until the wings cease to beat. Everything flows, flies, flees, flits, flutters to its end. But what's the alternative? Lack of change, stasis, no zygote, no gamic development, no you, no me, no earthlings to perceive the play of infinite forms. But that would be okay, too. As it is, Mother Nature and Father Time spawned endings, the ends as a means to move things along and keep the system refreshed and vital, and perhaps also to give lonely planet inhabitants a sense of urgency, and writers something to think about. Without an ending, being is less thought-provoking. The posthumous phase brings a lot to the party: no "post," no ghost. Recognizing Death as a life force, forcing awareness of the finite nature of the infinite forms, offers the greatest post-graduate lesson of all as this teaches you how to live. Miguel de Unanumo describes this real-world academic exercise in *The Tragic Sense of Life* (translated by J.E. Crawford Futch):

> Although this meditation upon mortality may soon induce in us a sense of anguish, it fortifies us in the end. Retire, reader, into yourself and imagine a slow dissolution of yourself—the light dimming about you—all things becoming dumb and soundless, enveloping you in silence—the objects that you handle crumbling away between your hands—the ground slipping from under your feet—your very memory vanishing as if in a swoon—everything melting away from you into nothingness and you yourself also melting

away—the very consciousness of nothingness, merely as the phantom harbourage of a shadow, not even remaining to you.

In the twilight of my life it's as clear to me as a cloudless summer mid-day that a "tragic sense of life" shadows humankind's existence on the lonely planet. It somehow seems in a way that we're zombies, mere phantom figures formed by chance of impure debris from lucky predecessor beings somehow purified by their experiments in existing, an experience quite unlike ours. From what I saw, a certain sadness darkly cloaks the cosmic experiment which by chance created the world out of nothing and which will eventually reduce it all to nothing. Within that vast macro fatal circle, each and every individual mortal thing which ever exists reflects the eventual fate of the entirety, which itself is mortal. Nothing led to something which in turn will end in nothing. Something ventured but nothing gained.

In *The Birth and Death of the Sun* George Gamow describes how the earth—for the time being caressed with warmth and light by the sun to nurture life—will darken and chill: "When all the available sources of sub-atomic energy will finally have been exhausted, our sun will begin its ultimate contraction." Although the life-giving brightness in the cold cosmos will for a time remain hot and luminous, after a further long period "our sun will turn into a giant lump of lifeless matter covered with eternal ice and surrounded by a system of frozen but still faithful planets." To this scenario Gamow adds more dark details, describing how the moon "will be so poorly illuminated by the dying sun that it will be practically invisible. The temperature of the earth will drop down to 200 degrees below the freezing point (-328° F.), making any kind of life on the earth's surface quite impossible. But all these inconveniences of darkness and cold will probably be of no importance to humanity, which...will have been burned to death by the increasing solar activity long before the ultimate contraction and thermal death commences."

In the meantime, however, earth-confined humanimals from generation to generation will grope and feel their ways through the dark woods, for now fire-powered by the sun's light and warmth. It's a truly fabulous experiences trekking through the woods, a strange and wonderful grove leading to the grave. There's nothing else like it down here, here down below if you believe in heaven or up here if you're a hell-raiser destined to be kept warm, or only just here if you doubt any after-life. Whatever you might believe or non-believe, the twirling pebble spins on through space, a random speck of cosmic Big Bang dust thrown into existence by the whirligig of time and chance. In the very nature of things, this plaything planet created by random natural forces is a sad sort of place. The fatal circle which encircles human existence confines it to a fleeting cosmic moment permeated with *saudade*, the very sound of the Portuguese word evoking its sense—yearning for an absent person or place or time. Yearn as we may, the "something" humankind seeks to explain everything eludes us. Life exists, but meaning's missing. More than just a brief gap year separates existence from significance: lack of meaning represents a permanent earthly condition. It's passing sad how all your hard-won experiences, experiments, accomplishments, thoughts, sensory impressions, memories, how your long-nurtured and treasured relationships with dogs, family, friends, spouses, lovers, colleagues, acquaintances, how the delights you've derived from favorite places, happenings, books, ice cream cones, Snoopy's wisdom and much, much more all come to an end. All. End.

It somehow seems contrary to reason and a proper order of things for someone, for everything, just to disappear, poof, like smoke. But in fact, non-being and disorder represent the natural order. There's really no reason that nature should be reasonable. Its purpose is not to provide logic but only to experiment, an enterprise with no purpose. Given the conditions for earthlings on the lonely planet, it's somewhat curious that those benighted mortals want to cling to existence. Once you get used to being alive you'd like it to go on longer, but no such luck. With something of a charmed life lived in your own way, you'd wanted it to go on and on, like the surrey with the fringe on top—continue to enjoy the relationships, the experiments and experiences, the adventures, the memories, just keep clip-clopping on without a stop. But it stops. The surrey with the fringe on top includes no fringe benefits which allow you to keep going. You weren't unrealistic: you didn't wish it would go on forever and never stop, like the surrey, because you knew in truth that the wingèd chariot would in the end outrun the surrey and bring you to the desert of vast eternity.

That desert destination is your destiny, and Everyman's. Those are the terms of engagement, the iron-clad rules which decree that someday you must disengage. Our bargain with existence demands a fixed-term contract with no renewal option. Although we know a last stop lurks ahead for both an individual and the entire system, how and why you—*mes semblables, mes frères, mes soeurs*—and I and all the rest began will ever remain a mystery. The random natural forces which somehow combined to produce each humanimal and the entirety resulted in a strange and unnatural sort of world. It's odd to exist, as a person and as a lively but lonely planet and as a forlorn cosmos. These chance experiments yielded a coldly mechanical and emotionless system, unfeeling and indifferent to all its infinite forms. One odd exception is the humanimal, a sentient part of the system endowed with emotions and feelings, perhaps in order to perceive and feel an existential sadness. This activates a humanistic reaction to the annihilating whirligig of a cold cosmic experiment.

Once thrown into life, through no fault of your own, you're confronted with how to react to a situation in which death has dominion. In light of the eventual darkness—the sun extinguished, twilights no longer passingly glimmering, eternal night, a lifeless and very lonely planet—it's a challenge to decide how a benighted earthling should live. I can't answer for anyone except myself, and in these pages now at their end I've described my response to the situation. My reminiscence represents my own extended answer to the question of how to react to the ineluctable realities of the world and also to the cosmic fatal circle which encircles earthlings. Concerned with the great existential questions of human existence, Gauguin wrote them on one of his Tahitian paintings: *D'où venons nous? Où allons nous? Que sommes nous?* Gauguin's painted words were just decorative graffiti. Who knows where we came from, where we're going, what we are. People want to know, but they never find out.

In Voltaire's *The Story of a Good Brahmin* the good man concedes, after spending his life searching for happiness, that he's learned almost nothing: "I do not know whence I came, whither I go, what I am nor what I shall become." Unlike the two Frenchmen, I never deemed it worth contemplating such unearthly ethereal metaphysical questions. I deemed it a waste of time to ponder those unanswerable matters. The question is not where we came from or where we're going or what we are. The only valid inquiry is to determine for one's-self, if you by chance happen to be thrown into existence, how to bridge the brief interval between your tenure as a terrestrial humanimal and the aftermath of eternal non-existence. Any transcendental meaning is irrelevant and anyway non-existent. Only we exist, very briefly, but not meaning. I avoided spending my limited earthly time on insoluble questions so many other earthlings in their time ventured to address. I restricted my adventures to more earthy matters. Only the immediate and personal puzzle of being as represented by my own existence concerned me. I contemplated how to get through the dark woods to the far side to arrive at the desert of eternity in the best way possible, given the fatal circles—mine and humankind's—which surrounded me. My story was that of Everyman, a variation on a theme. In the end, time and luck run out, for emperors and their underlings alike. On August 15, 1945, Emperor Hirorito conceded that, "The war situation has developed not necessarily to Japan's advantage," and yet a third Frenchman, Napoleon, said when he abdicated as emperor at Fontainebleu on April 4, 1814, "Events have turned against me." I know exactly how they felt, those diminished emperors, as soon I'll be forced to give up the empire of earthly existence for the desert of vast eternity. Empirical evidence shows that in that ghostly wasteland only Father Time and Mother Nature rule over their dominion of eternal oblivion. Like Everyman, in the end an emperor has no clothes, no empire, nothing. Death, not the deposed disposed ruler, shall have dominion. Events have turned agsinst me, and my life has developed not necessarily to my advantage. Now it's my turn, as soon I will turn into a ghost.

All life stories are ghost stories. You live and then you disappear, leaving some passing impressions for a few fleeting years recalled by those who once knew you, perhaps leaving behind some descendants, maybe leaving a few fond memories momentarily revived every so often by your family and close friends until they, in their turn, finally disappear into the desert.

In Ibsen's *Ghosts* Mrs. Alving says, "I am half inclined to think we are all ghosts." She was half right. Before long I will become a phantom as insubstantial and evanescent as the wispy curls of smoke which rose into the air and disappeared at Phantom Lake campfires when I was a boy a long lifetime ago. "How a long life grows ghostly toward the close/as any man dissolves in Everyman," wrote poet Howard Nemerov. I have now finished and I myself, facing the fate of Everyman, am nearly finished, soon to be forever a ghostly shadow, a phantom, a nothing. May luck be with you as you advance toward the desert of vast eternity. Reader, I wish you well and bid you farewell.

In 1895 Wilhelm Röntgen, who had just discovered X-rays, made an image of his wife Anna Bertha's hand wearing a ring. Upon seeing her ghostly skeletal fingers she said, "I have seen my death." In this book I have seen my death—and yours.

www.ingramcontent.com/pod-product-compliance
Lightning Source LLC
Chambersburg PA
CBHW071648160426
43195CB00012B/1389